Structural Design

Structural Design

A Practical Guide for Architects

James R. Underwood
and
Michele Chiuini

JOHN WILEY & SONS, INC.

New York   Chichester   Weinheim   Toronto   Singapore   Brisbane

Library of Congress Cataloging-in-Publication Data

Underwood, James R.
    Structural design: a practical guide for architects / James R.
  Underwood and Michele Chiuini
            p.   cm.
      Includes index.
      ISBN: 0-471-14066-X (cloth : alk. Paper)
      1. Structural design. 2. Architectural design. I. Chiuini,
  Michele. II. Title.
  TA658.U53 1998
  624.1'771--dc21   97-52009

Dedication and Acknowledgment

We would like to dedicate this undertaking to our families: our parents, wives, and children who have provided the opportunity to attempt this book through their support of our education and their sacrifices and understanding throughout its creation. Without their support and tolerance, it would never have been possible. We hope it will make them proud!

Special acknowledgment and thanks goes to Tanner Underwood, R.A., for the seemingly never ending effort on her part required to create the illustrations that help make it understandable.

CONTENTS

CONTENTS

SPECIAL SYSTEMS

27 Permanent wood foundations

PART THREE: CONCRETE

28 Materials and properties

29 Reinforced concrete in architecture

BENDING MEMBERS

30 Beams strength theory

31 Beam Design

PART FOUR: MASONRY

IN THE BEGINNING: THE PREMISE

One of the most important aspects of structural design for architects is the selection and configuration of structural systems. After a system is pragmatically conceived, individual members can be designed within that "system." Although we would like to be able to give you a set of "rules" associated with this decision, they simply don't exist. For each material, we will identify those considerations associated with that particular material. If you consider those issues, consult with a competent structural engineer, and remember that the structure and the architecture are inseparable partners, you can arrive at reasonable solutions. Finally, don't be fooled by the NINE decimal point answers your calculator will give you; if you start with faulty assumptions, you end with WRONG NINE decimal point answers.

Structural design books and manuals for the practical designer are few. They are also frequently written in a manner that leads a designer to believe that approaches to the design of various components of a structure are unique to each component. To help alleviate some of your anxiety, this book will focus on the idea that there are basically **three elements you can design in structures**: a bending member, a compression member, and a tension member. This is made slightly more complex by combining the options and adding **bending + compression** and **bending + tension**. This probably seems too simple, but as we go on, focus on this idea and you'll discover it's a fair representation.

To design these three or five, if you must, members, we'll also focus on the use of the two commandments of structures:

$$F = P/A \quad \text{and} \quad F = Mc/I$$

Most of the infinite equations presented in structural texts and manuals are elaborations or reinterpretations of these two simple formulas.

The level of math required is simple algebra and trigonometry. You need to solve a simple quadratic once in a while, using either the quadratic equation or completion of the square, whichever is the easier.

The most important thing to remember is that you've been dealing intuitively with structural principles all of your life; you walk, you carry things, you bend things, you break things. All we are going to do is quantify what's happening in these situations. The math is simple and logical; don't let it frighten you.

Our purpose is not to make you a structural engineer, but to give you an appreciation and understanding of the considerations that are necessary for the successful completion of any architectural project.

If you wish to be a complete designer, you must be able to integrate technical issues into your design vocabulary. You will ultimately be obligated to show that understanding and ability

through the registration examination. You will also be confronted by situations that will necessitate that you do simple structural calculations either for a small project or preliminary calculations for larger projects. It is assumed the reader knows statics. This understanding is the foundation for understaning sytems.

When you understand the approaches to structural design for small buildings, you should be able transfer that knowledge to larger situations. The principles are basically the same; the scale differs. With this knowledge, you can talk intelligently with your structural consultants. Remember, **you are responsible for the work of your consultants**; if you allow them to work in the shadow of an intellectual vacuum, **you are placing your future in their hands.**

We can classify structural systems into three general categories, each associated with a type of space. You will immediately notice that these systems are basically versions of the original wood and stone systems that are the foundation for what we do today with more sophisticated materials.

Frame

In its simplest form, a frame consists of two vertical poles with a horizontal pole spanning between them. The vertical poles are frequently pure compression elements; a pile of stones would work as well. The horizontal pole has several different types of forces acting on it (bending, shear, and deflection), requiring that it have both tensile and compressive capacity. Even taking this simple principle, you can understand why the columns on the Parthenon are so closely spaced. Stone has little tensile capacity, so it will only span short distances as a beam. The combination of close column spacing and wide capitals reduces the span to match the capacity of the stone. (The Golden Section wasn't the only driving force!)

Fig. P.1 Ise shrine

Architecturally, the most forceful use of the ancient wood frame can be found in early Japanese architecture. The best known example of indigenous Japanese architecture is the Ise shrine (Fig. P.1), originally constructed around the third century A.D. The wood columns are set in the ground, so they act as cantilevers for earthquake loads. In later times, the columns would rest on a stone base; consequently, the columns were not cantilevers and the earthquake resistance was ensured by a very elaborate "rigid" joinery at the connection of columns and beams, as illustrated in the eighth-century Toshodai-ji temple (Fig. P.2).

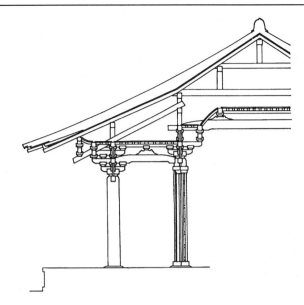

Fig. P.2 Toshodai-ji section

Arch and dome:

The vault created by a series of arches is an axial space that can theoretically extend indefinitely at the two ends, and at the same time has a well-defined edge at the spring of the arches. The modern arched structures have a formal analogy with the masonry frame of the Gothic cathedrals, which may well have been inspired by the interlacing of tree canopies, and a construction analogy with timber ships, which are upside-down vaults (Fig. P.3). If you were to take an arch and rotate it 180°, you would create an enclosed volume of space recognized as a dome. These domes can be cut and redefined as the half-dome forms we find in many masonry religious structures. Early third century B.C. Buddhist tombs, the Pantheon, the exhibition halls of Nervi, and the simple igloo of the Eskimo are all variations of an elementary compression structure.

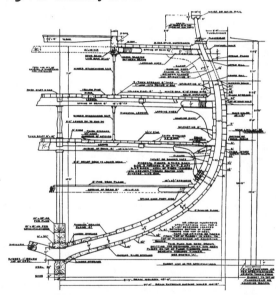

Fig. P.3 United States Frigate Constitution

One of the most commonly recognized masonry structures in the world is the Colosseum in Rome. It combines the arch systems of the Roman aqueducts and the round curves of domes to create a monument that seemingly will live forever. The form may have been both for view lines and to resist the forces of the ropes that once supported the removable roof, like many contemporary sports stadiums.

Fig. P.4 Roman Colosseum

Wall

Early stone buildings, log cabins, and modern panelized construction are all systems resulting in cellular spaces. These systems are still widely used in conventional or manufactured housing. A modular house can consist of several sections built in the factory and assembled on-site. All walls have the potential of becoming load-bearng, so the space is not as flexible as in the Japanese house, based on the frame, or as universal as that defined by the arch or dome. The wall has the advantage over the frame of combining in one panel element load-bearing and enclosure functions, making a rational use of material with modern construction technologies. It can be easily constructed in an area that has lots of stone or, as we learned from Plains settlers in American history classes, from great slabs of sod.

Fig. P.5 Maronry bearing wall housing complex in Venice, Italy.

These systems can teach us several valuable lessons. Look at indigenous materials; make use of those systems that require minimal manufacturing and are readily available. Consider the environmental context; structures of great thermal mass are excellent choices in areas of rapid and regular changes in temperature, whereas lighter structures that do not retain heat are better in areas that have consistently high temperatures. It is not an accident that most indigenous architecture is more environmentally sensitive than that which our "technology" can support. Don't be a salmon; work with nature; don't always swim upstream. We apologize to any salmon we might have offended.

Although these examples may include some domestic scale structures, their systems are the basis for what we do today at almost every scale. The John Hancock Building in Chicago is partially based on the structural idea of a farm gate: a cantilevered element supported from one end and "trussed" to make it light, yet strong (Fig. 14.7). Modern sports arenas are frequently large-scale examples of dome structures you might build with snow in your backyard in the winter or in other cases simply big "cold-air" balloons.

This book attempts to address structures both on a conceptual level and on a numerical level. We acknowledge that some types of structural design are more likely to be attempted by architects than others. For this reason, we discuss wood and steel in rather laborious detail and present reinforced concrete in a more simplified (believe it or not) version.

In most cases, you **can** do more than you probably **should** do. A good structural consultant will provide the architect with options based on "structural" considerations. It is your job to coordinate these options with the "architectural" considerations as well as those of the mechanical consultants. Structural consultants will ultimately provide the best solutions to the specifics of your design approaches, but a good architect has to be able to understand the engineer's language. Bring them in at the beginning of the design process and use their conceptual thinking, not just their calculators.

This text is intended to be used with the Manual of Steel Construction, published by the American Institute of Steel Construction (AISC); National Design Specifications for Wood Construction (NDS) and Design Values for Wood Construction: Supplement (NDS, 1997), published by the National Forest Products Association.

INTRODUCTION: UNDERSTANDING LOADS

1 Loads

In statics, loads are forces acting on structural components, and are represented as uniform or varying forces or points. In practice, they represent the weights of the building materials used to construct the building (Fig. 1.1), the weights of the people and equipment that will occupy the building, and the forces of nature that the building will be exposed to during its life.

Material weights are gravity loads which act down, surprise! **People and equipment** are primarily gravity loads, but in some instances may cause forces that act in some other direction. An example would be a piece of horizontally moving equipment that suddenly comes to a stop, causing a horizontal force to be induced in the structure.

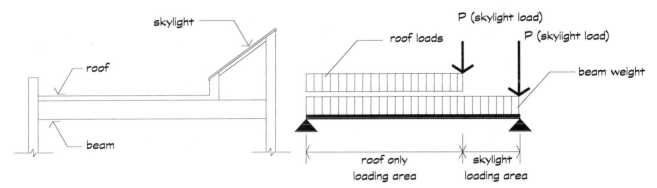

Fig. 1.1 Structural loading diagram of architectural condition

Wind forces are primarily horizontal, but can induce vertical forces when blowing over surfaces. Note that wind passing over an airplane wing causes the upward lift that keeps the plane in the air. Similar conditions can be induced in the roof structure of a building.

Earthquakes by contrast are wavelike forces that have both horizontal and vertical components; however, the horizontal force component is typically the most destructive of the two since most structures are designed to be primarily vertical load-carrying systems. Wind and earthquake loads are discussed in more detail in Chapter 17, at a point when you can better understand their implications for the systems you have designed.

The effect of these forces is to induce states of stress and deformation or deflection in the structure. Deflections are often the governing factors in the design of a structural system. Obviously, a structure fails when it collapses; however, excessive deflection that causes damage to finishes or other building components without collapse is also defined as a structural failure.

Building codes categorize these loads into two classifications: dead loads and live loads. Dead loads are those permanent loads generated by the constructional system. Live loads are those nonpermanent loads that are applied to the structure after it is completed. Some loads may be in either category, depending upon their **time of application.** It is essential to understand the construction sequence of the building and to design for deflection caused by live loads introduced after the construction is complete. For example, a typically permanent (**dead**) load such as an HVAC (Heating, Ventilating, Air Conditioning) unit should be considered a live load if installed after ceiling finishes are in place, since it would cause deflection of the ceiling/floor components similar to that created by snow on a roof or human occupancy of a level above. This may occur even if a building component is assumed to be in place prior to "finishes." A manufacturing delay, a labor dispute, a delivery problem, or even a design change may be responsible for an out-of-sequence installation that could have serious deflection implications.

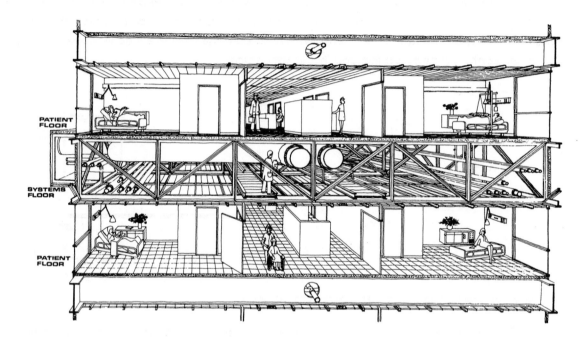

Fig. 1.2 Live and dead loads on Boston City Hospital

An objective of the building codes is to limit the deflection of structural members to the extent that they will not damage the connected nonstructural components or affect the functionality of the building. In the Uniform Building Code (UBC), limitations are imposed on deflections due to both dead and live loads (Table 1.1).

Table 1.1 Maximum allowable deflection for structural members

Type of member	Allowable LL deflection	Allowable LL + KDL deflection
Roof supporting plaster or floor member	L/360	L/240

Source: Adapted from Table No. 16-D, UBC, 1997 ed.

The value for K (a percentage of the DL to be used in deflection calculations to account for long-term permanent deflection) varies for each material discussed in this book. The value of K for steel is O. This means that the maximum allowable deflection is ONLY a function of LL and is limited to L/360 (1/360 of the span L).

This shouldn't suggest that dead loads don't cause deflection. The dead-load deflection of the structure isn't considered in some cases since it is compensated for during the construction process. For example, the ceiling finish that is (obviously) installed after the horizontal framing is enclosed will be installed "level." Any dead-load deflection that exists in the framing will be hidden by adjusting the finishes.

The load values to be used depend on the use or occupancy of the structure. Typically, loads are floor loads, roof loads, and wind loads acting on walls and roofs, and are given in lb/sq ft (or psf)[kN/m²]. For example, floor loading for offices is 50psf [2.39kN/m²], and for school classrooms 40psf [1.92 kN/m²]. In both cases, the buildings will have corridors or circulation spaces on each level that will have a live loading of 100psf [4.78kN/m²]. As this suggests, structures are subjected to a variety of live-loading conditions and the design must work for the worst-case scenario those loading conditions that cause the worst effect on the criteria being investigated. Why are we telling you this? No matter how well you understand the processes for designing individual members, if you begin with a misunderstanding of what loads must be resisted, how they will be applied, when they will be applied (in what combinations), and how the total system transfers these loads, your calculations will be no more than mathematical exercises.

How are loads determined?

1.1 Gravity loads

Dead loads are calculated by looking at the construction system of the building and calculating the actual weights of the materials. This may be difficult since preliminary calculations must be made during the design of the project, at a time when actual materials and systems have yet to be completely defined. Knowledge of a wide variety of alternative material systems is a great asset.

Probably the best technique for determining the dead load of a building system is to sketch a typical construction section of your projections (a roof or floor "sandwich") and determine the weights of the components from manufacturers' catalogues or from tables of standard weights (AISC Manual).

Of course, you will have to add any nonuniform components to this system sketch in the final calculations, i.e. walls, movable point loads, etc. Note that the UBC Live Load Tables have occupancy loading criteria, which include both uniform loads and movable concentrated loads.

Table 1.2 Live load requirements
(uniform and concentrated)

Occupancy category	Description	Uniform load type	Concentrated load type
Access floor systems	Office use Computer use	50psf 100psf	2,000lb[1] 2,000lb[1]
Armories		150psf	0
Assembly areas[2] (auditoriums/balconies)	Fixed seating Movable seating/other areas	50psf 100psf 125psf	0 0 0
Cornices and marquees		60psf[3]	0
Exit facilities[4]		100psf	0[5]
Garages	General storage/repair	100psf 50psf	?[6] ?[6]
Hospitals	Wards and rooms	40psf	1,000lb[1]
Libraries	Reading rooms Stack rooms	60psf 125psf	1,000lb[1] 1,500lb[1]
Manufacturing	Light Heavy	75psf 125psf	2,000lb[1] 3,000lb[1]
Offices		50psf	2,000lb[1]
Printing plants	Press rooms Composing/linotype	150psf 40psf	2,500lb[1] 0[5]
Residential[7]	Basic floor area Exterior balconies Decks Storage	40psf 60psf[3] 40psf[3] 40psf	0[5] 0 0 0
Restrooms[8]		50psf maximum	
Reviewing stands, grandstands, bleachers,		100psf	0
Roof decks	Same as area served		
Schools	Classrooms	40psf	1,000[1]
Sidewalks/driveways	Public access	250psf	?[6]
Storage	Light Heavy	125psf 250psf	
Stores		100psf	3,000[1]
Pedestrian bridges and walkways		100psf	

Source: Adapted from the UBC, 1997 (consult the UBC in any case).

1. Floors shall be designed for the specified loading of either uniform or concentrated, that will create the worst loading condition for any criteria: moment, shear, deflection, or any localized mode of failure. The specified concentrated load, as set forth, shall be placed upon any space 2-1/2ft [762mm] square. It is not necessary to combine the two loading conditions.

2. Assembly areas shall include such uses as dance halls, drill rooms, gymnasiums, playgrounds, plazas, terraces and similar occupancies that are generally accessible to the public.

3. When snow loads occur that are in excess of design conditions, the structure shall be designed to support the loads. These may be created by drift buildup or a greater snow design. The 1997 UBC specifies both additions and reductions as a result of roof geometries in Sec. 1614 Snow Loads.

4. Exit facilities shall include such uses as corridors serving an occupant load of 10 or more persons, exterior exit balconies, stairways, fire escapes, and similar uses.

5. Individual stair treads shall be designed to support a 300lb [1.33kN] concentrated load placed in a position that would cause maximum stress. Stair stringers may be designed for the uniform load specified in the table.

6. This one is fairly complicated!

   For general storage, the floor must be designed for either the uniform load or two or more concentrated loads spaced 5ft [1524mm] nominally on center. Each load shall be 40% of the gross weight of the maximum-sized vehicle to be accommodated.

   For private "pleasure-type" vehicles, the floor system shall be designed for either the uniform load or a single 2,000lb [8.9kN] load acting on an area of 20sqin [12,903mm²].

   If the storage area is also an open roof, additional loading combinations including snow must be considered in accordance with Sec. 1612.2 and 1612.3 of the UBC.

7. Residential occupancies include private dwellings, apartments, and hotel rooms.

8. Restroom loads shall not be less than the load for the occupancy with which they are associated, but need not exceed 50psf [2.4kN/m²].

Live loads are code-specified and are a result of testing and are somewhat subjective; consequently, they are usually conservative. For example, parking garages must support a code live load of 50psf [2.39kN/m²]. If a car weighs around 2500lb and covers an area of 14ft × 6ft = 84sqft, the corresponding distributed load on the floor is 30psf [1.44kN/m²]. If people and luggage are included in the car, the load increases. The code values are conservative since there are a wide range of exceptional conditions that could occur.

Table 1.3 Special live load requirements
(uniform and concentrated)

Occupancy category	Description	Uniform load type	Concentrated load type
Construction	Walkway	150psf	O
	Canopy	150psf	O
Reviewing stands, grandstands, bleachers, folding/telescoping seating	Seats and footboards (load in lb/ft)	120plf	see footnote 1
Stage accessories	Catwalks	40psf	O
	Followspot, projection, and control rooms	50psf	O
			O
Ceiling framing	Over stages	20psf	O
	All other locations	10psf[2]	O
Balcony railings and guardrails	Exit facilites with occupancy > 50 persons		50plf [3]
	Other than exit facilities		20plf [3]
	Components		25lb [4]
6. Handrails			see footnote 5
7. Fire sprinkler support		250lb[5]	

Source: Adapted from the UBC, 1997 (consult the UBC in any case).

1. Lateral sway bracing loads of 24plf [350N/m] parallel and 10plf [145.9N/m] perpendicular to seat and footboards.
2. Does not apply if the ceiling space does not have access above. Use in conditions where the attic areas above the ceiling are provided with access even if they do not have flooring.
3. A load in pounds per linear foot to be applied horizontally at right angles to the top rail.
4. Intermediate rails, panel fillers, and their connections shall be capable of withstanding a load of 25plf [1.2kN/m²] applied horizontally at right angles over the entire tributary area, including openings and spaces between rails. Reactions due to this loading need not be combined with those created by the load applied at the top rail (note 3).
5. The mounting of handrails shall be such that the complete handrail and supporting structure are capable of withstanding a load of at least 200lb [890N] applied in any direction at any point on the rail. These loads shall not be assumed to act cumulatively with note 4.

Roof live loads account not only for snow, wind, or people having access for maintenance, but also for water that might accumulate if drains become clogged. The code for northern Indiana specifies 30psf [1.44kN/m²] (Fig. 1.3) which is equivalent to 6in [52mm] of water or as much as 60in [1524mm] of snow. Water weighs 62.4pcf [9.81kN/m³], so every inch of water weighs 62.4pcf/12in/ft = 5.2psf/in [9.81N/m²/mm] of depth. Five feet (60in) of snow is unusual; however, it is not uncommon to find drifts of this depth, especially on roofs where building forms of different heights are adjacent to one another.

Fig. 1.3 Indiana Minimum Design Snow Loading
(from 1993 state building code)

Table 1.4 Roof live-load requirements (other than snow)
(uniform and concentrated)

Roof Slope	Method 1 (tributary loaded area in sqft for any structural member)			Method 2		
	0-200	201-600	600+	Uniform Load	Rate (%) reduction	Max. (%) reduction
Flat[1] or rise ≤ 4:12 Arch or dome with rise < 1/8 span	20psf	16psf	12 psf	20psf	0.08	40%[2]
Rise > 4:12 < 12:12 Arch/dome with rise ≥ 1/8 span to < 3/8 span	16psf	14psf	12psf	16psf	0.06	25%[2]
Rise 12:12 and greater Arch/dome with rise ≥ 3/8 span	12psf	12psf	12psf	12psf	None allowed	None allowed
Awnings except cloth covered[3]	5psf	5psf	5psf	5psf	None	None
Greenhouse, lath houses, and agricultural buildings	10psf	10psf	10psf	10psf	None allowed	None allowed

Source: Adapted from the UBC, 1997 (consult the UBC in any case).

1. A flat roof is any roof with a slope of less than 1/4:12 (2%). The live load for flat roofs is in addition to the ponding load required by Sec. 1611.7 of UBC.
2. Reduction = r(A - 150sqft); where r = rate of reduction and A = area of roof supported by the structural member. Note the maximum load reductions listed.
3. Defined by Sec. 3206 of UBC.

Wind loads, earthquakes, and some machinery cause dynamic forces on building structures. These are usually referred to as "lateral loads" and are treated in Chapter 17. The loads may have a considerable influence on the overall structure.

1.2 Tributary areas

Loads are transferred from nonstructural parts to the load-bearing structure of the building; structural members transfer the loads to each other until the forces eventually reach the foundations (Fig. 1.4).

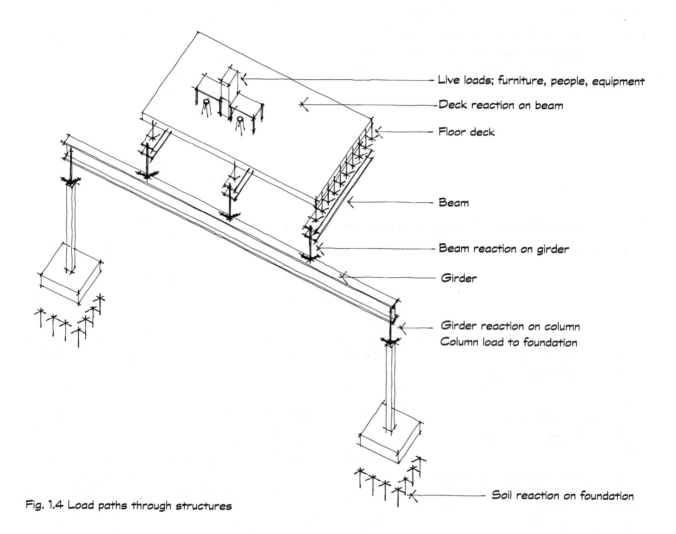

Live loads; furniture, people, equipment
Deck reaction on beam
Floor deck
Beam
Beam reaction on girder
Girder
Girder reaction on column
Column load to foundation
Soil reaction on foundation

Fig. 1.4 Load paths through structures

Understanding how these transfers take place is essential to assign the correct loads to all members and to design the connections between members. Location and type of load, connections and type of structural member affect the distribution of stress (axial, bending, shear, and torque) in the cross-section of the member. The assumptions made by the designer regarding the type of structural system and loads must match the reality of construction as closely as possible.

If members support large areas of floor or roof surfaces, the likelihood of the entire area being fully loaded with live load decreases as the area increases. In this case, the UBC allows you to reduce the live load applied to specific members in the system. The conservative position would be to apply the full live load to the entire system, but this conservatism will come back to haunt you when you begin to realize the cost of the system.

Dead loads do not have any area-reduction considerations.

1.3 Lateral loads: Wind

Lateral loads act horizontally and are typically dynamic. They push and pull and under the most extreme conditions can cause rhythmic loading, which amplifies their effect. Consider the rhythmic application of force that causes the amplitude of your motion to increase when pushing a child's swing. Wind, the most common lateral load, acting on sloped surfaces will create forces that may cause inward or outward movement. Two alternative methods of evaluating these forces are presented in detail in Chapter 17 .

1.4 Lateral loads: Earthquake

Earthquake loads are ground waves and have both vertical and horizontal components. We generally ignore the vertical component since the normal live load of the structure will account for this aspect. The horizontal force is another matter entirely. The horizontal magnitude of force is generally much greater than that created by wind and is more dangerously rhythmic in character. Specific load development and resistance techniques are presented in Chapter 17.

In the case of both wind and earthquake forces, the discussion is more relevant after you understand the considerations of the design of a variety of members and how these forces might affect the design of these members.

1.5 Loading conditions

It is a common misconception that if you use the maximum amount of load, the members will be designed for the maximum condition - au contraire!

The removal of load from a beam cantilever will actually increase the amount of moment the beam will experience as well as maximize the deflection between the supports. At the same time

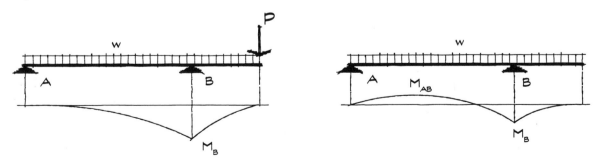

Fig. 1.5 Effect of load placement on a cantilever system

removal of load between the supports will increase cantilever deflection, but will cause no increase in the maximum moment in the cantilever (Fig. 1.5). To design this or any other bending member:

all loading conditions that will maximize the moment, shear and deflection must be considered!

No amount of "overdesigning" will compensate for inadequate consideration of the critical loading conditions.

PART ONE

STEEL

Kenzo Tange, Olympic Pool, Tokyo (photo by Michele Chiuini)

2 MATERIALS AND PROPERTIES OF STEEL

2.1 Structural properties of steel

The qualities that make steel a desirable building material are **strength** and **ductility**. Strength is measured relative to the yield strength, corresponding to the stress value where, with virtually no increase in loading, steel deformation continues to increase. This can be visualized by bending a piece of mild steel wire with your fingers. The elastic limit is quickly passed and, once the pressure of the fingers is released, the wire stops bending but the deformation remains. The wire is not broken and the steel has not lost its elastic properties. In fact, it can be bent back and straightened, at least theoretically, to its initial shape, with no loss of strength. Under repeated bends, the material will become hot, evidence that the steel is undergoing changes in its properties. Eventually, the wire will become brittle in the localized area and will break because it lost its ductility.

You can see visual evidence of a similar process if you crimp a cheap plastic cup and notice that the material changes from translucent to a milky white. The physical properties have been altered in this area by a process similar to strain hardening. We'll pursue some of the cases when you might use this characteristic to your advantage later.

The ability of the wire to "flow" under the constant stress created by your fingers and still maintain its strength is a measure of its **ductility**. The greater the ductility, the more a structure can adjust to high stresses without collapsing. Compare this behavior with a glass rod: Its strength in bending is substantial, but as you apply more and more load, it will snap suddenly. Glass displays strength but little ductility when compared to steel.

This structural behavior can be described in the stress-strain diagram of Fig. 2.1. This is a generalized diagram indicating the elastic limit, the yield point, and the failure point of a steel specimen. For all practical purposes, the elastic limit and the yield point occur at the same point; this first portion of the curve where the stress is directly proportional to the strain is called the **elastic range**. The stresses used in the design of steel members fall within this range in the Allowable Stress Design (**ASD**) method used in this book:

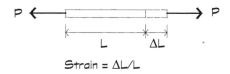

Strain = $\Delta L / L$

$$F = P/A$$

Pt. A = elastic limit, yield stress F_y

Pt. C = ultimate stress F_u

    (58ksi for A36, off diagram)

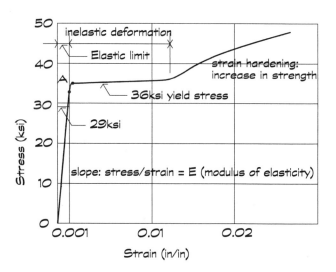

Fig. 2.1 Stress-strain diagram for A-36 steel

Several grades and types of structural steel are available today, whose properties depend on carbon content, alloys, and casting processes. Strength is the main property for designating steel, and varies from 32,000psi [220MPa] to 130,000psi [900MPa]. Steel is classified according to American Society for Testing of Materials (ASTM) specifications. Table 2.1 lists widely used range of steels with ASTM designation. The most common structural steel is A36, which has a yield stress of 36,000psi [250MPa].

Table 2.1 Structural steels for buildings

ASTM designation	Yield stress ksi	MPa	Available forms (plate thickness) Group per ASTM A6	Steel type
A36	36.0	250	All shapes, plates, bars	Carbon
A441	40.0	276	Plates and bars over 4 to 8in 102 to 203mm	High-strength
	50.0	345	Shapes group 1, 2 and small plates	Low-alloy
A572	42.0	290	All shapes, plates to 6in thick 152mm	
	50.0	345	All shapes, plates to 2in thick 51mm	
A588	50.0	345	All shapes, plates to 4in thick 102mm	Corrosion-resistant, high-strength

Source: Adopted from A.I.S.C., Manual of Steel Construction, 1991: Dimensions and Properties

The ratio of unit stress to unit deformation (strain) is called the modulus of elasticity, E and reflects the relative stiffness of a material. The modulus of elasticity of any material can be determined by calculating the slope of the elastic portion of the stress-strain diagram. The units of stress are psi or ksi and the units of strain are in/in, effectively psi/1, so E is expressed in psi [MPa]. All steel classifications have the same modulus of elasticity, E = 29,000,000psi [200,000MPa] regardless of the strength of the steel.

2.2 Allowable stress

The loading on the structural system will induce a state of stress in each member. As long as we ensure that the highest stress remains below the elastic limit, there will be no permanent deformation or collapse of the member (with the exception of compression members where the slenderness ratio is the controlling factor). Elastic deformation can be tolerated within limits, as discussed in Chapter 1. For each steel, according to the ASD procedure, we can assign an **allowable stress** typically well below the yield point. In the AISC specifications, the allowable stress is given for each type of stress (bending, shear, tension, etc.) as a percentage of the yield stress (Table 2.2). A36 steel has a yield point of 36ksi [250MPa], and in some cases, the allowable stress is taken as $F_b = 0.66F_y = 0.66(36ksi) = 23.76ksi$ (commonly taken as 24 ksi) [0.66(250MPa) = 165MPa].

Table 2.2 Allowable stress for steel members

Stress type	Symbol	Value [a]	AISC ref.
Tension (axial)	F_t	$0.60F_y$	5-40
Tension (net threaded area)	F_t	$0.33F_u$	4-3
Tension (plate, web tear)	F_t	$0.50F_u$	5-78
Compression	F_a	Variable (Kl/r, F_y)	5-42, 3-16
Shear	F_v	$0.40F_y$	5-49
Shear (plate, web tear)	F_v	$0.30F_y$	5-77
Shear (bolts)	F_v	See Table 1-D	4-5
Shear (welds, on effective area)	F_v	$0.30F_{weld}$	5-70
Bearing on bolts	F_p	$1.2F_u$	5-74
Bending, strong axis			
lateral support @ $\leq L_c$	F_b	$0.66F_y$	5-45
$L_u \geq$ unbraced L > L_c	F_b	$0.60F_y$	5-46
unbraced L > L_u	Variable		5-46, 2-146
Bending weak axis, bars, plates	F_b	$0.75F_y$	5-48

Source: Adapted from the AISC Manual

[a] Yield stress, F_y, From 32 to 100ksi [220 to 690MPa], depending on the grade of steel

[b] Ultimate stress, F_u, From 58 to 130ksi [400 to 900MPa], depending on the grade of steel

The AISC now supports a second structural design procedure for buildings called the LRFD (**L**oad and **R**esistance **F**actor **D**esign) method that is gaining increasing recognition in the codes and in the engineering profession. This is close to an ultimate-strength method, where the member is designed using stresses in the range where plastic deformation occurs. The expected loads (determining the "required strength") and the resistance of the structure (or "design strength") are factored to ensure that the yield is not actually exceeded.

LRFD can provide savings in material if serviceability aspects do not control. Since smaller sections will result, and higher stresses are considered for designing, deflection and buckling must receive special attention.

2.3 Standard shapes

Steel structures are frames made of columns and beams. Bracing in the form of diagonal members in the vertical or horizontal plane of structural bays may be necessary to resist lateral loads of wind and earthquakes.

Standard shapes are generally used in steel construction and their properties are described in AISC Manual of Steel Construction, Dimensions and Properties. The most common is the I shape that works as an efficient beam, since a major portion of the material is placed at the top and bottom of the web, where the bending stress is highest. This type of shape is classified by ASTM into W, M, S, and HP shapes. The difference between these is essentially the width of the flanges and the thickness of flanges and web. In Fig. 2.2, examples of W, M, S, and HP shapes with the same depth are shown. The W12X210 is very thick, weighs 210plf [*3065N/m*] and is intended to be used as a column (a lighter W12X22 is more likely to serve as a beam); S shapes have relatively narrow flanges and are typically used as joists.

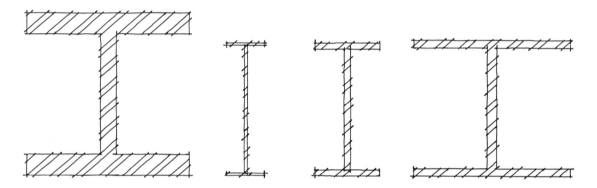

Fig. 2.2 W12X210, M12X11.8, S12X35, HP12X84

Other standard shapes are **channels**, **angles**, and **tees**. Tees are obtained by cutting W, M, and S shapes along the web. This group of shapes is commonly used in combination to fabricate horizontal and vertical framing systems. Double angles and double channels are common compression members for trusses and bracing elements (Fig. 2.3).

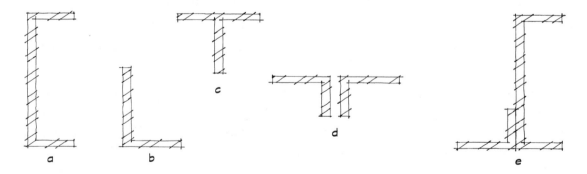

Fig. 2.3 a. Channel, b. Angle, c. Tee, d. Double angle, e. Built-up section; Channel and angle welded to form a lintel for masonry construction.

A third group of tubular shapes, called **steel pipe** (round) and **structural tubing** (rectangular), can be very effectively used for columns, beams, and truss construction (Fig. 2.4); unfortunately, they are almost twice as expensive as simple rolled shapes.

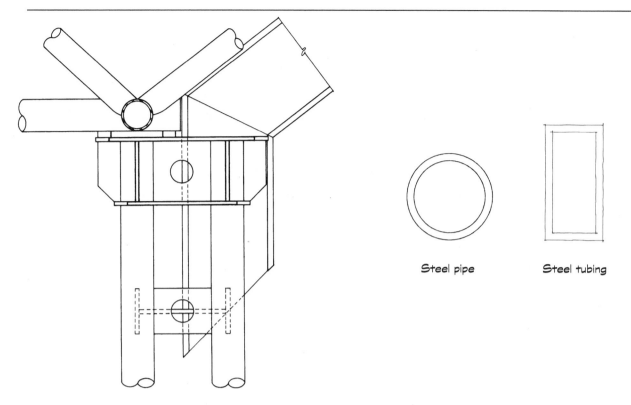

Fig. 2.4 Steel pipe used for exposed structure of O'Hare Airport, Chicago

Finally, **steel plates** are used for fabricating special members or to make gusset plates, stiffeners, and bearing plates. Plate dimensions are indicated in AISC 1-105 and 1-156 (Standard Mill Practice). All of these sections are classified as heavy hot-rolled shapes, with the exception of some pipes and tubing. There is a large variety of **light-gage** cold-formed structural members: the most common are the channel shapes used for relatively small structural members such as studs and purlins and the ribbed or corrugated sheets used for floor and roof decking (Fig. 2.5).

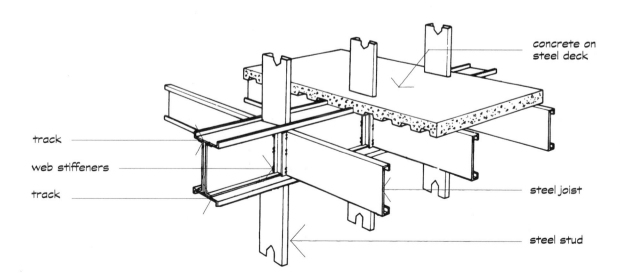

Fig. 2.5 Dale/Incor steel framing; load bearing interior wall

2.4 Fire considerations

Steel is not combustible under normal conditions, which would seem to make it a highly desirable material to use in high-risk fire situations. Unfortunately, although it will not support combustion, it does undergo significant strength changes beginning at 600°F [316°C]. At 1200°F [649°C], it has effectively lost all structural capacity. These temperatures are reached in almost any fire, causing structural deformation and collapses. It won't burn, but it will collapse, which is equally disastrous!

Most steel structures in urban areas must be protected by physical coverings (Fig. 2.6), by fire-suppression systems (sprinklers), or other means. Figure 2.7 illustrates a hollow structural system that utilizes water to carry heat away from the exposed structural steel.

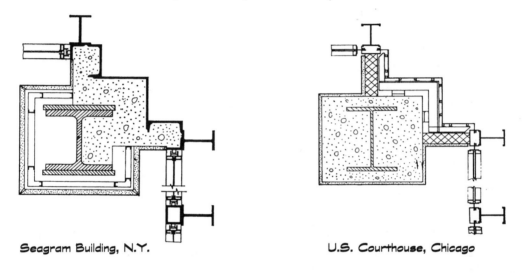

Seagram Building, N.Y.   U.S. Courthouse, Chicago

Fig. 2.6 Fireproofing techniques

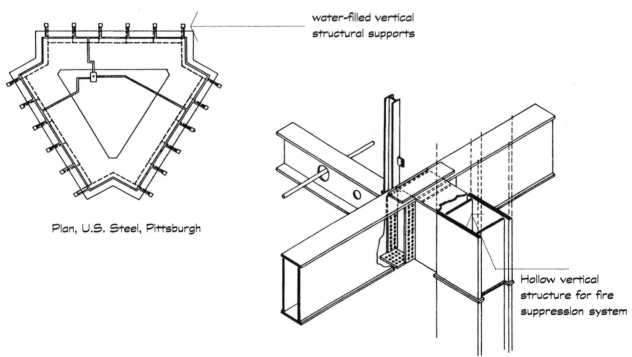

water-filled vertical structural supports

Plan, U.S. Steel, Pittsburgh

Hollow vertical structure for fire suppression system

Fig. 2.7 US Steel hollow fire resistive structural system

2.5 Surface finishes

Structural steel is subject to corrosion and must be painted or protected by some means. If not, the corrosion will eventually cause a significant loss of structural capacity due to the physical reduction of the section.

There are corrosion-resistant "weathering" steels that will create a very tight corroded surface that will act similar to paint and seal the inner material and protect it from further action. The unique feature of these steels is that if the surface is "scratched," it will heal itself by forming a new corroded surface. The con to this pro is that the structure will "bleed" a stain onto adjacent materials. The John Deere Headquarters utilizes this material to capture the rugged character of farm equipment. The building stands like a great piece of farm equipment on the open rolling Illinois landscape.

Fig. 2.8 John Deere Headquarters using Cor-Ten steel.

3 STRUCTURAL ELEMENTS AND SYSTEMS

$$F = P/A \quad \text{and} \quad F = Mc/I$$

Did you read the premise? Please do. You'll see that most of the virtually infinite number of formulas presented in structural texts and manuals are an elaboration or reinterpretations of these two simple formulas. Remember, there are only three basic member types you can design, and there are two additional types that are simply combinations of the first three, so here we go!

3.1 Member types

Beams: These are the probably the most common elements used in structures. They typically are horizontal, although they can be at any orientation. Small components, bearing plates, column base plates, stair treads, even the individual elements of Levelor blinds, are basically simple beams at a small scale. Beams are primary bending members and we must consider moment, shear, and deflection in all cases, even though all three will not be applicable in all beam situations. The decision to consider or not consider any of these three BIG ones must be a conscious one. Look at **each of them every time** and decide if they are relevant.

Columns: Although columns may not be used in every building situation (bearing walls may be substituted for a column), you'll see as we progress through the materials that even walls are designed with many of the same considerations that we use in column design. These are basically elements that carry pure compression loads regardless of physical orientation. The upper chords of many trusses, which may be horizontal or sloped, may be pure compression members and are technically "columns."

Tension members: These members are another example of a pure axially loaded member with the load being tension as opposed to compression in columns. This small distinction has a major effect on the geometric considerations of the member. Cross-sectional form is not a consideration with tension members, but it is a major consideration in compression components; we'll discuss that further at a later time. Again, the orientation of the member has no effect on its "type." Although tension members are frequently thought of as "hangers," in a vertical orientation, the lower chords of trusses are typically tension members in a horizontal orientation, and a catenary curve is an efficient way to develop horizontal spanning capacity with a tension member. Under any of these conditions, however, a tension member remains a pure axially loaded member.

Bending + compression or tension: These are the "bi-athletes" of structures. (The "tri-athletes" are these same members that have bending about **two** axes and compression or tension all at the same time - later, OK?) They must be able to simultaneously act as two different types of members.

A simple sloped roof joist used in most residential construction is a good example of Moment + Compression since it is a part of a "three-hinged arch" with a uniform roof load applied along its length. The ceiling joist in this same system, which resists the outward horizontal thrust of the joist while supporting a ceiling, is a perfect example of a Moment + Tension member.

Most structures, even simple ones as described for Moment + Compression or Tension, have all of these relatively simple elements combined with a system of support. This is a little scary since we often think of a small residence of such simplicity that it requires no engineering, yet the roof and ceiling joist are among the most complicated of all possible structural situations.

3.2 System selection

Don't be surprised when you see this section, as a reference, included in each material description. The issues are similar and reiteration will help you remember what you're looking for when you make system selection decisions.

Selection of a steel system involves a complex set of interrelated issues, so there is no single answer. You need to at least consider the following.

3.2.1 Spatial requirements

What are the volumetric requirements of the function that the space will house?

Is there a planning module or a leasing module that should be column-free or could be easily isolated? Parking garages, office structures, schools, hospitals, and retail establishments all have identifiable modules that are more desirable. Unfortunately, these are not absolute, with the possible exception of parking facilities, and must be identified during programming and schematic design stages.

3.2.2 Expansion

Is the spatial system likely to require frequent remodeling?

What is the likelihood of expansion of the system?

3.2.3 Integration of systems

Will the environmental/lighting systems require frequent change?

Does the spatial system require isolated volumes of space: for temperature, sound, security, utilities management, etc.?

Where will the environmental systems be located, potentially based on the region of the country where the building will be constructed i.e. supply conditioned air high where you need cooling primarily and low where you need heating primarily.

3.2.4 Ease of erection/construction

What is the availability of skilled labor?

Are material or component production facilities easily accessible?

3.2.5 Fire resistance

What are the fire rating requirements of the building as a whole?

What are the isolation characteristics of the zones of the building?

What are the patterns of access required for emergency egress?

3.2.6 Soil conditions

What is the nature of the soil on the site?

Chemical reactions with steel products?

Bearing capacities? Steel structures are relatively light structures, this may be an advantage for foundation design, and a disadvantage for thermal balancing. A steel structure may weigh slightly more than its equivalent in wood and significantly less than its concrete counterpart. This could be a major consideration if the building is supported by weak soils that would require an excessive investment in foundations.

3.2.7 Spans and loadings

Some common combinations of span/loading/mechanical integration characteristics for simple systems would include the following.

3.3 Low-rise frame systems

Skeleton frames composed of simple beams and columns can be cost-effective even for small buildings, when designed carefully following some basic rules for economy. The cost of the structure is due to :

3.3.1 Quantity (tons) of material utilized and its relative cost as a system

Use of high-strength steel (F_y = 50ksi)[345MPa], where the design is controlled by bending moment or shear, can lead to smaller members and cost savings, acknowledging that high-strength steel has a price/strength ratio 25% higher than mild steel (F_y = 36ksi)[250MPa]. This means that a grade 50 beam may use much less material for the same performance, assuming deflection does not control the design. In secondary members, less expensive steel, readily available from a fabricator's stock, could be more appropriate.

In most cases, it will be necessary for the architectural designer to look beyond the specific cost of the material to evaluate the cost of the entire system. Some of the most obvious considerations would include such items as fireproofing, finished appearance, finishing techniques, volume of space occupied by structural elements and its compatibility with other systems, ranging from mechanical ceilings to wall systems.

3.3.2 Fabrication of the members

The cost of fabrication is roughly the same whether the member is large or small. It is therefore convenient from a purely structural point of view to minimize the number of columns and beams by increasing the span as much as practical. In practice, the limit of the span is set by the following:

Limitations of depth of the horizontal structure:

> Very deep beams increase the height of the building and the volume to enclose and ultimately heat and cool, **for the life of the building**. Only in exoskeletal buildings like Crown Hall at the Illinois Institute of Technology does the depth of the roof structure not have interior implications. This increased depth might otherwise cause a conflict between the structure and continuity of the horizontal mechanical system. The downside of an exoskeletal system is that it creates a much larger exterior surface area to be maintained over the years. Every decision has pros and cons.

Limitation of deflection and vibration

Long spans are often controlled by these serviceability aspects. You'll remember from statics that shear is generally a function of L, moment is generally a function of L^2, and deflection is generally a function of L^4. This means that doubling the span doubles the shear, makes the moment 4X as large and creates 16X as much deflection. It's easy to see that an increase in span has serious implications on the depth of the structure.

Limitation of transportation:

The maximum length of the loads a truck can transport should be considered. Excessively long spans (those exceeding 60ft) [*18m*] will likely have to be assembled in the field from components that can be more easily transported.

In cases of structural/mechanical conflicts, a 2:1 bay proportion i.e. 40ft [*12m*] joist span with a 20ft to 30ft [*6m to 9m*] beam supporting the joist will result in a system that has beams and joist of approximately equal depth. This may not be a problem if horizontal mechanical runs are parallel to the deep beams. This could become a major factor if these systems are forced to cross one another, although we'll show you how you can penetrate a beam with a duct or pipe by reinforcing the beam or by selecting a penetration location where the hole will not be significant. Architects must look at the total picture of spatial systems and material integration, not just at the lightest structure.

3.3.3 Erection of the system

Bracing and moment-resisting connections are both expensive items and their use should be rationalized. For instance, moment-resisting frames might be used at spandrel beams only, located at the faces of the building, where they can be deeper without affecting floor height. A deeper beam makes rigid connections easier to develop since it increases the potential moment couple between compression and tensile forces. (Remember discussing moment couples in statics? Here they are!)

3.3.4 Miscellaneous considerations

Repetitive members reduce the cost of design, detailing, shop drawings and fabrication costs. Each component of the building will have a shop drawing produced by the fabricator for its construction. If you have 50% more unique members, you'll have 50% more shop drawings that the fabricator will be required to produce and 50% more that someone in your office or the engineer's will have to check and approve. This even has implications in the field. The more unique components that need to be stored and tracked, the more complicated the erection process becomes and the larger the staging areas become. KISS - keep it simple, ... uh, students!

Paint only members required by AISC specifications; members that will be fireproofed or covered with concrete need not be painted. A paint job as cheap as $0.15/sqft [$1.50/m²] multiplied by a steel surface of 200,000 sqft [*20,000m²*]can result in a savings of $30,000.00. This may seem to be an extremely large square footage for a structural system, but a single common medium-scale beam, a W16X40, has 4.89 sqft[*1.49m²/m*] of surface area per foot of length. This means a single 30ft [*9m*] W16X40 would have 146.6 sqft [*13.4m2*] of surface area to paint. This is the area of an 8-foot-high wall, 18.33 feet long, roughly the size of a small bathroom. **One beam equals one bathroom.** This adds up quickly even for a small-scale structure.

The cost of the structure should always be considered in the context of the building. A relatively expensive structural solution could result in savings on the overall building or on the life-cycle costs, due to gains in usable space or reductions in the cladding and finishing materials, or in simplification of the ductwork layout. The chapters dealing with beams and columns will help clarify these issues, although it should be clear that there is not a single simple solution and the answer is left to the creativity of the designer.

A typical framing plan for a low-rise commercial building is shown in Figure 3.1. The 30ft x 30ft [9mx9m] structural bay is likely to meet the economy criteria just illustrated.

W16x26 steel beams (F_y = 36ksi) at 10ft o.c. [3m]:

    3-1/4in [82mm] lightweight concrete topping on

    3in [76mm] metal deck

22K5 open-web steel joist at 3ft o.c. [0.9m]:

    3in concrete (f_c = 3.0ksi) [21MPa] on 9/16in [14mm] metal form

Open-web steel joist design is explained in Chapter 4. The two solutions are structurally sound and competitive in cost for a 50lb/sqft [2400N/m²] code live load. A choice should be made on the basis of construction (ducts and piping layout) and other architectural considerations.

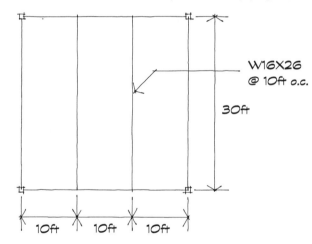

W16X26 @ 10ft o.c.

30ft

10ft   10ft   10ft

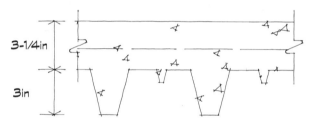

3-1/4in

3in

a. Mesh reinforced concrete on deep metal deck

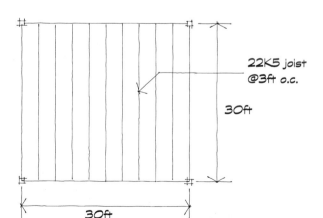

22K5 joist @3ft o.c.

30ft

30ft

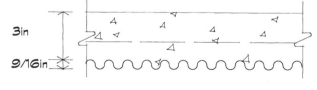

3in

9/16in

b. Mesh reinforced concrete on light metal deck

Fig. 3.1 Alternative floor framing systems

Roof construction differs from floor construction as a result of different loads and environmental considerations. The creation of slope for storm water drainage can be done in different ways, as shown in Figure 3.2. Each solution affects the structural design.

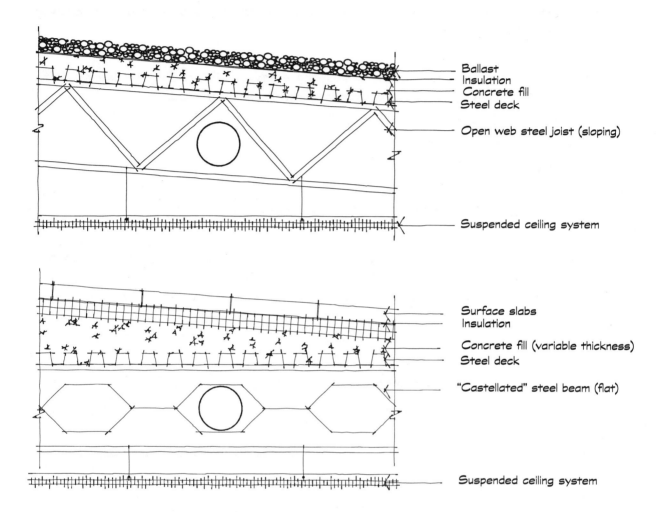

Fig. 3.2 Alternative methods of creating slope for typical roof framing systems

The floor over the joists is typically built using a steel deck or "centering" with a concrete fill that can act with the deck as a composite slab. The use of a concrete topping on a steel deck or form also can act as a composite design with the steel joists or beams as well, and the required steel sections can be reduced with significant savings. This does require additional erection cost to assure that they do actually act as a composite section. Although composite steel-concrete beam theory is beyond the scope of this book, this type of construction is mentioned here due to its increasing use in practice. The connection between the steel beam and the concrete topping is made possible by using steel shear studs welded to the top flange, or in some cases the web members of a joist are extended above the upper chord to create a connection. A major consideration with this second option is that the deck must be cut into short sections to fit between the joist and cannot be used in the more desirable three-span condition. The continuity over three beams or joists helps distribute the moment and deflection in the system, which potentially increases its capacity. Composite design is commonly used in highway bridge construction; you may have seen steel beams during construction that appear to have "railroad spikes" sticking out of their top flange. These are the shear studs just mentioned that anchor the concrete to the steel, creating a beam section that will include a portion of the concrete topping, as shown in Fig. 3.3. This technique is used in highway construction to minimize the differential distance between roadways. One foot of elevation difference between crossing roadways may mean a change in fill over several hundred feet.

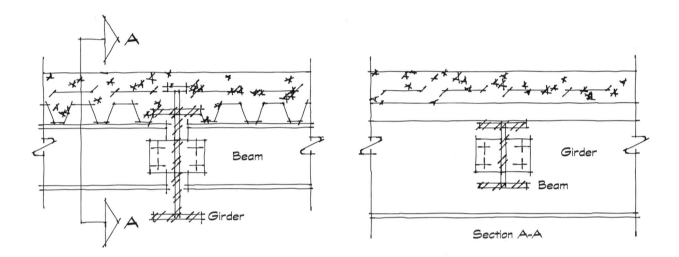

Fig. 3.3 Composite steel-concrete beams for floor structures

Additional advantages of concrete compared to lightweight steel systems are fireproofing, sound insulation, and, in some cases, reduced vibrations. Steel protection against fire is typically accomplished by shielding the steel with cement or gypsum board, through the use of a suspended ceiling. Fireproofing can be sprayed on beams and on roof decks to give the appropriate fire resistance. Where only 1 hour of fire protection between floors is required (based on UBC requirements according to building occupancy), such as in apartment houses and hotels, a thickness of 2.5in [63mm] lightweight concrete or 3.5in [89mm] standard concrete is sufficient for the floor slab. For ceiling and floor or roof slab construction, Table 43-C of the UBC should be consulted.

3.4 Medium and high-rise systems

Skeleton frames have been used for medium- and high-rise buildings since the end of the Nineteenth century. As described earlier, a frame consists essentially of columns, beams, and a bracing system. The bracing system, or more specifically the system adopted to resist lateral loads from wind and earthquakes, generates the real difference between steel-frame systems. This will be further discussed in Chapter 17, which deals with lateral loads. We can, however, easily understand the basic difference between a **braced frame** and a **moment-resisting frame** by looking at a simple portal (which could represent one bay of a larger system). In Fig. 3.4, the portal with hinged connections at the ends of the column is obviously unstable. We can (a) add bracing or (b) make at least one end of the columns fixed. In both cases, the statics are still simple, since the beam can be calculated as a simply supported member, and the columns are cantilevers subject to combined bending and compression. One very common technique used in

relatively small-scale buildings is to provide stability by utilizing the stair towers and/or the elevator shaft, all of which must be fire-resistive construction. Since this fire resistance is frequently
accomplished by the use of masonry or reinforced concrete enclosures, these potentially heavy
structures can also act as vertical tubes providing stability and resisting lateral loads in the
system.

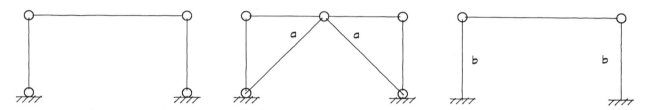

Fig. 3.4 Hinged portal stabilized with bracing (a) or moment-resisting connections (b)

Common examples of small- and medium-span one-story structures are the two-hinged frame and
the three-hinged frame, which are statically determinate structures. The "arch" type frames,
often employed in industrial buildings, are also called portal frames (Fig. 3.4). It is essential to
ensure structural continuity of the member where it forms a corner, or knee, between the
column and the horizontal or sloping roof member. This is done by creating a connection at this
joint that has the capacity to transfer any bending moment from the sloped member to the
vertical member, beam to column.

Low- and medium-rise buildings adopt these concepts, as illustrated in Fig. 3.5.

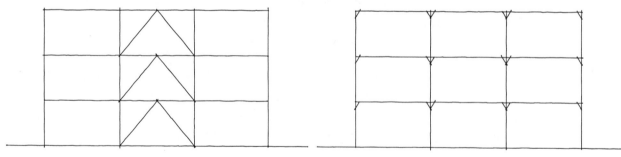

Fig. 3.5 Low-rise buildings using diagonal bracing or moment resisting connections

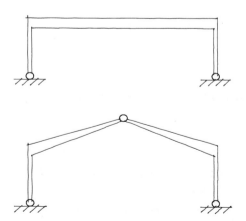
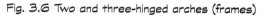
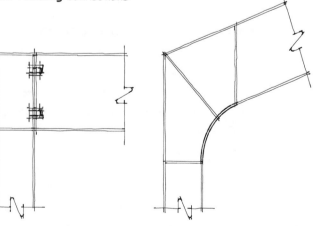

Fig. 3.6 Two and three-hinged arches (frames)

High-rise buildings sometimes use braced bays at each floor effectively forming a vertical truss, as in the John Hancock tower in Chicago. This building is like four large "gates" cantilevered from the ground (instead of from a post). The four "gates"/trusses on the faces of the building help provide lateral bracing to resist the wind loads. Once again, the diagonal bracing is designed to act only in tension; the "X's" always provide one tension diagonal regardless of the direction of the wind. Very tall buildings in seismic areas have more complex structural solutions; the most obvious is the use of several braced bents connected horizontally, every so many floors, by trusses (Fig. 3.7).

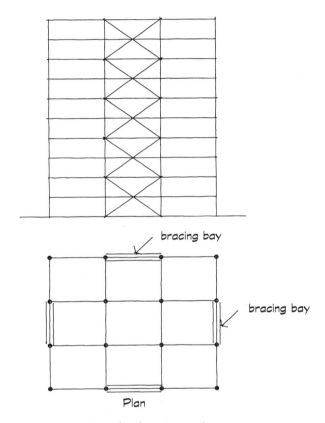

bracing bay

bracing bay

Plan

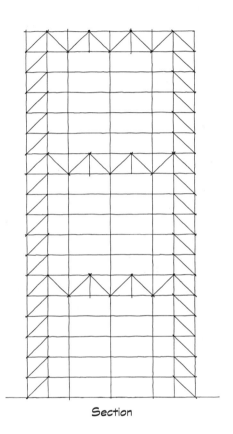

Section

Fig. 3.7 Braced bent (bay) and superframe concepts

Lateral support techniques include:

(a) Vertical circulation/mechanical cores providing interior lateral bracing similar to the trunk of a tree.

(b) Introduction of shear walls: solid/reinforced vertical planes concentrically placed and located at 90° to one another to provide lateral bracing in all directions. The plywood sheathing panels you may have noticed at the corners of wood stud residential construction are an example of this idea.

(c) Diagonal tie bracing in "X"'s to prevent lateral movement. This is similar to the rope ties used to hold up a tent, except they are placed on the interior of the structure instead of on the exterior.

(d) Trussed floors between braced bents (superframe).

(e) Pierced tube (extrarigid perimeter frame).

3.5 Architectural considerations

In the choice of the structural system for a building, considerations relative to the use and maintenance of the building and aesthetic considerations will play an important part. Possible ways of accommodating HVAC ductwork and other pipes are shown in Fig. 3.8. Horizontal ducts can travel within the beam depth through chases in the web and, in the case of trussed joists, between the two chords.

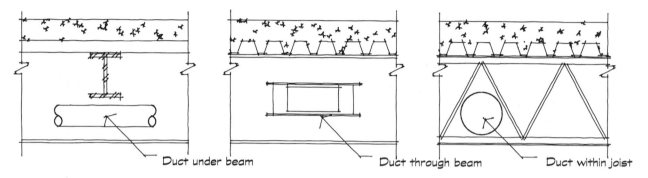

Duct under beam   Duct through beam   Duct within joist

Fig. 3.8 Mechanical/structural systems integration

The steel structure will generally be lighter and require less volume than reinforced concrete, providing more usable space. It is quick to erect and later modifications to the structure are easier. However, when fireproofing and maintenance are considered, it is not always possible to leave the steel exposed. In the Seagram building, Mies Van der Rohe used a steel frame encased in concrete, and finally clad in metal. The result is ultimately a sheep, in wolf's clothing, in sheep's clothing; it is a bit monomaniacal, but an appearance expressing the aesthetic nature of the frame without exposing it directly. Roche and Dinkeloo, in the Knights of Columbus building, have met the fire requirements by placing the exposed spandrel beams at a 5ft distance from the windows and by that technique avoiding placing fireproofing on the structural members (Fig. 3.9).

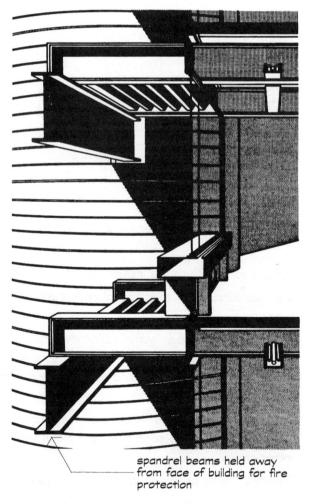

spandrel beams held away from face of building for fire protection

Fig. 3.9 Knights of Columbus building.

4 PRE-ENGINEERED SYSTEMS

4.1 Open-web steel joist

Steel members for beams and columns are produced in standard shapes, and, in some cases, a preliminary design can be carried out with the aid of tables. This is also true for a large number of proprietary structural systems (not just steel, but also wood and concrete). In addition to the ASTM (the American Society for Testing of Materials) shapes for beams and columns, there are other lighter shapes such as the "M" section (a rolled section of the same form as the standard "W" and "S" beams, but thinner). In addition, lightweight cold-formed shapes are used for secondary framing (steel "C" studs and plates), and for relatively light uniform floor and roof loads, open-web joist can be used. One type is shown in Fig. 4.1.

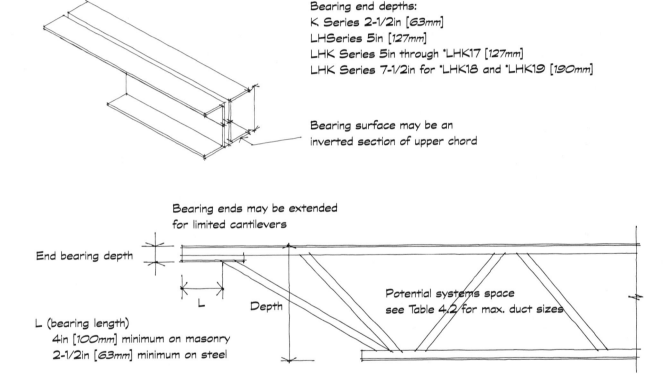

Bearing end depths:
K Series 2-1/2in [63mm]
LHSeries 5in [127mm]
LHK Series 5in through *LHK17 [127mm]
LHK Series 7-1/2in for *LHK18 and *LHK19 [190mm]

Bearing surface may be an
inverted section of upper chord

Bearing ends may be extended
for limited cantilevers

End bearing depth

L

Depth

Potential systems space
see Table 4.2 for max. duct sizes

L (bearing length)
   4in [100mm] minimum on masonry
   2-1/2in [63mm] minimum on steel

Fig. 4.1 Typical open-web steel joist characteristics

These sections are commonly referred to as "bar joist" because some of the original sections were constructed using bar stock for both web and flanges. Modern joist use alternative flange configurations consisting of light-gage angles or a formed section similar in configuration to angles. The web consists usually of a continuous round bar bent and welded to the flanges, yielding a member similar to a truss. These sections, for all practical purposes, are "generic" trusses. These joists are spaced from 24 to 48in [610-1220mm] apart, depending on the magnitude of load and the capacity of the steel decking or "centering" spanning between them. Special sections are manufactured that have the web members penetrating through and extending above the upper flange elements to allow for encasement and therefore composite structural action with a poured concrete deck.

Open-web joists have the convenience as well as the limitations of standardized components; the main limitation is that they are designed to work as simply supported beams. However, they are produced in different sizes and shapes to satisfy a variety of spans and roof pitches (Fig. 4.1). The strength of the steel used for the chords (F_y = 50ksi [350MPa] or higher), combined with a relative light weight, allow the use of these joists for spans over 200ft [60m]. The major characteristics are shown in Table 4.1.

Table 4.1 Major characteristics of joist series, F_y = 50ksi [350MPa]

Series	Depth		Maximum span	
K	8-30in	203 thru 762mm	60ft	18meters
LK	18-48in	457 thru 1219mm	96ft	29meters
DLK	52-72in	1321 thru 1829mm	144ft	44meters
SLK	80-120in	2032 thru 3048mm	240ft	73meters

Table 4.2 Maximum duct dimensions for K series joist [× 25.4mm]

Joist depth	Circular duct	Square duct	Rectangular duct
8 inches	5 inches	4X4 inches	3X8 inches
10 inches	6 inches	5X5 inches	3X8 inches
12 inches	7 inches	6X6 inches	4X9 inches
14 inches	8 inches	6X6 inches	5X9 inches
16 inches	9 inches	7-1/2X7-1/2 inches	6X10 inches
18 inches	11 inches	8X8 inches	7X11 inches
20 inches	11 inches	9X9 inches	7X12 inches
22 inches	12 inches	9-1/2X9-1/2 inches	8X12 inches
24 inches	13 inches	10X10 inches	8X13 inches
26 inches	15-1/2 inches	12X12 inches	9X18 inches
28 inches	16 inches	13X13 inches	9X18 inches
30 inches	17 inches	14X14 inches	10X18 inches

In all cases you should consult your joist manufacturer of current specifications.

Manufacturers catalogs may give different ranges of available spans, but all joists with the same series denomination have the same strength, i.e. they carry the same safe live and total loads with the code limitations for deflection. Camber is optional with the manufacturer, but when provided, the Steel Joist Institute recommends cambers between 1/1000 and 1/500 of the span. The upward camber (curve) helps compensate for the dead-load deflection.

The standard Steel Joist Institute specifications include the design of end supports; as an example, Table 4.3 illustrates those support minimum requirements for K series joists:

Table 4.3 End support minimum requirements for K joists

Support material	Extension of chord over support N	Width of bearing plate B
Masonry or concrete	4in [100mm]	6in [150 mm]
Steel	2.5in [63mm]	Not required

Notice in Fig. 4.2 that the ends are constructed of an inverted section of the upper chord. This provides a wider surface at the end bearing point to provide for connection or anchorage and to distribute the reaction over a larger surface (a snowshoe vs. a spike heel). Since these sections are commonly used with masonry construction, the end-bearing condition has been designed to be dimensionally compatible with block and brick coursing. Light K series joist are 2-1/2in [63mm] deep, and heavier LK and DLK series joist are 5in [127mm] deep. In most applications of masonry construction, a small bearing plate (4x6x1/4in) [100x150x6mm] with two welded anchors (#3 bars)[#10M] will be placed in the bond beam or bearing location by the masonry contractor. Although this may not be structurally necessary, in some situations it provides a consistently level surface to attach the joist. It is much easier to set the elevation of a small light plate properly than to struggle with a much heavier joist and shim it to the proper elevation. It also allows the steel erector to quickly place and "weld" the joist into final location.

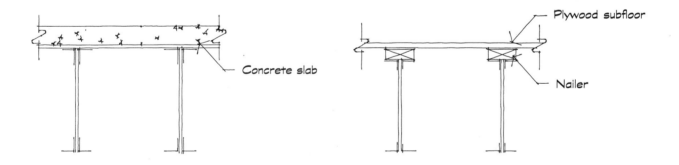

Fig. 4.2 Concrete and wood floor systems on open-web steel joist

The Steel Joist Institute has adopted a Standard Load Table (reproduced in this chapter, Table 4.4) for open-web steel joist "selection," where joist are designated with a code indicating the depth (in), the series (K, LK, DLK, etc.) and the chord designation in relative terms (1 to 12), i.e. a 12K4 joist is 12in [305mm] deep, belongs to the K series, and has a No. 4 chord. No actual physical dimension is associated with the chord designation, however, its properties must be such that regardless of the manufacturers configuration, it will meet the industry load-performance standards for a "12K4." This has some architectural significance: If the system is intended to be left exposed, some consideration may have to be given to the specific manufacturer of the joist to ensure that architectonic expectations are met.

The joist selection tables are based on several assumptions:

The maximum live load deflection is limited to L/360, although L/240 may be used in certain instances.

The maximum slope of the joist is limited to 1/2in per ft [42mm per m].

All loads, both live and dead, are uniformly distributed.

No joist may be used at a span exceeding 24 times its depth.

All loads are computed in lb/ft or plf [N/m]

The joist load table gives the allowable live load (for deflection of L/360) and the allowable total load (for moment or shear) for each joist in a variety of different spanning conditions. The criteria for choosing a particular joist may be:

1. Weight: If the cost of the joist is the only consideration, that cost will be determined by the total tonnage of steel involved.

2. Depth: If mechanical considerations will be best met by using a deeper joist so that systems can be integrated into the system as opposed to hanging them below the joist, potentially creating a deeper total system than necessary.

3. Standardization: If a wide variety of different sizes are specified for a project, the shop drawings, field storage and bridging installation complexity may all be increased, causing increased cost.

The joist load tables do not provide for any loading condition other than uniform. Unfortunately, most buildings have loadings other than this single uniform condition. To execute a design for other loading conditions, it is necessary to find the uniform total load (TL) and live load (LL) that will cause the same maximum conditions generated by the combined uniform and concentrated loads. This will give a relative accurate selection, but the manufacturer MUST be informed of any loadings other than uniform so appropriate local reinforcement of the truss may be made. This should be included as part of the construction documents. Note that since the TL values are based on the **worst-case scenario** of moment or shear, it may be necessary to evaluate more than one TL loading condition. Later, Example 4.2 will illustrate this technique.

Span "definition" for the purposes of this book.

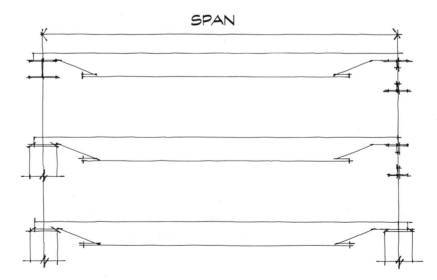

SPAN

Table 4.4 Standard load table   K - series open-web steel joist

The **BOLD** (first) figures give the TOTAL (live load + dead load) safe, uniformly distributed, load-carrying capacities in plf (pounds per linear foot) of K-series steel joist. The *ITALICIZED* (second) figures are the uniform live loads in plf that the joists will support within the limits of L/360 (1/360 of the span). This limitation is the standard for floors and flat roof systems. For a higher allowable deflection, i.e., L/240, the *ITALICIZED* values may be multiplied by 1.5. Other deflection criteria may be obtained by a similar proportioning method.
UNDER NO CONDITIONS MAY THE TOTAL LOAD CAPACITY OF THE JOIST BE EXCEEDED!   Table contains most common spans and sizes.

12-inch and 14-inch deep sections [305 and 356mm]

Joist	12K1	12K3	12K5	14K1	14K3	14K4	14K6
Wt. plf [a]	5.0plf	5.7plf	7.1plf	5.2plf	6.0plf	6.7plf	7.7plf
Span (ft) [b]							
12	*550/550*	*550/550*	*550/550*				
13	*550/510*	*550/510*	*550/510*				
14	*500/425*	*550/463*	*550/463*	*550/550*	*550/550*	*550/550*	*550/550*
15	*434/344*	*543/428*	*550/434*	*511/475*	*550/507*	*550/507*	*550/507*
16	*380/282*	*476/351*	*550/396*	*448/390*	*550/467*	*550/467*	*550/467*
17	*336/234*	*420/291*	*550/366*	*395/324*	*495/404*	*550/443*	*550/443*
18	*299/197*	*374/245*	*507/317*	*352/272*	*441/339*	*530/397*	*550/408*
19	*268/167*	*335/207*	*454/269*	*315/230*	*395/287*	*475/336*	*550/383*
20	*241/142*	*302/177*	*409/230*	*284/197*	*356/246*	*428/287*	*525/347*
21	*218/123*	*273/153*	*370/198*	*257/170*	*322/212*	*388/248*	*475/299*
22	*199/106*	*249/132*	*337/172*	*234/147*	*293/184*	*353/215*	*432/259*
23	*181/93*	*227/116*	*308/150*	*214/128*	*268/160*	*322/188*	*395/226*
24	*166/81*	*208/101*	*282/132*	*196/113*	*245/141*	*295/165*	*362/199*
25	*0/0*	*0/0*	*0/0*	*180/100*	*226/124*	*272/145*	*334/175*
26				*166/88*	*209/110*	*251/129*	*308/156*
27				*154/79*	*193/98*	*233/115*	*285/139*
28				*143/70*	*180/88*	*216/103*	*265/124*
29				*0/0*	*0/0*	*0/0*	*0/0*

[a] × 4.6N/m
[b] × 0.305m
Safe load in lb/ft × 14.6N/m

Table 4.4 Standard load table   K - series open-web steel joist
       16-inch deep sections [406mm]

Joist	16K2	16K3	16K4	16K5	16K6	16K7	16K9
Wt. plf	5.5plf	6.3plf	7.0plf	7.5plf	8.1plf	8.6plf	10.0plf
Span (ft)							
16	550/550	550/550	550/550	550/550	550/550	550/550	550/550
17	512/488	550/526	550/526	550/526	550/526	550/526	550/526
18	456/409	508/456	550/490	550/490	550/490	550/490	550/490
19	408/347	455/386	547/452	550/455	550/455	550/455	550/455
20	368/297	410/330	493/386	550/426	550/426	550/426	550/426
21	333/255	371/285	447/333	503/373	548/405	550/406	550/405
22	303/222	337/247	406/289	458/323	498/351	530/385	550/385
23	277/194	308/216	371/252	418/282	455/307	507/339	550/363
24	254/170	283/189	340/221	384/248	418/269	465/298	550/346
25	234/150	260/167	313/195	353/219	384/238	428/263	514/311
26	216/133	240/148	289/173	326/194	355/211	395/233	474/267
27	200/119	223/132	268/155	302/173	329/188	366/208	439/246
28	186/106	207/118	2249/138	281/155	306/168	340/186	408/220
29	173/95	193/106	232/124	261/139	285/151	317/167	380/198
30	161/86	180/96	216/112	244/126	266/137	296/151	355/178
31	151/78	168/87	203/101	228/114	249/124	277/137	332/161
32	142/71	158/79	190/92	214/103	233/112	259/124	311/147
33	0/0	0/0	0/0	0/0	0/0	0/0	0/0

Table 4.4 Standard load table   K - series open-web steel joist

18-inch deep sections [457mm]

Joist	18K3	18K4	18K5	18K6	18K7	18K9	18K10
Wt. plf	6.6plf	7.2plf	7.7plf	8.5plf	9.0plf	10.2plf	11.7plf
Span (ft)							
18	550/550	550/550	550/550	550/550	550/550	550/550	550/550
19	514/494	550/523	550/523	550/523	550/523	550/523	550/523
20	463/423	550/490	550/490	550/490	550/490	550/490	550/490
21	420/364	506/426	550/460	550/460	550/460	550/460	550/460
22	382/316	460/370	518/414	550/438	550/438	550/438	550/438
23	349/276	420/323	473/362	516/393	550/418	550/418	550/418
24	320/242	385/284	434/318	473/345	526/382	550/396	550/396
25	294/214	355/250	400/281	435/305	485/337	550/377	550/377
26	272/190	328/222	369/249	402/271	448/299	538/354	550/361
27	252/169	303/198	342/222	372/241	415/267	498/315	550/347
28	234/151	282/177	318/199	346/216	385/239	463/282	548/331
29	218/136	263/159	296/179	322/194	359/215	431/254	511/298
30	203/123	245/144	276/161	301/175	335/194	402/229	477/269
31	190/111	229/130	258/146	281/158	313/175	376/207	446/243
32	178/101	215/118	242/132	264/144	294/159	353/188	418/221
33	168/92	202/108	228/121	248/131	276/145	332/171	393/201
34	158/84	190/98	214/110	233/120	260/132	312/156	370/184
35	149/77	179/90	202/101	220/110	245/121	294/143	349/168
36	141/70	169/82	191/92	208/101	232/111	278/132	330/154
37	0/0	0/0	0/0	0/0	0/0	0/0	0/0

Table 4.4 Standard load table   K - series open-web steel joist
20-inch deep sections [508mm]

Joist Wt. plf Span (ft)	20K3 6.7plf	20K4 7.6plf	20K5 8.2plf	20K6 8.9plf	20K7 9.3plf	20K9 10.8plf	20K10 12.2plf
20	517/517	550/550	550/550	550/550	550/550	550/550	550/550
21	468/453	550/520	550/520	550/520	550/520	550/520	550/520
22	426/393	514/461	550/490	550/490	550/490	550/490	550/490
23	389/344	469/402	529/451	550/468	550/468	550/468	550/468
24	357/302	430/353	485/396	528/430	550/448	550/448	550/448
25	329/266	396/312	446/350	486/380	541/421	550/426	550/426
26	304/236	366/277	412/310	449/337	500/373	550/405	550/405
27	281/211	339/247	382/277	416/301	463/333	550/389	550/389
28	261/189	315/221	355/248	386/269	430/298	517/353	550/375
29	243/170	293/199	330/223	360/242	401/268	482/317	550/359
30	227/153	274/179	308/201	336/218	374/242	450/286	533/336
31	212/138	256/162	289/182	314/198	350/219	421/259	499/304
32	199/126	240/147	271/165	295/179	328/199	395/235	468/276
33	187/114	226/134	254/150	277/163	309/181	371/214	440/251
34	176/105	212/122	239/137	261/149	290/165	349/195	414/229
35	166/96	200/112	226/126	246/137	274/151	329/179	390/210
36	157/88	189/103	213/115	232/125	259/139	311/164	369/193
37	148/81	179/95	202/106	220/115	245/128	294/151	349/178
38	141/74	170/87	191/98	208/106	232/118	279/139	331/164
39	133/69	161/81	181/90	198/98	220/109	265/129	314/151
40	127/64	153/75	172/84	188/91	209/101	251/119	298/140
41	0/0	0/0	0/0	0/0	0/0	0/0	0/0

Table 4.4 Standard load table   K - series open-web steel joist
22-inch deep sections [559mm]

Joist	22K4	22K5	22K6	22K7	22K9	22K10	22K11
Wt. plf	8.0plf	8.8plf	9.2plf	9.7plf	11.3plf	12.6plf	13.8plf
Span (ft)							
22	550/548	550/548	550/548	550/548	550/548	550/548	550/548
23	518/491	550/518	550/518	550/518	550/518	550/518	550/518
24	475/431	536/483	550/495	550/495	550/495	550/495	550/495
25	438/381	493/427	537/464	550/474	550/474	550/474	550/474
26	404/338	455/379	496/411	550/454	550/454	550/454	550/454
27	374/301	422/337	459/367	512/406	550/432	550/432	550/432
28	348/270	392/302	427/328	475/364	550/413	550/413	550/413
29	324/242	365/272	398/295	443/327	532/387	550/399	550/399
30	302/219	341/245	371/266	413/295	497/349	550/385	550/385
31	283/198	319/222	347/241	387/267	465/316	550/369	550/369
32	265/180	299/201	326/219	363/242	436/287	517/337	549/355
33	249/164	281/183	306/199	341/221	410/261	486/307	532/334
34	235/149	265/167	288/182	321/202	386/239	458/280	516/314
35	221/137	249/153	272/167	303/185	364/219	432/257	494/292
36	209/126	236/141	257/153	286/169	344/201	408/236	467/269
37	198/116	223/130	243/141	271/156	325/185	386/217	442/247
38	187/107	211/119	230/130	256/144	308/170	366/200	419/228
39	178/98	200/110	218/120	243/133	292/157	347/185	397/211
40	169/91	190/102	207/111	231/123	278/146	330/171	377/195
41	161/85	181/95	197/103	220/114	264/135	314/159	359/181
42	153/79	173/83	188/96	209/106	252/126	299/148	342/168
43	146/73	165/82	179/89	200/99	240/117	285/138	326/157
44	139/68	157/76	171/83	191/92	229/109	272/128	311/146
45	0/0	0/0	0/0	0/0	0/0	0/0	0/0

Table 4.4 Standard load table   K - series open-web steel joist
24-inch deep sections [610mm]

Joist	24K4	24K5	24K6	24K7	24K8	24K9	24K10	24K12
Wt. plf	8.4plf	9.3plf	9.7plf	10.1plf	11.5plf	12.0plf	13.1plf	16.0plf
Span (ft)								
24	520/516	550/544	550/544	550/544	550/544	550/544	550/544	550/544
25	479/456	540/511	550/520	550/520	550/520	550/520	550/520	550/520
26	442/405	499/453	543/493	550/499	550/499	550/499	550/499	550/499
27	410/361	462/404	503/439	550/479	550/479	550/479	550/479	550/479
28	381/323	429/362	467/393	521/436	550/456	550/456	550/456	550/456
29	354/290	400/325	435/354	485/392	536/429	550/436	550/436	550/436
30	331/262	373/293	406/319	453/353	500/387	544/419	550/422	550/422
31	310/237	349/266	380/289	424/320	468/350	510/379	550/410	550/410
32	290/215	327/241	357/262	397/290	439/318	478/344	549/393	549/393
33	273/196	308/220	335/239	373/265	413/289	449/313	532/368	532/368
34	257/179	290/201	315/218	351/242	388/264	423/286	502/337	516/344
35	242/164	273/184	297/200	331/221	366/242	399/262	473/308	501/324
36	229/150	258/169	281/183	313/203	346/222	377/241	447/283	487/306
37	216/138	244/155	266/169	296/187	327/205	356/222	423/260	474/290
38	205/128	231/143	252/156	281/172	310/189	338/204	401/240	461/275
39	195/118	219/132	239/144	266/159	294/174	320/189	380/222	449/261
40	185/109	208/122	227/133	253/148	280/161	304/175	361/206	438/247
41	176/101	198/114	216/124	241/137	266/150	290/162	344/191	427/235
42	168/94	189/106	206/115	229/127	253/139	276/151	327/177	417/224
43	160/88	180/98	196/107	219/118	242/130	263/140	312/165	406/213
44	153/82	172/92	187/100	209/110	231/121	251/131	298/154	387/199
45	146/76	164/86	179/93	199/103	220/113	240/122	285/144	370/185
46	139/71	157/80	171/87	191/97	211/106	230/114	272/135	354/174
47	133/67	150/75	164/82	183/90	202/99	220/107	261/126	339/163
48	128/63	144/70	157/77	175/85	194/93	211/101	250/118	325/153
49	0/0	0/0	0/0	0/0	0/0	0/0	0/0	0/0

Table 4.4 Standard load table K - series open-web steel joist
26-inch deep sections [660mm]

Joist	26K5	26K6	26K7	26K8	26K9	26K10	26K12
Wt. plf	9.8plf	10.6plf	10.9plf	12.1plf	12.2plf	13.8plf	16.6plf
Span (ft)							
26	542/535	550/541	550/541	550/541	550/541	550/541	550/541
27	502/477	547/519	550/522	550/522	550/522	550/522	550/522
28	466/427	508/464	550/501	550/501	550/501	550/501	550/501
29	434/384	473/417	527/463	550/479	550/479	550/479	550/479
30	405/346	441/377	492/417	544/457	550/459	550/459	550/459
31	379/314	413/341	460/378	509/413	550/444	550/444	550/444
32	356/285	387/309	432/343	477/375	519/407	549/431	549/431
33	334/259	364/282	406/312	448/342	488/370	532/404	532/404
34	315/237	343/257	382/285	422/312	459/338	516/378	516/378
35	297/217	323/236	360/261	398/286	433/310	501/356	501/356
36	280/199	305/216	340/240	376/263	409/284	486/334	487/334
37	265/183	289/199	322/221	356/242	387/262	460/308	474/315
38	251/169	274/184	305/204	337/223	367/241	436/284	461/299
39	238/156	260/170	289/188	320/206	348/223	393/262	449/283
40	227/145	247/157	275/174	304/191	331/207	393/243	438/269
41	215/134	235/146	262/162	289/177	315/192	374/225	427/256
42	205/125	224/136	249/150	275/164	300/178	356/210	417/244
43	196/116	213/126	238/140	263/153	286/166	339/195	407/232
44	187/108	204/118	227/131	251/143	273/155	324/182	398/222
45	179/101	194/110	217/122	240/133	261/145	310/170	389/212
46	171/95	186/103	207/114	229/125	250/135	296/159	380/203
47	164/89	178/96	199/107	219/117	239/127	284/149	369/192
48	157/83	171/90	190/100	210/110	229/119	272/140	353/180
49	150/78	164/85	183/94	202/103	220/112	261/131	339/169
50	144/73	157/80	175/89	194/97	211/105	250/124	325/159
51	139/69	151/75	168/83	186/91	203/99	241/116	313/150
52	133/65	145/71	162/79	179/86	195/93	231/110	301/142
53	0/0	0/0	0/0	0/0	0/0	0/0	0/0

Table 4.4 Standard load table   K - series open-web steel joist
28-inch deep sections [711mm]

Joist	28K6	28K7	28K8	28K9	28K10	28K12
Wt. plf	11.4plf	11.8plf	12.7plf	13.0plf	14.3plf	17.1plf

Span (ft)

Span	28K6	28K7	28K8	28K9	28K10	28K12
28	548/541	550/543	550/543	550/543	550/543	550/543
29	511/486	550/522	550/522	550/522	550/522	550/522
30	477/439	531/486	550/500	550/500	550/500	550/500
31	446/397	497/440	550/480	550/480	550/480	550/480
32	418/361	466/400	515/438	549/463	549/463	549/463
33	393/329	438/364	484/399	527/432	532/435	532/435
34	370/300	412/333	456/364	496/395	516/410	516/410
35	349/275	389/305	430/333	468/361	501/389	501/389
36	330/252	367/280	406/306	442/332	487/366	487/366
37	312/232	348/257	384/282	418/305	474/344	474/344
38	296/214	329/237	364/260	396/282	461/325	461/325
39	280/198	313/219	346/240	376/260	447/306	449/308
40	266/183	297/203	328/222	357/241	424/284	438/291
41	253/170	283/189	312/206	340/224	404/263	427/277
42	241/158	269/175	297/192	324/208	384/245	417/264
43	230/147	257/163	284/179	309/194	367/228	407/252
44	220/137	245/152	271/167	295/181	350/212	398/240
45	210/128	234/142	259/156	282/169	334/198	389/229
46	201/120	224/133	248/146	270/158	320/186	380/219
47	192/112	214/125	237/136	258/148	306/174	372/210
48	184/105	206/117	227/108	247/139	294/163	365/201
49	177/99	197/110	218/120	237/130	282/153	357/193
50	170/93	189/103	209/113	228/123	270/144	350/185
51	163/88	182/97	201/106	219/115	260/136	338/175
52	157/83	175/92	193/100	210/109	250/128	325/165
53	151/78	168/87	186/95	203/103	240/121	313/156
54	145/74	162/82	179/89	195/97	232/114	301/147
55	140/70	156/77	173/85	188/92	223/108	290/139
56	135/66	151/73	166/80	181/87	215/102	280/132
57	0/0	0/0	0/0	0/0	0/0	0/0

Table 4.4 Standard load table   K - series open-web steel joist
30-inch deep sections [762mm]

Joist	30K7	30K8	30K9	30K10	30K11	30K12
Wt. plf	12.3plf	13.2plf	13.4plf	15.0plf	16.4plf	17.6plf
Span (ft)						
30	550/543	550/543	550/543	550/543	550/543	550/543
31	534/508	550/520	550/520	550/520	550/520	550/520
32	501/461	549/500	549/500	549/500	549/500	549/500
33	471/420	520/460	532/468	532/468	532/468	532/468
34	443/384	490/420	516/441	516/441	516/441	516/441
35	418/351	462/384	501/415	501/415	501/415	501/415
36	395/323	436/353	475/383	487/392	487/392	487/392
37	373/297	413/325	449/352	474/374	474/374	474/374
38	354/274	391/300	426/325	461/353	461/353	461/353
39	336/253	371/277	404/300	449/333	449/333	449/333
40	319/234	353/256	384/278	438/315	438/315	438/315
41	303/217	335/238	365/258	427/300	427/300	427/300
42	289/202	320/221	348/240	413/282	417/284	417/284
43	276/188	305/206	332/223	394/263	407/270	407/270
44	263/176	291/192	317/208	376/245	398/258	398/258
45	251/164	278/179	303/195	359/229	389/246	389/246
46	241/153	266/168	290/182	344/214	380/236	380/236
47	230/144	255/157	277/171	329/201	372/226	372/226
48	221/135	244/148	266/160	315/188	362/215	365/216
49	212/127	234/139	255/150	303/177	347/202	357/207
50	203/119	225/130	245/141	291/166	333/190	350/199
51	195/112	216/123	235/133	279/157	320/179	343/192
52	188/106	208/116	226/126	268/148	308/169	336/184
53	181/100	200/109	218/119	258/140	296/159	330/177
54	174/94	192/103	209/112	249/132	285/150	324/170
55	168/89	185/98	202/106	240/125	275/142	312/161
56	162/84	179/92	195/100	231/118	265/135	301/153
57	156/80	173/88	188/95	223/112	256/128	290/145
58	151/76	167/83	181/90	215/106	247/121	280/137
59	146/72	161/79	175/86	208/101	239/115	271/130
60	141/69	156/75	169/81	201/96	231/109	262/124

4.1.1 Bridging

As in any bending member, the ability to sustain the calculated moment is dependent on the lateral-stability considerations of the system. The means of providing this lateral stability in joist systems is through the use of "bridging." Bridging may be either horizontal or diagonal, and its size depends on the slenderness ratio (L/r) of the unbraced length to the least radius of gyration.

$$\text{Horizontal bridging: } L/r \leq 300; \text{ Diagonal bridging: } L/r \leq 200$$

Although either of these systems will provide lateral stability to the lower chord of the joist, assuming the upper chord is held in line by attachment to the decking material, diagonal bridging will serve an additional purpose. Every project can be expected to be exposed to loads other than "uniform" ones at some point during its life. Diagonal bridging will transfer some portion of any concentrated load or localized overload to an adjacent joist whereas horizontal bridging will not. Remember, a horizontal member has no potential vertical component of force, so it cannot transfer loads until excessive potentially destructive deflection has already occurred. This shouldn't be construed as a recommendation to use the diagonal bracing for overloads, but to acknowledge the fact that it will provide some additional strength that horizontal bridging won't. The diagonal bridging requirement, $L/r \leq 200$ can be rewritten in the form:

$$r \geq L/200$$

Any section that will provide the adequate minimum r value will work, but the lightest and least expensive sections are typically light equal-leg angles. Although round hollow tube sections have the best minimum r value at the lightest weight, they are expensive to manufacture, and it is difficult to reconcile their round geometry with the orthogonal geometry of the joist to install them. If you have relatively long cross-bridging, the section can be pinned at the center and the unbraced length L based on one half of its actual length. You will often see this done as a constructional convenience as well. The resulting X shape is easily installed as one piece. Since it is pinned at the center, it can be "closed" to a straight line for shipping and storage and rotated out to its X shape when it is finally put in place.

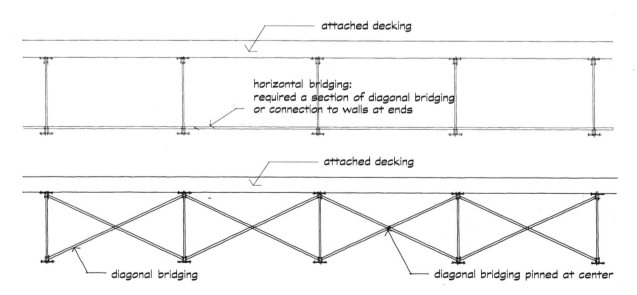

Fig. 4.3 Diagonal bridging alternatives

When selecting the bridging, you'll note that the angles have three "r" values; r_x, r_y, and r_z. For angles, the r_z value will typically be the minimum. You should also notice that since r is a function of the area and the moment of inertia of a section, a heavier section may actually have a lesser r value than a light one. This sometime leads you to select a more costly section simply because its r value is closer to your design value. **Select the lightest section that meets your requirements,** not just the minimum r, (r_z), closest to your value.

Table 4.5 Bridging requirements

Section	1 Row	2 Rows	3 Rows	4 Rows*	5 Rows*
1	up thru 16ft	16ft+ thru 24ft	24ft+ thru 28ft		
2	up thru 17ft	17ft+ thru 25ft	25ft+ thru 32ft		
3	up thru 18ft	18ft+ thru 28ft	28ft+ thru 38ft	38ft+ thru 40ft	
4	up thru 19ft	19ft+ thru 28ft	28ft+ thru 38ft	38ft+ thru 48ft	
5	up thru 19ft	19ft+ thru 29ft	29ft+ thru 39ft	39ft+ thru 50ft	50ft+ thru 52ft
6	up thru 19ft	19ft+ thru 29ft	29ft+ thru 39ft	39ft+ thru 51ft	51ft+ thru 56ft
7	up thru 20ft	20ft+ thru 33ft	33ft+ thru 45ft	45ft+ thru 58ft	58ft+ thru 60ft
8	up thru 20ft	20ft+ thru 33ft	33ft+ thru 45ft	45ft+ thru 58ft	58ft+ thru 60ft
9	up thru 20ft	20ft+ thru 33ft	33ft+ thru 46ft	46ft+ thru 59ft	59ft+ thru 60ft
10	up thru 20ft	20ft+ thru 37ft	37ft+ thru 51ft	51ft+ thru 60ft	
11	up thru 20ft	20ft+ thru 38ft	38ft+ thru 53ft	53ft+ thru 60ft	
12	up thru 20ft	20ft+ thru 39ft	38ft+ thru 53ft	53ft+ thru 60ft	

Span × 0.305m

"Section" refers to the last digit(s) of the joist designation,

i.e., a 16K7, 18K7, 20K7, 22K7, 24K7, 26K7, etc., are all section 7.

Note:

*When four or five rows of bridging are required, the row nearest midspan of the joist must be diagonal bridging with bolted connections at chords and bridging intersection.

In some circumstances, additional bridging may be required to compensate for uplift forces. Consult the final manufacturer for specific recommendations regarding their joist in all cases.

DO NOT APPLY POINT LOADS TO ANY JOIST WITHOUT CONSULTING THE MANUFACTURER. Although the joist may be sufficiently strong, additional local members may be necessary to carry local overstresses.

4.2 Steel decks

Steel decks are produced in a variety of shapes corresponding to the three basic construction methods: composite floor/roof deck, permanent form, and roof deck (Fig. 4.4). Composite floor decks have deformations on the sides of the ribs designed to interlock with the cast-in-place concrete topping. The steel deck works as at least a portion of the reinforcement of the concrete slab; additional reinforcement is generally required for shrinkage of the concrete. During construction, the deck provides a form that can carry the full load of the concrete before curing without shoring. In long deck spans, shoring could be required to reduce deflection during construction, before the concrete cures. The manufacturers' tables indicate the maximum unshored span for each type.

Permanent-form decks are primarily used to support the load of concrete during casting, occasionally with some temporary shoring. In this case, the concrete may be designed to work as a reinforced concrete slab without any composite action with the deck, which is simply left in place as a nonremovable form.

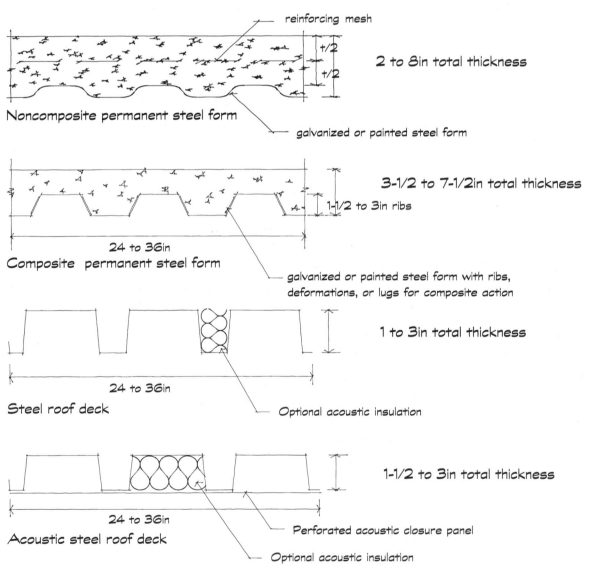

reinforcing mesh

2 to 8in total thickness

$t/2$

$t/2$

Noncomposite permanent steel form

galvanized or painted steel form

3-1/2 to 7-1/2in total thickness

1-1/2 to 3in ribs

24 to 36in

Composite permanent steel form

galvanized or painted steel form with ribs, deformations, or lugs for composite action

1 to 3in total thickness

24 to 36in

Steel roof deck

Optional acoustic insulation

1-1/2 to 3in total thickness

24 to 36in

Acoustic steel roof deck

Perforated acoustic closure panel

Optional acoustic insulation

Fig. 4.4 Typical structural decking configurations and types

Roof decks are connected to the roof joists or purlins and support an insulated roofing system, usually of lightweight construction, or have adequate internal voids to provide space for additional insulation.

In the first two systems, the concrete slab may or may not act in composite action with the steel joists or beams. "Composite" and permanent-form decks can be used to build composite joists and beams: A composite deck simply means that the deck works with the concrete topping, but the latter, permanent-form deck, may or may not be designed to act with its concrete topping.

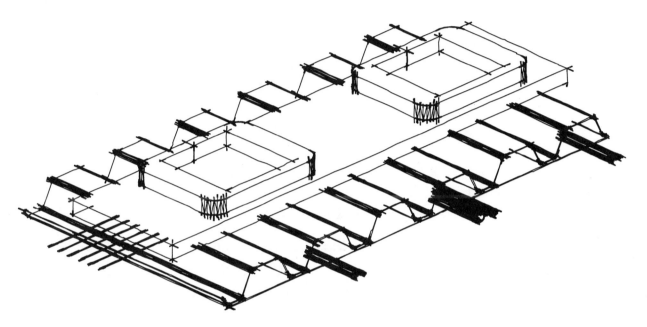

Fig. 4.5 Cellular deck for systems integration

Electrical and communication systems can be run under/within the floor using cellular floor deck units. This system provides virtually unlimited flexibility in the planning of office building floor space. Generally, the cellular deck units are intermixed with regular composite metal deck units to provide underfloor duct runs at regular intervals (Fig. 4.5). Cellular deck units cause some structural complications for fireproofing; trench headers require additional filler beams or joists.

Steel deck thickness is given as **gage**: Increasing gage corresponds to decreasing thickness, as shown in Tables 4.2 and 4.3. The tables assume a dead load of 10psf [480N/m²] for the deck and a typical built-up roof with standard aggregate. Live-load deflection does not exceed 1/240 of the span. You should notice that the safe loads are given for one-, two-, or three-span decks. It is more efficient to lay the deck in longer panels to save labor if a crane is available on site. (A 30ftx3ft [9mx0.9m] roof deck panel, 16 gage, weighs 314lb [142 kg] and is obviously too heavy to erect by hand.) This is also structurally convenient, since multiple spans reduce the maximum moments and deflections between supports.

Table 4.6 Typical roof deck

Single-span condition

Safe TOTAL loading in psf (pounds per square foot) [× 47.88N/m²]

Type (ga)	Span 5'-0"	Span 5'-6"	Span 6'-0"	Span 6'-6"	Span 7'-0"	Span 7'-6"	Span 8'-0"	Span 8'-6"	Span 9'-0"	Span 9'-6"	Span 10'-0"
22	93psf	72psf	58psf	48psf	40psf	35psf	30psf	27psf	24psf	22psf	20psf
20	117psf	91psf	72psf	59psf	49psf	42psf	36psf	32psf	26psf	26psf	23psf
18	163psf	126psf	99psf	80psf	66psf	56psf	48psf	41psf	36psf	32psf	29psf
16	210psf	160psf	125psf	101psf	83psf	69psf	59psf	51psf	44psf	39psf	35psf

[Span × 0.305m]

Double-span condition

Safe TOTAL loading in psf (pounds per square foot) [× 47.88N/m²]

Type (ga)	Span 5'-0"	Span 5'-6"	Span 6'-0"	Span 6'-6"	Span 7'-0"	Span 7'-6"	Span 8'-0"	Span 8'-6"	Span 9'-0"	Span 9'-6"	Span 10'-0"
22	102psf	84psf	71psf	60psf	52psf	45psf	40psf	35psf	31psf	28psf	25psf
20	125psf	104psf	87psf	74psf	64psf	56psf	49psf	43psf	39psf	35psf	31psf
18	168psf	138psf	116psf	99psf	85psf	74psf	65psf	58psf	52psf	46psf	42psf
16	212psf	175psf	147psf	125psf	108psf	94psf	83psf	73psf	65psf	58psf	52psf

[Span × 0.305m]

Triple-span condition

Safe TOTAL loading in psf (pounds per square foot) [× 47.88N/m²]

Type (ga)	Span 5'-0"	Span 5'-6"	Span 6'-0"	Span 6'-6"	Span 7'-0"	Span 7'-6"	Span 8'-0"	Span 8'-6"	Span 9'-0"	Span 9'-6"	Span 10'-0"
22	127psf	105psf	88psf	75psf	65psf	56psf	50psf	44psf	39psf	35psf	31psf
20	157psf	130psf	109psf	93psf	80psf	70psf	61psf	54psf	48psf	43psf	39psf
18	209psf	173psf	145psf	124psf	107psf	93psf	82psf	70psf	61psf	53psf	47psf
16	265psf	219psf	184psf	157psf	135psf	118psf	102psf	87psf	75psf	65psf	57psf

[Span × 0.305m]

Table 4.7 Typical composite floor deck

Single unshored span condition

5-1/2in [140mm] slab standard-weight concrete (145pcf) [23KN/m³]

Safe SUPERIMPOSED* loading in psf (pounds per square foot) [× 47.88N/m²]

(Superimposed load is that load which it will support in addition to its own weight.)

Type(ga)	Shoring(max)	Span 9'-0"	Span 9'-6"	Span 10'-0"	Span 10'-6"	Span 11'-0"	Span 11'-6"	Span 12'-0"	Span 12'-6"	Span 13'-0"	Span 13'-6"
22	8'-3"	140psf	120psf	105psf	90psf	80psf	70psf				
20	11'-8"	165psf	150psf	135psf	120psf	110psf	105psf	90psf	75psf	70psf	
18	12'-7"		185psf	170psf	156psf	140psf	130psf	120psf	105psf	95psf	85psf
16	13'-3"				185psf	170psf	155psf	140psf	130psf	125psf	110psf

[Span × 0.305m]

Single unshored span condition

6-1/2in [165mm] slab standard-weight concrete (145pcf) [23KN/m³]

Safe SUPERIMPOSED* loading in psf (pounds per square foot) [× 47.88N/m²]

(Superimposed load is that load which it will support in addition to its own weight.)

Type (ga)	Shoring(max)	Span 9'-0"	Span 9'-6"	Span 10'-0"	Span 10'-6"	Span 11'-0"	Span 11'-6"	Span 12'-0"	Span 12'-6"	Span 13'-0"	Span 13'-6"
22	7'-0"	180psf	160psf	145psf	120psf	105psf	90psf	80psf	70psf		
20	11'-0"			175psf	155psf	140psf	125psf	115psf	105psf	95psf	80psf
18	11'-11"				195psf	180psf	165psf	150psf	140psf	130psf	115psf
16	12'-7"						195psf	180psf	165psf	150psf	140psf

[Span × 0.305m]

Example 4.1

Open web steel joist design

A roof system spanning 35ft is located in northern Indiana. The UBC specifies a **30psf snow live load** in northern Indiana.

A typical roof construction might consist of the following:

Membrane roofing 1.0psf

4in average tapered rigid insulation 6.0psf

Steel deck (2ft-4ft span) 1.0psf

Estimated joist weight:

35ft span would have a min. 18in joist

an average 18in joist weight = 9.0plf

If the joist is spaced @ 3ft-0in o.c.

9.0plf/3ft 3.0psf

Suspension system 1.0psf

1/2in gypsum ceiling 2.0psf

Mechanical system estimates should also be included, a heavy sprinkler/drain piping running parallel to a joist or pair of joists is especially critical.

Miscellaneous ductwork/electrical 1.0psf

Total live load .... 30.0psf x 3ft o.c. = 90plf

Total dead load 15.0psf

Total live load + dead load .... 45.0psf x 3ft o.c. =135plf

Use the joist load tables to select the "best" section:

(a) At 35ft, a 18K4 joist carries 179plf TL and 90plf LL.

Live load: Deflection controls and the weight is 7.2plf.

(b) At 35ft, a 20K3 joist carries 166plf TL and 96plf LL.

Live load once again controls and the weight is 6.7plf

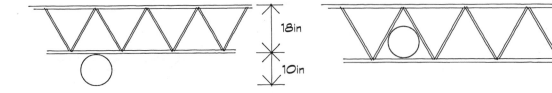

Either of these two choices might be "best" depending on the ability to integrate mechanical systems into the joist space. Even a 22K4 at 8.0plf, which is both deeper and heavier than the previous two selections, could be "best".

By comparing the two systems, the "deeper and heavier" joist actually saves 6inches of building height per floor, which does not need to be:

(a) initially enclosed

(b) heated and cooled for the life of the building

(c) maintained both inside and out; painted, waterproofed, etc.

In a 16-story solution this amounts to a full extra 8ft floor "built, but producing no income". Consequently, to make informed decisions about structural selections, you have to look beyond simplistic considerations of weight or depth and determine the impact on the entire system.

For the purposes of this example we will select the 18K4 (7% heavier initially, but 11% shallower **FOREVER**).

18K4 joist @ 3ft-0in o.c.

Either horizontal or diagonal bridging could be used, but for the system to work, **bridging must be specified**. The diagonal bridging is more costly, but will transfer some overloads to adjacent joists in an overload situation. The horizontal bridging is cheaper and easier to install, but will not distribute any overloads; it simply provides lateral support. We would suggest that for the small added expense, the diagonal bridging is the best choice.

Bridging length $= \sqrt{36^2 + 18^2}$ = 40.24in

Maximum L/r for diagonal bridging = 200

$\quad$ 40.24in/r = 200

r min. = 40.24in/200 = 0.2012in

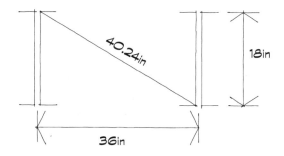

A 1-1/8x1-1/8x1/8in angle has a minimum r_z = 0.221in. This is the lightest section.

Based on a 35ft span, and a chord designation of "4" (18K"4"), Table 4.5 indicates that three rows of bridging at 1/4 points (equally spaced) must be used.

Solution: 18K4 joist @ 3ft-0in o.c. with

$\quad$ 3 rows of 1-1/8x1-1/8x1/8in diagonal bridging at 1/4 points of span

The open-web steel joist tables do not provide for any loading condition other than uniform. Unfortunately, most buildings have loadings other than this uniform condition; consequently, to do a preliminary design it is necessary to find the single uniform equivalent (TL and LL) that would cause the same maximum conditions as the more complex loading condition.

Example 4.2

Open-web joist design with nonuniform loading

A floor system for an office building has a joist span 30ft

UBC specifies 50psf uniform load or a 2000lb point load at any point applied over a 2-1/2ft square. These loads do not have to be considered simultaneously. Evaluate the alternative maximum conditions and use the larger.

Span 30ft:

Joist at 4ft-0in o.c.

Uniform DL 35.2psf
Uniform LL 50psf
or 2000lb applied at any location

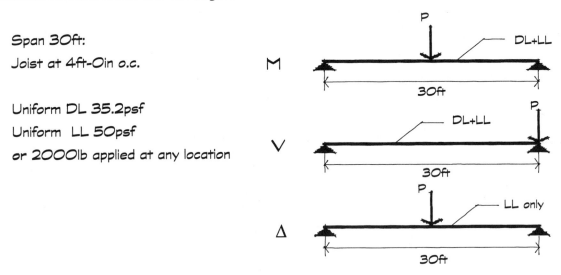

An easy way to evaluate this situation is to compare the maximum values for each of the criteria and determine which load is critical in each case.

Alternative loadings

1. Uniform: 50.0psf LL × 4ft-0in o.c. = 200plf   (live load)

                    35.2psf DL   (dead load)

    Total uniform: 85.2psf × 4ft-0in o.c.   = **340.8plf**   (total load)

2. Combined load: 35psf DL × 4ft-0in o.c.   = 140.8plf   (dead load)

                    Alternative point live load = 2000lb   (live load)

First trial loading condition:

uniform maximums:

    uniform LL + uniform DL

    Maximum moment   = $wL^2/8$ = 340.8plf×(30ft)²/8 = **38.340ftlb**

    Maximum shear   = $wL/2$ = 340.8plf×30ft/2 = **5112lb**   Maximum

Maximum $\Delta = \dfrac{5wL^4}{384EI} = \dfrac{5\left(\frac{200plf}{12in/ft}\right)\left(30ft \times \frac{12in}{ft}\right)^4}{384(29,000,000psi)I} = \dfrac{125.7}{I}$

Second trial loading condition:

combined maximums:

Concentrated LL + uniform DL

Maximum moment: using a shear diagram

$\dfrac{3112lb + 1000lb}{2} \times 15ft = 30840ftlb$

Maximum moment: assuming a full 2000lb at center

Uniform load conditions control moment design.

Maximum shear: using a shear diagram

Maximum shear = 4112lb

assuming a full 2000lb at end

Uniform load conditions control shear design.

Maximum deflection condition: using concentrated LL only

$\Delta = \dfrac{PL^3}{48EI} = \dfrac{2000lb(30ft \times 12in/ft)^3}{48(29,000,000psi)I} = \dfrac{67.0}{I}$

Uniform load conditions control deflection design.

At this point, we find that the controlling condition is uniform loading and we would proceed with the example using the same technique as in Example 4.1. (Don't forget to include bridging or it will not work!)

This is unfortunately not the "worst case" since we did not have to develop "equivalent uniform loads" so, . . .

Example 4.3

Open-web steel joist design with complex loading conditions

Loading is an office space next to a corridor with a permanent partition separating the two.

Span of 26ft: LL = 50psf office loading

Joist at 3ft-0in o.c.:   LL = 100psf corridor loading

DL = 36.3psf construction materials

DL = 64plf partition loading

By using the adjacent shear diagram:

  Maximum moment:

  0.5(13.29ft)(3441.3lb) = 22,871ftlb

  Maximum shear: 4082.1lb

  Maximum deflection: OOPS!

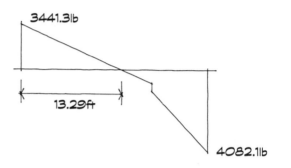

Since we have a complex LL condition, we will combine formulas. The different loading conditions will create maximum deflection at different locations, so if we directly add the maximum values together, we will obtain a conservative answer. If we ultimately find that deflection is **not** the controlling criterion, the additional precision of lengthier calculations is unnecessary.

$$\Delta_1 = \frac{5wL^4}{384EI}$$

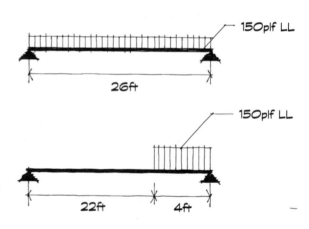

Rather than take a much more complex approach to the uniform corridor load, it is being combined into a single point load applied at the corridor wall. This is a very conservative assumption, but if deflection is ultimately not the controlling criterion, more precision is a waste of time and effort. So . . . .

$$\Delta_2 = \frac{Pab(a+2b)\sqrt{3a(a+2b)}}{27EIL}$$

Solving the combination of these two equations:

$$\Delta_1 = \frac{5\left(\frac{150plf}{12in/ft}\right)(26ft \times 12in/ft)^4}{384(29,000,000psi)I} = \frac{53.18}{I}$$

The total corridor load is 50psf × 3ft-0in o.c. = 150plf.

At half of the load being placed at the corridor wall,

150plf x a 2ft-0in corridor = 300lb

$$\Delta_2 = \frac{300lb(48in)(264in)[48in + (2\times264in)]\sqrt{3(48in)[48in + (2\times264in)]}}{27(29,000,000psi)(26ft\times12in/ft)I} = \frac{2.58}{I}$$

$$\Delta_{total} = \Delta_1 + \Delta_2 = \frac{55.76}{I}$$

Now we can convert these complex loading conditions into "equivalent uniform" conditions that would create these same maximum values.

Maximum moment: $wL^2/8 = 22,871ftlb$ $\qquad w = \frac{22,871ftlb\times8}{(26ft)^2} = 270.6plf$

Maximum shear: $wL/2 = 4082lb$ $\qquad w = \frac{4,082lb\times2}{26ft} = 314.0plf$

Maximum deflection: $= \frac{5wL^4}{384EI} = \frac{55.76}{I}$

Since the I values would be the same, we can cancel out this term and solve for w:

$$w = \frac{55.76\times384\times29,000,000psi}{5\times(26ft\times12in/ft)^4} = 13.1pli$$

OOPS, our units are not consistent with the other criteria:

w = 13.1lb/in×12in/ft = 157.3lb/ft

The controlling uniform loading criteria for this problem are

w = 314lb/ft TL

w = 157lb/ft LL

We now select a joist just as we did in Example 4.1

The alternatives are

18K4 at 7.2plf

20K3 at 6.7plf

Once again, we must evaluate the total system to determine whether less weight or less depth or more mechanical space is most desirable. We will "arbitrarily" select an 18K4 and design the necessary bridging.

Bridging length = $\sqrt{(18in)^2 + (36in)^2} = 40.25in$

Maximum L/r = 200; therefore $r_{min.}$ = (40.25in)/200 = 0.201in

Use 1in × 1in × 1/8in L's with minimum r_z = 0.196in (within our 5% error margin) **or**

Use 1-1/8in × 1-1/8in × 1/8in L's with minimum r_z = 0.221in (clearly works!)

Solution:

Use 18K4 joist @ 3ft-0in o.c. with

2 rows of 1in × 1in × 1/8in cross bridging at 1/3 points of span.

Although this design would work for the specified load conditions, the fabricator MUST be notified of any concentrated loads on the joist system. These point loads, if applied at some location other than a joint, could cause local buckling of the upper or lower (wherever applied) chord members. The joist manufacturer will introduce an additional web member to support the chord at this point if necessary.

These examples are typical uses of open-web steel joist. A less traditional use would be as extremely tall window wall mullions as in the Oakland Alameda Stadium (Fig. 4.6) where they span vertically approximately 70ft along the concourses to support the glass enclosure, yet remain light and visually unobtrusive. These obviously do not meet the maximum of 1/2in/ft slope. Their load is predominantly lateral, from the wind, which is perpendicular to their axis, the normal loading scenario. In this case, they would need to be checked for axial-load capacity as well: A consideration not generally required in other uses.

Fig. 4.6 Oakland Alameda Stadium concourse

The following problems will give you an opportunity to exhibit your new-found skills and to determine if you need to go back and take another look at the chapter. Good luck!

Problem 4.1

A floor system is built with open-web steel joists @ 4ft o.c. spanning 36ft and carries a live load of 70psf and a dead load of 30psf (including an assumed joist weight of 14lb/ft). Find the most economical K-series joist.

Solution:

Distributed LL = 70psf(4ft) = 280plf

Distributed TL = 100psf(4ft) = 400plf

28K8 has TL = 406plf, LL = 306plf, weight = 12.7plf OK

30K7 has TL = 395plf < 400plf, but it is close; weight = 12.3plf, possible choice.

Too easy!

Problem 4.2

A roof system with a 36ft span carries a skylight in the middle, causing two symmetrical 300lb point loads (DL) in addition to the roof distributed load. Roof LL = 20psf, DL = 22psf. Find the most economical open-web joist (you may ignore deflection in this problem).

Solution:

TL: Maximum M = 26,304ftlb, Maximum V = 2988lb

Equivalent loads for V:

$V = wL/2$; $w = 2V/L$; $w = 2(2988lb)/36ft = 166plf$

w (shear) = 166plf

Equivalent loads for M:

$M = wL^2/8$; $w = 8M/L^2$; $w = 8(26,304ftlb)/(36ft^2) = 163.4plf$

w (moment) = 162.4plf

Required safe loads:

TL = 166plf + assumed joist weight 10plf = **176plf**

18K5 carries TL = 191plf, LL = 92plf (but not a consideration in this case)

and the weight = 7.2plf, the lightest.

METRIC VERSIONS OF ALL EXAMPLES AND PROBLEMS

Example M4.1

Open-web steel joist design

A roof system spanning 10.8m is located in northern Indiana. The UBC specifies a 1440N/m² snow live load in northern Indiana.

A typical roof construction might consist of the following:

Membrane roofing 48N/m²

100mm average tapered rigid insulation 288N/m²

Steel deck (0.6m-1.2m span) 48N/m²

Estimated joist weight:

10m span would have a minimum 457mm joist

An average 457mm joist weight = 131N/m

The joist is spaced @ 900mm o.c.,131N/m/0.9m = 145N/m²

Suspension system 48N/m²

12mm gypsum ceiling 96N/m²

Mechanical system estimates should also be included; a heavy sprinkler/drain piping running parallel to a joist or pair of joists is especially critical.

Miscellaneous ductwork / electrical 48N/m²

Total live load ....(1440N/m²)(0.9m) = 1296N/m

Total dead load 289N/m

Total live load + dead load 1585N/m

Use the joist load tables to select the "best" section:

a. At 11 m, a 18K4 joist carries 2612N/m TL and 1313N/m LL.

Live load: . . . . deflection controls and the weight is 61.3N/m.

b. At 11 m, a 20K3 joist carries 2422N/m TL and 1401N/m LL.

Live load once again controls and the weight is 97.8N/m.

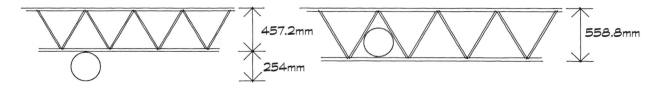

Either of these two choices might be "best" depending on the ability to integrate mechanical systems into the joist space. Even a 22K4 at 117N/m, which is both deeper (559mm) and heavier than the previous two selections, could be "best."

By comparing the two systems, the "deeper and heavier" joist actually saves 100mm of building height per floor, which does not need to be:

(a) initially enclosed

(b) heated and cooled for the life of the building

(c) maintained both inside and out; painted, waterproofed, etc.

In a 30-story solution this amounts to a full extra 3m floor "built, but producing no income." Consequently, to make informed decisions about structural selections, you have to look beyond simplistic considerations of weight or depth and determine the impact on the entire system.

For the purposes of this example we will select the 18K4 (7% heavier initially, but 11% shallower **FOREVER**).

$$18K4 \text{ joist at } 900mm \text{ o.c.}$$

Either horizontal or diagonal bridging could be used, but for the system to work, **bridging must be specified.** The diagonal bridging is more costly, but will transfer some overloads to adjacent joist as in an overload situation. The horizontal bridging is cheaper and easier to install, but will not distribute any overloads; it simply provides lateral support. We would suggest that for the small added expense diagonal bridging is the best choice.

Bridging length $= \sqrt{(914mm)^2 + (457mm)^2}$

Bridging length $= 1022mm$

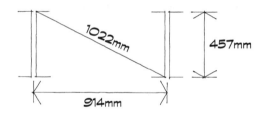

Maximum L/r for diagonal bridging $= 200$

$(1022mm)/r = 200$

r minimum $= (1022mm)/200 = 5.11mm$

A 29mm × 29mm × 3.2mm angle has a minimum $r_z = 5.61mm$. This is the lightest section.

Based on a 10.8m span, a chord designation of four (18K"4") Table 4.5 indicates that three rows of bridging at 1/4 points (equally spaced) must be used.

Solution: 18K4 joist at 900mm o.c. with

3 rows of 29mm × 29mm × 3.2mm diagonal bridging at 1/4 points of span

Example M4.2

Open-web joist design with nonuniform loading

A floor system for an office building has a joist span 9.0m.

The UBC specifies 2400N/m² uniform load or a 9000N point load at any point applied over a 0.76m square. These loads do not have to be considered simultaneously. Evaluate the alternative maximum conditions and use the larger.

Span 9m: Joist at 1.2m o.c.

Uniform DL 1685N/m²
Uniform LL 2400N/m²
or 9000N applied at any location.

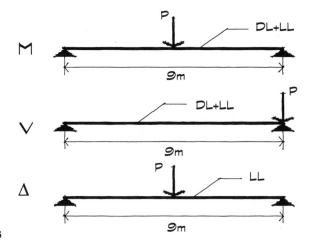

An easy way to evaluate this situation is to compare the maximum values for each of the criteria and determine which load is critical in each case.

Alternative loadings

1. Uniform: 2400N/m² LL × 1.2m o.c. = 2880N/m (live load)
                    1685N/m² DL   (dead load)
    Total uniform: 4085N/m² × 1.2m o.c.   = 4902N/m (total load)

2. Combined load: 1685N/m² DL × 1.2m o.c. = 2022N/m (dead load)
                    Alternative point live load = OOON   (live load)

First trial loading condition:

uniform maximums:

uniform LL + uniform DL

Maximum moment = $wL^2/8$ = (4902N/m)(9m)²/8 = **49,632Nm**

Maximum shear = $wL/2$ = (4902N/m)(9m)/2 = **22,059N**   Maximum

Maximum $\Delta = \dfrac{5wL^4}{384EI} = \dfrac{5(2.88\text{N/mm})(9000\text{mm})^4}{384(200,000\text{MPa})I} = \dfrac{1.23(10^9\text{mm}^5)}{I}$

Second trial loading condition:

combined maximums:

Concentrated LL + uniform DL

Maximum moment: using a shear diagram

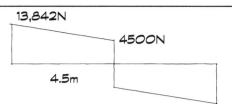

Maximum moment: assuming full 9000N at center

$$\frac{13842 + 4500N}{2}(4.5m) = 41,270Nm$$

Uniform load conditions control moment design.

Maximum shear: again using a shear diagram

Maximum shear = 17,719N

assuming full 9000N at 0.38m from end

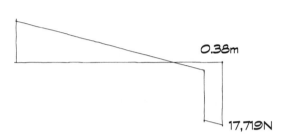

Uniform load conditions control shear design.

Maximum deflection condition: using concentrated LL only

$$\Delta = \frac{PL^3}{48EI} = \frac{9000N(9000mm)^3}{48(200,000MPa)I} = \frac{0.68(10^9 mm^5)}{I}$$

Uniform load conditions control deflection design.

At this point, we find that the controlling condition is uniform loading and we would proceed with the example using the same technique as in Example M4.1. (Don't forget to include bridging or it will not work!)

This is unfortunately not the "worst case" since we did not have to develop "equivalent uniform loads" so,....

Example M4.3

Open-web steel joist design with complex loading conditions

Loading is an office space next to a corridor with a permanent partition separating the two.

Span of 7.9m   LL = 2400N/m² office loading

Joist at 0.9m   LL = 4800N/m² corridor loading

                       DL = 1742N/m² construction materials

                       DL = 934N/m partition loading

R_A = 15,246N R_B = 20,228N

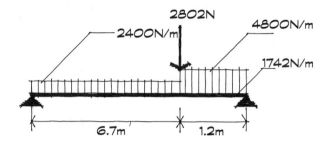

By using the adjacent shear diagram:

M = 0 at x = (15,246N)/(2,400+1,742)N/m² = 3.68m

Maximum moment:

(15,246N)(3.68m)/2 = 28,059Nm

Maximum shear: 20,228N

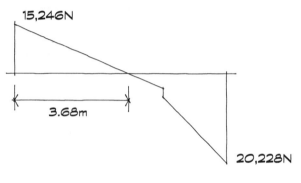

Maximum deflection: OOPS!

Since we have a complex LL condition, we will combine formulas. The different loading conditions will create maximum deflection at different locations, so if we directly add the maximum values together, we will obtain a conservative answer. If we ultimately find that deflection is not the controlling criterion, the additional precision of lengthier calculations is unnecessary.

$$\Delta_1 = \frac{5wL^4}{384EI}$$

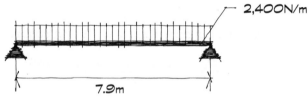

Rather than take a much more complex approach to the uniform corridor load, it is being combined into a single point load applied at the corridor wall. This is a very conservative assumption, but if deflection is ultimately not the controlling criterion, more precision is a waste of time and effort. So. .

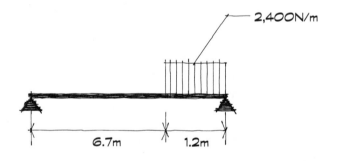

$$\Delta_2 = \frac{Pab(a+2b)\sqrt{3a(a+2b)}}{27EIL}$$

Solving the combination of these two equations:

$$\Delta_1 = \frac{5(7200 \times 10^{-6}N/mm^2)(7900mm)^4}{384(200,000MPa)I} = \frac{1.825 \times 10^6}{I}$$

The corridor LL is (4800N/m²)(0.9m)(1.2m) = 5184N.

At half of the load being placed at the corridor wall,

P = (5184N)/2 = 2592N, with a = 1200mm and b = 6700mm

$$\Delta_2 = \frac{51.707 \times 10^6 \text{mm}^5}{I}$$

$$\Delta_{total} = \Delta_1 + \Delta_2 = \frac{53.532 \times 10^6 \text{mm}^5}{I}$$

Now we can convert these complex loading conditions into "equivalent uniform" conditions that would create these same maximum values.

Maximum moment: $wL^2/8 = 28,059 \text{Nm}$ $w = \frac{8(28,059\text{m})}{(7.9\text{m})^2} = 3,597\text{N/m}$

Maximum shear: $wL/2 = 20,228\text{N}$ $w = \frac{20228\text{N/m}}{7.9\text{m}} = 5,121\text{N/m}$

Maximum deflection: $= \frac{5wL^4}{384EI} = \frac{53.532 \times 10^6 \text{mm}^5}{I}$

Since the I values would be the same, we can cancel out this term and solve for w:

$$w = \frac{384(200,000\text{MPa})(53.532 \times 10^6 \text{mm}^5)}{5(7900\text{mm})^4} = 2,111\text{N/m}$$

The controlling uniform loading criteria for this problem are
$$w = 5,121\text{N/mTL}$$
$$w = 2,111\text{N/m LL}$$
We now select a joist just as we did in Example M4.1.

The alternatives are:

18K4 @ 32N/m

20K3 @ 30N/m

Once again, we must evaluate the total system to determine whether less weight or less depth or more mechanical space is most desirable. We will "arbitrarily" select an 18K4 and design the necessary bridging.

Bridging length $= \sqrt{(457\text{mm})^2 + (914\text{mm})^2} = 1022\text{mm}$

Maximum L/r = 200; therefore $r_{min.} = 1022\text{mm}/200 = 5.11$

Use 25mm × 25mm × 3.2mm L's

with minimum r_z = 4.98mm (within our 5% error margin) or

Use 25mm × 25mm × 3.2mm L's with minimum r_z = 5.61mm (clearly works!)

Solution: Use 18K4 joist at 0.9m o.c. with

2 rows of 25mm × 25mm × 3.2mm cross bridging at 1/3 points of span.

Problem M4.1

A floor system is built with open-web steel joists at 1.2m o.c. spanning 11m and carries a live load of 33,60N/m² and a dead load of 1,440N/m². Assuming a joist weight of 62N/m, find the most economical K-series joist.

Solution:

Distributed LL = (3,360N/m²)(1.2m) = 4,032N/m

Distributed TL = (4,800N/m²)(1.2m) = 5,760N/m

28K8 has TL = 5,925N/m, LL = 4,466, weight = 181N/m OK

30K7 has TL = 5,765N/m, LL = 4,714N/m; weight = 55N/m, possible choice.

Problem M4.2

A roof system with an 11m span carries a skylight in the middle, causing two symmetrical 1,334N point loads (DL) in addition to the roof distributed load. Roof LL = 960N/m², DL = 1,056N/m². Find the most economical open-web joist (you may ignore deflection in this problem).

Solution:

TL: Maximum M = 35,684Nm, Maximum V = 13,291N

Equivalent loads for V:

V = wL/2; w = 2V/L; w = 2(13,291N)/11 m = 2,416N/m

Equivalent loads for M:

M = wL²/8; w = 8M/L²; w = 8(35,684Nm)/(11m)² = 2,359N/m

Required safe loads:

TL = 2416N/m + assumed joist weight 50N/m = 2,466N/m

18K5 carries TL = 2787N/m, LL = 1,343N/m (but not a consideration in this case), and the weight = 34N/m is the lightest.

---

5 STEEL BEAMS

5.1 Beam theory

Beams are structural components that are subjected to loads or components of loads that act perpendicular to their axes. Although the most common use of beams is in a horizontal orientation with gravity loads acting on them, they may be in any orientation. A vertical sign becomes a bending member when subjected to horizontal wind loads. In this situation, the member will be subjected to loads that act both perpendicular (wind) and parallel (its own weight) to its axis. If one of the force components is very small, its effect may be disregarded. Bending members in general, and in this case **steel beams specifically, are designed based on at least three criteria:**

1. Shear: Use $F = P/A$ or $F_v = V/A$. The physical property A (area of web = thickness of web X depth of section) is used to select appropriate members. The actual shear formula is $F_v = VQ/Ib$; however, for steel W and I sections, the $F_v = V/A$ approximation is quite accurate.

    F_v is the allowable shear stress based on material/code (AISC).

    V is the actual shear load at the point of evaluation, usually the maximum.

    A is the web area of the section, typically, the value you're attempting to determine.

2. Moment: Use $F_b = Mc/I$. Since $I/c = S$ (section modulus), we usually change this to $F_b = M/S$, with S, the section modulus, being the physical property used for selection.

    F_b is the allowable bending stress based on material/code (AISC).

    M is the actual moment at the point of evaluation, usually the maximum.

    S is the section modulus of the section, typically, the value you're attempting to determine.

3. Deflection: Any number of "formulas" or combinations of formulas from the AISC Manual can be used todetermine the required moment of inertia to ensure that the live-load deflection does not exceed code-specified criteria. I is the value you will be seeking when using these formulas. Look carefully at the definitions of all of the values used in these formulas and their associated units. The most common mistake in these calculations is a units error. The three design values, required physical properties A_w, S_x, and I, obtained from these calculations, are material/building code-dependent.

The governing building code for the project will specify for a particular material with an allowable F_y, the yield strength of steel; F_b, the maximum allowable bending stress, as a function of F_y, to be used in moment/ bending calculations; F_v, the maximum allowable shear stress, also a function of F_y, to be used in shear calculations; and Δ, the maximum allowable deflection based on specific criteria for each structural component (which subsequently is used, with E, the modulus of elasticity of the material, to determine the required moment of inertia).

An additional consideration that will be relevant is the constructional condition of lateral support. A bending member will attempt to "twist or rotate" at certain levels of bending stress. This twisting is a result of the combination of the compression flange/zone trying to buckle sideways and the tension flange/zone attempting to remain in a straight line. The net result of the sideways movement of one flange and the upward movement of the other is a twisting or rotating motion. Unfortunately, these critical bending stress points are a function of stability, which is related to the specific geometry of the beam; consequently, all beams have different "critical" lateral-support dimensions (spacing between points of lateral support). These tendencies are compensated for by introducing reductions in the allowable F_b.

Initially, we'll deal with those beam conditions that have "full lateral support," making no stress adjustments necessary. Don't worry about what the criteria are for "full lateral support," we'll explain them when we get to an appropriate point.

As mentioned, the flexure formula $F_b = Mc/I$ can be written as $F_b = M/S$, where $S = I/c$. S is defined as the section modulus; since the moment of inertia (I) and the distance from the neutral axis to the extreme fiber (c) are both constant for a particular section, S is also a constant for that section. It's just a simplified method of selecting members based on two variables at the same time.

You should note that in the AISC manual there are two tables that define section modulus; they can be easily confused if you're not alert. One is for S_x, the elastic section modulus, and the other is for Z_x, the plastic section modulus. Be sure you use the **elastic tables for S_x**. You may note in the AISC Manual under each individual section that there is a listing for S_x and S_y. Typically, we use S_x for bending calculations. However, in some instances where a section may be loaded or used on its side, the S_y value may be appropriate. A simple sketch of your section under load (Fig. 5.1) will help make this clear.

Steel W (wide-flange) sections use their material very efficiently: They develop high section modulus values and high I values without much material. This is reflected in relatively low cost and explains why they are the most common beam sections. Other sections, however, may be more appropriate under certain conditions where you are looking at architectural as well as structural considerations.

5.2 Beams in structural systems

Beams as horizontal members are among the most commonly designed elements of structure; however, any member that is subject to moment is a bending member, and although its orientation may create a wider variety of loading conditions, the basic premise is the same. It must resist moment, shear, and deflection in every case. Occasionally, the criteria may be inconsequential, but this should be a considered decision, not an omission.

Beams may be known by alternative names, depending on their location in a system and the material that they use. Basic floor joists are really beams, although spaced closely together and

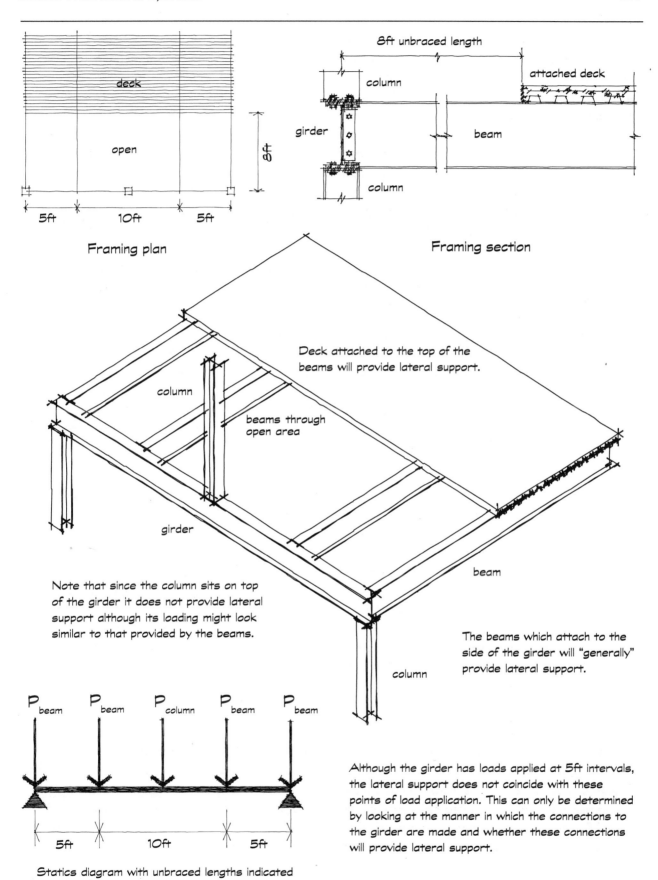

deck

open

8ft

5ft   10ft   5ft

Framing plan

8ft unbraced length

column

girder

attached deck

beam

column

Framing section

Deck attached to the top of the beams will provide lateral support.

column

beams through open area

girder

beam

Note that since the column sits on top of the girder it does not provide lateral support although its loading might look similar to that provided by the beams.

column

The beams which attach to the side of the girder will "generally" provide lateral support.

P_{beam} P_{beam} P_{column} P_{beam} P_{beam}

5ft   10ft   5ft

Statics diagram with unbraced lengths indicated

Although the girder has loads applied at 5ft intervals, the lateral support does not coincide with these points of load application. This can only be determined by looking at the manner in which the connections to the girder are made and whether these connections will provide lateral support.

Fig. 5.1 Axonometric sketch of construction system to determine unbraced lengths and load paths.

generally lightly loaded. A roof joist, although sometimes called a rafter, is still a beam (although it may carry a small compression load). A beam that supports other beams is commonly called a **girder**; while still a beam, its loads are generally much greater and it may support major portions of the structure. Transfer beams are "big girders" or "major" beams that generally pick up loads from columns in cases where they cannot continue vertically through the building. The ultimate confusion might be associated with "plate girders," or large built-up sections used as beams. These plate girders have a technical definition based on the height-to-thickness ratio of their webs and have different stress values for both bending and shear. At the other end of the spectrum are purlins and girts, which are light-rolled "C" or "Z" shapes used in prefab steel building construction for roof and wall members, respectively. Don't be confused by the name; a rose is a rose is a beam is a rafter is a girder is a girt, and so on. Although the variety of unique names in some cases defines the beam's role, it is also occasionally confusing. We'll try to be clear. **The bottom line is that they are all bending members and they are distinguished primarily by the manner in which they are loaded, ultimately, a minor statics consideration.**

As a final example, a simple handrail is a beam with a complicated life. It must resist a lateral (horizontal) load of 50plf [730N/m] or a single point load of 200lb [890N] applied in any direction. This means that this simple handrail must be a beam about both axes simultaneously. At the same time, the vertical supports (balusters) are both columns supporting axial load and beams resisting the potential lateral loads of the handrail. The end condition for these cantilevered beams is crucial to the stability and continued life of both the inhabitants and the structural system.

In steel structures, simply supported beams are very common, since the bolted connections used are generally regarded, at least for the purpose of simplifying calculations, as **pinned** connections or perfect hinges. This is obviously not a very accurate assumption, since any bolted connection will provide some moment resistance either by friction or by a couple corresponding to a moment reaction. This potential connection-induced moment is a prime concern in connection design. It is important that connection angles in framed beam connections be **as light as possible** while still meeting all connection criteria. This will allow the angle to flex more easily under normal beam deflection and not transfer moment into a column or beam as a stiffer or more extended angle might. For similar reasons, loads should be transferred to adjacent members as close to the centroid of that member as possible. This reduces the eccentricity, which also leads to unintended moments (Fig. 5.2).

This moment reaction is difficult to quantify and is generally disregarded. This is a conservative procedure, since the end moments reduce the maximum moment and deflection in the beam. You should be aware, however, that even partial moment resistance can be considered in refining the design of beams to achieve economy of construction. In our examples, both bolted and welded connections will be considered as hinges as long as the connection is made exclusively **at the web of the beam**. As a generalization, if the flanges of a beam are connected to the adjacent member, beam or column, it is likely that some significant moment transfer will occur.

Another important consideration is the method or points of application of loads. If a beam is loaded from only one side, a potential torsional force may be induced. To avoid this possibility, a system should be designed so that the loads are applied to the centriod of any member. This is true not only of beams, but of columns as well. In fact, we will see later that tension splices, truss connections, and framed beam connections should all be designed with the least possible amount of eccentricity. This means that the centriodal axis of the members should coincide or should be as close as possible. The shortest distance between two points is a straight line and this is also the most efficient line of load transfer.

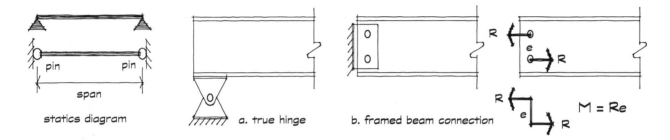

Fig. 5.2 Diagram of a simply supported beam with the fixed connection symbol at the ends. Although a hinge can be built (a), in practice beams in buildings are bolted (or welded) to columns (b) and the simplest two bolt framed beam connection will provide a moment reaction.

The support of overhangs is a particularly delicate matter. If the overhang can be developed by simply extending the framing members beyond their support point, we have no problem. Unfortunately, this means we can have overhangs only in one direction. To create overhangs in a direction perpendicular to the direction of span, a technique similar to basic residential roof construction is used. In this case, the overhanging members are continued over the support back into the adjacent framing generally an amount equal to two to three times the overhang. This accounts for the fact that the overhang will have some point load at the end from a railing, parapet or at least surface finishes that wouldn't normally exist. If you look at houses under construction, you will almost always see this technique. **The worst possible solution is to bracket the overhang off the side of a single beam, creating a massively eccentric loading condition.** Although it's possible to design for this torsional force, it can be solved more easily and inexpensively with a more considered structural framing solution.

Simply supported beams subject to gravity loads are statically determinate and easy to design. Remember that the V, M and Δ formulas can be used for a combination of point loads and distributed loads. In other words, the shear, moment, or deflection value at any point along the beam under investigation due to a load condition can be **added** to the corresponding values **at the same point** produced by another load. This is useful in situations such as the one shown in Fig. 5.3, where a point load is applied asymmetrically on a simple beam. You should note that although you might use the maximum values for each of these two loading conditions, since the maximums do not occur at the same point, they will give a conservative value that will be greater than the actual maximum created by the combined loading. This is a useful technique for deflection, which can become quite cumbersome to calculate. Actual values for moment are easily and more accurately determined by simply using the combined shear diagram for the loading condition and calculating the area under the shear diagram to the point of maximum moment (zero shear).

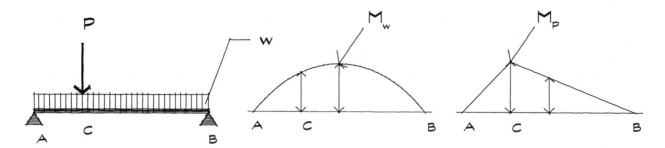

Fig. 5.3 Maximum deflection can be calculated separately for each load and the results combined for a **conserva-tive** approximation of the actual value.

Beam design tables are provided in most statics textbooks and in the AISC Manual. These tables give formulas to compute values of support reactions, shear, moment, and deflection at any point for various types of beams and loads, including statically indeterminate beams. A common type of statically indeterminate beam is shown in Fig. 5.4; these are continuous beams over three or more supports. The familiar statics method of equations of equilibrium is not sufficient to solve for the support reactions, but once the values of maximum shear, moment and deflection are calculated using the tabulated formulas, the continuous beam can be designed. The problem with indeterminate structures is not the design of the members themselves, but in determining the loads that must be resisted. Examples of continuous beams would be a roof girder supported

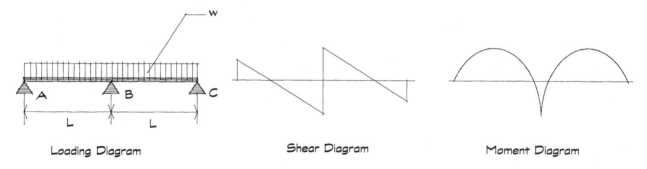

Loading Diagram   Shear Diagram   Moment Diagram

Fig. 5.4 Continuous beam diagrams

by a series of columns, a handrail, or a floor deck spanning across several joists.

Using our understanding of how cantilevers reduce interior moment and deflection, we could create a series of cantilevers with hanging spans in a continuous system to maximize the use of materials. This combination of beams is referred to as a Gerber beam and is simple to calculate and construct, making it cost-effective for long repetitive spans (Fig. 5.5). Shear or moment diagrams are used to optimize the location of splices in the system of beams. In the continuous beam of Fig. 5.4, the moment diagram of the continuous beam has two points of zero moment in the central bay. It is logical to locate any necessary splice at these points in order to avoid the design of an expensive moment-resisting joint. The statics are simplified also, since the system is reduced to two overhangs with a simply supported beam in the middle Fig 5.6. This technique retains most of the efficiency of the more statically complex system without the math complexity. Deflection will change somewhat since the hanging span can be a section of smaller cross-section, typically providing less moment of inertia. Since the hanging span also can be somewhat shallower than the adjacent cantilevered beams, this could provide the opportunity to integrate mechanical cross passages into the structural system.

A very common type of statically indeterminate beam is the one with fixed ends, representing the condition of a moment-resisting frame. The calculation of a frame takes into account the fact that rigid connections will generally allow for some rotation of the end of the beam, and the magnitude of the moment at the ends will basically depend on the relative rigidity of the column, as shown in Figs. 5.8 and 5.9. Given equal loads, a very deep and rigid beam supported on slender columns will display a behavior more similar to pinned ends than the opposite case of a flexible beam on very rigid supports. In the latter beam, the ends are subject to a stronger bending moment, whereas in the former, the bending moment is bigger at midspan. Of course, we're assuming a perfectly rigid connection in both cases, which is not a realistic situation.

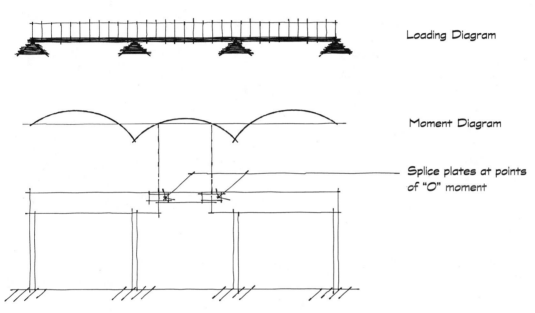

Loading Diagram

Moment Diagram

Splice plates at points of "O" moment

Fig. 5.5 Gerber beam as a statically determinate equivalent to a indeterminate continuous span

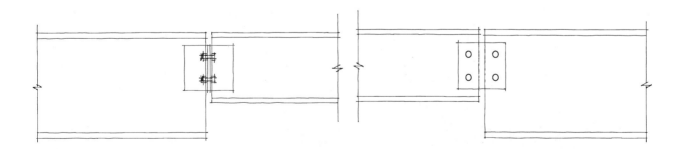

Fig. 5.6 Alternative non-moment transfer connections for Gerber beam systems. The primary concern is to avoid flange continuity and to transfer shear loads from web to web only.

This concept is used to calculate the actions on multistory frames, where beams and columns are effectively subject to bending moments applied at the ends.

Fig. 5.7 Gerber beams stacked into a multistory building

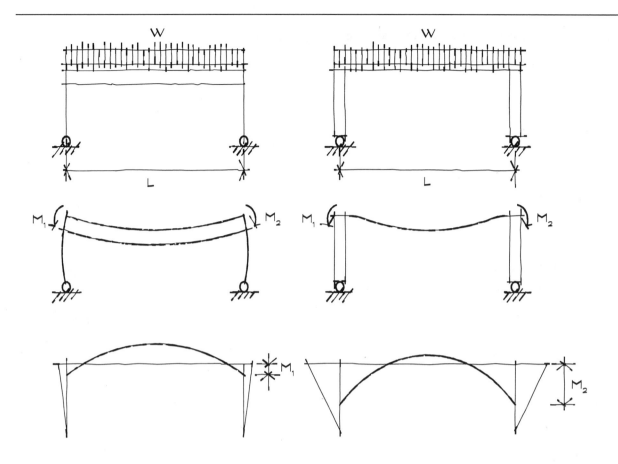

Fig 5.8 Moments are generated by gravity loads at the end of beams in a frame relative to the rigidity of the columns. (The free body diagrams do not show the vertical reactions.)

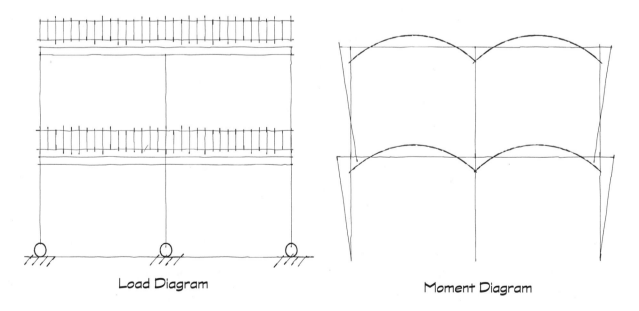

Load Diagram   Moment Diagram

Fig. 5.9 Statically indeterminate multistory frame with rigid connections and corresponding moment diagrams

Enough talk, let's do an example and see how it all works.

Example 5.1
Simple "fully laterally supported" beam

Span: 26ft-0in   LL = 500plf = 0.50klf

                                     DL = 150plf = 0.15klf

+ beam weight estimated at 40plf   = 0.04klf

Total load (LL + DL + beam wt.) = 690plf   = 0.69 klf

A-36 Steel, F_y = 36,000ksi

Allowable F_b = 0.66F_y AISC Manual, p.5-45 $F_b = Mc/I = M/S_x$

Allowable F_v = 0.40F_y AISC Manual, p.5.49 $F_v = V/A$

E = 29,000,000ksi $\Delta_{LL} = L/360$

                                                         $\Delta_{TL} = L/240$

Since the TL is less than 1.5 X the LL, L/240 will not control.

1. Required section modulus (S_x), based on maximum moment:

$$S_x = M/F_b = \frac{(58.31\text{ftk})(12\text{in/ft})}{(0.66)(36.0\text{ksi})} = 29.16\text{in}^3$$

2. Required web area (A_w):

$$A_w = \frac{V}{F_v} = \frac{8.97\text{k}}{0.4(36.0\text{ksi})} = 0.62\text{in}^2$$

3. Required moment of inertia (I) :

Maximum allowable deflection (Δ) =(26ft)(12in/ft)/360 = 0.867in.

$$I_x = \frac{5WL^4}{384E\Delta_{max}} = \frac{5\left(\frac{500\text{plf}}{12\text{in/ft}}\right)[(26\text{ft})(12\text{in/ft})]^4}{384(29,000,000\text{psi})(0.867\text{in})} = 204.5\text{in}^4$$

This deflection formula can be found on page 2-296 of the AISC Manual. Contrary to the Manual description of units, if you always use **pounds and inches**, you will be less prone to making units errors. This is especially true when you move between steel and wood where the load units generally change from kips (steel) to pounds (wood).

4. **First, make a preliminary selection from the section modulus table (AISC Manual, p. 2-12). The most "economical" sections, by weight, are listed in boldface type.**

W14X22: S_x = 29.0in³   (within 5%; may be acceptable)

W12X26: S_x = 33.4in³   (clearly works for moment)

5. **Second,** check the moment of inertia (I_x) of the preliminary sections. Check this requirement next, since it requires no calculations.

   I_x required = 204.5in⁴

   W14X22: I = 199in⁴   (within 5%; may be acceptable)

   W12X26: I = 204in⁴   (just meets I requirements. it works!)

6. **Third,** check the web area (A) for minimum requirements.

   Check this requirement last, since it requires some minimal calculations. In fact, this may not necessitate actual calculations if the value is clearly greater than the requirements.

   W14X22: t_w (web thickness) = 0.230in

              d (web depth or height) = 13.74in

              $A_w = t_w \times d$ = 0.230in (13.74in) $\geq$ 0.62 in² (it works!)

   W12X26: t_w = 0.230in

              d = 12.22in

              $A_w = t_w \times d$ = 0.230in (12.22in) $\geq$ 0.62 in² (it works!)

Based on all three criteria, either preliminary section works (if you accept a ± 5% margin of error). The W12X26 is shallower but slightly heavier. Based on life-cycle costing considerations the cost of this initial extra weight could be recovered by decreasing the building volume.

Select W12x26 as the solution.

You will note a previously unused column of values listed in the section modulus table labeled: L_c and L_u. For the W12X26, L_c = 6.9ft and L_u = 9.4ft. This means that to use full F_b = 0.66F_y, this particular beam must have lateral support at intervals not exceeding 6.9ft (for A-36 steel). The other figure, L_u, is the maximum spacing of lateral support to use a reduced F_b value (again for A-36 steel). We'll discuss other conditions of lateral support next.

Example 5.2

Beam without "full lateral support" potential

Assume 7ft unbraced length.

    A-36 steel, F_y = 36,000 ksi

    All. F_b = 0.66F_y AISC Manual

    All. F_b = 0.40F_y AISC Manual

    E = 29,000,000ksi

Assume beam weight = 30 plf

Maximum V = 13.1 k

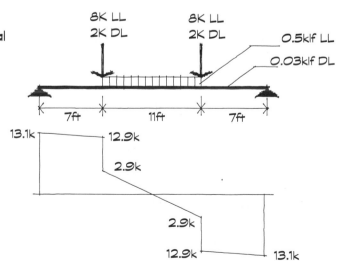

Maximum M = $\dfrac{(13.1k + 12.9k)}{2}$(7ft)   = 91.0ftk

$\qquad\qquad\quad$ + .5 × 5.5ft (2.9k)   =   8.0ftk

$\qquad\qquad\qquad\qquad\qquad$ Total M = 99.0ftk

Maximum Δ_{LL} = L/360 = (25ft×12in/ft)/360 = 0.833in

Since lateral stability considerations vary with every beam selection, initially assume full lateral support and make revisions after preliminary selection if necessary.

1. Required Section Modulus (S_x) , based on maximum moment:

$$S_x = M/F_b = \frac{(99.0ftk)(12in/ft)}{(0.66)(36.0ksi)} = 49.5in^3$$

2. Required Web area (A_w):

$$A_w = V/F_v = \frac{(13.1k)}{(0.40)(36.0ksi)} = 0.91in^2$$

3. Required Moment of inertia (I) :

$\quad$ Max. All. Deflection (Δ) = = 0.833in

$\quad$ Sketch the live-loading conditions:

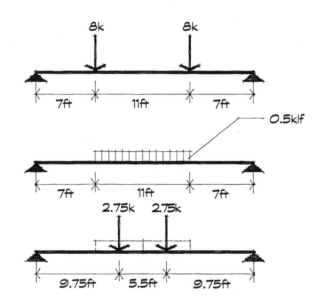

$\Delta_1 = \dfrac{Pa}{24EI}(3L^2 - 4a^2)$

Δ_2 = No formula exists in AISC Manual

$\Delta_2 = \dfrac{Pa}{24EI}(3L^2 - 4a^2)$

(a roughly equivalent loading condition)

$\Delta_T = \Delta_1 + \Delta_2$ or $I_T = I_1 + I_2$

$$I_1 = \frac{(8000lb)(84in)}{24(29,000,000psi)(0.833in)}[3(300in)^2 - 4(84in)2] = 280.2in4$$

$$I_2 = \frac{(2750lb)(117in)}{24(29,000,000psi)(0.833in)}[3(300in)^2 - 4(117in)^2] = 119.4in^4$$

I_{total} required = 399.6in⁴

4. Make a preliminary selection using S_x tables from the AISC Manual.

> W14X34: S_x = 48.6in^3 (within 5%, it may work)
>
> check L_c = 7.1ft $\geq$ 7.0ft (actual unbraced length)

This beam has full lateral support; consequently, our original assumption that allowed us to use F_b = 0.66F_y is valid.

> W18X35: S_x = 57.6in^3 (works based on original assumption)
>
> check L_c = 6.3ft $\leq$ 7.0ft (cannot use F_b = 0.66F_y; must reduce allowable stress: see formula F1-5 AISC Manual, p. 5-46)
>
> check L_u = 6.7ft $\leq$ 7.0ft (cannot use F_b = .60F_y; must use tables on p. 2-175 or formula F1-3, on p. 5-46 both in the AISC Manual)

Tables are much simpler and quicker to use.

Looking at "allowable moments in beams," p.2-171, AICS Manual, we see that a W18X35 will carry approximately 102ftk at an unbraced length of 7.0ft since this exceeds our require ment of 99.0ftk, the beam will work for the moment, but at actual bending-stress value below 0.60F_y.

5. Check I for W18X35 = 510in^4 OK $\geq$ 399.7in^4

> W14x34 = 340in^4 inadequate, $\leq$ 399.7in^4 required.

6. Check A_w for W18X35

> t_w = .30in
>
> d = 17.70in A_w = $t_w \times$ d $\geq$ .91in^2 required.

Select W18X35 as the solution.

You'll notice that we've used the "most economical" section (**boldfaced**) selections even when the beam does not have full lateral support. This will not always produce the lightest section as it will with laterally supported beams, but it is still a good place to start.

Try the following problems and see if you need to look at the chapter again!

Problem 5.1

Find the lightest A-36 steel beam that can carry a bending moment of 110ftk. Assume full lateral support, and ignore deflection, and shear.

Solution:

From the S_x tables, W18X35 has a moment capacity M_r = 114.0ftk and weighs 35plf.

Problem 5.2

An existing building has a W24X250 grade-50 girder installed between two columns. A retrofit is planned that requires a maximum bending moment on this beam of 1596ftk. Find out if the girder is adequate, assuming a maximum unbraced length of 10ft.

Solution:

From the S_x tables, W24X250 has L_c = 11.8ft and M_r = 1770ftk, so it's OK!

Problem 5.3

A steel beam with a 20ft span is designed to be simply supported at the ends on columns, and to carry a floor system made with open-web joists at 4ft o.c. (assume this spacing as the unbraced length). The joists span 28ft and frame into the beam from both sides. Using grade-50 steel, select the most economical wide-flange section for the beam. Floor loads are 50psf LL and 14.5 DL. Assume, for the initial calculations, a beam weight of 40plf.

Solution:

The load applied to the beam by each joist equals

(50psf)(14ft × 4ft)	= 2,800lb (LL)
(14.5psf)(14ft × 4ft)	= 812lb (DL)
Total load	= 3,612lb

At a 4ft o.c. spacing, there would be four joists equally spaced on the beam. The two joists that would sit directly over the columns will cause no moment, shear, or deflection in the beam, but would be included in any column or bearing calculations. Since the loads are all symmetrically placed, the easiest statics solution is to simply distribute 50% to each side.

Maximum shear = 2(3,612lb) + (40plf)(20ft/2) = 7624lb

Maximum moment = 3(3,612lb)(4ft) + (400lb)(20ft)/4 = 45,344lb

Maximum deflection = (20ft × 12in/ft)/360 = 0.67in

Moment requirements:

Assuming (initially) full lateral support:

$$S_x = \frac{M}{0.66F_y} = \frac{45,344\text{ftlb}}{0.66(50,000\text{psi})} = 16.49\text{in}^3$$

Shear requirements:

$$A = \frac{V}{F_v} = \frac{7,624\text{lb}}{0.4(50,000\text{psi})} = 0.38\text{in}^2$$

Required I_x for deflection:

The loading condition of four equally spaced point loads is not one of those listed in the beam diagrams and formulas portion of the AISC Manual, but if you look at the very beginning of these tables, you'll find a page that converts simple equally spaced point loads into simple formulas. For this case, the deflection formula would be:

$$\Delta = \frac{0.063PL^3}{EI} = \frac{0.063(2,800lb)(20ft \times 12in/ft)^3}{29,000,000psi(I)}$$

The load P used in this formula is the LL reaction from each joist calculated before.

For an allowable deflection of 0.67in and solving for I, this formula becomes

$$I = \frac{0.063(20ft \times 12in/ft)^3}{29,000,000psi(0.67in)} = 125.5in^4$$

From the S_x tables, W12X16 has S_x = 17.1in^3 OK, L_c = 2.9ft < 4ft NO.

Next try, W6X25, S_x = 16.7in^3 OK, L_c = 5.4ft OK, but is it the cheapest?

Next try, W10X17, S_x = 16.2in^3 OK (within 5%) L_c = 3.6ft < 4ft, but L_u = 4.4ft.

This section won't work at full lateral support, but has extra capacity, so it may work using "partial" criteria. Check S_x using a reduced allowable bending stress of $0.60F_y$ = 0.60(50ksi) = 30ksi. The new required S_x can be calculated as:

S_x = 16.49in^3(0.66/0.60) = 18.14in^3 > 16.2in^3 given by the W10X17, which is thus inadequate.

Try W12X19: S_x = 21.3in^3 OK!!! This shape looks excessive, but we have to satisfy the deflection criteria. A glance at the chart of allowable moments (AISC Manual, p. 2-210) shows that W10X19, M14X18, and W8X21 are possible options for moment with an unbraced length of 4ft, but should be checked for deflection.

The shapes satisfying both moment and deflection requirements are the M14X18 and the W12X19, which weigh about the same. We choose the W12X19 and check it for shear.

Required A_w = 0.38in^2

W12X19 A_w = 0.235in(12.16in) = 2.86in^2 OK

Use W12X19 !!!! Fini!!

METRIC VERSIONS OF ALL EXAMPLES AND PROBLEMS

Example M5-1

Simple "fully laterally supported" beam

Span: 7.93m   LL = 7297N/m = 7.297kN/m

                              DL = 2189N/m = 2.189kN/m

                    + beam weight estimated = 0.584kN/m

              Total load (LL + DL + beam wt.) = 10.07kN/m

A-36 Steel, F_y = 250 MPa

Allowable F_b = 0.66 F_y AISC Manual, p.5-45 $F_b = Mc/I = M/S_x$

Allowable F_v = 0.40 F_y AISC Manual, p.5-49 $F_v = V/A$

E = 200,000 MPa Δ_{LL} = L/360

                                                        Δ_{TL} = L/240

      Since the TL is less then 1.5 X the LL, L/240 will not control.

1. Required section modulus (S_x), based on maximum moment:

$$S_x = M/F_b = \frac{79,100,000Nmm}{0.66(250MPa)} = 479,000mm^3$$

2. Required web area (A_w):

$$A_w = V/F_v = \frac{40,000N}{0.4(250MPa)} = 400mm^2$$

3. Required moment of inertia (I) :

$$\text{Max. All. Deflection } (\Delta) = \frac{(7.93m)(1000mm/m)}{360} = 22mm$$

$$I_x = \frac{5wL^4}{384E\Delta_{max}} = \frac{5(7297N/m)(1/1000mm/m)(7930mm)^4}{384(200,000MPa)(22mm)} = 85.39 \times 10^6 mm^4$$

4. **First,** make a preliminary selection from the section modulus table (AISC Manual p. 2-12). The most "economical" sections, by weight, are listed in **boldface type.**

    W360X33 : S_x = 29.0in³ = 475x10³mm³   (within 5% may be acceptable)

    W310X39 : S_x = 33.4in³ = 547x10³mm³   (clearly works for moment)

5. **Second,** check the moment of inertia (I_x) of the preliminary sections. Check this requirement next, since it requires no calculations.

        I_x required = 85.39x10⁶mm⁴

W360X33: I = 199in⁴ = 82.83x10 mm⁴ (within 5%, may be acceptable)

W310X39: I = 204in⁴ = 84.91x10⁶mm⁴ (just meets I requirements; it works!)

6. Third, check the web area (A) for minimum requirements.

Check this requirement last, since it requires some minimal calculations. In fact, this may not necessitate actual calculations if the value is clearly greater than the requirements.

W360X33: t_w (web thickness) = 5.842mm

d (web depth or height) = 349mm

$A_w = t_w$ d = (5.842mm)(349mm) = 2039mm² > 400mm²   (it works!)

W310X39: t_w = 5.842mm

d = 310mm

$A_w = t_w$ d = (5.842mm)(310mm) = 1813mm² > 400mm²   (it works!)

Based on all three criteria either preliminary section works (if you accept a 5% margin of error). The W310X39 is shallower but slightly heavier. Based on life-cycle costing considerations, the cost of this extra initial weight could be recovered by decreasing the building volume.

Select W310X39 as the solution.

You will note a previously unused column of values listed in the section modulus table labeled: L_c and L_u. For the W310X39, L_c = 2.10m and L_u = 2.87m. This means that to use full $F_b = 0.66F_y$ this particular beam must have lateral support at intervals not exceeding 2.1m (for A-36 steel). The other figure, L_u, is the maximum spacing of lateral support to use a reduced F_b value (again, for A-36 steel). We'll discuss other conditions of lateral support next.

Example M5-2

Beam without "full lateral support" potential

Assume 2.134m unbraced length.

A-36 Steel, F_y = 250 MPa

All. $F_b = 0.66 F_y$ AISC Manual, p. 5-45

All. $F_b = 0.40 F_y$ AISC Manual, p. 5-49

E = 200,000MPa

Assume beam weight = 438N/m

Maximum V = 58,270N

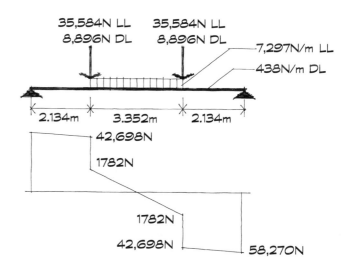

Maximum M $= \dfrac{(58,270N + 42698N)}{2}(2.134m) = 107,733Nm$

$$+ \dfrac{(1782N)(3.352/2m)}{2} \quad = \underline{1493Nm}$$

Total M $\quad = 134,127Nm$

Maximum Δ_{LL} = L/360 = (7620 mm)/360 = 21 mm

Since lateral stability considerations vary with every beam selection, initially assume full lateral support and make revisions after preliminary selection if necessary.

1. Required section modulus (S_x), based on maximum moment:

$$S_x = M/F_b = \dfrac{(134,127Nm)(1000mm/m)}{0.66(250MPa)} = 813 \times 10^3 mm^3$$

2. Required web area (A_w):

$$A_w = V/F_v = \dfrac{(58,270N)}{0.4(250MPa)} = 583mm^2$$

3. Required moment of inertia (I) :

Maximum allowable deflection Δ = 21mm.

Sketch the live-loading conditions:

$$\Delta_1 = \dfrac{Pa}{24EI}(3L^2 - 4a^2)$$

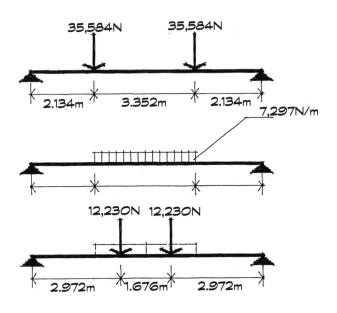

no formula exists in the AISC Manual

$$\Delta_2 = \dfrac{Pa}{24EI}(3L^2 - 4a^2)$$

(a roughly equivalent loading condition)

$\Delta_T = \Delta_1 + \Delta_2$ or $\quad I_T = I_1 + I_2$

$$I_1 = \dfrac{(35,584N)(2134mm)}{24(200,000MPa)(21mm)}[3(7620)^2 - 4(2134)^2] = 117.5 \times 10^6 mm^4$$

$$I_2 = \dfrac{(12230N)(2972mm)}{24(200,000MPa)(21mm)}[3(7620)^2 - 4(2972)^2] = 50.1 \times 10^6 mm^4$$

I_{total} required = $117.5 \times 10^6 mm^4 + 50.1 \times 10^6 mm^4 = 167.6 \times 10^6 mm^4$

4. Make a preliminary selection using S_x tables from the AISC Manual.

W360X51: $S_x = 48.6 in^3 = 796 \times 10^3 mm^3$ (within 5%, it may work)

check L_c = 2164mm > 2134mm   (actual unbraced length.)

This beam has full lateral support; consequently, our original assumption that allowed us to use $F_b = 0.66F_y$ is valid.

W460X52: $S_x = 57.6 in^3 = 944 \times 10^3 mm^3$ (works based on original assumption)

check L_c = 1920mm < 2134mm

(cannot use $F_b = 0.66F_y$, must reduce allowable stress; see formula F1-5 AISC Manual, p.5-46)

check L_u = 2042mm < 2134mm

(cannot use $F_b = 0.60F_y$, must use tables on p. 2-175 or formula F1-3 p. 5-46, both in the AISC Manual)

Tables are much simpler and quicker to use.

Looking at "allowable moments in beams," p. 2-271 we see that a W18X35 will carry approximately 138kNm at an unbraced length of 2134mm. (Actually, you'll need to convert your metric values to ftk and ft to use the table.) Since this exceeds our requirement of 134kNm, the beam will work for moment, but at an actual bending stress value below $0.60F_y$.

5. Check I for W460X52 = $510 in^4 = 212.3 \times 10^6 mm^4$; OK > $167.6 \times 10^6 mm^4$

W360X51 = $340 in^4 = 141.5 \times 10^6 mm^4$; inadequate, < $167.6 \times 10^6 mm^4$ required.

6. Check A_w for W460X52

t_w = 7.6mm

d = 449.6mm $A_w = t_w d = 3426 mm^2 > 583 mm^2$ required.

Select W460X52 as the solution.

You'll notice that we've used the "most economical" section (**boldface**) selections even when the beam does not have full lateral support. This will not always produce the lightest section as it will with laterally supported beams, but it is still a good place to start.

Problem M5.1

Find the lightest A-36 steel beam that can carry a bending moment of 149kNm. Assume full lateral support, and ignore deflection and shear.

Solution:

From the S_x tables, W460X52 has a moment capacity M_r = 154.6kNm and weighs 511N/m.

Problem M5.2

An existing building has a W610X372 grade-50 (345MPa) girder installed between two columns. A retrofit is planned that requires a maximum bending moment on this beam of 2165kNm. Find out if the girder is adequate, assuming a maximum unbraced length of 3.05m.

Solution:

From the S_x tables, W610X372 has L_c = 3.60m and M_r = 2401kNm, so it's OK!

Problem M5.3

A steel beam with a 6.1m span is designed to be simply supported at the ends on columns, and to carry a floor system made with open-web joists at 1.22m o.c. (assume this spacing as the unbraced length). The joists span 8.54m and frame into the beam from both sides. Using grade-50 (345 MPa) steel, select the most economical wide-flange section for the beam. Floor loads are 2400N/m² LL and 694N/m² DL. Assume, for the initial calculations, a beam weight of 600N/m.

Solution:

The load applied to the beam by each joist equals

$$(2400N/m^2)(4.27m \times 1.22m) = 12,503N \text{ (LL)}$$
$$(694N/m^2)(4.27m \times 1.22m) = \underline{3,615N} \text{ (DL)}$$
$$\text{Total load} = 16,118N$$

At a 1.22m o.c. spacing, there would be four joists equally spaced on the beam. The two joists that would sit directly over the columns will cause no moment, shear, or deflection in the beam, but would be included in any column or bearing calculations. Since the loads are all symmetrically placed, the easiest statics solution is to simply distribute 50% to each side.

Maximum shear = 2(16,118N) + (600N/m)(6.1m/2) = 34067N

Maximum moment = 3(16,118N)(1.22m) + (1830N)(6.1m) /4 = 61782Nm

Maximum deflection = (6,100mm)/360 = 17mm

Moment requirements:

Assuming (initially) full lateral support:

$$S_x = \frac{M}{0.66F_y} = \frac{(61,782)(1000Nmm)}{0.66(345MPa)} = 271 \times 10^3 mm^3$$

Shear requirements:

$$A = \frac{V}{F_v} = \frac{34,067N}{0.4(345MPa)} = 247mm^2$$

Required I_x for deflection:

The loading condition of four equally spaced point loads is not one of those listed in the beam diagrams and formulas portion of the AISC Manual, but if you look at the very beginning of these tables, you'll find a page that converts simple equally spaced point loads into simple formulas. For this case, the deflection formula would be:

$$\Delta = \frac{.063PL^3}{EI} = \frac{.063(12,503N)(6,100mm)^3}{200,000MPa(I)}$$

The load P used in this formula is the LL reaction from each joist calculated before.

For an allowable deflection of 17 mm and solving for I, this formula becomes

$$I = \frac{.063(12,503N)(6,100mm)^3}{200,000MPa(17mm)} = 52.6\times10^6 mm^4$$

From the S_x tables, W12X16 has S_x = 17.1in³ = 280x10³mm³ OK, L_c = 0.88m < 1.22m NO. Next try, W6X25, S_x = 16.7in³ = 274x10³mm³ OK, L_c = 1.65m OK, but is it the cheapest? Next try, W10X17, S_x = 16.2in³ = 266x10³mm³ OK (within 5%), L_c = 1.0m < 1.22m, but L_u = 1.34m.

This section won't work at full lateral support, but has extra capacity, so it may work using "partial" criteria. Check S_x using a reduced allowable bending stress of $0.60F_y$ = 0.60(345MPa) = 207MPa. The new required S_x can be calculated as:
S_x = 271x10³(0.66/0.60) = 298x10³mm³ > 266x10³mm³ given by the W10X17, which is thus inadequate.

Try W12X19: S_x = 21.3in³ = 349x10³mm³ OK!!! This shape looks excessive, but we have to satisfy the deflection criteria. A glance at the chart of allowable moments (AISC Manual, p. 2-210) shows that W10X19, M14X18 and W8X21 are possible options for moment with an unbraced length of 1.22m, but should be checked for deflection.

The shapes satisfying both moment and deflection requirements are the M14X18 and the W12X19, which weigh about the same. We choose the W12X19 and check it for shear.

Required A_w = 247mm²
W12X19 A_w = 6mm(320mm) = 1920mm² OK

Use **W12X19** !!!! Fini!!

6 LATERAL STABILITY OF BEAMS

6.1 Conditions of stability

As suggested in the previous chapter, the design of a beam is based on assumptions associated with its lateral stability. If a beam has "full" lateral support, as the previous ones were assumed to have, the full allowable bending stress, $0.66F_y$, may be utilized in calculating its required physical properties. However, this is not always the case.

Fortunately, only one of the previous properties, S (section modulus), is affected by the presence or absence of lateral support. This is because only the allowable bending stress that is used to calculate section modulus is affected.

You will see this same issue discussed in column design. Members under compression (columns) or elements of members under compression (beams) tend to buckle under high loads. That buckling can be resisted if the member or component has sufficient lateral support. That's what the entire issue of beam support addresses. This was the reason we provided lateral support or bridging to open-web steel joist. Joists are very slender for their depth and will twist under load. Wide-flange shapes will potentially experience the same "twisting," but since they have wide horizontal plates on the top and bottom (flanges), they tend to have more resistance to lateral movement. I-beams have narrower flanges and, as you would now expect, tend to buckle more easily.

Higher-yield-stress steels and the resulting comparatively lighter sections have led to the identification/classification of certain sections as being either compact or noncompact. (AISC Manual, Sec. B5.1, p. 5-35). The limiting factor for the definition of a compact section is the width-to-thickness ratio of the compression flange and the compressed area of the web in bending. The shape, the laterally unsupported length of the span, and the grade of steel are also determinants of the allowable bending stress, F_b.

You may have noted that the section modulus tables have additional information beyond the section modulus for F_y = 36ksi [250MPa] and F_y = 50ksi [345MPa] steel. Two columns, labeled L_c and L_u, are included. The notation L_c defines the maximum unbraced length of the compression flange for which the maximum allowable stress for a compact symmetrical shape, $F_b = 0.66F_y$, may be used. The notation L_u defines the maximum unbraced length of the compression flange for which the maximum allowable bending stress can be taken as $F_b = 0.60F_y$.

For unbraced lengths greater than L_u, AISC Manual Tables 2-149 through 2-211 are recommended. For the purists, actual specific allowable stresses can be calculated using formulas specified in Chapter F of the AISC Manual (pp. 5-45 through 5-48).

THE RULES TO REMEMBER

If the unbraced length of a beam is less than or equal to L_c,

$$F_b = 0.66F_y$$

If the unbraced length of a beam is greater than L_c but less than or equal to L_u,

$$F_b = 0.60F_y$$

If the unbraced length of a beam is greater than L_u. Proceed to AISC Manual, Tables 2-149 through 2-211.

The use of these AISC tables requires an understanding of the concept of moment capacity or resisting moment (M_R).

There's really nothing new here, simply a reinterpretation one of our basic formulas, $F_b = Mc/I$. You'll remember that we have previously rewritten this formula in the form $I/c = M/F_b$, with I/c defined as S (section modulus). This is the most convenient form to use to design a beam, but we can also write the formula in the form $M_r = F_bS$: the moment capacity (resisting moment) of a beam is equal to its specific allowable stress F_b times its section modulus S. As long as the moment capacity of a beam at its specific unbraced length is at least equal to our design moment, the beam works (for moment). At unbraced lengths equal to or less than L_u, the section modulus table is probably the most efficient way to find the economical section, but at lengths greater than L_u, the tables are essential.

Calculations associated with shear and deflection are unaffected by considerations of lateral stability. The following examples (Figs. 6.1a and 6.1b) illustrate alternative lateral-support conditions. They take you through calculations for a beam that either does not have "full" lateral support or for which we cannot be sure as to its lateral-support limitations initially. Remember, the problem arises in that we don't know the criteria until we've selected a section and we cannot select a section until we know the criteria.

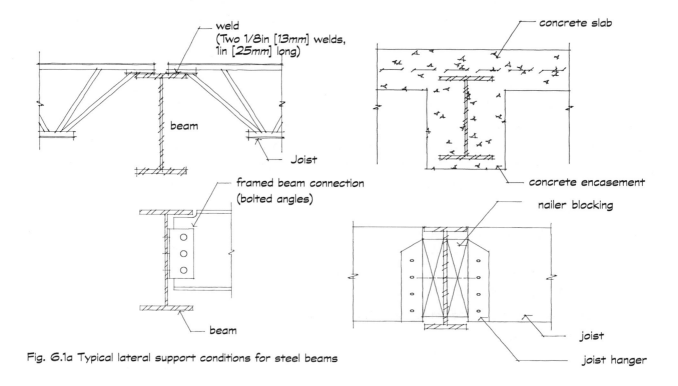

Fig. 6.1a Typical lateral support conditions for steel beams

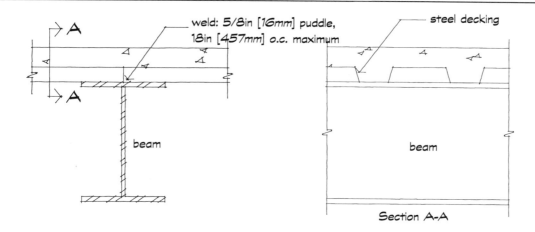

Fig. 6.1b Typical lateral support conditions for steel beams

Some common sense might help at this point. Look at the AISC Manual section modulus tables, (p. 2-7 through 2-13). Notice the range of values of L_c and L_u for the economical sections. Even for a very light section such as a **W12X19 [W310X28]**, the value for L_c is greater than 4ft [1.2m], so initial decisions can be made as to the "likelihood" of a condition having full, or partial, or critical lateral support with some confidence.

Example 6.1

A 30ft girder supports two concentrated loads, 21k LL and 15.6k DL, each at the 1/3 points of the span. There is no lateral support for the girder other than that provided by the connections of the beams. Assuming a girder weight of 60plf, select the most economical wide-flange shape using 50ksi steel.

Although the section has an unbraced length of 10ft (between the ends and the point loads, and in the midspan portion), we will initially assume "full lateral support."

Support reactions: Since we have a simple span with symmetrical loading, we can simply add all the loads and place 1/2 on each end.

$$R = 21k + 15.6k + (0.06klf)(15ft) = 37.5k$$

$$V_{max} = 37.5k$$

By using the shear diagram, the maximum moment equals

$$M_{max} = \frac{(37.5k + 36.9k)}{2}(10ft) + \frac{(0.3k)(5ft)}{2}$$

$$372ftk + .75ftk = 372.75ftk$$

$$M_{max} = 372.75ftk$$

$$\Delta_{max} = L/360 = 30ft(12in/ft)/360 = 1in$$

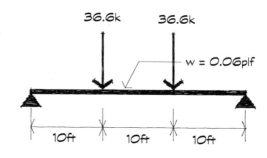

Section properties

Required section modulus:

$$F_b = 0.66F_y = 0.66(50\text{ksi}) = 33\text{ksi}$$

$$S_x = \frac{(372.75\text{ftk})(12\text{in/ft})}{33\text{ksi}} = 135.5\text{in}^3$$

Required shear area:

$$F_v = 0.4F_y = 0.4(50\text{ksi}) = 20\text{ksi}$$

$$A_v = \frac{37.5\text{k}}{20\text{ksi}} = 1.875\text{in}^2$$

Required I_x for deflection:

Since the LL condition is a simple one for which a formula exists, a simple substitution works.

$$\Delta = \frac{Pa}{24EI}(3L^2 - 4a^2)$$

$$\Delta = \frac{(21\text{k})(120\text{in})}{24(29,000\text{ksi})(1\text{in})}[3(360\text{in})^2 - 4(120\text{in})^2] = 1199.2\text{in}^4$$

Check S_x first, since our original assumption must be verified.

W21X68 has $S_x = 140\text{in}^3 >$ required 135.5in^3

$L_c = 7.4\text{ft}$ and $L_u = 8.9\text{ft}$

This means that the section does not have even partial lateral support with L = 10ft.

One option is to look in the section modulus table for a beam with a section modulus of 135.5in³ and an L_c = 10ft or more.

W14X90 works, or if we are willing to look at beams with an L_c < 10ft, but an L_u of 10ft or more, we can find other workable options. To do this, we must recalculate the required S_x based on "partial lateral support" with $F_b = 0.60F_y = 30\text{ksi}$.

Using a simple proportion will solve this one:

$$S_x = 135.5\text{in}^3\left(\frac{0.66}{0.6}\right) = 149.05\text{in}^3$$

W18X76 provides $S_x = 146\text{in}^3$ (within 5%) and $L_u = 13.7\text{ft}$, or

W21X83 provides $S_x = 171\text{in}^3$ and $L_u = 10.9\text{ft}$, or

W16X89 provides $S_x = 155\text{in}^3$ and $L_u = 18\text{ft}$

We could also use the "allowable moments in beams" tables on page 2-200 of the AISC Manual.

(Be sure to note the F_y listed in the upper right-hand corner of each page.)

Look for a moment capacity close to the maximum moment of our beam. In our case M_R must be at least 372ftk. Two parameters, 10ft and 372ftk, identify a point in the chart. The 10ft line is indicated with small crosses (unbraced lengths are noted at the top and bottom of the page). Mark this point with a pencil. If you imagine a rectangle with this point at the upper right-hand corner, any beam that has a line that passes outside that rectangle to the right works. If you find a dashed line, this means there will be a more economical section (if weight is your only consideration) and you should look further. If, however, you wish to use a beam of a specific depth, a dashed line may be the one you're looking for.

Some choices that this process indicates are:

W24X76: M_R = 372ftk out to an unbraced length of 13.25ft. Another way of looking at it is that this beam will carry M_R = 422ftk at exactly 10ft of unbraced length (you have to go to the preceding pages of the charts to do this for this shape). Obviously, from these tables, you can see that as the unbraced length increases beyond L_u, the moment constantly decreases. This reflects that the allowable bending stress changes continuously for unbraced lengths above L_u.

You will notice from the chart that for a value M_R = 484ftk, the graph is flat out to an unbraced length of 8ft, with a white dot at 8ft: this means M_R is constant out to the unbraced length (L_c), corresponding to "full lateral support" with F_b = 0.66F_y, and the diagram drops off to M_R = 440ftk at an unbraced length (L_u) of 8.5ft, corresponding to "partial lateral support" with F_b = 0.6F_y. In both cases, however, it is simpler to use the section modulus tables directly.

We could design a beam for a situation in which the unbraced length is not where the moment is maximum. In this case, even though the unbraced length might exceed the allowable from the section modulus table, the beam could still work if the bending stress is sufficiently low. One way to do this is to utilize the allowable moment charts.

Example 6.2

The beam illustrated has a span of 20ft: The 12ft section on the right is laterally braced and carries a distributed load of 1.5klf LL + 2.25klf DL. The 8ft section on the left is unbraced and carries only the beam dead laod. There is also a point dead load of 6k. Design this beam using A36 steel and assuming the beam weight is 40plf.

Statics:

R_A = 17.5k

R_B = 34.3k

V = "0" at 9.05ft from B

M_{max} = 155.2ftk

M_C = **138.8ftk** at 8ft from B. This is the maximum moment in the unbraced portion.

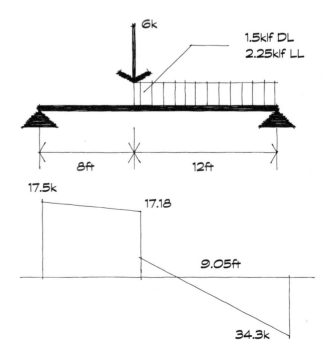

Required section properties (assuming full lateral support):

$$S_x = \frac{(155.2ftk)(12in/ft)}{0.66(36ksi)} = 78.4in^3$$

Required web area:

$$A_w = \frac{34.3k}{0.4(36ksi)} = 2.398in^2$$

Allowable deflection = (20ft)(12in/ft)/360 = 0.67in for live load.

Since we have no precise formula for this situation, we could initially check it with a single point load = (1.5klf)(12ft) = 18k equal to the uniform load and located at the centriod of the uniform load (6ft).

This would be case 8 from the formulas:

$$I = \frac{Pab(a+2b)\sqrt{3a(a+2b)}}{27EL\Delta}$$

Note that we have solved the formula for our unknown = I.

$$I = \frac{(18k)(72in)(168in)[72in + 2(168in)]\sqrt{3(72in)[72in + 2(168in)]}}{27(29,000ksi)(240in)(0.67in)} = 209.45in^4$$

Required I_x = 209.45in^4

From the section modulus charts, W21X44 has S_x = 81.6in^3, but L_c = 6.6ft and L_u = 7.0ft. However the maximum moment in the unbraced section of 8.0ft is only 138.8ftk.

We could check the "allowable moment" charts (AISC Manual p. 2-170) and check the capacity of W21X44 at an 8.0ft unbraced length.

The maximum moment at 8.0ft = 141.4ftk, which is above our 138.8ftk, so it WORKS (for moment)!

I_x = 843in^4 > 209.45in^4 required for deflection and provides plenty of extra capacity, so a more exact deflection calculation would be a waste of time.

Answer:

Use W21X44 A-36 steel.

___

Problem 6.1

A 50ksi steel beam has an unbraced length in the midspan portion of 12ft. The maximum moment is 333ftk. Basing your choice on moment only, select the cheapest (lightest) W shape.

Answer:

W24X68

Problem 6.2

Find an alternative section for Problem 6.1 using A-36 steel.

Answer:

W24X84

Metric tables are not currently available for these examples and problems.

7 SUPPORTS

7.1 Bearing plates for beams and columns

The reaction at the support results in localized stresses in the beam end and on the bearing structure. (One of Newton's observations is that every action has an equal and opposite reaction.) Beams that are supported on masonry or reinforced-concrete walls usually rest on steel bearing plates, both to distribute the load over an adequate bearing area and to simplify the construction process. The connection between the plate and the support, as well as between the plate and the beam, is discussed in Section 7.2. The function of the plate is to

(a) distribute the load transferred from the beam to the vertical structure in a uniform manner

(b) provide a contact area sufficient to maintain the bearing pressure below the allowable stress (allowable bearing pressure) of the support material

(c) create a level support area for the beam end and to allow for the compensation of any construction errors in the bearing elevation

As with many connection types, column and base plates transfer vertical loads and occasionally horizontal loads while also providing for constructional tolerances and thermal expansion. These requirements have to be met with special support design and materials (Fig. 7.1). Steel-to-steel contact provides horizontal reaction by friction proportionally to the load perpendicular to the surface. (Remember: Frictional force is equal to the load acting on the surface times the coefficient of friction of the contact area.) Friction may be decreased by wind uplift on a floor structure or by other materials, paint, or even brass plates in older structures that are introduced into the contact planes. To compensate for these forces, anchorage is usually provided for both beam ends (Fig. 7.1).

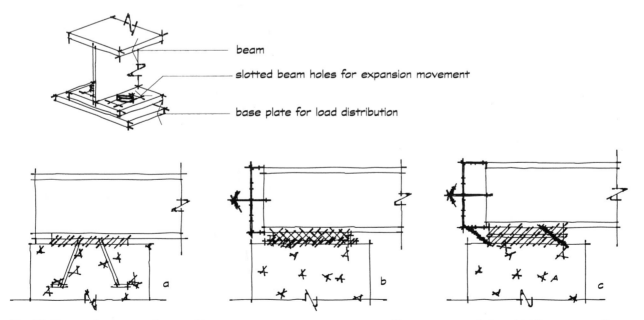

beam

slotted beam holes for expansion movement

base plate for load distribution

Fig. 7.1 Types of supports for steel beams on concrete or masonry. a. Beam on a steel plate (rigidly connected), b. Teflon pad with lateral restraints (i.e. slotted bolt holes) c. Steel-reinforced neoprene pad

On very long beams, especially in bridge construction, one end is left free, or supported on a rocker, to expand or contract. This movement is acknowledged in buildings with slotted holes in at least one of the beam attachment points. This provides for a lesser amount of movement than a rocker; however, building structures generally experience less thermal expansion than bridges since they are protected from the extremes of the environment. Special supports that allow movement in a controlled way can be built with Teflon pads or Teflon-coated plates, which reduce friction, or neoprene pads, which provide an elastic support. Teflon is a unique material that has a lower coefficient of static friction than dynamic friction. This means it will move without the initial force buildup that is experienced with other materials. This property greatly reduces the jerky movement experienced by force buildup and release. At one time, brass plates were used for this same purpose between steel bearing surfaces to reduce friction and make more effective expansion joints. Neoprene and other elastomeric products can be reinforced by steel plates for very large loads. Alternate layers of steel and neoprene with replaceable lead plugs are sometimes used as earthquake dissipators.

A more "architectural" solution of beam supports on concrete was used in the Knights of Columbus Tower by Roche and Dinkeloo (Fig. 7.2). The main girders are W36X194 Cor-ten, a "weathering" steel, spanning over 70ft [21m] between concrete towers. The supports are created by inserting steel boxes in the concrete; the beam end has a clearance of 4 inches [102mm] with the steel face and rests on a steel plate fixed with adhesive on a steel box. When the load on concrete or masonry is very large, the required bearing plate may be so wide and thick that other solutions are necessary. In these cases, the beam end may be placed on a grillage of short beams similar to a grillage foundation.

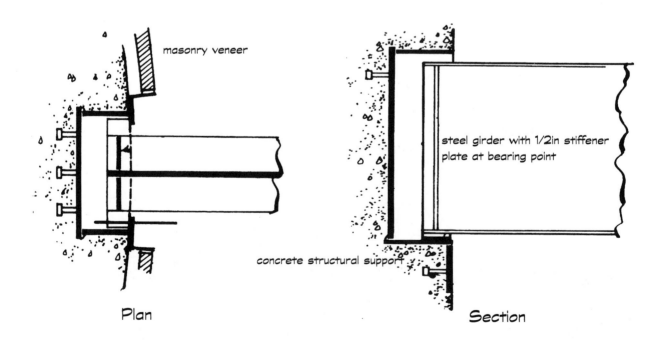

masonry veneer

steel girder with 1/2in stiffener plate at bearing point

concrete structural support

Plan

Section

Fig.7.2 Detail of support of steel girders on steel boxes set in concrete on Knights of Columbus Building, New Haven, Connecticut

7.2 Design of bearing plates

The nature of structures is to accumulate load and transfer it down through the building to ultimately be supported by the soil. This means that at every bearing point, there generally will be a relatively high load to be transferred and absorbed. When a strong material (steel) rests on a weaker material (concrete or masonry), the load must be distributed out over a large area. An excellent example of this principle is evident in a snowshoe. Its purpose is to take your weight and distribute it out over a large enough area so the snow will be capable of supporting you.

In the case of a bearing plate, the pressure on the contact area between the plate and vertical structure is not generally uniform, but is higher under the beam web, as illustrated in Fig. 7.3. The design procedure, however, generally adopts a uniformly distributed force on the bearing area, as represented by $F_p = P/A$, where

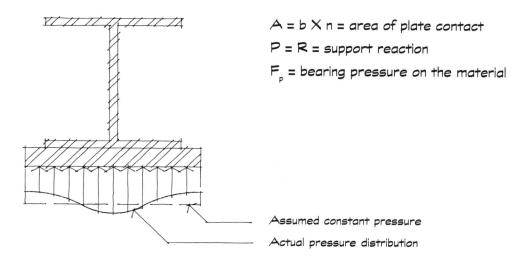

$A = b \times n =$ area of plate contact

$P = R =$ support reaction

$F_p =$ bearing pressure on the material

Assumed constant pressure

Actual pressure distribution

Fig. 7.3 Pressure distribution under bearing plates

Some allowable bearing pressures on masonry, to be used in the absence of more specific data, are listed in Table 7.1.

Table 7.1 Allowable stresses for beam supports

Material	Allowable bearing pressure F_p (ksi)	MPa
Masonry		
Sandstone and Limestone	0.40 ksi	2.80 MPa
Brick in cement mortar	0.25 ksi	1.70 MPa
Solid brick,		
S mortar, $f_m = 4500$ psi	0.34 ksi	2.30 MPa
Concrete		
On full area of support	$0.35 f_c'$	
On less than full area of support	$0.35 f_c'$ or $(A_2/A_1)^{1/2} \le 0.70 f_c'$	

Source: AISC Manual p. 5-79.

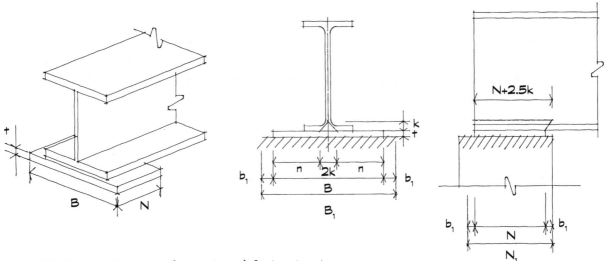

Fig. 7.4 Reference dimensions (nomenclature) for bearing plates

The allowable pressure on concrete (where f_c' = concrete yield strength; AISC Manual, Sec. J9, p. 5-79) is a function of the location of the plate relative to the edges of the support. Typical situations are illustrated in Fig. 7.5. In the case of a plate not on an edge of the support, the concrete under the plate is "contained," or supported, by the surrounding concrete, which allows us to increase the allowable concrete resistance. The larger the area A_2 relative to the plate size, the higher the allowable design pressure. According to the formula, area A_2 cannot be taken larger than four times A_1. If the plate occurs at the edge of the wall or column, even on one side only, this containment effect is not present and the allowable F_p value is the same as for a plate on the full area of concrete support, or $0.35f_c'$.

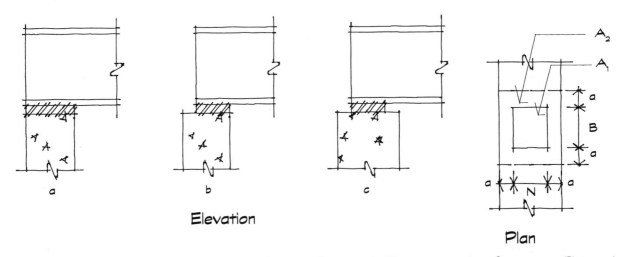

Fig. 7.5 Concrete support options a. Plate on full area of support, b. Plate on one edge of support, c. Plate on less than full area of support: a = minimum distance between plate and support edge.

The typical bearing plate (or column base plate for that matter) is designed as a double cantilever fixed at the interior support point. The uniform load is generated by the bearing pressure under the plate. This concept is similar to the one used to design foundations (Fig. 7.6), which is just another example of a device whose purpose is to take load from a very strong material and distribute it to a material of lesser strength.

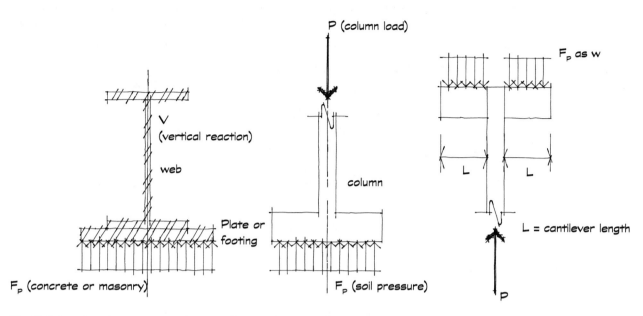

Fig. 7.6 Bearing plate as a cantilever and its analogy with foundation systems in general

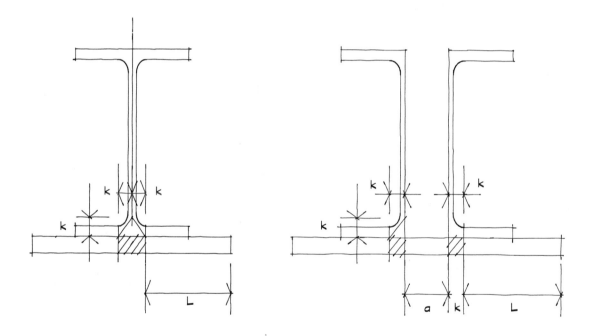

Fig. 7.7 Design cantilever length for W shapes and double channels. For the channels, spacing "a" must be < 2.4L.

Now for the general form of the design procedure:

A bearing plate or a column base plate is simply a beam consisting of a solid rectangular section bent about its weak axis. The basic design procedure begins with $F = Mc/I$, just as we might with any other beam. Deflection calculations are typically not a consideration with bearing plates since the dimensions are very small and consequently only negligible deflection will be developed. Shear is also not a problem since a solid rectangle is 100% "web." There are situations of solid rectangular sections (illustrated in the examples) that do require deflection calculations.

First, we must distribute the load over a large enough area so that we do not exceed the allowable strength of the bearing material. This area is calculated by using $F_p = P/A$, or in this case:

$$A_p = R/F_p$$

where A_p = BN (see Fig. 7.4) = area of the plate in contact with the bearing surface

R = reaction of the beam

F_p = allowable bearing pressure on the supporting material

The plate is designed as a beam cantilevering out of the bottom of the beam it supports, using the beam formula

$$F_b = Mc/I$$

If the plate is a simple cantilever, Maximum moment = $wL^2/2$

The plate is a rectangular section, bent about its weak axis, with c = t/2, and I = $bh^3/12$, or if we use a 1 inch typical strip through the plate: I = 1in(t)^3/12 = t^3/12

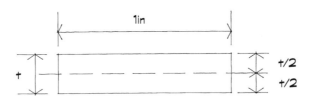

The length of the cantilever is equal to half the plate length less the web toe of the beam (Fig. 7.4), or

$$L = B/2 - k$$

where k is the distance from the top toe of the fillet to the bottom of the flange; values for k for each beam may be found in the AISC Manual, Sec. 1. Substituting these quantities in the beam formula, $F_b = Mc/I$, the bending stress becomes

$$F_b = \frac{(wL^2)\frac{t}{2}}{\frac{t^3}{12}} = \frac{3wL^2}{t^2}$$

which, solved for t, gives the **design formula:**

$$t = \sqrt{\frac{3wL}{F_b}}$$

where F_b = allowable bending stress for a solid rectangular section (a plate) = $0.75F_y$ (see ASD, 5.48, Sec. F2.1). This formula can be found in the AISC Manual, p. 2-142.

Note that this is the allowable stress to be used for weak axis bending of plates, solid rectangular bars, and I shapes (Fig. 7.8).

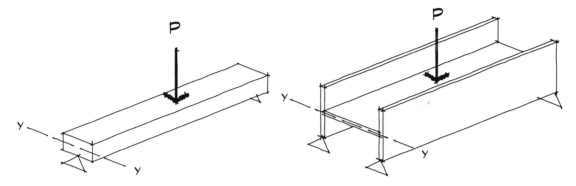

Fig. 7.8 Weak axis bending situations

Example 7.1

Beam bearing plate:

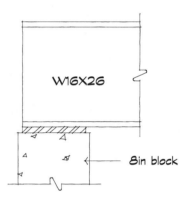

Beam W16X26

A36 steel

k = 1-1/16in = 1.06in

$F_b = 0.75F_y$ AISC Manual, Sec. F2.1, p. 5-48

Maximum reaction = 30.0k from beam statics analysis

8in thick concrete-block wall bearing: The UBC does not specifically address concrete block bearing capacities so we'll use the brick value for bearing capacity of 0.25ksi. Other values are possible based on the block strength and mortar type. (see masonry design) With an 8in block wall, the actual thickness is 7-5/8in, so a 7in plate dimension perpendicular to the wall would be constructionally convenient.

Use the previously developed formula, derived from F = Mc/I :

$$t_p = \sqrt{\frac{3f_pL^2}{0.75F_y}}$$

f_p = actual bearing pressure

For small-scale structures, this value will frequently be significantly smaller than the allowable F_p. This results from the bearing plate being large enough to provide constructional clearance for anchor bolts.

1. Find the bearing area required:

$$F_p = P/A; \quad A = P/F_p \qquad A = (30.0k)/(0.25ksi) = 120in^2$$

2. Determine the plate dimensions:

$(120in^2)/(7in \text{ maximum}) = 17.1in$ minimum width   Use 18in

Plates typically will be specified in full inch dimensions.

3. Determine the actual bearing pressure:

$$f_p = \frac{P}{A} = \frac{30.0k}{(7in)(18in)} = 0.238ksi$$

4. Determine the maximum plate cantilever length:

$$L = (18in)/(2) - k = 9in - 1.06in = 7.94in$$

5. Execute the formula:

$$t_p = \sqrt{\frac{3(0.238ksi)(7.94in)^2}{(0.75)(36.0ksi)}} = 1.29in \quad \text{Use 1-3/8in plate}$$

Answer:

Use a 7in x 18in x 1-3/8in bearing plate.

Either the masonry bearing points should be a grouted bond beam or the cores of the block should be grouted full (solid) a minimum of three courses below the bearing point to distribute the load over a larger number of ungrouted blocks. Screen or a fine hardware cloth is generally placed in the lowest mortar joint to keep the grout in place.

Example 7.2

A variation of a "solid rectangular section bent about its weak axis"

A 10in x 44in solid steel plate is used as a stair tread supported 1in from each end (similar to the Saarinen stairs at the GM Tech Center).

LL = 100psf (UBC "exit facility")
DL = 20psf   AISC Manual, p. 1-111
1/2in estimated thickness

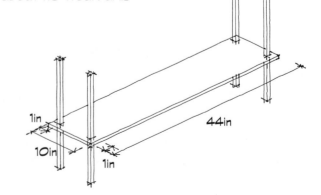

Total load = 120psf × 10in/(12in/ft)= 100plf = 100plf/12in = 8.3pli

Since this is not a cantilever condition, we cannot use the specialized bearing plate formula. Consequently, we go back to "old faithful": F = Mc/I.

$$\text{Maximum moment} = \frac{WL^2}{8} = \frac{(8.3\text{pli})(42\text{in})^2}{8} = 1,830\text{inlb}$$

$$\text{Maximum shear} = \frac{WL}{2} = \frac{(8.3\text{pli})(42\text{in})}{2} = 174.3\text{lb}$$

Maximum allowable Δ = L/360 = (42in)/360 = 0.12in

Using $F_b = Mc/I$

$F_b = 0.75F_y$

$c = t/2$

$I = bt^3/12$

1. Find the required thickness based on moment:

$$F_b = \frac{(1,830\text{inlb})(t/2)}{b(t^3)/12} = \frac{(1,830\text{inlb})(t)}{2} \times \frac{12}{b(t^3)} = \frac{(1,830\text{inlb})(t)}{b(t^3)}$$

$$t = \sqrt{\frac{(1,830\text{inlb})6}{(10\text{in})(0.75)(36,000\text{psi})}} = 0.20\text{in thick}$$

2. Find the required thickness based on deflection:

   Maximum allowable deflection = 0.12in

$$I_{req.} = \frac{5(8.3\text{pli})(42\text{in})^4}{384(29,000,000\text{psi})(0.12\text{in})} = 0.097\text{in}^4$$

$I = bt^3/12$; $t^3 = I(12/b) = .097\text{in}^4(12/10\text{in})$; $t^3 = 0.11\text{in}^3$; $t = 0.48\text{in}$

3. Find the required thickness based on shear:

$$F_v = \frac{P}{A} \qquad A = \frac{P}{F_v} = \frac{(174.3\text{lb})}{0.4(36,000\text{psi})} = 0.012\text{in}^2$$

Obviously, almost anything will provide 0.012in² of area.

Answer:

Use a 10in × 44in × 1/2in thick tread section.

Elaboration:

Let's look at an alternative solution of a channel to replace the flat plate. The only criterion that would change would be the approach to the section modulus:

$$F_b = M/S_y$$

The AISC Manual (Sec. F2.1, p. 5-48) suggests that we can use the same F_b for I and W sections. While C sections are not mentioned, we'll assume $F_b = 0.75F_y$.

$$S_y = (1830 \text{inlb})/0.75(36,000\text{psi}) = .068\text{in}^3$$

The moment of inertia and web area requirements remain unchanged.

Check a MC10X8.4

$$S_y = 0.118\text{in}^3 \quad \text{OK}$$
$$I = 0.112\text{in}^4 \quad \text{OK!}$$

Compare the weight of this channel section, 8.4plf, to that of the 1/2in x 10in flat plate, 17.0plf. The channel obviously makes more efficient use of material, especially when you compare the controlling criterion, deflection. The lightest standard channel section that will work is roughly 50% lighter and has 87% as much deflection as the plate. But, will it look as "cool"? This is your "architectural" decision.

7.3 Column base plates

Columns also require additional bearing area to transfer high loads to relatively weaker materials. The column end may be planed for direct transfer of load to the base plate; however, you should provide sufficient weld material to transfer the load regardless of the end condition. For large (thick flanged) columns, planing is the most reasonable solution. The column base plate is generally supported by four double-nutted anchor bolts to provide a means for both adjusting the "plumb" of the column and making minor adjustments in the support elevation (Fig. 7.9). The double nuts, one above and one below the plate on each bolt, simplify this process. The space under the plate, typically a minimum of 1 inch, can then be filled ("grouted") with a nonshrinking grout material (Fig. 7.10). This grout overcomes the natural tendency of cement-based materials to shrink, causing the bolts to take more load. Nonshrinking grout may have metal filings as a portion of the aggregate. These metal filings will "rust" or expand during the curing process, overcoming natural shrinkage. Unfortunately, they will create a typical "rust" stain on any adjacent material if exposed to weather.

The anchor bolts connect the plate to the foundation and prevent the column from overturning during construction. They also can be designed to resist any horizontal loads that the column may experience. In some cases, they may even be designed to create moment connections between the column and the foundation. This means the anchor bolts may be in tension under some loading conditions. Although anchor bolts may be either cast in place or drilled into the foundation at a later time, the cast-in-place are by far the most desirable and most common. The bolts are generally 3/4in [19mm or M20] in diameter and have a 4in [100mm] end hook to provide extra resistance to tensile loads that may inadvertently occur during construction.

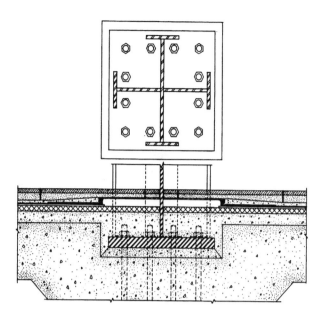

Fig. 7.9 Detail of Mies Van der Rohe column. Berlin Art Gallery

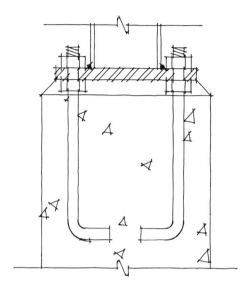

Fig. 7.10 Grouted base plate

Base plates wider than the column dimensions, as previously described, are designed as cantilevers. When the column loads are high and bearing surfaces are relatively weak, brackets may be used between the column and the base plate to reduce plate thickness. These brackets reduce the cantilever lengths and brace the plate (Fig. 7.11). Extremely heavy column loads can be transferred to the concrete foundations via layers of beams placed side by side in alternating perpendicular layers. This is called "grillage footing" and was originally used for the construction of tall buildings in Chicago; steel rails were originally used before I-beams became available (Fig. 7.12).

7.4 Design of column base plates

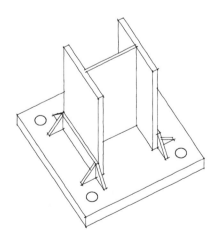

Fig. 7.11 Bracketed column/plate connection

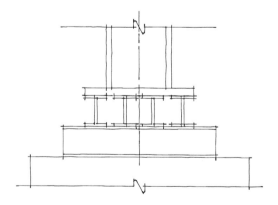

Fig. 7.12 Grillage footing connection

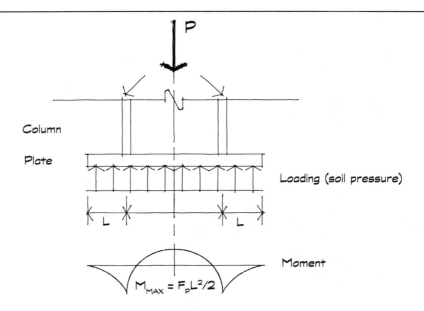

Fig. 7.13 Moment distribution in a bracketed column base plate

The design of column base plates follows the same basic procedure as that of bearing plates. Enough contact area (A) is provided between the column base and the bearing material to ensure that the allowable bearing pressure (F_p) is not exceeded.

The AISC Manual provides a description of the design procedure (p. 3-106 through 3-109), and uses two formulas:

$$t_p = 2m\sqrt{f_p/F_y} \quad \text{and} \quad t_p = 2n\sqrt{f_p/F_y}$$

where:

t_p = the thickness of the column base plate

m = the cantilever length of the plate measured from the top or bottom flange to the edge of the plate

n = the cantilever length of the plate measured from the sides of the column to the edge of the plate

f_p = the actual bearing pressure under the base plate

F_y = the yield stress of the steel being used in the plate design

These "new" formulas are redundant!

First, there is no reason to calculate two thickness. If you look at the formulas, there is only one variable: the cantilever length (m or n). A clever designer will only calculate t_p based on the longest cantilever, either m or n. Second, these are not new formulas; they are simply alternative forms of the formula used for designing bearing plates:

$$t_p = \sqrt{3wL^2/F_b}$$

where

$w = f_p$

L = cantilever length (m or n)

$F_b = 0.75\, F_y$

Substituting:

$$t_p = \sqrt{3f_p m^2/0.75F_y} = \sqrt{4f_p m^2/F_y} = 2m\sqrt{f_p/F_y}$$

Same formula, alternative form . . . . why bother?

The only potential difference in the design procedure concerns the determination of the cantilever length. In column base plate design, the load is assumed to be placed on the base plate on an area having a width of 0.80 of the column width and a depth equal to 0.95 of the column depth. The small flange ends are not considered to provide as rigid a support as the broad flange width, accounting for the difference between the 0.80 and 0.95. Figure 7.14 illustrates this "support" area.

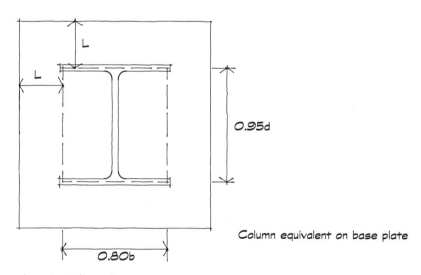

Column equivalent on base plate

Fig. 7.14 Column/base plate adjusted contact dimensions

Example 7.3

Column base plate design

W12X58 column connected to an 18in x 18in reinforced concrete pier:

P = 293k   A36 steel

d = 12.19in

b_f = 10.01in $f_{c'}$ = 3,000psi concrete strength

                                  f_c allowable = $0.35f_{c'}$ = 1050psi

$$t_p = \sqrt{\frac{3f_p L^2}{0.75 F_y}}$$

1. Find the required bearing area:

$$F_p = P/A ; \quad A = P/F_p = 293k/1.05ksi = 279in^2$$

2. Determine base plate size:

$$A = \sqrt{279in^2} = 16.7in \times 16.7in$$ use a 17in square base plate. This will allow a 1/2in chamfer on all sides of the top of the pier.

3. Determine the actual f_p :

$$F = P/A ; f_p = \frac{(293,000lb)}{(17in)(17in)} = 1.01ksi$$

4. Determine the column "support" dimensions:

$b_f = 10.01in \times 0.80 = 8.008in$

$d = 12.19in \times 0.95 = 11.58in$

$$\text{Maximum cantilever} = \frac{(17in - 8.01in)}{2} = 4.49in$$

5. Calculate the plate thickness:

$$t_p = \sqrt{\frac{3(1.01ksi)(4.49in)^2}{0.75(36ksi)}} = 1.50in$$ use 1-1/2in plate

Answer:

Use a 17in × 17in × 1-1/2in column base plate

Note that this is almost exactly the same procedure used in bearing-plate design. The primary difference is in the determination of the cantilever length.

METRIC VERSIONS OF ALL EXAMPLES

Example M7.1

Beam bearing plate:

Beam W410X39,

250MPa steel

k = 27mm

$F_b = 0.75F_y$ AISC Manual, Sec, F2.1, p. 5-48

Maximum reaction = 133.44KN from beam statics analysis

208mm thick concrete-block wall bearing: The UBC does not specifically address block bearing, so we'll use the brick value for bearing capacity of 1.7MPa. Other values are possible based on the block strength and mortar type (see masonry design). With a 203mm nominal block wall, the actual thickness is 194mm, so a 180mm plate dimension perpendicular to the wall would be constructionally convenient.

Use the previously developed formula, derived from F = Mc/I :

$$t_p = \sqrt{\frac{3fpL^2}{0.75Fy}}$$

f_p = actual bearing pressure

For small-scale structures, this value will frequently be significantly smaller than the allowable F_p. This results from the bearing plate being large enough to provide constructional clearance for anchor bolts.

1. Find the bearing area required:

$F_p = P/A;$ $A = P/F_p$ $A = (133,440N)/(1.7MPa) = 78,494mm^2$

2. Determine the plate dimensions:

$(78,494mm^2)/(180mm) = 436mm$ minimum width   Use 440mm

Plates typically will be specified in full millimeter dimensions.

3. Determine the actual bearing pressure:

$$F_p = P/A = \frac{(133,440N)}{(180mm)(440mm)} = 1.68N$$

4. Determine the maximum plate cantilever length:

$$L = 440mm/2 = 220mm - k = 220mm - 27mm = 193mm$$

5. Execute the formula:

$$t_p = t_p = \sqrt{\frac{3(1.68N)(193mm)^2}{.75 \times 250MPa}} = 31.64mm \quad \text{Use 32mm plate}$$

Answer:

Use a 180mm x 440mm x 32mm bearing plate.

Either the masonry bearing points should be a grouted bond beam or the cores of the block should be grouted full (solid) a minimum of three courses below the bearing point to distribute the load over a larger number of ungrouted blocks. Screen or a fine hardware cloth is generally placed in the lowest mortar joint to keep the grout in place.

Example M7.2

A variation of a "solid rectangular section bent about its weak axis"

A 250mm x 1120mm solid steel plate is used as a stair tread supported 25mm from each end (similar to the Saarinen stairs at the GM Tech Center).

LL = 4800N/m² (UBC "exit facility")
DL = 960N/m² AISC Manual, p 1-111
12mm estimated thickness

Total load = (5760N/m²)(0.25m)/1000mm/m = 1.44N/mm

Since this is not a cantilever condition, we cannot use the specialized bearing plate formula. Consequently, we go back to "old faithful": F = Mc/I.

Maximum moment = $WL^2/8 = \frac{(1.44N/mm)(1070mm)^2}{8} = 206{,}082Nmm$

Maximum shear = $WL/2 = \frac{(1.44N/mm)(1070mm)}{2} = 770N$

Maximum allowable $\Delta = L/360 = (1070mm)/360 = 3mm$

Using $F_b = Mc/I$:

$$F_b = 0.75F_y$$
$$c = t/2$$
$$I = bt^3/12$$

1. Find the required thickness based on moment:

$$F_b = \frac{(206,082 Nmm)(t/2)}{b(t)^3/12} = \frac{(206,082 Nmm)(t)}{2} \times \frac{12}{b(t)^3} = \frac{(206,082 Nmm)(6)}{b(t)^2}$$

$$t = \sqrt{\frac{(206,082 Nmm)(6)}{250(0.75)(250 MPa)}} = 5.14mm$$

2. Find the required thickness based on deflection:

Maximum allowable deflection = 3mm

$$I_{req.} = \frac{5(1.44 N/mm)(1070 mm)^4}{384(200,000 MPa)(3mm)} = 40,962 mm^4$$

$$I = bt^3/12 \quad ; \quad t^3 = Ix12/b = 40,962 mm^4 \times (12/250mm); \quad t^3 = 1966 mm^3; \quad t = 12.5mm$$

3. Find the required thickness based on shear:

$$F_v = P/A \quad ; \quad A = P/F_v = \frac{(770N)}{0.4(250 MPa)} = 7.7 mm^2$$

Obviously, almost anything will provide $7.7mm^2$ of area.

Answer:

Use a **250mm × 1120mm × 13mm tread section.**

Elaboration:

Let's look at an alternative solution of a channel to replace the flat plate. The only criterion that would change would be the approach to the section modulus:

$$F_b = M/S_y$$

AISC Manual (Sec. F2.1, p. 5-48) suggests that we can use the same F_b for I and W sections. Although C sections are not mentioned, we'll assume $0.75F_y$.

$$S_y = (206,082 Nmm)/0.75(250 MPa) = 1099 mm^3$$

The moment of inertia and web area requirements remain unchanged.

Check a MC250X12.5

$$S_y = 4.44 mm^3 \quad OK!$$
$$I_y = 0.137 \times 10^6 mm^4 \quad OK! \quad 0.041 \times 10^6 mm^4$$

Compare the weight of this channel section, 12.5kg/m, to that of the 13mm x 250mm flat plate, 25kg/m. The channel obviously makes more efficient use of material, especially when you compare the controlling criterion, deflection. The lightest standard channel section that will work is both 50% lighter and has only 30% (0.041/0.137) as much deflection as the plate. But, will it look as "cool"? This is your "architectural" decision.

Example M7.3

Column base plate design

W310X86 Column connected to a 450mm x 450mm reinforced-concrete pier:

$P = 1300kN$ 250MPa steel

$d = 310mmn$

$b_f = 254mm$ $f_{c'} = 21MPa$ concrete strength

                       f_c allowable $= 0.35 f_{c'} = 7.35MPa$

$$t_p = \sqrt{\frac{3 f_p L^2}{0.75 F_y}}$$

1. Find the required bearing area:

$$F_p = P/A ; \quad A = P/F_p = 1,300,000N / 7.35MPa = 176,870mm^2$$

2. Determine the base plate size:

$$A = \sqrt{(176,870mm)^2} = 420mm$$

Use a 430mm square base plate. This will allow a 10mm chamfer on all sides of the top of the pier.

3. Determine the actual f_p :

$$F = P/A ; f_p = 1,300,000N / (430mm \times 430mm) = 7.03MPa$$

4. Determine the column "support" dimensions:

$b_f = 254mm \times 0.80 = 203mm$

$d = 310mm \times 0.95 = 294mm$

$$\text{Maximum cantilever} = \frac{(430mm - 203mm)}{2} = 113.5mm$$

5. Calculate the plate thickness:

$$t_p = \sqrt{\frac{3(7.03MPa)(113.5mm)^2}{0.75(250MPa)}} = 38mm \quad \text{Use 38mm plate}$$

Answer:

Use a 430mm x 430mm x 38mm column base plate.

Note that this is almost exactly the same procedure used in bearing-plate design. The primary difference is in the determination of the cantilever length.

8 WEB YIELDING AND CRIPPLING

8.1 Localized failure of components

Members with concentrated loads applied to the flanges must have a web thickness or a bearing length great enough to satisfy web yielding and web crippling. (AISC Manual, Sec. K1.1. p. 5-80). When pairs of stiffeners are provided on opposite sides of the web, at concentrated loads, and extend at least half the depth of the member, local flange bending and local web yielding need not be checked (Fig. 8.1).

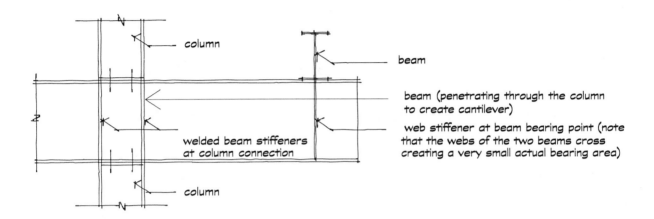

Fig. 8.1 Beam stiffeners at the support and at points of concentrated load on the flange.

A common use of this principle can be found at almost any social gathering of college students. After drinking a can of beverage, it is common practice to fold the can and smash it before disposing of it. (Probably to achieve less volume of trash, nothing macho.) If you look at the process carefully, it is a perfect example of web "yielding." For example, if you hold the can only by each end and try to fold it (bend it), you'll find it difficult, if not impossible. However, by supporting the ends in your palms and pressing your fingers into the top side of the can, you can push the top down, collapsing the sides (web) of the can, reducing the effective depth of the section, and easily bend it. This is done in one smooth motion, which hides the use of the structural principle. Now parties will never be the same.

In more architectural situations, stiffeners can be placed in pairs at unframed ends or at points of concentrated loads on the interior of beams, girders, or columns (AISC Manual, Sec. K1) to resist this local failure of the web. The thickness of the stiffeners cannot be less than t/2, where t is the thickness of the plate or flange delivering the concentrated load. The stiffeners will not automatically prevent sidesway web buckling or rotation of the flange, including the tension flange, if the beam is not braced at that point.

Bearing stiffeners can be provided if the compressive stress at the web toe of the fillets resulting from concentrated loads exceeds $0.66F_y$ (Sec. K1.3). Alternatives are to weld a plate to the web in order to increase its thickness or to increase the bearing length so a greater area of the web is accepting the applied load.

Web yielding should be considered when a large load is applied to a small area of the beam such as at the end supports or at a point of application of an interior load (remember the beverage can?). Yielding is clearly distinct from crippling, as the AISC Manual requires separate checks with different formulas. Yielding is failure of the web in pure compression and it is therefore checked with a direct stress formula, F = P/A. The AISC specifications require that at the end of a member, the compressive stress on the web be limited to $0.66F_y$:

$$F = \frac{P}{A} = \frac{R}{t_w(N + 2.5k)} \leq 0.66F_y$$

This formula assumes a distribution of the concentrated load on a length of web depending on the distance k of the toe from the flange, or fillet, where the web area is at the minimum (Fig. 8.3). The horizontal area of the web at this point is

$$A = t_w \, (N + 2.5k) \qquad \text{(at the end of the beam)}$$
$$A = t_w \, (2.5k + N + 2.5k) \quad \text{(at an interior point of load application)}$$

where t_w = web thickness

N = length of support plate transferring the point load to the flange of the beam

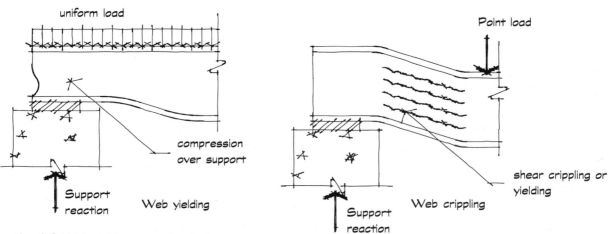

Fig. 8.2 Web yielding vs. web crippling

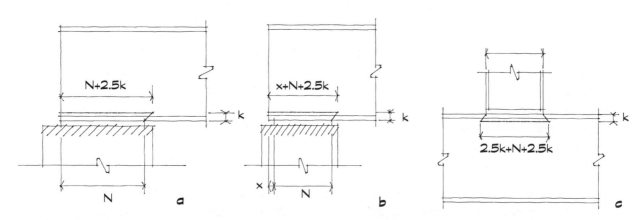

Fig. 8.3 Area of web subjected to concentrated load: (a) support at the end of the beam, (b) support at distance x from the end of the beam, (c) point load on interior of the beam

In intermediate situations, the area can be calculated as [Fig. 8.3(b)]

$$A = t_w (X + 2.5k + N)$$

You'll note that in each case, the value of A is simply the total length of the bearing times the web thickness.

Example 8.1

Web yielding at the point of concentrated load

A-36 steel: W12X26

k = 7/8in = 0.875in

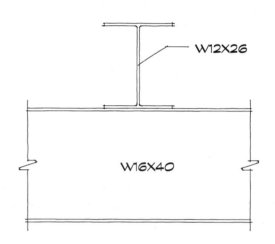

W16X40

t_w = 0.305in

t_f = 0.505in

d = 16.01in

k = 1-3/16in = 1.19in

Reaction of W12X26 on W16X40 = 23k

1. Determine the load application area:

    N = 2 (k) = 2 (.875 in) = 1.75 in

    This application dimension could be increased by using web stiffeners to support the cantilevered flanges of the W12X26. We'll use the bearing-plate technique to determine the direct-load-transfer location.

2. Calculate the web-yielding stress (in the W16X40) at the point of load application:

    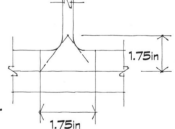

$$F_p = \frac{P}{A} = \frac{23k}{[(2.5)1.19in + 1.75in + (2.5)1.19in](.305in)} = 10.05ksi.$$

F_{all} = 0.66F_y = 24ksi This condition is workable.

This is more or less a worst-case scenario for web yielding and web crippling. Unless you have extremely heavy loads applied to a small dimension parallel to the web, this **should** not be a problem. Should, however, is no guarantee. . . . Check it if you have any doubts.

Problem 8.1

A W30X124 beam, A36 steel, is bearing on a 5in-wide steel plate and the support reaction is 98k.

Check the beam end for web yielding.

Solution:

This problem could be checked a variety of ways since we have all of the criteria. One of the simplest would be to determine the actual stress and compare it to the allowable stress.

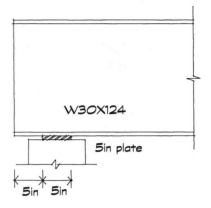

W30X124

5in plate

5in   5in

t_w = 0.585in; k = 1-11/16in = 1.69in

A_{web} = [2.5(1.69in) + 5in + 2.5(1.69in)](0.585in) = 7.86in²

Using F = P/A;

Actual web-yielding stress (98k)/(7.86in²) = 12.47ksi

Allowable stress = $0.66F_y$ = 0.66(36ksi) = 24ksi   OK!

Problem 8.2

A W44X198 girder, 50ksi steel, in the illustration has a pinned support (similar to the ones used in bridges) for a reaction of 245k. What is the minimum length N of the support plate in order to be safe against web yielding?

Solution:

In this case, we have basically the same problem, but need to solve for the only variable in the formula, N.

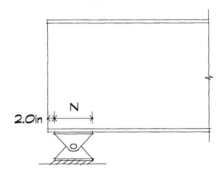

N

2.0in

t_w = 0.709in; k = 2.-1/8in = 2.125in

Using F = P/A in the form A = P/F

Required A = (245k)/(0.66)(50ksi) = 7.42in²

A = t_w(L) = (0.709in)(2.0in + N + 2.5k)

Notice that in this case we used a distribution of only 2.0in on the end instead of 2.5k since 2.5k of distribution did not exist. This is why we use the basic formula of F = P/A instead of specialized formulas which don't ALWAYS work.

Setting the two A's equal to one another:

7.42in² = 0.709in[2.0in + N + 2.5(2.125in)]

7.42in² = 0.709in(7.31in + N)

10.46in = 7.31in + N

N = 3.15in

Problem 8.3

A W30X90 transfer girder supports a column bolted on the top flange. The dimension N of the base plate is 8in, as shown in the illustration on the following page. The load from the column is 210kips. Determine if there are yelding problems in the girder web. Use A-36 steel. If the yielding stress is excessive, what are the alternate solutions?

Solution:

t_{web} = 0.47in; k = 1-5/16in = 1.312in

A_{web} = [2.5(1.312in) + 8in + 2.5(1.312in)](0.47in) = 6.84in²

Use F = P/A

Actual web-yielding stress = (210k)/(6.84in²) = 30.7ksi

Allowable web yeilding stress = 0.66(36ksi) = 24ksi

We have a problem!

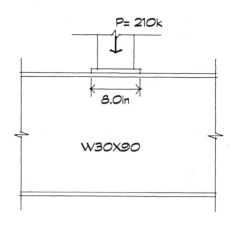

Alternative solutions to this problem:

1. Lengthen the connection plate to load a greater portion of the girder web.

2. Weld a plate on the girder web to create a thicker web, which provides more area.

3. Use stiffeners on each side of the web. This solution is a common one especially in statically indeterminate structures, as in Gerber-beam situations. Whereas the technical solution is quite complex, the most common solution is to provide four stiffener plates, two on each side, which align with the flanges of the column. This effectively allows the column load to pass through or into the beam without negative effect.

Problem 8.4

A W18X35 beam is supported on a W24X76 beam without any additional plate. Assuming the flanges of both shapes are insufficiently thick to distribute the concentrated load at this point, check the web-yielding stress in both beams. Reaction = 37.3k. Use A-36 steel.

Solution:

For the reaction force of the W24X76 "up" into the W18X35:

W24X76: t_w = 0.44in; k = 1-7/16in = 1.437in;

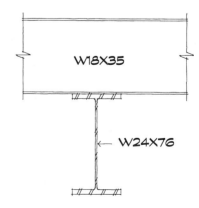

If we use the assumption of bearing plates, the support length (N) for the beam will be 2k = 2.87in.

W18X35: t_{web} = 0.3 in; k = 1 -1/8in = 1.125in

A_{web} = [2.5(1.125in) + 2.87 in + 2.5(1.125in)](0.3in) = 2.54 in²

Stress in web = (37.3k)/(2.54in²) = 14.7ksi

14.7ksi < [0.66(36ksi) = 24 ksi] OK!

Use a similar solution for the W24X76 verification:

The support length (N) from the W18X35 would be

2k = 2(1.125in) = 2.25in

A_{web} = [2.5(0.937in) +2.25in + 2.5(0.937in)](0.44in)

= 3.05in²

At this point, we can compare these numbers to the previous calculation and we see immediately that it works, but for the sake of example, here it is.

(37.3k)/(3.05in²) = 12.22ksi < 24ksi OK!

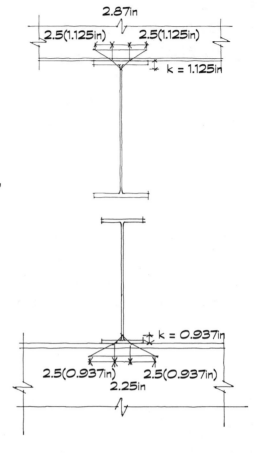

If a problem did occur, solutions similar to the ones suggested in the previous problem would be appropriate.

REMEMBER

These are all really quite simple problems based on F = P/A. As always, don't allow yourself to be confused by the special formulas developed in the AISC Manual and engineering texts.

METRIC VERSIONS OF ALL EXAMPLES AND PROBLEMS

Example M8.1 Web yielding at the point of concentrated load

    250MPa steel

    W310X39

        $k = 22mm$

    W410x60

        $t_w = 7.75mm$

        $t_f = 12.8mm$

        $d = 407mm$

        $k = 30mm$

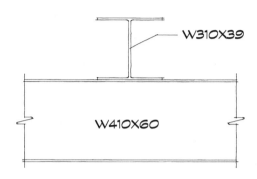

Reaction of W310X39 on W410X60 = 102,304N

1. Determine the load application area:

$N = 2(k) = 2(22mm) = 44mm$

This application dimension could be increased by using web stiffeners to support the cantilevered flanges of the W310X39. We'll use the bearing-plate technique to determine the direct-load-transfer location.

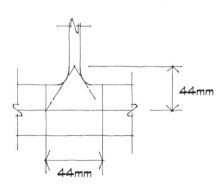

2. Calculate the web-yielding stress (in the W410X60) at the point of load application:

$$F = \frac{102,304N}{[2.5(30mm) + 44mm + 2.5(30mm)](7.75mm)} = 68MPa$$

$F_{all} = 0.66F_y = 165MPa$ This condition is workable.

This is more or less a worst-case scenario for web yielding and web crippling. Unless you have extremely heavy loads applied to a small dimension parallel to the web, this should not be a problem. Should, however, is no guarantee. . . . Check it if you have any doubts.

Problem M8.1

A W760X185 beam, 250MPa steel, is bearing on a 127mm-wide steel plate and the support reaction is 436kN. Check the beam end for web yielding.

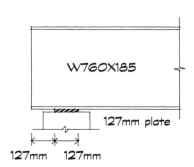

Solution:

Using F = P/A:

Actual web yielding stress = 86MPa

Allowable stress = 0.66F$_y$ = 0.66(250MPa) = 165MPa   OK!

Problem M8.2 There is no metric equivalent.

Problem M8.3

A W760X134 transfer girder supports a column bolted on the top flange. The dimension N of the base plate is 200mm, as shown in the illustration. The load from the column is 934kN. Determine if there are yelding problems in the girder web. Use 250MPa steel. If the yielding stress is excessive, what are the alternate solutions?

Solution:

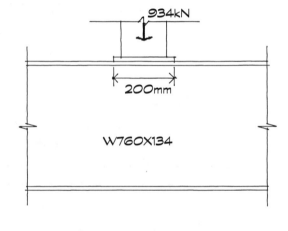

t$_{web}$ = 11.90mm; k = 33mm

A$_{web}$ = [2.5(33mm) + 200mm+ 2.5(33mm)](11.9mm)
       = 4343mm^2

Use F = P/A

Actual web yielding stress =

   (934,000N)/(4343mm^2) = 215MPa

Allowable web yielding stress =

   0.66(250MPa) = 165MPa

   We have a problem!

Alternative solutions to this problem:

1. Lengthen the connection plate to load a greater portion of the girder web.

2. Weld a plate on the the girder web to create a thicker web, which provides more area.

3. Use stiffeners on each side of the web. This solution is a common one especially in statically indeterminate structures, as in Gerber-beam situations. Whereas the technical solution is quite complex, the most common solution is to provide four stiffener plates, two on each side, which align with the flanges of the column. This effectively allows the column load to pass through or into the beam without negative effect.

Problem M8.4

A W460X52 beam is supported on a W610X113 beam without any additional plate. Assuming the flanges of both shapes are insufficiently thick to distribute the concentrated load at this point, check the web yielding stress in both beams. R = 165.9kN; 250MPa steel.

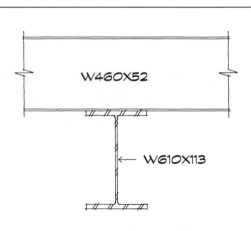

Solution:

For the reaction force of the W610X113 "up" into the W460X52:

W610X113: k = 37mm

If we use the assumption of bearing plates, the support length (N) for the beam will be

2k = 2(37mm) = 74mm

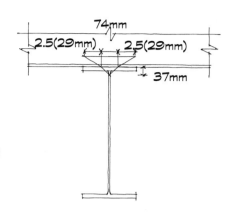

W460x52:

t_{web} = 7.62mm; k = 29mm

A_{web} = [2.5(29mm) + 74mm + .5(29mm)](7.62mm)

= 1669mm²

Stress in web = 165,900N/1669mm² = 99MPa

99MPa< [0.66(250MPa) = 165MPa] OK!

Use a similar solution for the W610X113 verification:

The support length (N) from the W460X52 would be

2k = 2(29mm) = 58mm

A_{web} = [2.5(37mm) + 58mm+ 2.5(37mm)](11.2mm)

= 2722mm²

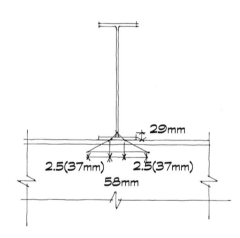

At this point, we can compare these numbers to the previous calculation and we see immediately that it works, but for the sake of example, here it is.

(165,900N)/(2722mm²) = 61MPa < 165MPa OK!

If a problem did occur, solutions similar to the ones suggested in the previous problem would be appropriate.

9 BUILT-UP BEAMS

9.1 Built-up sections

Built-up sections are used on the most ordinary buildings in the form of lintels. To create both enclosure and structural strength, a flat bottom plate combined with a C or S section or with a pair of angles is commonly used for support of a double-wythe masonry wall. This minimizes the thermal transfer that would occur through a solid precast-concrete lintel or bond beam section and hides the lintel.

Fig. 9.1 Palace of Labor, Turin, Italy. Steel umbrella structures on concrete columns

In the work of Pier Luigi Nervi, structures are frequently shaped to visibly represent the forces that exist within them. In the Palace of Labor in Turin (Fig. 9.1), the exhibition hall is roofed with steel cantilevers projecting out from concrete columns. The materials used are consistent with the primary forces they carry; steel for the cantilevered bending members and concrete for the major compressive structure. The cantilevers are deeper near the supports and reduce in cross-section toward the free end, clearly demonstrating the variation in bending stress. The vertical plates welded to the webs are stiffeners that resist the buckling of the webs as their height-to-thickness ratio gets large near the column supports.

Similar concepts are found in the design of the Renault Center by Foster Associates (Fig. 9.2). The beams in this case are part of a truss system. They have circular holes in the web primarily for aesthetic reasons. The holes near the supports are stiffened with pipe sections welded to the hole edge, to prevent buckling at points of concentrated loads.

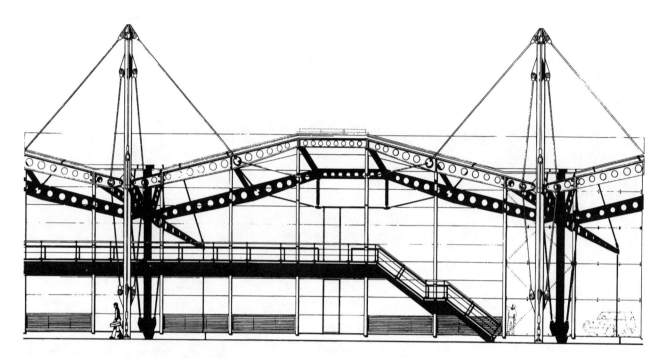

Fig. 9.2 Renault Center at Swindon, England

When rolled steel shapes are used to form built-up sections such as those illustrated in Fig. 9.3, their section properties may/should be recalculated. Sections are often built up to increase the moment capacity of an existing shape by creating more moment of inertia and consequently more section modulus. This can be done by welding a "cover" plate to the flanges. This solution is sometimes necessary when a hole is required in a flange, to compensate for the loss of area in tension or compression. Cover plates generally need to be added to the flanges for only a portion of the length, as determined by the moment diagram. Built-up beams such as in Fig. 9.3(d) are sometimes necessary for technical reasons (e.g., to accommodate vertical plumbing runs) or can be designed for architectural reasons.

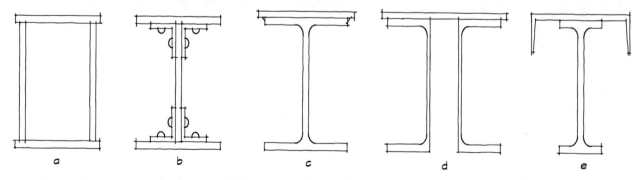

Fig. 9.3 Built-up plate girders (for large loads): a. welded box configuration, b. riveted I shape, c. Cover plates can be welded or bolted to flanges to increase load capacity or conpensate for flange holes, d. Channels connected by a plate, e. Combination section using channel and S joists.

A channel connected to the W shape (e) increases I in the x and y directions. This makes the beam more stable laterally and is often used when vertical loads are combined with lateral forces such as in beams for crane rails. The properties of some common combination sections are tabulated in the AISC Manual, pages 1-83 through 1-90.

Where two or more rolled beams or channels are used side by side to form a flexural member, they should be connected together at intervals of not more than 5ft [1.5m]. Through-bolts and separators (spacers) are permitted, provided that in beams having a depth of 12in [305mm] or more, no fewer than two bolts are used at each separator location (AISC Manual, F6, p. 5-50). Without these connections, it would be very difficult to ensure that the beams were both carrying their equal share of the load. Excessive loading from framing from one side could potentially overload that half if the beams were not forced to act together. The separators or spacers hold the sections apart and provide a greater moment of inertia and section modulus relative to the "weak" axis.

"Castellated" beams can be considered a type of built-up beams, since they are fabricated by cutting an I or W shape along the web (Fig. 9.4). Because of the "castellated" profile of the cut, the two halves can be offset and welded back together to form a deeper beam section. The cuts in the web may be ground if a more refined architectural finish is desired.

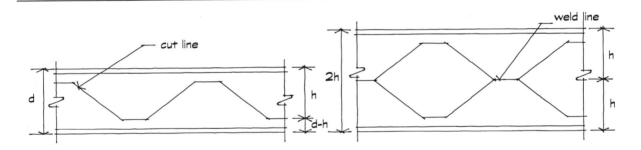

Fig. 9.4 Construction procedure for a castellated beam

Example 9.1

The load of a simply supported beam is 2klf DL and 1.8klf LL, so the total load is 3.8klf (not including an estimate for the weight of the beam). The beam is 30ft long and is fully laterally supported. If the overall depth is limited to 18in, design a section using A-36 steel.

Total load W = 3.8klf + 0.2klf beam estimate = 4klf

Determine the required section properties of the beam.

Maximum M = $wL^2/8$ = (4.0klf)(30ft)^2/8 = 450ftk

Maximum V = $wL/2$ = (4.0klf)(30ft)/2 = 60k

Maximum LL Deflection = L/360 = (30ft × 12in/ft)/360 = 1in

Although we would calculate the area and the moment of inertia as well, we'll limit our concerns to moment initially. (When we calculate the properties for the built-up section for moment, we will also be finding the moment of inertia, which can be evaluated later for deflection.)

Required $S_x = M/F_b$ $M = 450ftk \times 12in/ft = 5400ink$

$\quad\quad F_b = 0.66F_y = 0.66(36ksi) = 23.7ksi = 24ksi$

$\quad\quad S_x = M/F_b = (5400ink)/24ksi = 225in^3$

From the S_x tables (AISC Manual, p. 2-10), the lightest section with maximum S under 18in is W18X50 with a d = 17.99in. Unfortunately, the S_x is only $88.9in^3$, which is insufficient. We can use a smaller shape (albeit a heavier one) and weld a cover plate on each flange.

Shape	S_x	d	I
W16X100	$175in^3$	17in	$1490in^4$
W14X120	$190in^3$	14.5in	$1380in^4$
W14X109	$173in^3$	14.4in	$1240in^4$

Try W16X100: I = $1490in^4$; d = 16.97in; S_x = $175in^3$. This shape does not meet the section modulus requirements, but we can increase its moment capacity by welding a cover plate on the flanges (Fig. 9.3). By using a plate thickness of 1/2in, the depth of the built-up beam becomes

$\quad\quad d = 17in + 2(.5in) = 18in$ (as required by the problem statement).

At this point, we can calculate the required I value for moment concerns. Remember F = Mc/I, I/c = S. Therefore $I_{req.} = S_{req.}(c)$, or in our case:

$\quad\quad I_{req.} = 225in^3(9in) = 2025in^4$

At this point it's worth calculating the I value required for deflection to determine which will control the design of the section.

$\quad\quad I_{required} = \dfrac{5wL^4}{384E\Delta};$ for live loads, the allowable Δ = 1in, therefore:

$\quad\quad I_{required} = \dfrac{5\left(\frac{1800lb/ft}{12in/ft}\right)(30ft \times 12in/ft)^4}{384(29,000,000psi)(1in)} = 1131in^4$

From these calculations, it's easy to see that the moment requirements govern the design.

Find the required section properties of the plates using the parallel axis theorem; $I = \Sigma(I_o + Ad^2)$.

Required I of the plates: = total I - beam I = $2,025in^4$ - $1,490in^4$ = $535in^4$

The distance "y" from the neutral axis to the centroids of the plates is:

$$y = 9in - (0.5in)/2 = 8.75in$$

We can now determine the width of plate B that yields the necessary I.

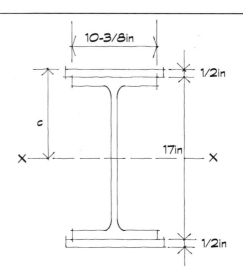

Each plate has a moment of inertia of $I = I_o + Ad^2$. Therefore, the moment of inertia of the plates is:

$$I = 2 ["O" + .5inB(8.75in)^2] = 535in^4$$

The moment of inertia of a rectangle less than 1 inch high is a verrrry small number, so we'll ignore it.

Solving for B: $B = 535in^4/2(0.5in)(8.75in)^2 = 6.89in$

Use the 7.0in plate.

Check shear requirements:

Required $A_w = P/F_v = 60k/0.4(36ksi) = 4.17in^2$

   Actual web area = 0.5in x 18in = 9.0in²   OK!

9.2 Connection and length of cover plates

Cover plates are typically welded, but may be bolted or riveted to the flanges. These connections must be designed to carry the horizontal shear that would exist at that point. A deck of cards can illustrate this consideration. If you hold a deck of cards by the edges, you can easily bend them, but if you glue the cards together, it increases the strength of the cards enormously. Without the cards being "locked" together to act as one section, their full potential connot be developed. (Laminated wood beams?) The same is true of individual structural pieces. The following illustrates how this force would be calculated.

In the previous example, the web-shear formula has been applied using the full depth of the built-up section; this assumes that the cover plates are solidly connected to the W shape by welds. At this connection, the shear stress is completely resisted by the welds; this is a **horizontal shear stress** that can be computed with a variation of the general shear formula:

$$P_v = VQ/I$$

where P_v = horizontal shear load calculated in units of lb or kips/inch

   V = shear value at the point of investigation

      (note that this value changes as you move along the beam)

   Q = statical moment of the cover plate about the neutral axis

      (also can be written as   or)

   I = total moment of inertia of the **entire cross-section** about the neutral axis

In this case, P_v is the load for which any connector at this location must be designed.

When a cover plate is designed to provide additional moment capacity at a point of maximum moment, as in Example 9.1, we know that this moment capacity is not required for the full length of the beam, but there are points where the beam without the plates has a moment capacity higher than the required bending moment at those points (Fig. 9.5). These locations can be easily obtained by evaluating the moment diagram for each loading condition.

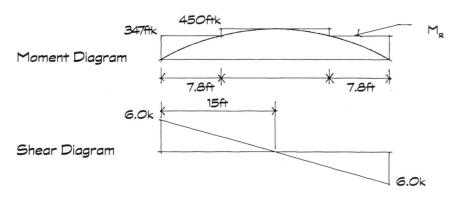

Fig. 9.5 Length of cover plates for the beam in Example 9.1

At the points where the moment capacity of the beam equals or exceeds the value of the required moment, the cover plates can be theoretically discontinued; these are the "theoretical cutoff points." However, the stress transfer from a flange to a cover plate is gradual, and the plates must be extended to a region of lower stress. The extension length beyond the theoretical cutoff points is called the terminating length, and in the absence of specific calculations, this should not be less than two times the width of the cover plate.

The design of cover plates introduces the notion that a steel beam could be built up around the moment and shear diagrams in order to maximize its efficiency and reduce the weight of steel. Steel areas could be eliminated where understressed and added as stresses increase according to the moment diagrams. In fact, a steel beam selected for maximum moment is an inefficient structure, since as we move away from the point of maximum moment the flanges are underused (although they do provide lateral support for both bending and moment of inertia for deflection). At the same time, the depth required to resist the maximum moment in a simply supported beam gives an excessive web area at that point, since that is a point of **zero** shear. Ideally, a beam could be designed with a variable cross-sectional area with the web area corresponding to the variation of the shear diagram and the moment of inertia (variable depth and flange area) corresponding to the moment diagram. A great idea, but not easy to manufacture. We'll address this later when we discuss the conceptual basis for trusses.

Example 9.2

Find the required length of the cover plates for the beam of Example 9.1. The plates can be cut off where the bending moment is equal to the actual moment capacity. M_r (from the section modulus tables) = 347ftk for the W16X100 without cover plates. Knowing that the moment at any point on a beam is equal to the area of the shear diagram to that point, we can easily determine the point at which the moment would equal 347ftk.

With a maximum reaction of 60k and a rate of change of 4klf the area of the shear diagram would equal:

$$A = \frac{V + V_x}{2}(X) = 347ftk$$

$$347ftk = \frac{60k + (60k - 4klf(Xft))}{2}(Xft)$$

$$347ftk = [30k + 30k - 2klf(Xft)](Xft) = [60k - 2klf(Xft)](Xft)$$

$$347ftk = 60Xftk - 2X^2ftk$$

$$2X^2 - 60X + 347 = 0$$

$$x^2 - 30X + 173.5 = 0$$

This is a quadratic equation, which can be written as

$$x = \frac{-(-30) \pm \sqrt{-(30)^2 - 4(1)(173.5)}}{2(1)} = \frac{30 \pm \sqrt{900 - 694}}{2} = \frac{30 \pm 14.35}{2}$$

This yields two possible solutions: $x_1 = 22.17ft$ and $x_2 = 15ft - 7.17ft = 7.83ft$.

In this case since the rate of change of the shear diagram is uniform over the entire span, both answers are correct. If distances are measured from the left end, the moment is equal to 347ftk at both 22.1ft and 7.83ft. Therefore, the plate must be 30ft - 2(7.83ft) = 14.34ft long. By adding the terminating length = 2(7.0in) = 14in = 1.17ft, the total length of the plates is 14.34ft + 1.17ft = 15.5ft. Make the 7in x 1/2in cover plates 16ft long, centered at midspan.

9.3 Holes in the web and in the flanges

Beams designed for maximum moment or deflection almost always have a redundant web area. This can be exploited when it is necessary to cut holes in the web for ducts or piping. On the other hand, whenever it is necessary to cut holes larger than the shear will allow, it is necessary to reinforce the web with additional plates. In general, it is convenient to select a point on the beam where holes can be cut without having to reinforce the web.

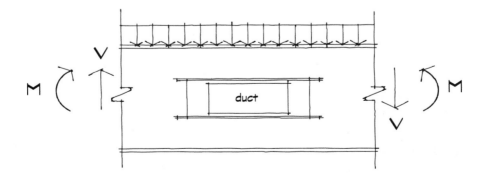

Fig. 9.6 Stiffened hole in web around duct penetration

Where the hole is cut in the web, the beam must have sufficient I_x and web area for M, Δ, and V requirements. Whenever we modify the shape, all of the usual procedures apply for the modified beam design: It is like designing a different beam for the same loading condition.

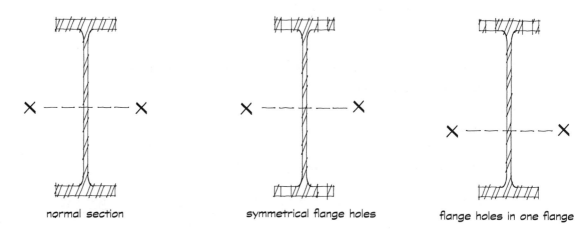

normal section   symmetrical flange holes   flange holes in one flange

Fig. 9.7 Effect of flange holes on location of neutral axis.

Often, holes are cut in a web in such a way that the resulting shape is locally inadequate. It is common to reinforce the shape with vertical or horizontal stiffeners, as illustrated in Fig. 9.6.

If the flanges of a beam have holes for bolts, rivets, or piping, the net area of the cross-section through a group of holes is obviously less than the gross cross-sectional area. When the holes are symmetrical with respect to the centroidal neutral axis X-X of the cross-section, no shift of this axis occurs. If, however, the holes are not symmetrical as described, the neutral axis is shifted away from the holes. The new moment of inertia is calculated as the difference between the gross i of the beam relative to the new neutral axis and the I of the cross-section of the holes about that same axis. The procedure is used on both symmetrical and asymmetrical sections, such as illustrated in Figs. 9.3(c), (d), and (e) and in Example 9.3. In these cases, the new location of the neutral axis must be calculated in order to find I and ultimately S. The change in I will have little effect on the deflection of the beam since it affects only a small proportion of the entire beam and deflection is an accumulative change in dimension. Flexural stress, by comparison, is a maximum value that must not be exceeded at any point on the beam.

Example 9.3

For the same beam of Example 9.2, a 12in square hole must be cut in the web. Find the minimum distance from the support where this can be done. (Note that the shape of the hole has nothing to do with the problem; only the removed cross-section is important).

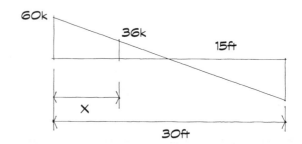

If a 12in hole is cut out, the area of web remaining is

$$A_w = (17in - 12in)(0.5in) = 2.5in^2$$

The shear capacity of the web with the hole is:

$$V_r = f_v A_w = (14.4ksi)(2.5in^2) = 36k$$

The distance X from the support where V = 36k can be found by looking at the shear diagram:

The maximum shear value is 60k

The rate of change on this diagram is 4klf

Therefore, if we're looking for the point where the shear is 36k, we can write

$$60k - (4klf)(X) = 36k$$
$$X = 6ft$$

The hole can be cut at any location beyond 6ft from the support.

Example 9.4

A hole must be cut in the web of a W30X116 grade-50 beam, as shown in the illustration. At that point, M = 830ftk and V = 0. Check if the section with the hole is adequate.

From the AISC Manual, W30X116 has the following section properties:

$d = 30in;$ $\quad t_w = 0.565in;$ $\qquad I_x = 4930in^4$

Find the new position of neutral axis (N.A.)

Moment of the areas: $Ay = (A_{section})y_1 - (A_{hole})y_2$

$A_{section} = 34.2in^2$
$A_{hole} = (10in)(.565in) = 5.65in^2$

$A = 34.2in^2 - 5.65in^2 = 28.55in^2$

$y_1 = 30in/2 = 15in$
$y_2 = 16in + 5in = 21in$
$$y = \frac{34.2in^2(15in) - 5.65in^2(21in)}{28.55in^2} = 13.8in$$

(distance of the N.A. from axis A-A)

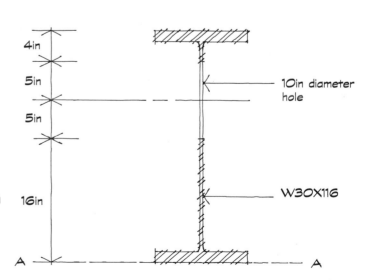

Moment of inertia of the composite section:

$I = \Sigma(I_o + Ay^2)$

W30X116: $I_o = 4930in^4$; $A = 34.2in^2$; $y = 15in - 13.8in = 1.2in$

$I_o + Ay^2 = [4930in^4 + 34.2in^2(1.2in)^2] = 4979in^4$

Hole: $I_o = t_w h^3/12 = .565in(10in)^3/12in = 47in^4$
             $A = t_w h = .565in(10in) = 5.65in^2$
             $y = 16in - 13.8in = 2.2in$
             $I_o + Ay^2 = 47in^4 + (5.65in^2)(2.2in)^2 = 74in^4$

W30X116 with hole: $I_x = 4979in^4 - 74in^4 = 4905n^4$
                    $c = 30in - 13.8in = 16.2in$
                    $S_x = 4905/16.2 = 302.8in^3$
                    $M_R = 33ksi(302.8in^3) = 9992ink$
                    $M_R = 9992ink/12in/ft = 833ftk > 830ftk$ OK!

Example 9.5

A W12X26 beam of $F_y = 50ksi$ steel has two 13/16in holes for 3/4in bolts in both flanges. At that section, the beam has a bending moment of M = 72ftk. Is the section adequate?

Area of the holes in a flange: $A_h = 2(13/16in)(0.38in) = 0.62in^2$

For the holes on two flanges,
$I_h = 2(Io + Ad^2) = 2(``O" + 0.62in^2 [(12.22in - 0.38in)/2]^2) = 43.46in^4$

Net moment of inertia $I_{net} = 204in^4 - 43.46in^4 = 160.5in^4$

Required $I = Mc/F_b = \dfrac{(72ftk)(12in/ft)(12.22in/2)}{0.66(50ksi)} = 159.9in^4$ OK!

OK, we've gone through a discussion of altered sections. At this point you should try some problems to see if you can execute the procedures we've outlined. GOOD LUCK!

Problem 9.1

The beam shown in the illustration is built up with a C10X15.3 welded to a W14x30. Determine the combined moment of inertia I_x.

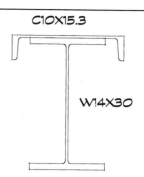

C10X15.3

W14X30

Answer:

$I_x = 420.3 \text{in}^4$

Problem 9.2

This built-up section consists of an S10X25.4 with two 12in x 0.5in plates welded to the flanges. Calculate the moment of inertia of the sections and find the allowable distributed live load giving a maximum deflection = L/360 for the 12ft span. (Based on deflection only.) Use A36 steel.

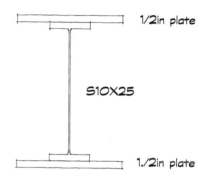

1/2in plate

S10X25

1./2in plate

Answer:

$I = 455 \text{in}^4$; $w = 943 \text{pli} = 11.3 \text{klf}$

Problem 9.3

A handrail is supported with vertical double angles (two unequal leg angles, short legs back to back) 4ft o.c. welded to the floor edge, supporting a plate (to receive a non-structural wooden handrail). The space in between these vertical supports is closed with glass that does not give any structural support to the railings. The handrail must resist a horizontal load of 100plf applied at the handrail plate and have a max. deflection = L/360. Use A36 steel. Select the appropriate L shape.

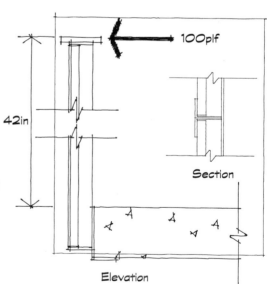

100plf

42in

Section

Elevation

Answer:

L - 3-1/2in x 3in x 5/16in has $I = 3.17 \text{in}^4$ > required I;
$S = 1.44 \text{in}^3$ > required S; $A = 3.87 \text{in}^2$ > required A.

Problem 9.4

Design a beam to span 60ft, laterally braced, with a DL of 200plf and LL 400plf. The maximum depth is limited to 36in. As shown in the illustration, the beam is built up with two round pipes welded to a steel plate web. The circular holes in the web have a diameter of 18in for ductwork. Use A36 steel. (Illustrated on the following page, but don't look at the answer yet!)

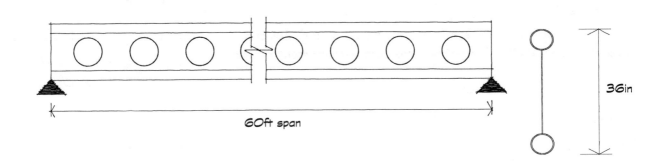

Answer:

Required I of flanges = 2455in⁴; required A_w = 1.25in². Use 6in standard-weight pipes and 3/16in web plate with 18in holes; total depth = 36in.

METRIC VERSIONS OF ALL EXAMPLES AND PROBLEMS

Example M9.1

The load of a simply supported beam is 29,188N/m DL and 26,269N/m LL, so the total load is 55,457N/m (not including an estimate for the weight of the beam). The beam is 9.15m long and is fully laterally supported. If the overall depth is limited to 460mm, design a section using 250MPa steel.

Total load W = 55,457N/m + 3000N/m beam estimate = 58,454N/m

Determine the required section properties of the beam.

Maximum M = $wL^2/8$ = 58,454N/m(9.15m)^2/8 = 611,739Nm

Maximum V = WL/2 = 58,454N/m(9.15m)/2 = 267,427N

Maximum LL deflection = L/360 = 9150mm/360 = 25mm

Although we would calculate the area and the moment of inertia as well, we'll limit our concerns to moment initially. (When we calculate the properties for the built-up section for moment, we will also be finding the moment of inertia, which can be evaluated later for deflection.)

Required S_x = M/F$_b$ = $\dfrac{(611,739\text{Nm})(1000\text{mm/m})}{0.66(250\text{MPa})}$ = 3707x10^{3}mm^3

The strongest shape with d = 460mm is W460X106 with S_x = 2080x10^{3}mm^3, which is not sufficient. We will assume in our example that problems of availability limit our choice for the lightest section under 460mm to a W410X149, with I = 619x10^{6}mm^4, d = 431mm, and S_x = 2870x10^{3}mm^3. This shape is not meeting the requirements, but we can increase its moment capacity by welding a cover plate on the flanges (Fig. 9.3). By using a plate thickness of 12mm, the depth of the built-up beam becomes

d = 431mm + 2(12mm) = 455mm satisfying the problem requirements.

At this point, we can calculate the required I value for moment concerns. Remember that F = Mc/I, and I/c = S. Therefore I$_{req.}$ = S$_{req.}$(c) or in our case:

I$_{req.}$ = 3707x10^{3}mm^3(455mm/2)= 843x10^{6}mm^4

At this point, it's worth calculating the I value required for deflection to determine which will control the design of the section.

I$_{required}$ = $\dfrac{5wL^4}{384E\Delta}$; for live loads, the allowable Δ = 25mm, therefore:

I$_{required}$ = $\dfrac{5(29188\text{N/m}/1000\text{mm/m})(9150\text{mm})^4}{384(200,000\text{MPa})(25\text{mm})}$ = 533x10^{6}mm^4

From these calculations, it's easy to see that the moment requirements govern the design.

Find the required section properties of the plates using the parallel axis theorem; $I = \Sigma(I_o + Ad^2)$.

Required I of the plates = total I - beam I = $843 \times 10^6 mm^4$ - $619 \times 10^6 mm^4$ = $224 \times 10^6 mm^4$

The distance y from the neutral axis to the centroids of the plates is:
$$y = (455mm - 12mm)/2 = 221.5mm$$

We can now determine the width of plate B that yields the necessary I.

Each plate has a moment of inertia of $I = I_o + Ad^2$. Therefore the moment of inertia of the plates is

$$I = 2[0 + B(12mm)(221.5mm)^2] \quad = 224 \times 10^6 mm^4$$

The moment of inertia of a rectangle less than 25mm high is a verrrrrry small number, so we'll ignore it. Solving for B:

$$B = \frac{224 \times 10^6 mm^4}{2[12mm(221.5mm)^2]} = \quad 190mm \quad \text{Use 190mm x 12mm plate.}$$

Check shear requirements: Required $A_w = P/F_v = (267,427N)/0.4(250MPa) = 2674mm^2$

$$\text{Actual web area} = 14.9mm(455mm) = 6779.5mm^2 \qquad OK!$$

Example M9.2

Find the required length of the cover plates for the beam of Example M9.1. The plates can be cut off where the bending moment is equal to the actual moment capacity.

$$M_r = S_x F_b = \frac{(2870 \times 10^3 mm^3)165MPa)}{1000mm/m} = 473,550Nm$$

for the W410X149 without cover plates. Knowing that the moment at any point on a beam is equal to the area of the shear diagram to that point, we can easily determine the point at which the moment would equal 473 kNm.

With a maximum reaction of 267,427N and a rate of change of 55,457N/m, the area of the shear diagram equals:

$$A = \frac{V + V_x}{2}(X) = 473550 \, Nm$$

$$473{,}550 \, Nm = \frac{267{,}427N + [(267{,}427N - 55{,}457N/m(X)]}{2}X$$

where X is the distance from the left support in meters. Solving for X:

473,550Nm = [133,713N + 133713N - (27,728N/m)X]X

473,550Nm = (267,427N)X - (27,728N/m)X²

(27,728N/m)X² - (267,427N)X + 473,550Nm = 0

X² - 9.64X + 17.08 = 0

This is a quadratic equation, which can be written as:

$$X = \frac{-(-9.64) \pm \sqrt{-(9.64)^2 - 4(1)(17.08)}}{2(1)} = \frac{9.64 \pm \sqrt{92.93 - 68.31}}{2}$$

This yields two possible solutions: x_1 = 7.3m, and x_2 = 2.3m.

In this case, since the rate of change of the shear diagram is uniform over the entire span, both answers are correct. If distances are measured from the left end, the moment is equal to 473,550Nm at both 7.3m and 2.3m. Therefore, the plate must be 9.15m - 2(2.3m) = 4.55m long.

Adding the terminating length = 2(190mm) = 380mm = 0.38m, the total length of the plates is 4.55m + 0.38m = 4.93m. Make the 190mm x 12mm cover plates 5m long, centered at midspan.

Example M9.3

For the same beam of Example M9.2, a 300mm square hole must be cut in the web. Find the minimum distance from the support where this can be done. (Note that the shape of the hole has nothing to do with the problem; only the removed cross-section is important.)

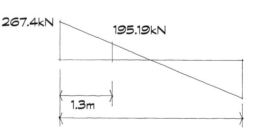

If a 300mm hole is cut out, the area of web remaining is

$$A_w = (431mm - 300mm)(14.9mm) = 1952mm^2$$

The shear capacity of the web with the hole is

$$V_r = f_v A_w = 0.4(250MPa)(1952mm^2) = 195{,}190N$$

The distance X from the support where V = 195,190N can be found by looking at the shear diagram:

> The maximum shear value is 267,427N
>
> The rate of change on this diagram is 55,457N/m

Therefore, if we're looking for the point where the shear is 267,427N, we can write:

$$267,427N - 55,457N/m(X) = 267,427N$$
$$X = 1.3m$$

The hole can be cut at any location beyond 1.3m from the support.

Example M9.4

A hole must be cut in the web of a W760X173 grade 345MPa beam, as shown in the illustration. At that point, M = 1,126,000Nm and V = 0. Check if the section with the hole is adequate.

From the AISC manual, W760X173 has the following section properties:

$d = 762mm$; $t_w = 14.4 mm$; $I_x = 2060 \times 10^6 mm^4$

Find the new position of neutral axis (N.A.)

Moment of the areas: $Ay = (A_{section})y_1 - (A_{hole})y_2$

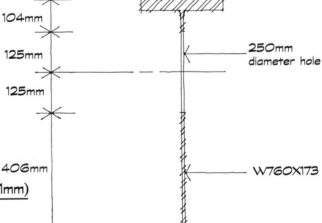

$A_{section} = 22100mm^2$

$A_{hole} = (250mm)(14.4mm) = 3600mm^2$

$A = 22100mm^2 - 3600mm^2 = 18500mm^2$

$y_1 = 762mm/2 = 381mm$

$y_2 = 406mm + 125mm = 531mm$

$$y = \frac{(22100mm^2)(381mm) - (3600mm^2)(531mm)}{18500mm^2}$$

$$= 352mm$$

(distance of N.A. from axis A-A)

Moment of inertia of the composite section:

$I = \Sigma(I_o + Ay^2)$

W760X173: $I_o = 2060 \times 10^6 mm^4$; $A = 22100mm^2$; $y = 381mm - 352mm = 29mm$

$I_o + Ay^2 = 2060 \times 10^6 mm^4 + 22100mm^2(29mm)^2 = 2079 \times 10^6 mm^4$

Hole: $I_o = t_w h^3/12 = 14.4mm(250mm)^3/12 = 18.7\times10^6 mm^4$

$A = t_w h = 14.4mm(250mm) = 3600mm^2$

$y = 406mm - 352mm = 54mm$

$I_o + Ay^2 = 18.7\times10^6 mm^4 + (3600mm^2)(54mm)^2 = 29\times10^6 mm^4$

W760X173 with hole: $I_x = 2060\times10^6 mm^4 - 29\times10^6 mm^4 = 2050\times10^6 mm^4$

$c = 762mm - 352mm = 410mm$

$S_x = 2050\times10^6 mm^4/410mm = 5000\times10^3 mm^3$

$M_R = 0.66(345MPa)(5000\times10^3 mm^3)/1000mm/m = 1,138,500Nm$

$1,138,500Nm > 1,126,000Nm$ OK!

Example M9.5

A W310X39 beam of F_y = 345MPa steel has two 20mm holes for 19mm bolts in both flanges. At that section, the beam has a bending moment M = 98,000Nm. Is the section adequate?

$I_x = 84.8\times10^6 mm^4$; d = 310mm; t_f = 9.7mm.

Area of the holes in a flange: $A_h = 2(20mm)(9.7mm) = 388mm^2$

For the holes on two flanges:

$$I_h = 2(I_o + Ad^2) = 2\left["O" + 388mm^2\left(\frac{310mm - 9.7mm}{2}\right)^2 \right] = 17.5\times10^6 mm^4$$

Net moment of inertia $I_{net} = 84.8\times10^6 mm^4 - 17.5\times10^6 mm^4 = 67.3\times10^6 mm^4$

Required $I = Mc/F_b = \dfrac{98,000Nm(1000mm/m)(310mm/2)}{0.66(345MPa)} = 66.7\times10^6 mm^4$ OK!

Problem M9.1

The beam shown in the illustration is built up with a C250X23 welded to a W360X45.

Determine the combined moment of inertia I_x.

C250X23

W360X45

Answer:

$I_x = 175\times10^6 mm^4$

Problem M9.2

This built-up section consists of an S250X38 with two
305mmx12mm plates welded to the flanges. Calculate the
moment of inertia of the sections and find the allowable
distributed live load giving a maximum deflection = L/360
for the 3.66m span. (Based on deflection only.)
Use 250MPa steel.

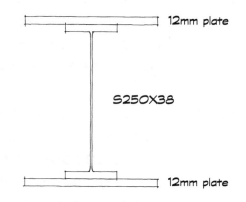

Answer:

I = 189×10^6mm^4; w = 13.8kN/m

Problem M9.3

A handrail is supported with vertical double angles (two
unequal leg angles, short legs back to back) 1.2m o.c.
welded to the floor edge, supporting a plate (to receive a
nonstructural wooden handrail). The space in between
these vertical supports is closed with glass that does not
give any structural support to the railings. The handrail
must resist a horizontal load of 1460N/m applied at the
handrail plate and have a maximum deflection = L/360.
Use 250MPa steel. Select the appropriate L shape.

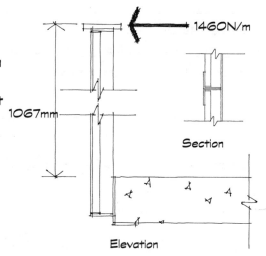

Answer:

L 89mm x 76mm x7.9mm

Problem M9.4

Design a beam to span 18.3m, laterally braced, with a DL of 2920N/m and a LL of 5840N/m.
The maximum depth is limited to 914mm. As shown in the illustration, the beam is built up with
two round pipes welded to a steel plate web. The circular holes in the web have a diameter of
457mm for ductwork. Use 250MPa steel.

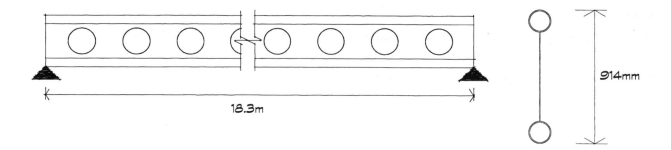

Answer:

Required I of flanges = 1022×10^6mm^4; required A$_w$ = 806mm^2. A possibility is to use 152mm
standard-weight pipes and a 5mm web plate with 457mm holes; total depth = 914mm.

10 COLUMNS

10.1 Column theory

Columns are axially loaded members in compression. The compressive stress formula is used to calculate the stress, $F_a = P/A$, or to find the required section area, $A = P/F_a$, where F_a is the allowable compressive stress. F_a varies depending on the slenderness ratio of the column (AISC Manual, p. 5-42). Slenderness is a critical issue in column design with any material. Its influence is felt every time you attempt to punch a straw through the top of a drink box or even to simply penetrate the plastic cover on a "take out." To punch the hole, you might initially hold the straw at the top and with dismay, watch it buckle before it punches a hole. To make the straw "stronger," you hold it closer to its base and the same straw can easily apply enough force to penetrate the top. It's the same straw, made of the same material, but it will hold more load (or in this case apply more force) if it's shorter (has a smaller slenderness ratio). Clearly, F_a is not only a function of the material strength, but also of the properties of the cross-section of the column, of its height, and of its end conditions. With simple rectangular shapes like wood columns, the slenderness ratio is the unbraced length divided by the dimension of the least side. Steel columns have more complex shapes and the radius of gyration is used. The radius of gyration is defined as $r = I/A$, and it is therefore a dimension in inches like the side of a wood column. The slenderness ratio is kL/r, where kL is the effective length of the column (units of inches or mm) and r is the least radius of gyration of the cross-section (units of inches or mm). The AISC Specification (AISC Manual, p. 5-135) gives recommended design values for effective column length depending on end conditions or restraints (essentially fixed or pinned) and the ability of the system to resist lateral displacement.

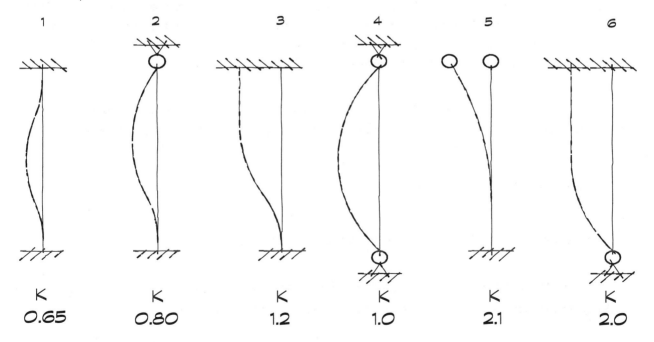

Fig. 10.1 Recommended K values for typical end conditions.
Case 4 is the base to which everything is related.

Example 10.1

The two columns (adjacent) have the same height, but different effective lengths.

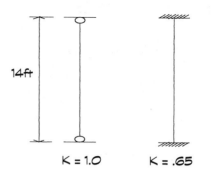

The first column is case (4) of Fig. 10.1. Therefore, its effective length is

KL = 1.0(14ft)(12in/ft) = 168in

The second column is case (1); its effective length is

KL = 0.65(14ft)(12in/'ft) = 109.2in

Example 10.2

The adjacent frame has no diagonal bracing in this plane, and its lateral stability is ensured by fixed-end connections with the foundations; a rigidly connected beam prevents rotation of the top of the column. Lateral motion of the beam can take place, as shown by dashed line; therefore the column is case (3): KL = 1.2(14ft) = 16.8ft, or 201.6in

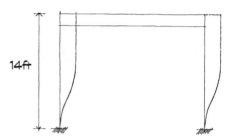

Symmetrical shapes such as W, S, and C have in general two different radii of gyration, r_x and r_y, for the x and y axes; for angles, an additional value is given with respect to the oblique z axis, as was shown in bridging for open-web steel joists. The best columns (the ones using steel most efficiently) have the same radius of gyration in all directions: round or square tube sections. Tubes and pipes however, are significantly more expensive than other rolled shapes, and it is often more economical to use a W shape even if it has unequal r values. Columns are also "built-up" with the attempt to create equal radii of gyration in all directions.

The AISC Specifications (AISC Manual, p. 5-42) classify columns by comparing their slenderness ratio KL/r to a coefficient C_c. Two values of F_a are given correspondingly for what we could call "short" and "long" columns.

$$C_c = \sqrt{\frac{2\pi^2 E}{F_y}}$$

where E is the modulus of elasticity and F_y is the yield stress.

1. When the largest KL/r < C_c, F_a is given by AISC formula E2-1 ("short").
2. When the largest KL/r > C_c, F_a is given by AISC formula E2-2 ("long").

You will notice that the value of F_a in those formulas depends on F_y, KL/r, and E. In other words, given a type of steel, the allowable compressive stress F_a is reduced as the slenderness ratio increases, since F_y is multiplied by a factor smaller than 1.0. In practice, it is not necessary to calculate these formulas every time, since F_a values are tabulated in the AISC Allowable **Stresses for Compression Member,** (AISC Manual, pp. 3-16 and 3-17). The maximum slenderness ratio is 200, and you should notice from the tables that the steel yield strength has no effect at slenderness ratios above 120. The design procedure for a column of a given height and with a given load P is made very rapid using the values in the allowable axial load tables in the AISC Manual (pp. 3-19 through 3-105). In the following example, a column is designed using these allowable axial load tables. (These tables do not presently exist for metric values, but F_a can be converted into metric by multiplying the value in ksi by 6.894MPa.)

Example 10.3

An A36 steel column with an unbraced length of 12ft has to support a load of 148k, K = 1.0. From the AISC Manual, (p. 3-31). For KL = 1.0(12ft) we find 149k allowable load corresponding to a W8X31. . . . That's it, it works, we're done!

We could verify these table values by utilizing F = P/A or P = FXA.
From the AISC Manual (p. 3-31): W8X31 has A = 9.13in² and r_y = 2.02in,
so KL/r = 1.0(144in)/(2.02in) = 71.3, which we will round conservatively to 71.
From the AISC Manual, (p. 3-16): For KL/r = 71, F_a = 16.33ksi

P = 16.33ksi × 9.13in² = 149.1k > 148k. The slight variation from the design tables is accounted for by rounding off in both.

Checks are not necessary when selecting shapes from the tables. However, the check in the example illustrates the procedure to follow if the allowable load of a specific section is not listed in the table.

Example 10.4

Find a circular column that can be used in place of the W8X31 of Example 10.3.

From the AISC Manual, (p. 3-36): KL = 12ft An 8in standard steel pipe has an allowable load of 155k. The weight is 28.55plf, lighter than W8X31. Even though this section is lighter, differences in manufacturing will make the pipe section more expensive.

You'll notice that the allowable axial load tables do not contain all the W shapes listed in the AISC Manual, Section 1. For instance, W12's below W 12X40 [*W310's below W310X60*]are not in the column tables, whereas in the dimensions table (p. 1-28), smaller W12's can be found. This is because the lighter sections are not intended to be used as columns but as beams, and their r with respect to Y-Y is much smaller than the other, so they are inefficient as columns. The AISC Manual, (Table 2, p. 1-8), classifies shapes in five groups; groups 4 and 5 are generally contemplated for application as columns. The heaviest shape is W14X730 [*W360X1086*] and the lightest W12X210 [*W310X313*]. Obviously, this list does not exclude the use of other shapes, such as the W12's mentioned earlier, when calculations indicate their acceptability.

Compression members take many forms: columns in a multistory building are generally wide flanges, and truss members are angles, tubes, pipes, and wide flanges. We can find columns everywhere, since they are one of the three basic structural elements. An internal combustion engine has a "connecting rod" that connects the piston to the crankshaft. This connecting rod/column on one stroke is compressed down by the explosion of the air/fuel mixture and is a column. The piston is next pushed back up by the rotation of the crankshaft, and on the next cycle it is pulled down by the crankshaft and draws a new supply of air and fuel into the combustion chamber to be compressed (column again) and then ignited. The rod shape is much like a wide flange that has a strong axis in one direction, the one free to rotate on the crankshaft, and a weak axis, the one parallel to the crankshaft, which is restrained against rotation and has a K value of 0.5. This variation in end condition explains why the rod has two different moments of inertia and is not shaped like a pipe.

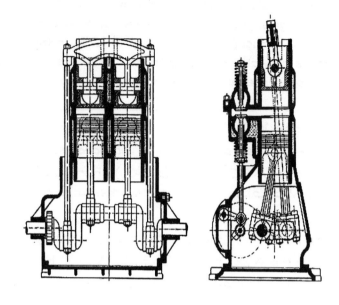

Fig. 10.2 Small internal combustion engine section

10.2 Built-up columns

There are occasions when the column load is either above or below the range of typical columns and consequently no options are listed in the AISC Manual, or a specific architectural configuration is desired that is not a standard shape. In these cases, it is possible to design a composite configuration from several different approaches.

One approach would be to combine columns listed in the tables. If this is done, and the column sections are not attached together at intermediate points, the table values simply can be added. This is typically not a good approach, however, since these same column sections, if attached at intermediate levels, will have a significantly different slenderness ratio as a group than as individual components. Consequently, these attached column sections will have a much greater load capacity than the sum of the individual parts.

An alternative approach is to combine a set of shapes: angles, channels, or even wide-flange sections, in a manner that would give a configuration similar to a tube, an ideal column section. The tube is ideal because it has no weak axis. To carry this idea one step further, the composite section could be either literally a tube or mathematically a tube. A "mathematical" tube would be a section that has equal slenderness ratios about both major axes and attempts to maximize the moment of inertia about both major axes. Mies van der Rohe's columns at Crown Hall, Illinois Institute of Technology, and many other locations are examples of this idea. In this case, the section consists of apparently intersecting wide flanges, creating a cruciform section.

Example 10.5

The column shown adjacent is built with two C10X30 channels back to back, spaced 8in by a continuous 1/2in plate. The column length is 28ft, and K is 1.0. Find the allowable axial load with 36ksi steel and 50ksi steel.

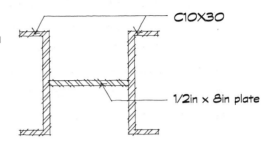

C10X30

1/2in × 8in plate

The allowable load can be found from the tables of allowable stresses for compression members in the AISC Manual (Part 3: Column Design), based on the slenderness ratio KL/r. For this special shape, we do not know which axis is stronger, so we have to calculate both I_x and I_y.

$$I_x = I_{x\ channels} + I_{x\ plate} = 2(103in^4) + \sim 0 = 206in^4$$
$$I_y = I_{ychannels} + I_{y\ plate}$$

$$I_{ychannels} = 2(I_{yo} + Ad^2) = 2[3.94in^4 + 8.82in^2(4in + 0.65in)^2] = 389in^4$$

where 8.82in² is the area of a single channel.

$$I_{y\ plate} = bh^3/12 = (0.5in)(8in)^3/12 = 21in^4$$

$$I_y = 389in^4 + 21in^4 = 410in^4 > I_x,$$

so X-X is the weak axis, and it controls the buckling. The allowable stress will depend on the slenderness about X-X, which is necessarily greater than that about Y-Y.

$$A = 2(8.82in^2) + (0.5in \times 8in) = 21.64in^2$$

$$r_x = \sqrt{\frac{I_x}{A}} = \sqrt{\frac{206in^4}{21.64in^2}} = 3.08in$$

$$KL/r_x = (28ft)(12in/ft)/3.08in = 109 < 200 \text{ OK}$$

From the table for 36ksi steel, $F_a = 11.81ksi$, thus, $P_a = (11.81ksi)(21.64in^2) = 255k$
From the table for 50ksi steel, $F_a = 12.57ksi$, thus, $P_a = (12.57ksi)(21.64in^2) = 272k$
The shift from 36 to 50ksi steel increases the load bearing capacity by less than 7%, although you would probably pay 25% more for the stronger material.

XStrong 12in pipe, 36ksi   w = 65.4plf
9in × 9in × 9/16in square tube, 46ksi   w = 61.8plf lightest,

     but since a more expensive grade of steel, maybe not the cheapest.

10.3 Columns with unequal unbraced lengths

Columns can have different effective lengths (KL) in the X-X and Y-Y axes. This can be the result of at least two different cases:

1. They can have different end conditions with respect to the X-X and Y-Y axes.
2. Their axes are braced at different intervals relative to the X-X and Y-Y axes.

In all cases the column will fail about or relative to the axis with the largest slenderness ratio, KL/r.

Example 10.6

The illustrated column is rigidly connected at the base and at the joint with the girder with respect to the X-X axis; therefore K = 1.2 for this axis. The Y-Y axis of the column is braced and simply pinned both top and bottom, so K = 1.0.

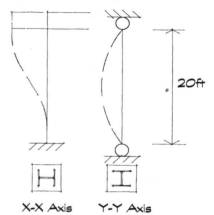

X-X Axis   Y-Y Axis

The design load is 72k. From the AISC Manual, (p. 3-31), a W8X31 can support 95k, for KL = 20ft based on Y-Y axis failure which means that KL_y/r_y is larger than KL_x/r_x. The next lightest section is a W8X28 with an allowable load of 56kips, which is not adequate. Since our column has different KL values for each axis, we must check the X-X axis to verify which is the controlling factor.

Y-Y: $KL_y/r_y = 1.0(20ft)(12in/ft)/2.02in = 118.8$
X-X: $KL_x/r_x = 2.0(20ft)(12in/ft)/3.47in = 138.3$

Therefore the X-X axis is the controlling failure axis for this particular section, end, and loading condition. Our use of the tables is incorrect since they are developed based on the premise that the Y-Y axis is the controlling failure axis.

So, let's start at the beginning and develop an approach.

If KL_x/r_x is = KL_y/r_y: . . . . The tables work.

> Then either or both axes are controlling and we could use the tables, which assumes the the Y-Y axis controls.

if KL_x/r_x is < KL_y/r_y: . . . . The tables work.

> Then the Y-Y axis controls and the tables are the easiest way to select a column section.

if KL_x/r_x is > KL_y/r_y: . . . . The tables do not work.

> Then the X-X axis controls and the tables do not work without some mathematical compensation.

Using any of these formulas, we can approach them from a slightly different direction.

We could rewrite the previous criteria as follows:

If $KL_x/r_x > KL_y/r_y$, then $KL_x > KL_y (r_x/r_y)$ or $KL_x / (r_x/r_y) > KL_y$.

If $KL_x/(r_x/r_y) = KL_y$: . . . . The **tables work**; neither axis controls.

If $KL_x/(r_x/r_y) < KL_y$: . . . . The **tables work**; Y-Y axis controls.

If $KL_x/(r_x/r_y) > KL_y$: . . . . The **tables don't work**; X-X axis controls.

In the third case, however, the value $KL_x/(r_x/r_y)$ could be considered as an "equivalent" KL_y and this value could be used in the tables as though it was an actual KL_y and load capacities for this value would be correct. Note that you may actually change column choices as long as their r_x/r_y values remain the same. If this value changes, however, you will have to recalculate the equivalent KL_y value and reevaluate the controlling factors.

Yes, it's a bit confusing, but an example will help you see how to do it. So . . . .

Example 10.7

Design a column section for the following conditions of lateral support.

"A" is a pinned connection.

"B" is a pinned connection.

"C" is not truly a pinned or a fixed connection.

We will use a conservative pinned assumption and assume A-36 steel.

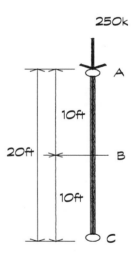

1. First assume that the Y-Y axis is the mode of failure, as we normally do, and check that assumption for the selected alternative sections.

   By using $K = 1.0$ and an unbraced $L_y = 10$ft: $KL_y = 10.0$ft. From the AISC Manual Column Design Tables, we find at least three possibilities:

   W8X48 carries 249k at 10.0ft.

   W10X49 carries 268k at 10.0ft.

   W12X50 carries 254k at 10.0ft.

2. Check the equivalent length of L_x of each section:

   W8X48: $r_x/r_y = 1.74$ (this value is located at the bottom of the column tables)

   $$\frac{KL_x}{\frac{r_x}{r_y}} = \frac{1.0(20.0\text{ft})}{1.74} = 11.49\text{ft} \geq 10.0\text{ft}$$

   (L_x controls, X-X axis failure)

This section is controlled in this case by X-X axis failure. The following alternative is possible:

A section could be selected from the tables if it **satisfies the following:**

      (a) a section that carries 250k at 11.49ft

      (b) has a r_x/r_y value of 1.74 or greater

A W8X58 carries 288k at 11.5ft (interpolated value between 293k and 283k), which is a workable section. **This is one option.**

3. Check the equivalent length of the W10X49:

W10X49: r_x/r_y = 1.71

$$\frac{KL_x}{\frac{r_x}{r_y}} = \frac{1.0(20.0ft)}{1.71} = 11.7ft \geq 10ft$$

(L_x controls, X-X axis failure)

A W10X49 carries 258k at 11.7ft (interpolated value).

In this case, the original section works due to its original "extra" capacity at a 10.0ft unbraced length. **This is a second option.**

4. Check the equivalent length of the W12X50

W12X50: r_x/r_y = 2.64

$$\frac{KL_x}{\frac{r_x}{r_y}} = \frac{1.0(20.0ft)}{2.64} = 7.57ft \leq 10.0ft$$

(L_y controls, "Y-Y" axis failure)

Since our original assumption was OK, we can simply use this section. **This is a third option.**

Problem 10.1

An 18ft column pinned at both ends has to carry an axial load of 355k. Design the column with (a) wide flange, (b) pipe, and (c) square tubing. Use all types of steel listed and find the lightest column.

Answer:

W12X79, 36ksi   w = 79plf

W12X65, 50ksi   w = 65plf

The following examples cannot be compared for cost by weight alone, since their manufacturing process is roughly 100% more expensive than a rolled shape: wide flange, I section, or angles.

XStrong 12in pipe, 36ksi: w = 65.4plf

9in x 9in x 9.16in square tube, 46ksi   w = 61.8plf

The square tube is the lightest, but since it is a more expensive grade of steel, it may not be enough lighter to compensate for the added unit cost of the steel.

Problem 10.2

A compression member of a truss is made with a pair of 3in x 2in x 1/4in angles (unequal legs), back to back, with long vertical legs, and 36ksi steel. The angles are attached at the connections by 3/8in thick plates. The unbraced length is 8ft in any direction; and, as in most trusses, you should consider the ends pinned. What is the allowable axial load in compression?

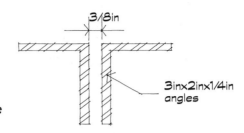

3/8in

3inx2inx1/4in angles

Answer:

27k controlled by the Y-Y axis

Problem 10.3

Design a 20ft tall column , braced on the Y-Y axis at 10ft, to carry an axial load of 210k. Use 36ksi steel.

Answer:

W10x45:

P_a = 216k; $KL_x/(r_x/r_y)$ = 1.0(20ft)/2.15 = 9.3ft < 10ft The Y-Y axis controls.

Problem 10.4

Design a 28ft column built up with two 6in standard pipes, structurally connected by lacing, so that they work as a unit for axial load. The pipes are spaced 18in apart, the Y-Y axis is braced at 7ft intervals, and the X-X axis is unbraced for the full 28ft. As-

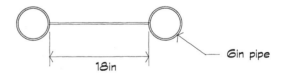

6in pipe

18in

sume pinned connections even at the 7ft bracing points. The column has to support an axial load of 200k. Is the design adequate if you use 36ksi steel?

Answer:

r_x = 9in, r_y = 2.24in

KL_x/r_x = 1.0(28ft)(12in/ft)/9in = 37.33

KL_y/r_y = 1.0(7ft)(12in/ft)/2.24in = 37.50 controls;

F_a = 19.42ksi ; P_a = 19.42ksi(2x5.58in^2) = 216.7k

Problem 10.5

The column shown is 24ft tall and is braced top and bottom about the X-X axis and has K bracing at 12ft for the Y-Y axis. It is built with a W10X45, A36 steel. Does this condition support an axial load of 210k?

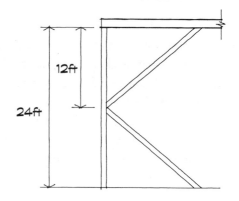

Solution:

With KL_y = 12ft, the W10X45 carries 216k. It works if the slenderness about X-X is not greater than about Y-Y. We can check this immediately by adjusting the KL_y ratio with r_x/r_y of 2.15. Since $KL_x/(r_x/r_y)$ = 1.0(24ft)/2.15 = 11.16ft < KL_y = 12ft, Y-Y axis controls, and the choice of W10X45 is correct.

Problem 10.6

A column has to carry an axial load of 210k. The column is 24ft tall (for a factory building) and is braced (pinned) at the ends about the X-X axis by the building frame. A double series of beams is inserted at 8ft and 16ft to support the exterior cladding panels, and these act as braces for the columns about the Y-Y axis. Find the most economical wide-flange shape for this column with A36 steel. You may once again assume K = 1.0 at all connection points (a conservative approach).

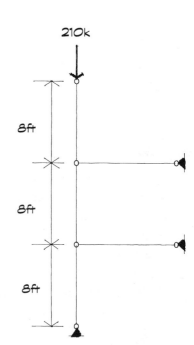

Solution:

KL_x = 1.0(24ft), KL_y = 1.0(8ft)
Select a column from the tables assuming KL_y controls. We find W10X39 as the lightest shape (laod capacity = 213k),
      with r_x/r_y = 2.16.
To test our assumption that the Y-Y axis controls check the equivalent X-X axis length:
$KL_x/(r_x/r_y)$ = 1.0(24ft)/2.16 = 11.1ft > 8ft
      OOPS!   X-X axis controls!

The choice is not valid, since the tables are designed for Y-Y axis failure. However, we can use the tables for the equivalent X-X axis length.
with $KL_x/(r_x/r_y)$ = 11.1ft = KL_y, we look for KL_y = 11.1ft:
      W10X45 is OK, as it carries 223.2k > 210k.
Use a W10X45, A36 steel.

We obtain this value by interpolating the table values between 10 and 11ft.
      224k - (224k - 216k)(0.1) = 224k - 0.8k = 223.2k

METRIC VERSIONS OF ALL PROBLEMS AND EXAMPLES

Example M10.1

The two columns (adjacent) have the same height, but different effective lengths.

The first column is case (4) of Fig. 10.1. Therefore, its effective length is

KL = 1.0(4.27m)(1000mm/m) = 4270mm. The second column is case (1); its effective length is

KL = 0.65(4.27m)(1000mm/m) = 2775mm.

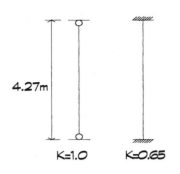

Example M10.2

The adjacent frame has no diagonal bracing in this plane, and its lateral stability is ensured by fixed-end connections with the foundations; a rigidly connected beam prevents rotation of the top of the column. Lateral motion of the beam can take place as shown by dashed line; therefore, the column is case (3): KL = 1.2(4270mm) = 5124mm.

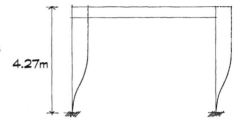

Example M10.3

A 250MPa steel column with an unbraced length of 3658mm has to support a load of 658kN. K = 1.0. Find an adequate W shape.

Since there is no metric table giving allowable axial loads for KL values, the problem cannot be solved with this procedure. However, we can check the load-bearing capacity of the W8X31 [W200X46] indicated in Example 10.3.

A = 5890mm²

KL/r = 1.0(3658mm)/51mm = 71.72 rounded conservatively to 72

From the AISC Manual (p. 3-16): For KL/r = 72, F_a = 16.33ksi = 6.894(16.33) = 112.6MPa.

P = (112.6MPa)(5890mm²) = 663kN > 658kN OK!

Example M10.4

Find a circular column that can be used in place of the W200X46 of Example M10.3.

From the AISC Manual (p. 3-36): KL = 3658mm A 203mm standard steel pipe has an allowable load of 689kN. The weight is 42.5kg/m, lighter than W200X46. Even though this section is lighter, differences in manufacturing will make the pipe section more expensive.

Example M10.5

The column shown here is built with two C250X45 channels back to back, spaced 203mm by a continuous 12mm plate. The column length is 8.54m, K = 1.0. Find the allowable axial load with 250MPa steel and 345MPa steel.

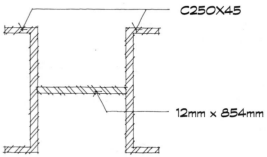

The allowable load can be found from the tables of allowable stresses for compression members in AISC Part 3 - Column Design, based on the slenderness ratio Kl/r. For this special shape we do not know which axis is stronger, so we have to calculate both I_x and I_y.

$$I_x = I_{x \, channels} + I_{x \, plate} = 2(42.9 \times 10^6 mm^4) + \sim 0 = 85.8 \times 10^6 mm^4$$

$$I_y = I_{ychannels} + I_{y \, plate}$$

$$I_{ychannels} = 2(I_{yo} + Ad^2) = 2[1.64mm^4 + 5690mm^2(101.5mm + 16.5mm)^2] = 158.4 \times 10^6 mm^4$$

where 5690mm² is the area of a single channel.

$$I_{y \, plate} = bh^3/12 = (12mm)(203mm)^3/12 = 8.4 \times 10^6 mm^4$$

$I_y = 158.4 \times 10^6 mm^4 + 8.4 \times 10^6 mm^4 = 166.8 \times 10^6 mm^4 > I_x$, so X-X is the weak axis, and it controls the buckling. The allowable stress will depend on the slenderness about X-X, which is necessarily greater than about Y-Y.

$$A = 2(5690mm^2) + (12mm \times 203mm) = 13816mm^2$$

$$r_x = \sqrt{\frac{I_x}{A}} = \sqrt{\frac{85.8 \times 10^6 mm^4}{13816mm^2}} = 78.8mm$$

$KL/r_x = 8534mm/78.8mm = 108 < 200$ OK

From the table for 250MPa steel, $F_a = 81.4$MPa, thus $P_a = 81.4MPa(13816mm^2) = 1125kN$
From the table for 345MPa steel, $F_a = 86.7$MPa, thus $P_a = 86.7MPa(13816mm^2) = 1197kN$

The shift from 250 to 345MPa steel increases the load-bearing capacity by less than 7%, while you would probably pay 25% more for the stronger material.

Example M10.6

The illustrated column is rigidly connected at the base and at the joint with the girder with respect to the X-X axis, therefore K = 1.2 for this axis. The Y-Y axis of the column is braced and simply pinned both top and bottom, so K = 1.0.

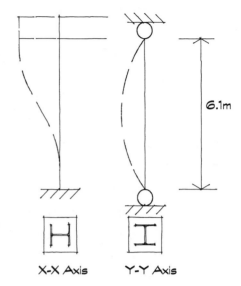

6.1m

X-X Axis   Y-Y Axis

The design load is 320kN. From the AISC Manual (p. 3-31), a W200X46 can support 423kN, for KL = 6.1m based on Y-Y axis failure, which means that KL_y/r_y is larger than KL_x/r_x. The next lightest section is a W200x42 with an allowable load of 249kN, which is not adequate. Since our column has different KL values for each axis, we must check the X-X axis to verify which is the controlling factor.

Y-Y: KL_y/r_y = 1.0(6100mm)/51mm = 119.6
X-X: KL_x/r_x = 2.0(6100mm)/87.9mm = 138.8

Therefore, the X-X axis is the controllng failure axis for this particular sectiion, end, and loading condition. Our use of the tables is incorrect since they are developed based on the premise that the Y-Y axis is the controlling failure axis.

Example M10.7

Design a column section for the following conditions of lateral support.

"A" is a pinned connection.

"B" is a pinned connection.

"C" is not truly a pinned or a fixed connection.

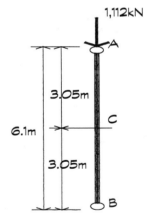

1,112kN

3.05m

6.1m

3.05m

We will use a conservative pinned assumption and assume 250MPa steel.

1. First assume that the Y-Y axis is the failure axis, as we normally do, and check that assumption for the selected alternative sections.
   By using K = 1.0 and an unbraced L_y = 3.05m, KL_y = 3.05m.

   From the AISC Manual Column Design Tables, we find at least three possibilities:
   W200X71 carries 1107kN at 3.05m
   W250X73 carries 1192kN at 3.05m
   W310X74 carries 1130kN at 3.05m

2. Check the equivalent length of L_x of each section:

> W200X71: $r_x/r_y = 1.74$ (this value is located at the bottom of the column tables)
>
> $$\frac{KL_x}{\frac{r_x}{r_y}} = \frac{1.0(6.1m)}{1.74} = 3.5m > 3.05m \quad (L_x \text{ controls, "X-X" axis failure})$$

This section is controlled in this case by X-X axis failure, and the following alternative is possible:

A section could be selected from the tables if it **satisfies** the following:

> (a) a section that carries 1112kN at 3.5m
>
> (b) has a r_x/r_y value of 1.74 or greater

A W200X86 carries 1281kN at 3.51 which is a workable section. **This is one option.**

3. Check the equivalent length of the W250X73

> W250x73: $r_x/r_y = 1.71$
>
> $$\frac{KL_x}{\frac{r_x}{r_y}} = \frac{1.0(6.1m)}{1.71} = 3.57m > 3.05m \quad (L_x \text{ controls, "X-X" axis failure})$$

A W250X73 carries 1148kN at 3.57m (interpolated value).

In this case, the original section works due to its original "extra" capacity at a 3.05m unbraced length. **This is a second option.**

4. Check the equivalent length of the W310X74:

> W310x74: $r_x/r_y = 2.64$
>
> $$\frac{KL_x}{\frac{r_x}{r_y}} = \frac{1.0(6.1m)}{2.64} = 2.31m < 3.05m \quad (L_y \text{ controls, "Y-Y" axis failure})$$

Since our original assumption was OK, we can simply use this section. **This is a third option.**

Problem M10.1

An 5.486mm column pinned at both ends has to carry an axial load of 1579kN. Design the column with (a) wide flange, (b) pipe, (c) square tubing. Use all types of steel listed and find the lightest column.

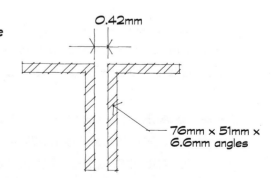

0.42mm

76mm x 51mm x 6.6mm angles

Answer:

W310X117, 250MPa	w = 117kg/m
W310X97, 345MPa	w = 97kg/m

The following examples cannot be compared for cost by weight alone, since their manufacturing process is roughly 100% more expensive than a rolled shape: wide flange, I section, or angles.

XStrong 310mm pipe, 250MPa   w = 97.4kg/m

254mmx254mmx12.7mm tube, 317MPa   w = 93kg/m

The square tube is the lightest, but since it is a more expensive grade of steel, it may not be enough lighter to compensate for the added unit cost of the steel.

Problem M10.2

A compression member of a truss is made with a pair of 76mm x 51mm x 6.6mm angles (unequal legs), back to back, with long vertical legs, and 250MPa steel. The angles are attached at the connections by 9.5mm thick plates. The unbraced length is 2.4m in any direction; and, as in most trusses, you should consider the ends pinned. What is the allowable axial load in compression?

Answer:

120kN controlled by Y-Y axis

Problem M10.3

Design a 6.10m tall column , braced on the Y-Y axis at 3.05, to carry an axial load of 934kN. Use 250MPa steel.

Answer:

W250X67

Problem M10.4

Design a 8.54m column built up with two 152mm standard pipes, structurally connected by lacing, so that they work as a unit for axial load. The pipes are spaced 457mm apart, the Y-Y axis is braced at 2.135m intervals, and the X-X axis is unbraced for the full 8.54m. Assume pinned connections even at the 2.135m bracing points. The column has to support an axial load of 890kN; is the design adequate if you use 250MPa steel?

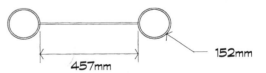

Answer:

KL_y/r_y = 37.50 controls;

F_a = 134MPa;

P_a = 134MPa(2x3600mm²) = 964.8kN >890kN OK.

Problem M10.5

The column shown is 7.32m tall and is braced top and bottom about the X-X axis and has K bracing at 3.66m for the Y-Y axis. It is built with a W250X67, 250MPa steel. Does this condition support an axial load of 934kN?

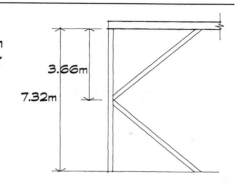

Answer:

A W250X67 has an allowable load of 961kN > 934kN;
for KL_y = 3.66m.
The efffective length X-X is $KL_x/(r_x/r_y)$ = 1.0(7.32m)/2.15 = 3.40m < 3.66m; Y-Y axis controls.

Problem 10.6

A column has to carry an axial load of 934kN. The column is 7.32m tall (for a factory building) and is braced (pinned) at the ends about the X-X axis by the building frame. A double series of beams is inserted at 2.44 and 4.88m to support the exterior cladding panels, and these act as braces for the columns about the Y-Y axis. Find the most economical wide-flange shape for this column with 250MPa steel. You may once again assume K = 1.0 at all connection points (a conservative approach).

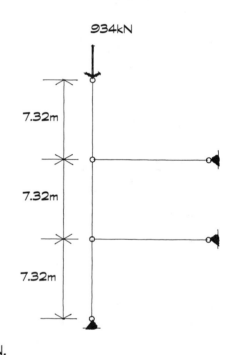

Solution:

KL_x = 1.0(7.32m), KL_y = 1.0(2.44m)
W10X39 [W250X49] has P = 213k = 947kN
with r_x/r_y = 2.16
$KL/(r_x/r_y)$ = 1.0(7.32m)/2.16 = 3.38m >2.44m
X axis controls! Look up K_yL = 11.1ft
W10X45 [W250X67] carries 223.2k = 993.2kN > 934kN.

Use W250X67 metric (based on X-X axis failure).

11 FABRICATED COLUMNS

11.1 General discussion

Columns can be fabricated from a variety of other shapes to satisfy unique architectural criteria, address specific structural problems, or to alter the load capacity of a standard column shape.

Since a major consideration in the load-carrying capacity of any column is the material distribution within the section, built-up sections start with the objective of "balanced" strength. A standard wide flange would be made significantly stronger by the introduction of material parallel to the weak (Y-Y) axis. Plates across the open-flange sides will be a quick, economical solution. In any case, assuming the column has equal unbraced lengths for each axis, an approximation of a hollow tube will create the greatest capacity with the least material. Mies van der Rohe's famous cruciform columns (Fig. 11.1) made from I or W sections and structural "T's" (in reality, one half of an I or W shape) and the Foster Associates spaced column at the Sainsbury Center for the Arts are good examples of the tube analogy. Although they have internal elements (the webs of the sections), these constitute only a small proportion of the column's material and the composite is still quite efficient.

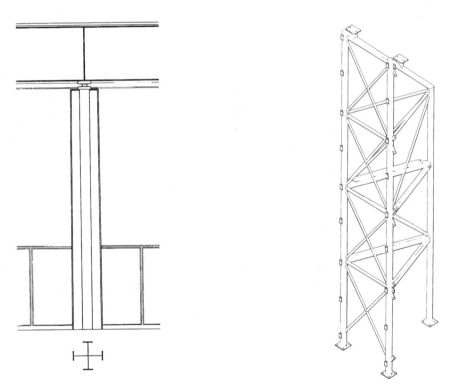

Fig 11.1 Typical Miesian built-up column and Sainsbury Center for the Arts spaced column

11.2 Design

Initially, it might seem that the design of built-up sections is an exercise in trial and error, but a little logical reasoning and the use of some of the design tables should allow us to get a close approximation on the first try.

Example 11.1

The adjacent column is 11ft-3in high and must carry a load of 73k.
Assume R = 1 and F_y = 36ksi

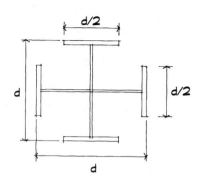

The AISC Manual tables (p.3-35 through 3-52) contain the
allowable loads on round and rectangular tube columns. Although
these tables are for a higher F_y than we're using, they will at
least give us an idea of where to start.

In the AISC Manual (p. 3-42 and 3-43), we find that for an
effective length of KL = 11ft, and a load of 73k, the potential sections are as follows:

Shape	Thickness	Load Capacity	Weight
4x4	5/16in = 0.3125in	72k	14.83plf
6x6	3/16in = 0.1875in	93k	14.53plf

Note that even here less material more advantageously placed will carry significantly more
load.

If we wish to use an I or W section to provide the same material at the same location, we need
to select a section that has a flange area approximately equal to the material provided by one of
the faces, or "flanges," of the tube. The other consideration is one of constructional/architectural
requirements. Obviously, selecting a 6in deep section with a 6in wide flange will not yield a
cruciform section; you'll end up with a square tube. Adequate space will be required between the
ends of the flanges to provide access for welding the sections together and to apply finishes. A
4in wide flange would be a starting point. The flange would have the same area (+ or -) as the
tube, so we might look for 6in x 0.1875in = 1.125in² This would equate to a 0.28in thick flange,
4in wide (1.125in²/4in = 0.28in). Since we're just looking for an approximation, anything around
.28in will be acceptable. You'll want to remember that these comparisons are for 46ksi steel in
tubes, whereas we're only using 36ksi steel, but at the same time, our configuration will have
some additional steel due to the webs of our crossed sections, so we only have a good guess.

Looking in the AISC Manual tables (p. 1-32), we find:

Shape	Flange b_f	Flange t	A	I_x	I_y
W6X16	4.03in	0.405in	4.74in²	32.1in⁴	4.43in⁴
W6X12	4.00in	0.280in	3.55in²	22.1in⁴	2.99in⁴
W6X9	3.94in	0.215in	2.68in²	16.4in⁴	2.19in⁴

(Although these values for A, I_x, and I_y are not necessary at this point, we'll be using them for the
next calculations; so we save a little time later if we record them now.)

Geometrically, they could all work, but the W6X16 provides an excess of material, whereas the
W6X12 and W6X9 are reasonable starts. We'll try W6X12.

To calculate r for the cruciform section, we use the parallel-axis theorem: $I = (I_o + Ad^2)$.

We could proceed with the calculation by accumulating the values for all parts, but a quicker method is to use BIG parts.

This will allow us to calculate I for the total section by simply adding I_x and I_y values.

$$I_{revised} = 22.1in^4 + 2.99in^4 = 25.1in^4$$

$$r = \sqrt{\frac{25,1in^4}{2 \times 3.55in^2}} = 1.88in$$

Since the cruciform section minimum r is twice that of r_y of an individual W6X12, the slenderness ratio is only one half as high. This means we're using the material in the section much more efficiently and can develop a higher design stress.

KL/r = 1.0(11.25ft)(12in/ft)/1.88in = 71.8

We now look at the allowable stress tables (AISC Manual, p. 3-16, A36 steel).

F_a can be interpolated for the 71.8 value to get a precise number, but to get a first look, we can just use a conservative value associated with 71.

   KL/r = 71, and therefore F_a = 16.33ksi

The approximate capacity using $F_a = P/A$, $P = F_a \times A$:

   P = (16.33ksi) × 2(3.55in) = 115.94k >> 73k,

   so it's probably worth trying a smaller section.

By using the W6X9 following the same procedure:

W6X9	Cruciform column section
A = 2.68in²	A = 2(2.68in²) = 5.36in²
I_x = 16.40in⁴	$I = I_x + I_y$ = 18.59in⁴
I_y = 2.19in⁴	$r = \sqrt{\dfrac{I}{A}} = 1.86in$

   KL/r = 11.25ft(12in /ft)/1.86in.= 72.6

A quick look says at KL/r = 73, and therefore F_a = 16.12ksi

   $P = F_a \times A$ = 16.12ksi × 5.36in² = 86.40k > 73k,

      we're done!

If we wanted to interpolate the exact F_a value, we would do this:

$KL/r = 72$, and therefore $F_a = 16.22$ksi

$KL/r = 72.6$, no value for $F_a = $?

$KL/r = 73$, and therefore $F_a = 16.12$ksi

$Kl/r = 72.6$ is 60% of the way between 72 and 73, so the change in value will be 60% of the difference:

$$16.22\text{ksi} - 16.12\text{ksi} = .10\text{ksi}$$

$$0.10\text{ksi} \times 0.6 = .06\text{ksi}$$

Therefore, at $KL/r = 72.6$, $F_a = 16.22\text{ksi} - 0.06\text{ksi} = 16.16\text{ksi}$

and the precise load capacity is

$$P = F_a \times A = 16.16\text{ksi} \times 5.36\text{in}^2 = 86.62\text{k}$$

Example 11.2

If architectural considerations were not a factor, for example, a hidden column, we might just reinforce a W shape. Note that this will probably be more expensive than simply using the tube. Labor is expensive! Unfortunately, it may be necessary in a retrofit situation.

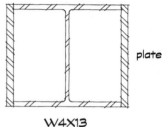

plate

W4X13

To design an alternative to the column in Example 11.1, a 4in section that would approximate the 4in tube might be a W4X13. We could then weld plates on the open sides, resulting in a tube equivalent.

The 4in tube section has 5/16in thick walls, so the 4in plate sections that we add should approximate this dimension (5/16in = 0.3125in) more or less, depending on the flange thickness of the 4in W section we use.

Since the W4X13 has flanges 0.345in thick, we might try a slightly thinner side plate. (We would have quite a bit of extra area with the thicker flanges and the web.) You'll note in the figure that the plates are slightly farther from the axis, so they'll have a larger I value.

W4X13

$A = 3.83\text{in}^2$

$I_x = 11.3\text{in}^4$ $I_y = 3.86\text{in}^4$

We calculate I_x and I_y for the combined shape:

$I_x = I_x \text{(W4x13)} + I_x \text{(plates)}$

$$I_x = 11.3\text{in}^4 + 2\left(\frac{bh^3}{12}\right) = 11.3\text{in}^4 + 2\left[\frac{\left(.25\text{in}(4\text{in})^3\right)}{12}\right] = 13.97\text{in}^4$$

$I_y = I_y$ (W4X13) $+ I_y$ (plates)

I_y (W4X13) $= 3.86\text{in}^4$

I_y (plates) $= (I_o + Ad^2)$

$I_o = 0.0\text{in}$ (for all practical purposes)

$A = 4\text{in} \times .25\text{in} = 1.0\text{in}^2$

$d = 4.06\text{in}/2 + .25\text{in}/2 = 2.15\text{in}$

$I_y = 3.86\text{in}^4 + 2[0.0\text{in}^4 + 1.0\text{in}^2 (2.15\text{in})^2] = 13.10\text{in}^4$

I_y is the controlling axis and will be used in KL/r calculations:

$$r_y = \sqrt{\frac{13.10\text{in}^4}{3.83\text{in}^2 + 2(.25\text{in} \times 4\text{in})}} = 1.499\text{in} = 1.5\text{iin}$$

KL/r $= 1.0(11.25\text{ft} \times 12\text{in/ft})/1.5\text{in} = 90$

KL/r $= 90$, therefore $F_a = 14.20\text{ksi}$

The allowable load capacity of the built-up section is

$P_a = F_a \times A = 14.20\text{ksi} \times 5.83\text{in}^2 = 82.78\text{k}, > 73\text{k}$

Finis!

Problem 11.1

Design a column 12ft tall, with pinned ends, made with two channels (C shapes), back to back, and with 36ksi steel. The two channels are to be adequately connected with spacers at regular intervals to work as a single column. The axial load is 80k.

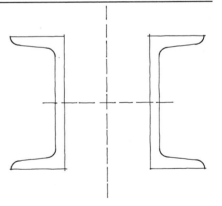

Solution:

Compare with available standard shapes:

5in × 5in × 1/4in; $P_a = 87\text{k}$ 4in × 4in × 3/8in; $P_a = 74\text{k}$

Try a 5in channel, for example, C5X9:

$$r_x = \sqrt{\frac{I}{A}} = \sqrt{\frac{8.9\text{in}^4}{2.64\text{in}^2}} = 1.83\text{in}$$

$KL_x/r_x = 144\text{in}/1.83\text{in} = 78.7 \ldots\ldots F_a = 15.5\text{ksi}$

$P_{ax} = 15.5\text{ksi}(2 \times 2.64\text{in}^2) = 81.8\text{k} > 80\text{k}$, OK for the X-X axis; now we can make the Y-Y axis work too.

If we make r_y the same (i.e., $I_x = I_y$), then $Pa_y = Pa_x$ and all will be swell. So

$I_y = I_x = 2(8.9in^4) = 17.8in^4$, (and we can also write:

$I_y = 2(I_o + Ad^2) = 17.8in^4$; solving for d:

$$Ad^2 = \frac{I_y}{2} - I_o$$

$$2.64in^2 d^2 = \frac{17.8in^4}{2} - 0.632in^4$$

$$2.64in^2 d^2 = 8.268in^4$$

$$d = \sqrt{\frac{8.268in^4}{2.64in^2}} = 1.77in$$

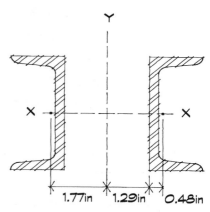

d is the distance of the centroid of each channel from the Y-Y axis that gives $I_y = I_x$.

The minimum spacing between the channels back, as you can see from the figure, is

$s = 2(d - x) = 2(1.77in - 0.48in) = 2.58in$, make it 3in for convenience of fabrication, so d > 1.77in, and without calculations, we know now that the X-X axis controls (weak axis); thus $P_a = 81.8k$, as previously calculated!

Problem 11.2

A 30ft column is fabricated with two W30X148 welded together, as in the illustration. Find the allowable axial load in compression with 36ksi steel.

Solution:

X-X axis: $KL_x/r_x = 1.0(30ft)(12in/ft)/12.4in = 29$

Y-Y axis: $I_y = 2(I_{oy} + Ad^2) = 2[227in^4 + 43.5in^2(10.5in/2)^2]$

Y-Y axis: $I_y = 2852in^4$

$$r_y = \sqrt{\frac{2852}{2(43.5)}} = 5,74 \le 12.4 \quad \text{so Y-Y controls, as:}$$

$KL_y/r_y = 1.0(30ft)(12in/ft)/5.72in = 63$, therefore $F_a = 17.14$

$P_a = 17.14ksi(2 \times 43.5in^2) = 1491k$

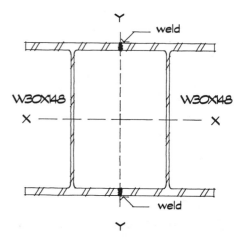

Problem 11.3

You are working at Foster Associates, London, and Norman asks you to design a fabricated column with four pipes, spaced 14in, 36ksi steel, 45ft tall, with an axial load of 450k. Find the adequate column dimension and pipe size.

Solution:

(Note that this is one of several possible solutions depending on the size of the individual pipes selected.)

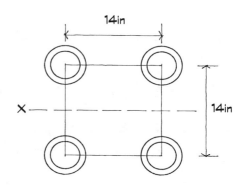

For these end conditions: K = 0.8, and KL = 0.8(45ft) = 36ft

Comparable square tube: 14in x 14in x 1/2in, A = 26.4in², P_a = 482k > 450k

Note that tubes are 46ksi steel, so a 36ksi column with the same section properties would carry a smaller load.

Equivalent pipe area = 26.4in²/4 = 6.6in²

Try with extrastrong 6in pipes spaced 14in:

A = 4(8.40in²)

I = 4[40.5in⁴ + (8.40in²)(7in)²] = 1808in⁴

$$r = \sqrt{\frac{1808in^4}{4(8.40in^2)}} = 7.33in$$

KL/r =1.0(36ft)(12in/ft))/7.33in = 59 < 200 OK

F_a = 17.53ksi

P_a = 17.53ksi(4x8.40in²) = 589k > 450k OK

Norman will complain that the column is overdesigned, so you can try with a smaller overall size or a smaller/lighter pipe section with this same column size.

Problem 11.4

A square column is built up with four 2-1/2inx2-1/2inx3/16in angles. Assuming the 15 foot long angles are laced (attached together) to act as a unit, and the steel is 36ksi, what axial load would the column carry?

At what intervals would laces (diagonal bracing between angles, as in the illustration) be provided to ensure that the angles act as a unit?

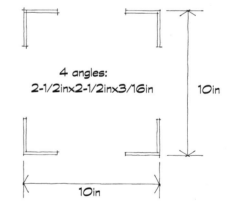

4 angles:
2-1/2inx2-1/2inx3/16in

10in

10in

Solution:

Angle section properties:

A = 0.902in²

I_x = 0.547in⁴

d = 5in - 0.694in = 4.306in

Column properties:

I = 4(I_o + Ad²) = 4[0.547in⁴ + 0.902in²x(4.306in)²]
  = 69.01in⁴

A = 4(0.902in²)= 3.608in²

$$r = \sqrt{\frac{69.01in^4}{3.608in^2}} = 4.37in$$

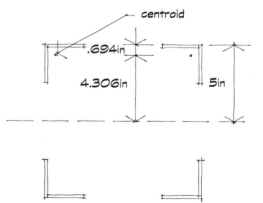

centroid

.694in

4.306in

5in

KL/r = 1.0(15ft)(12in/ft)/4.37in = 41.1, therefore F_a = 19.1ksi

P_a = 19.1ksi(4x.90in²) = 68.9k

Since the load is axial, each angle must carry P = 68.9k/4 = 17.22k, thus the working stress is:
F_a = P/A = 17.22k/.902in² = 19.1ksi
as we should have guessed! So working the tables in reverse:

F_a = 19.1ksi, requires KL/r = 41.1 which solved for KL,
with K = 1.0:

L = 41.1(.495in)/1.0 = 20.3in

where r_z = 0.495in is the minimum radius of gyration of the angle, giving the maximum spacing for the laces. Use laces at 20in o.c.

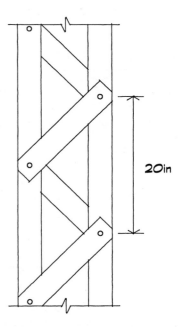

20in

METRIC VERSIONS OF ALL EXAMPLES AND PROBLEMS:

Example M11.1

The adjacent column is 3430mm high and must carry a load of 325kN.

Assume R = 1.0

F_y = 250MPa

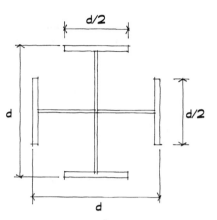

The AISC Manual tables (p.3-35 through 3-52) contain the allowable loads on round and rectangular tube columns. Although these tables are for a higher F_y than we're using, they will at least give us an idea of where to start.

In the AISC Manual (p. 3-42 and 3-43), we find that for an effective length of KL = 11ft = 3353mm, and a load of 325kN, the potential sections are as follows:

Shape	Thickness	Load capacity	Weight
4X4 = 102X102mm	7.9mm	320kN	22.1kg/m
6X6 = 152X152mm	4.8mm	414kN	21.6kg/m

Note that even here less material more advantageously placed will carry significantly more load.

If we wish to use an I or W section to provide the same material at the same location, we need to select a section that has a flange area approximately equal to the material provided by one of the faces, or "flanges," of the tube. The other consideration is one of constructional/architectural requirements. Obviously, selecting a 152mm deep section with a 152mm wide flange will not yield a cruciform section; you'll end up with a square tube. Adequate space will be required between the ends of the flanges to provide access for welding the sections together and to apply finishes. A 102mm wide flange would be a starting point. The flange would have the same area (+ or -) as the tube, so we might look for 152mm(4.8mm) = 730mm². This would equate to a 7.15mm thick flange, 102mm wide. (730mm²/102mm = 7.15mm) Since we're just looking for an approximation, anything around 7.5mm will be acceptable. You'll want to remember that these comparisons are for 317MPa steel in the tubes, whereas we're only using 250MPa steel, but at the same time, our configuration will have some additional steel due to the webs of our crossed sections, so we only have a good guess.

Looking in the AISC Manual tables (p. 1-32), we find:

Shape	Flange b_f	Flange t	A	I_x x10⁶	I_y x10⁶
W150X24	102mm	10.3mm	3060mm²	13.4mm⁴	1.83mm⁴
W150X18	102mm	7.1mm	2290mm²	9.19mm⁴	1.26mm⁴
W150X14	100mm	5.5mm	1730mm²	6.84mm⁴	0.912mm⁴

(Although these values for A, I_x, and I_y are not necessary at this point, we'll be using them for the next calculations; so we save a little time later if we record them now.)

Geometrically, they could all work, but the W150X24 provides an excess of material, whereas the W150X18 and W150X14 are reasonable starts. We'll try W150X18.

To calculate r for the cruciform section, we use the parallel- axis theorem: $I = (I_o + Ad^2)$.

We could proceed with the calculation by accumulating the values for all parts, but a quicker method is to use BIG parts.

This will allow us to calculate I for the total section by simply adding the I_x and I_y values.

$$I_{revised} = 9.19 \times 10 mm^4 + 1.26 \times 10^6 mm^4 = 10.45 \times 10^6 mm^4$$

$$r = \sqrt{\frac{10.45 \times 10^6 mm^4}{2(2290 mm^2)}} = 47.77mm$$

Since the cruciform section minimum r is twice that of r_y of an individual W150X18, the slenderness ratio is only one half as high. This means we're using the material in the section much more efficiently and can develop a higher design stress.

KL/r = 1.0(3430mm)/47.77mm = 71.8

We now look at the allowable stress tables (AISC Manual, p. 3-16, A36 steel). F_a can be interpolated for the 71.8 value to get a precise number, but to get a first look, we can just use a conservative value associated with 71.

KL/r = 71......... F_a = 112.58MPa

The approximate capacity using F_a = P/A,   P = F_aA

P = 112.58MPa(4580mm²) = 515,616N = 516kN >> 325kN,

so it's probably worth trying a smaller section.

By using the W150X14 following the same procedure:

W150X14   Cruciform column section

A = 1730mm²   A = 2(1730mm²) = 3460mm²

I_x = 6.84×10⁶mm⁴ $I = I_x + I_y = 7.752 \times 10^6 mm^4$

I_y = 0.912×10⁶mm⁴

$$r = \sqrt{\frac{I}{A}} = 47.33m$$

KL/r = 3430mm/47.33mm= 72.5

A quick look says at KL/r = 73, F_a = 111.13MPa

$P = F_a A = 111.13MPa(3460mm^2) = 384,514N = 384kN > 325kN,$

We're done!

If we wanted to interpolate the exact F_a value, we would do this:

$KL/r = 72$, therefore $F_a = 111.82MPa$

$KL/r = 72.5$, is not listed $F_a = ?$

$KL/r = 73$, therefore $F_a = 111.13MPa$

$KL/r = 72.6$ is 50% of the way between 72 and 73, so the change in value will be 60% of the difference:

$111.82MPa - 111.13MPa = 0.69MPa$

$0.69MPa(0.50) = 0.35MPa$

Therefore, at $KL/r = 72.5$, $F_a = 111.82MPa - 0.35MPa = 111.47MPa$

and the precise load capacity is

$P = F_a \times A = 111.47MPa(3460mm^2) = 385,686N = 386kN.$

Example M11.2

If architectural considerations were not a factor, for example, a hidden column, we might just reinforce a W shape. Note that this will probably be more expensive than simply using the tube. Labor is expensive! Unfortunately, it may be necessary in a retrofit situation.

To design an alternative to the column in Example M11.1, a 102mm section that would approximate the 102mm tube might be a W100X19. We could then weld plates on the open sides, resulting in a tube equivalent.

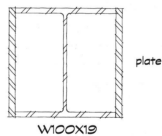

plate

W100X19

The 102mm tube section has 7.9mm thick walls, so the 102mm plate sections that we add should approximate this dimension (102mmx7.9mm) more or less, depending on the flange thickness of the nominal 100mm W section we use.

Since the W100X19 has flanges 8.8mm thick, we might try a slightly thinner side plate. (We would have quite a bit of extra area with the thicker flanges and the web.) You'll note in the figure that the plates are slightly farther from the axis, so they'll have a larger I value.

W100x19: $d = 106mm$, $b_f = 103mm$

$A = 2470mm^2$

$I_x = 4.75 \times 10^6 mm^4$ $I_y = 1.60 \times 10^6 mm^4$

We calculate I_x and I_y for the combined shape:

$I_x = I_x$ (W100X19) $+ I_x$ (plates)

$I_x = 4.75 \times 10^6 mm^4 + 2(bh^3/12) = 4.75 \times 10^6 mm^4 + 2 [(6mm(106mm)^3/12)] = 5.94 \times 10^6 mm^4$

$I_y = I_y$ (W100X19) + I_y (plates)

I_y (W100X19) = $1.60 \times 10^6 mm^4$

I_y (plates) = $(I_o + Ad^2)$

$I_o = 0.0$ in.

A= 6mm(103mm) = $618mm^2$

d = 103mm/2 + 6mm/2 = 54.5mm

$I_y = 1.60 \times 10^6 mm^4$ + $2[0.0 in^4 + 618 mm^2 (54.5mm)^2] = 5.38 \times 10^6 mm^4$

I_y is the controlling axis and will be used in KL/r calculations:

$$r = \sqrt{\frac{5.38 \times 10^6 mm4}{2470 + 2(636mm^2)}} = 37.92mm$$

KL/r = 1.0(3430mm)/37.92mm = 90

KL/r = 90, therefore F_a = 97.89MPa

The allowable load capacity of the built-up section, $P_a = F_a \times A$

A = $2470mm^2 + 2(636mm^2) = 3642mm^2$

Thus, $P_a = F_a A$ = 97.89MPa($3742mm^2$) = 366,304N = 363kN > 325kN.

Finis!

Problem M11.1

Design a column 3660mm tall, with pinned ends, made with two channels (C shapes), back to back, and with 250MPa steel. The two channels are to be adequately connected with spacers at regular intervals to work as a single column. The axial load is 356kN.

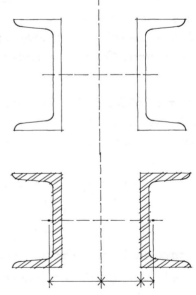

Solution:

Compare with L 127mm x 127mm x 6.4mm, P = 387kN

L 102mm x 102mm x 9.5mm, P = 329kN

Try 127mm channel e.g. C130X13:

$$r_x = \sqrt{\frac{I}{A}} = \sqrt{\frac{3.70 \times 10^6 mm^4}{1700mm^2}} = 46.65mm$$

KL_x/r_x = (3660mm)/(46.65mm) = 78.5 F_a = 107 MPa

P = 107MPa(2)(1700mm) = 363,783N = 364kN > 356kN

Use two C130X13, back to back, spaced 75mm, which give approximately $P_a = 364kN$.

Problem M11.2

A 9140mm column is fabricated with two W760X220s welded together as shown in the illustration. Find the allowable axial load in compression with 250MPa steel.

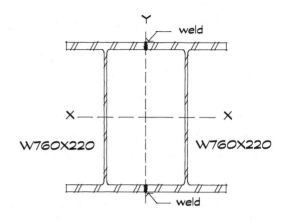

W760X220   W760X220

Solution:

X-X axis:

$KL_x/r_x = 1.0(9140mm)/(315mm) = 29$

Y-Y axis:

$I_y = 2[94.4\times10^6mm^4 + 28,100mm^2(266mm/2)^2]$

$\quad = 1,183\times10^6mm^4$

$r_y = \sqrt{\dfrac{1,183\times10^6mm^4}{2(28,100mm^2)}} = 145.1mm < 315mm$

$KL_y/r_y = 1.0(9140mm)/(145.1mm) = 63,\ F = 118.16MPa$

$P_a = 118.16MPa(2)(28,100mm^2) = 6640kN$

Problem M11.3

You are working at Foster Associates, London, and Norman asks you to design a fabricated column with four pipes spaced 356mm, 250MPa steel, 13.72m tall, with an axial load of 2000kN. Find the adequate column dimension and pipe size.

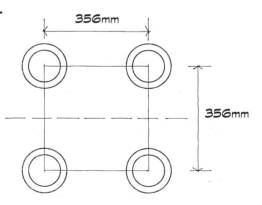

356mm

356mm

Solution:

$K = 0.8\ KL = 0.8(13.72m) = 10.98m$

Comparable square tube: 356mm X 356mm x 15.9mm

$A = 20,900kN > 2000kN$ OK!

Tubes are 317MPa > 250MPa

Equivalent pipe $A = (20,900mm^2)/4 = 5225mm^2$

Try: XStrong 152mm pipe spaced at 356mm

$A = 4(5,420mm^2)$

$I = 4[16.8mm^4 + 5,420mm^2(178mm)] = 687\times10^6mm^4$

$r = \sqrt{\dfrac{687\times10^6mm^4}{4(5420mm^2)}} = 77 < 200$

KL/r = 1.0(13,720mm)/(178mm) = 77 < 200 OK!

F = 108.2MPa

P = (108.2MPa)(4)(5420mm²) = 2345kN > 2000kN OK!

Use four Extra Strong 152mm pipes giving a total of 2345kN

Problem M11.4

A square column, 4.6m long, is built up with four 64mm ×
64mm × 4.8mm angles. Assuming the angles are laced (attached together) to act as a unit, and the steel is 250MPa,
what axial load would the column carry?

At what intervals would laces (diagonal bracing between
angles, as in the illustration) be provided to ensure that the
angles act as a unit?

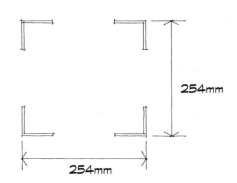

Solution:

Angle properties:

$\quad$ A = 582mm²

$\quad$ I = 0.233×10⁶mm⁴

$\quad$ d = 127mm - 17.8mm = 109.2mm

I = 4[0.233×10⁶mm⁴ + 582mm²(109.2mm)²]

$\quad$ = 28.7×10⁶mm⁴

A = 4(582mm²) = 2328mm²

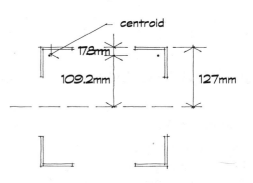

$$r = \sqrt{\frac{28.7 \times 10^6 mm^4}{2328 mm^2}} = 111mm$$

KL/r = 1.0(4,600mm)/(111mm) = 41.4

F = 136.6MPa

P = (136.6MPa)(2328mm²) = 318kN

Each angle P = (318kN)/4 = 79.5kN

F_a = P/A = (79.5kN)/(582mm²) = 136.6MPa

F = 136.6MPa, therefore, KL/r = 41.4

Solved for KL:

L = 41.4(12.6mm)/1.0 = 518mm

where 12.6mm = r_z

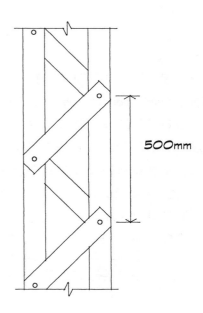

Maximum interval of lacing = 518mm.

Use laces @ 500mm o.c.

12 TENSION MEMBERS

12.1 Types of tension members

Fig. 12.1 Pompidou (Beaubourg) Center, Paris

Steel structural systems almost always include tension members in the form of frame bracing if not after completion, during construction. Bracing is not always visible in the completed building, particularly bracing for floor diaphragms, but in some examples, it is exploited as an architectural feature of the elevation. Some well-known examples are the Pompidou (Beaubourg) Center in Paris (Fig. 12.1), Terminal 1 of O'Hare Airport and the John Hancock Tower, both in Chicago.

In the Pompidou, the cross-bracing rods are very slim, so that when the lateral loads act in one direction, the rod in compression will slightly bend and will not offer any resistance (Fig. 12.2). The load therefore will be taken by the diagonal in tension BC. BC is then de-signed for the calculated tension, and diagonal AD will be designed similar to (equal in this case) BC for loads acting in the opposite direction.

Suspension systems use tension members to support beams of floors and roofs. Tension members may also be necessary to brace columns or compression mem-bers (Fig. 12.3). Multistory buildings can be built with floors suspended from a trussed structure located

Fig 12.2 Tensile bracing at Pompidou Center

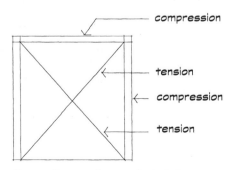

Force diagram for tensile bracing

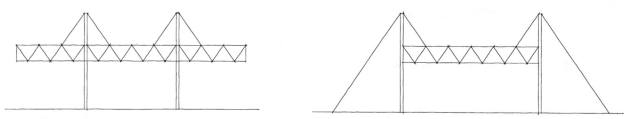

Fig. 12.3 Suspension members providing support only for a long-span structure or providing both support and lateral bracing for long-span structures

above roof level or within the wall surfaces of the facades. Cable systems can be used either to support a roofing membrane or a series of floors (Fig. 12.4). Cable-supported structures are very efficient (structurally), but end connections are expensive and somewhat difficult to achieve. An approach similar to that used for your shoe strings (a tensile system) is efficient because it uses a very long "cable," laced back and forth, with only two terminators. Cables are obviously a bit harder to tie! The most common uses of tension members are probably in truss design and bracing.

Fig. 12.4 Federal Reserve Bank, Minneapolis, Minn. under construction

Roof systems often require tension members; the oldest examples are the iron chains used to resist tensile forces at the base of masonry domes and arches in Fourteenth-century Italy. These were necessary to balance the horizontal thrust at the spring line of the arch at a time when columns and walls were becoming more slender. (Remember, a dome is an arch rotated 360°.) Frank Lloyd Wright used a chain "ring" to support the octagonal roof above his first studio in Oak Park, Illinois. A system of umbrellas or trees has been designed by Norman Foster for Stanstead Airport, London (Fig. 12.5). The size of the members reflect the type of compressive or tensile stress. Even your home likely uses a three-hinged arch consisting of roof "rafters" and a ceiling joist tensile "tie."

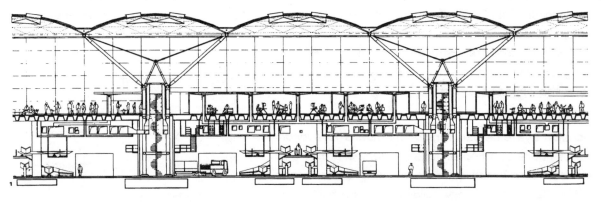

Fig. 12.5 Stanstead Airport terminal, London.

A three-hinged arch supported on walls or columns is in fact an elementary truss composed of two compression chords and one tension rod (chord). Many other types of elementary trusses, such as the one conceived by Foster, are common (Fig. 12.6). Trusses will be discussed in Chapter 14.

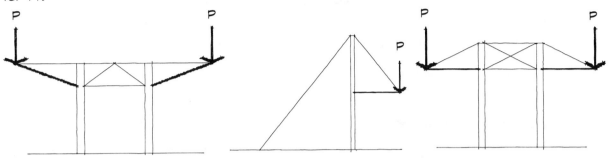

Fig. 12.6 Some simple tension schemes

12.2 Basic design

Tension members are calculated with the direct stress formula giving tensile stress:

$$F_a = P/A$$

where P is the tensile load, and A is the **net** area of the tension member; this, solved for A_n, becomes the design formula where stress is replaced by the allowable tensile stress F_a:

$$A_n = P/F_a$$

For members subject to pure axial tension (forces acting through the centroidal axis), the allowable stress is (AISC Manual, p. 5-40)

$$F_a = 0.6F_y \text{ for the gross section}$$
$$F_a = 0.5F_u \text{ for the net section}$$

where F_y is the yield stress for the selected steel, and F_u is the ultimate stress. F_u values are given in the AISC Manual (Table 1, p. 1-7).

The net section is the area of material remaining at any point of reduced cross-section, such as at the threads on a bolt or rod or around holes in a plate.

Tension members are used in different types of structural systems: three-hinged arches, trusses, and bracing of columns are the most common examples (Fig. 12.7). Various shapes are used, ranging from plates and angles to solid round bars (rods). In some cases, tension members may be subject to additional stresses: bending and/or thermal stresses. The ceiling joist in a "three-hinged arch" residential roof system has both tension, as part of the three-hinged-arch action, and bending, as a result of ceiling finishes and required LL.

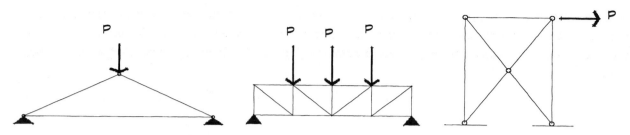

Fig. 12.7 Common uses of tensile members

12.3 Connection of tension members

Tension plates or platelike sections are connected at their ends with welds or bolts. Round tension bars (rods) have threaded ends connected by turnbuckles or clevises. Bolt holes and threaded sections reduce the cross-sectional area of the member and that reduction must be accounted for. The AISC Manual (p. 5-40) specifies that the allowable stress may be increased at these limited length locations.

$$F_a = 0.5F_u \text{ (on the net cross-section)}$$

For A36 steel, this means that $F_a = 0.6(36\text{ksi}) = 21.6\text{ksi}$ for the gross area, and $F_a = 0.5(58\text{ksi}) = 29\text{ksi}$ for the net (threaded or bolted) "tensile stress" area in tension. At the connection, the allowable stress is closer to the yield stress (80% of F_y, therefore still within the elastic limit); this allows for slightly larger deformations potentially to take place, which is generally tolerable over a **short length** of the connection part. It would be inappropriate to use this value if the entire rod was threaded, since the net section and the gross section would be the same. In this case, the more conservative gross value (for the long threaded cross-section) should be used to determine load capacity.

Example 12.1

A rectangular plate of A36 steel, 1in thick, must carry a tensile load of 150k. Its gross cross-sectional area must be

$$A_g = 150\text{k}/0.6(36\text{ksi}) = 6.94\text{in}^2$$

Therefore, its required width would be $B = (6.94\text{in}^2)/1.0\text{in} = 6.94\text{in}$ or 7in.

If this same plate was connected to a member using a single 2in-diameter bolt with a 2-1/16in or 2.06in hole, the plate would need to be

$$A_n = (150\text{k})/0.5(58\text{ksi}) = 5.17\text{in}^2$$

and with a 1in thickness, it would be 5.17in wide. When we add the required bolt hole to this dimension, we get 5.17in + 2.06in = 7.23in

This means that the plate could be 7in at locations where there are no holes and 7.25in where there are holes. This explains the "dog-bone-shaped" tie members we sometimes see in old truss bridges. The same principles and requirements exist today, but labor costs make it cheaper to simply build the plate using the largest dimension along its entire length and call it done.

Example 12.2

The bracing of a row of columns (pinned at both ends) shown below must resist a wind load, transferred from the end wall to column AC, of 900plf. Design the bracing with rod and turn-buckle.

$P = 900plf\ (18ft/2) = 8100lb$

Length of diagonal bracing: $\sqrt{(18ft)^2 + (20ft)^2} = 26.9ft$

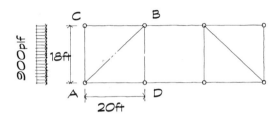

Equilibrium of joint B:

$$\frac{P_{ab}}{P_{cb}} = \frac{26.9ft}{20.0ft}$$

$$\frac{P_{ab}}{8.1k} = \frac{26.9ft}{20.0ft}$$

$$P_{ab} = \frac{(8.1k)(26.9ft)}{20.0ft} = 10.9k$$

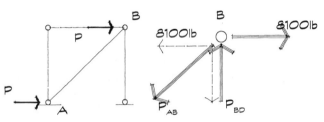

Required gross area: $A_g = (10.9k)/0.6(36ksi) = 0.5in^2$

Required net area: (tensile stress) $A_n = (10.9k)/0.5(58ksi) = 0.38in^2$

From the AISC Manual (p. 4-147), a rod with diameter D = 7/8in has

Gross area: $A_g = A_D = 0.601in^2$

Net area: $A_n = 0.462in^2$ ("tensile stress area")

for turnbuckles safe working loads are listed in the AISC Manual (p. 4-149):

A 1-1/8in turnbuckle has a safe load of 11.6k

A 1-1/8in "upset" end (simply a larger section of threaded rod welded to the base rod) will have to be welded to the ends of the 7/8in rod for the turnbuckle connection.

12.4 Suspension structures

Long-span structures are becoming more commonplace and tensile structures are being employed more frequently to respond to this need. They are not really new: the Coliseum (the one in Rome) at one time had a tensile "roof" to provide shade. We can look at suspension bridges for analogies; large cables are supported by towers/pylons and the roadway is suspended from these cables. San Francisco's Golden Gate Bridge is a classic example of this system. In buildings, the cables typically support a roof of a single story, although a potentially tall single story building. Some well-known examples of these lightweight tensile roofs are New York's Madison Square Garden and Washington's Dulles Airport. The Garden is a circular building with a diameter of about 420ft [127m], spanned by 48 radial cables. The cables are anchored to an outer concrete "compression" ring and are tied at the center of the roof by a steel tensile ring. If you can visualize a horizontal bicycle wheel, you'll have a good idea of how it works. You might also appreciate the use of materials reflecting their classic properties: concrete for the compression ring and steel for the tension ring. In Dulles Airport, the heavy concrete support towers lean outward against the tension of the cables to provide stability. Their geometric configuration maximizes their "overturning" moment inducing tension in the roof cables

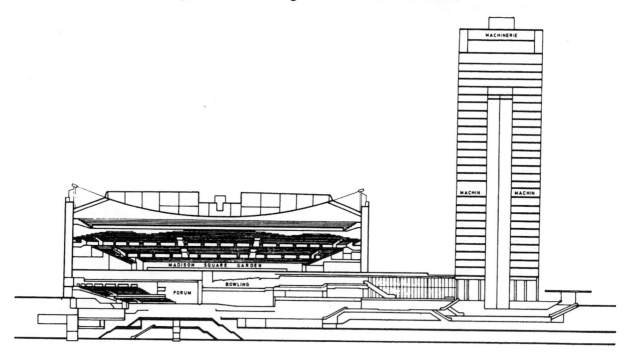

Fig. 12.8 Madison Square Garden tensile roof

The bridge concept has been applied in a more creative way to support the floors of the Federal Reserve Bank building in Minneapolis. Massive towers at each end of the building, containing vertical circulation and mechanical systems, provide vertical support with a deep truss spanning between them to resist the inward horizontal force of the suspended tensile cable. Floors are suspended from both the tensile cable and from the truss.

The configuration of the cables of a bridge or a roof is determined by the loads. Unlike an arch, which can maintain a specific shape, a cable is a flexible structure subjected to pure tension. Since tensile forces follow the line of least resistance, the load will determine the ideal shape. This can be seen by taking a string, holding it between your hands, and letting it sag. Now place a load along the string at any point and you will see the system adjust itself to the most efficient form. Gaudi's Colonia Guell Chapel and Saarinen's St. Louis Arch are both basically compression structures whose geometries were investigated using an inverted tensile configuration to optimize their final form.

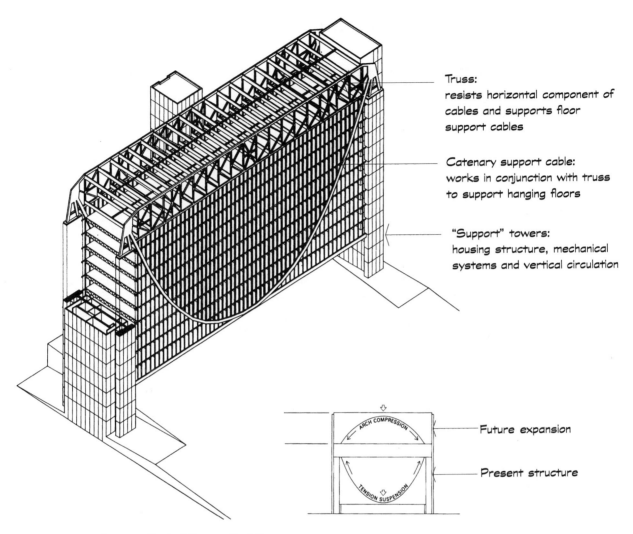

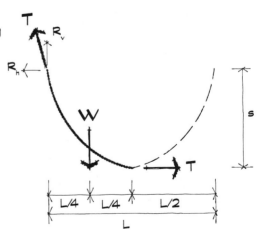

Truss:
resists horizontal component of
cables and supports floor
support cables

Catenary support cable:
works in conjunction with truss
to support hanging floors

"Support" towers:
housing structure, mechanical
systems and vertical circulation

Future expansion

Present structure

Fig. 12.9 Federal Reserve Bank, Minneapolis, Minn.

The stress in a cable is always parallel to the tangent to the axis of the cable and is consequently in pure tension. **Cables are not designed to support bending stresses!** A single cable is a statically determinate structure and we can find its tensile force at any point (which will be its tension at every point) given a particular load configuration. In Madison Square Garden, each cable can be represented by the diagram in Fig. 12.10. From simple equations of equilibrium, we can find the components of the support reaction R:

Horizontal reaction: $R_h = T$

Vertical reaction: $R_v = W$

Fig. 12.10 Cable diagram for the "Garden"

where W is the total roof load on the half cable.

Summation of moments about the high end of the cable results in

$$T_h(s) = W\left(\frac{1}{2}\right)\frac{L}{2}$$

$$T_h = \frac{W}{s}\left(\frac{1}{2}\right)\frac{L}{2}$$

where L = total span of the cable

      s = cable sag (vertical dimension) at the point of tangency

The answer we are looking for is tension T in the cable, which also equals the horizontal force component that must be resisted at the point of support.

Example 12.3

Find the size of the cables required to support the roof of Madison Square Garden using a span of 420ft, a point load P of 145k at the 1/3 point, and a sag of 32ft. (Since the plan is radial, the load will be roughly the area supported by a triangle.)
Use 100ksi steel.

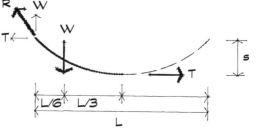

Summing the moments about one end of the cable:

$$T_h(s) = W\frac{L}{6} \text{ or,}$$

$$T_h = \frac{WL}{6s} = \frac{145k(420ft)}{6(32ft)} = 317k$$

The resultant force in the cable would be:

$$T = \sqrt{(317k)^2 + (145k)^2} = 348k$$

We can now design the cable (or any other tensile member) to carry 348k.
Required cable gross area:

$$A_g = \frac{P}{F_t} = \frac{T}{F_t} = \frac{348k}{0.6(100ksi)} = 5.8in^2$$

From the AISC Manual (Threaded Fasteners, p. 4-147), we can find the size member that would have a gross area of 5.8in² (or more). This table is obviously not for a "cable," but the cross-sectional area of a circle remains the same regardless of its composition.

Use a 2-3/4in diameter cable. This calculation would also need to consider the type of connection system used, but the basic cable design is complete.

Although cable structures use materials very efficiently, their lightness can also become a problem. If you look at classic tensile structures by Frei Otto or even by a spider, you'll see that they not only have cables that define surface, but have "tie downs" that keep the lightweight structure from vibrating or "fluttering" in the wind. If you transport your mattress to school and cover it with a sheet or tarp to keep it clean, you may have noticed that if the sheet is not well secured, it will "flutter and flap" in the wind until it tears itself to pieces. The mattress is strong enough to resist basic loads, but the dynamic movement caused by wind can destroy it. In a cable system, these stabilizers are frequently statically indeterminate.

Alternative methods of resisting these forces include using mass on the surface to create enough load to stabilize the structure. Dulles Airport uses this technique. You can also see it on power lines (a mass, or series of masses, located along the cable to keep it from vibrating). You could also use two tensile systems braced against each other. Your bicycle wheel uses this system to maintain its shape under 360 different loading configurations. This system is used architecturally in the Municipal Auditorium in Utica, New York. Yet another technique is to preload the cables, as we might find in a pneumatic structure with a system of cables that are held in place by internal pressure. The Indianapolis RCA Dome uses this technique.

If you want to really understand tensile structures, sit for a while in a hammock and contemplate the strange conditions of structure and architecture.

Problem 12.1

The tie rod of a three-hinged arch has to support the thrust of 1.67k, due to gravity roof loads. Calculate the size of the steel rod using 36ksi steel and select an appropriate turnbuckle. If the length of the rod is 48 ft, what will be its elongation under full load?

Solution:

F_n = 0.5(58ksi) = 29ksi net (tensile stress) area

F_g = 0.6(36ksi) = 21.6ksi gross area

Required A_n = P/F_n = 1.67k/29ksi = 0.057in^2 net area

Required A_g = P/F_g = 1.67k/21.6ksi = 0.077in^2 gross area

A 3/8in diameter rod has a net (tensile stress) A_n = 0.078in^2, A_g = 0.11in^2 gross area: OK!
A 1/2in turnbuckle has safe working load = 2.2k > 1.67k; use an upset end on a 3/8in rod or use a 1/2in rod over the entire distance. The addition of the upset ends to a 3/8in rod will offset the cost savings of the smaller rod, so the best alternative is probably to use a 1/2in rod over the entire length.

The expected elongation in the rod can be determined by using the stress-strain diagram and remembering that E, the modulus of elasticity of any material, is the ratio of the stress to the strain within the elastic limit.

$$E = \frac{F}{\delta} = \frac{Stress}{Strain} = \frac{\frac{P}{A}}{\frac{\Delta L}{L}} \text{ , or } \frac{\Delta L}{L} = \frac{P}{AE} \text{, or } \Delta L = \frac{PL}{AE}$$

Deformation of rod:

$$\Delta L = \frac{1.67k}{0.20in(29,000ksi)}[48ft(12in/ft)] = 0.165in$$

Problem 12.2

A tension member of a truss has a tension of 43k and is made with a pair of equal leg angles back to back. There is an 11/16in diameter bolt hole in the vertical legs of both angles (to accommodate a 5/8in bolt) to connect to a hanger and avoid excessive sag. Find the appropriate angle size using A441 grade 50 steel.

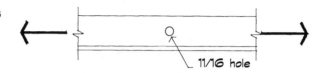

11/16 hole

Solution:

A_g = 43k/0.6(50ksi) = 1.43in² gross area

A_n = 43k/0.5(70ksi) = 1.23in² net area

A pair of L - 2inx2inx3/16in has A = 1.43in²

(Note that a 5/8in bolt has a hole 1/16in larger or 11/16in.)

Area of hole = 3/16inx11/16in = 0.13in²

Total area required: A = 1.23in²+ 0.13in² = 1.36in² < gross area. OK!

Use L - 2inx2inx3/16in.

Problem 12.3

A cross brace is built with double L - 5in x 3in x 3/8in angles that have to support a tension of 120k. Is this section adequate? The angles are adequately welded at the center of the cross by an octagonal plate, as shown. Find the required thickness of the center plate with 36ksi steel. You may assume that all loads are pure tension and no eccentricity exists in the plate.

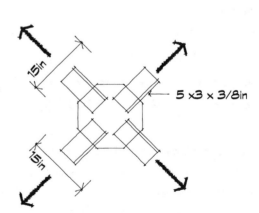

5 x3 x 3/8in

Solution:

Required A = (120k)/0.6(36ksi) = 5.55in²

The given angle has A = 5.86in² OK!

At the end of the angles, the plate has a width of approximately 2in, so A = (2.0in)(t) and the minimum thickness is t = (5.55in²)/8.0in = 0.69in.   Use 11/16in plate (t = 0.69in).

METRIC VERSIONS OF ALL EXAMPLES AND PROBLEMS

Example M12.1

A rectangular plate of 250MPa steel, 25mm thick, must carry a tensile load of 667kN. Its gross cross-sectional area must be

$$A_g = 667000N/.6(250MPa) = 4447mm^2$$

Therefore, its required width would be $B = (4447mm^2)/(25mm) = 178mm$

If this same plate was connected to a member using a single 51mm diameter bolt with a 52mm hole, the plate would need to be

$$A_n = (667000N)/.5(400MPa) = 3335mm^2$$

and with a 25mm thickness, it would be 133mm wide. When we add the required bolt hole to this dimension, we get 133mm + 52mm = 185mm.

This means that the plate could be 178mm at locations where there are no holes and 185mm where there are holes. This explains the "dog-bone-shaped" tie members we sometimes see in old truss bridges. The same principles and requirements exist today, but labor costs make it cheaper to simply build the plate using the largest dimension along its entire length and call it done.

Example M12.2

The bracing of a row of columns (pinned at both ends) shown below must resist a wind load, transferred from the end wall to column AC, of 13.13kN/m. Design the bracing with rod and turnbuckle.

P = 13.13kN/m(5.486m/2) = 36kN

Length of diagonal bracing $= \sqrt{(5.486m)^2 + (6.096m)^2} = 8.201m$

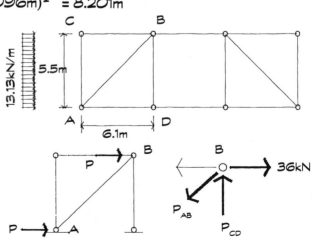

Equilibrium of joint B:

$$\frac{P_{ab}}{P_{cb}} = \frac{8.201m}{6.096m}$$

$$\frac{P_{ab}}{36kN} = \frac{8.201m}{6.096m}$$

$$P_{ab} = \frac{36kN(8.201m)}{6.096m} = 48.43kN$$

Required gross area: $A_g = 48.43kN/0.6(250MPa) = 0.32mm^2$

Required net area (tensile stress): $A_n = (48.43kN)/0.5(400MPa) = 0.24mm^2$

From the AISC Manual (p. 4-147), a rod with diameter D = 22mm has

> Gross area: $A_g = A_D = 380mm^2$
>
> Net area: $A_n = 298mm^2$ ("tensile stress area")

for turnbuckles safe working loads are listed in the AISC Manual (p. 4-149):

> A 28mm turnbuckle has a safe load of 51.6kN.

A 28mm "upset" end (simply a larger section of threaded rod welded to the base rod) will have to be welded to the ends of the 22mm rod for the turnbuckle connection.

Example M12.3

Find the size of the cables required to support the roof of Madison Square Garden using a span of 128m, a point load P of 645kN at the 1/3 point, and a sag of 9.8m. (Since the plan is radial, the load will be roughly the area supported by a triangle.) Use 690MPa steel.

Summing the moments about one end of the cable:

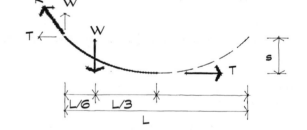

$$T_h(s) = W\frac{L}{6} \text{ or,}$$
$$T_h = \frac{WL}{6s} = \frac{645kN(128m)}{6(9.8m)} = 1404kN$$

The resultant force in the cable would be:

$$T = \sqrt{(1404kN)^2 + (645kN)^2} = 1545kN$$

We can now design the cable (or any other tensile member) to carry 1545kN.

Required cable gross area:

$$A_g = \frac{P}{F_t} = \frac{T}{F_t} = \frac{1545000kN}{0.6(690MPa)} = 37.37mm^2$$

From the AISC Manual (Threaded Fasteners, p. 4-147), we can find the size member that would have a gross area of 3737mm² (or more).

Use a 70mm diameter cable. This calculation would also need to consider the type of connection system used, but the basic cable design is complete.

Problem M12.1

The tie rod of a three-hinged arch has to support the thrust of 7.43kN, due to gravity roof loads. Calculate the size of the steel rod using 250MPa steel, and select an appropriate turnbuckle. If the length of the rod is 14.63m, what will be its elongation under full load?

Solution:

$F_n = 0.5(400\text{MPa}) = 200\text{MPa}$ (net cross-section)

$F_g = 0.6(250\text{MPa}) = 150\text{MPa}$ (gross cross-section)

Required:

$A_g = (7,430\text{N})/(200\text{MPa}) = 37.15\text{mm}^2$

$A_n = (7,430\text{N})/(150\text{MPa}) = 49.53\text{mm}^2$

9.5mm diameter rod has a net $A_n = 50.3\text{mm}^2$ and a gross $A_g = 71\text{mm}^2$

12.7mm turnbuckle load capacity = 9.27kN > 7.43kN

$$\Delta L = \frac{7,430\text{N}}{(126.6\text{mm}^2)(200,000\text{MPa})}(14,630\text{mm}) = 4.3\text{mm}$$

Deformation of the rod: $\Delta L = 4.3\text{mm}$

Problem M12.2

A tension member of a truss has a tension of 191kN and is made with a pair of equal leg angles back to back. There is an 17mm diameter bolt hole in the vertical legs of both angles (to accommodate a 16mm bolt) to connect to a hanger and avoid excessive sag. Find the appropriate angle size using A441 grade 345MPa steel .

Solution:

$A_g = (191,000\text{N})/0.6(345\text{MPa}) = 923\text{mm}^2$

$A_n = (191,000\text{N})/0.5(483\text{MPa}) = 791\text{mm}^2$

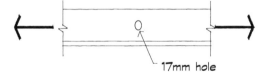

17mm hole

L - 51mm × 51mm s 4.8mm has an A = 922mm²

(a 16M bolt has a 17mm hole)

$A_{hole} = (4.8\text{mm})(17\text{mm}) = 81.6\text{mm}^2$

Required total area, $A_T = 791\text{mm}^2 + 81.6\text{mm}^2 = 873\text{mm}^2 < A$ OK!

Use a pair of L - 51mm × 51mm × 4.8mm.

Problem M12.3

A cross brace is built with single L- 127mm × 76mm × 9.5mm angles that have to support a tension of 534kN. Is this section adequate? The angles are adequately welded at the center of the cross by an octagonal plate, as shown. Find the required thickness of the center plate with 250MPa steel. You may assume that all loads are pure tension and no eccentricity exists in the plate.

Solution:

Required A = 534,000N/0.6(250MPa) = 3,560mm²

The given angle has an A = 3690mm² OK!

Plate width = 200mm, so A = 200mm(t)

t = 3560mm²/200mm = 17.8mm

Use an 18mm plate

13 COMBINED AXIAL LOADING AND BENDING

13.1 Basic considerations and procedure

Although the subject of combined axial loading and bending may seem to be beyond the scope (or interest?) of architects, this is a very common structural condition encountered in almost every building you design. The roof of your home, if sloped, has joists or rafters that have both bending, due to the loads acting directly on them, and axial compression loads from their part in the triangulation of the "three-hinged" arch of the roof system. (Even if you have roof trusses, the same structural condition exists.) The three-hinged arch is the most elementary of all trusses, consisting of only three basic members. The lower "chord," which you recognize as a ceiling, has both bending, from the ceiling construction and the minimum of 10psf LL the code requires, and tension to keep the side walls from being pushed out by the roof joist. The other common application of combined axial loading and bending is in stair stringers, those members that support the treads and span from level to level or landing to landing. Any member, and there are many, that has components of force acting at an angle to its axis will experience this dreaded combination.

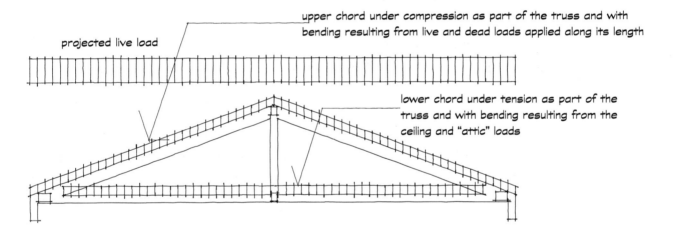

upper chord under compression as part of the truss and with bending resulting from live and dead loads applied along its length

projected live load

lower chord under tension as part of the truss and with bending resulting from the ceiling and "attic" loads

Fig. 13.1 Simple truss with the most complex combined axial loading combinations

An easy metaphor to allow us to understand how these systems work, as well as what sorts of limitations result from these combinations, would be that of an athlete. A decathlon sprinter might run a 100 meters in just under 10 seconds; that would be an excellent time. This same athlete might dead lift (just lift a weight off the ground) 450lb [200kg], again an excellent effort. Now just imagine how quickly that same athlete could run the 100 meters carrying the 450lb [200kg].

Contrary to the assertions of sports commentators, one can give only 100%, not 110% of one's resources. Structural members have the same limitations placed on them. If a column used 100% of its potential strength as an axially-loaded member, it has nothing left to resist any bending stresses. Consequently, combined axial loading and bending is simply a matter of allocating the total resources of a particular structural element in such a way as to ensure that those resources will not be exceeded. Just as we have previously!

The formula for combined stresses is

$$\frac{f_a}{F_a} + \frac{f_{b_x}}{F_{b_x}} + \frac{f_{b_y}}{F_{b_y}} \leq 1.0$$

The formula looks intimidating at first, but we'll see that it is just a straightforward combination of simple parts:

The expression f_a/F_a simply represents the percentage of the member's capacity that is being utilized as an axially loaded member. This could be written as f_t/F_t, if the member were subjected to tensile forces, or f_c/F_c, if the member were in compression.

In either case, the individual components of the solution are relatively simple:

f_t is the **actual tensile stress** = P/A.

F_t is the **allowable tensile stress** = $0.60F_y$.

f_c is the **actual compressive stress** = P/A.

F_c is the **allowable compressive stress** (a function of KL/r, found in the tables in the AISC Manual, p. 3-17).

But, if you remember that the total term represents the percentage of the member's capacity that is being utilized, you could also use the column design tables. By looking up the allowable capacity of a particular section and comparing the actual load to that value, you could arrive at the same percentage value with almost no calculations; i.e., (P required)/(P allowable) is the percentage of capacity utilized as a column.

Similarly, the second and third terms, f_{b_x}/F_{b_x} and f_{b_y}/F_{b_y}, represent the percentage of the capacity of the member that is being utilized as a beam (about either the X-X or the Y-Y axis). Usually, we will be using only the bending component about the X-X axis since this is the orientation that has the most bending strength.

f_b (the **actual bending stress**) = M (actual) $\times c/I$ = M (actual)/S_x

F_b (the **allowable bending stress**) = $0.66F_y$ or $0.60F_y$, depending on its lateral support.

Again, if you realize that the term represents the percentage of capacity utilized as a beam, we could compare the S_x required to the S_x provided and arrive at the percentage capacity utilized as a beam;

i.e., (S_x required)/(S_x provided) is the percentage of capacity utilized as a beam.

This explains the basic idea of the formulas. There is another factor that comes into play in compression plus bending. If a member is subjected to bending forces, it will bend (what a revelation!). That bending creates a small amount of eccentricity relative to the axis of the member, and that eccentricity when multiplied by the axial load that acts along the original axis of the member causes additional moment, which referred to as a secondary bending moment. That secondary bending moment causes additional lateral deflection, and that creates yet additional moment, causing more lateral deflection, and, potentially, on . . . . . . and on . . . . untill failure.

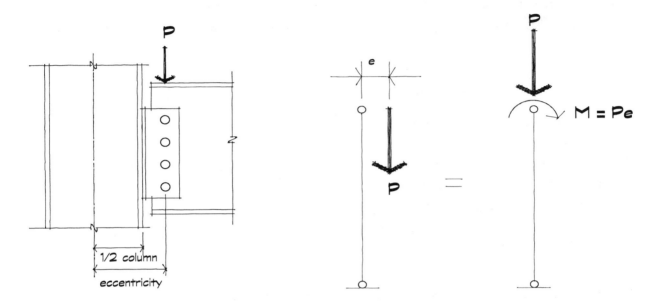

Fig. 13.2 Secondary moment development resulting from eccentricity in compression members

The code specifies that if you are utilizing less than 0.15 or 15% of the member's capacity as a column ($f_a/F_a < 0.15$), then you do not need to consider any secondary moment effects. If, however, you exceed this 0.15 value, then you must include some factor to acknowledge its effect. This factor is called an **AMPLIFICATION** factor, and it is represented in the formulas by the letter A. The two versions of the combined stresses formulas are as follows:

For tension + bending:
$$f_t/F_t + f_b/F_b \leq 1.0$$

For compression + bending:
$$f_a/F_a + A\, f_b/F_b \leq 1.0$$

The Amplification factor A is in reality a formula:

$$A = \frac{C_m}{1 - \dfrac{f_a}{F'_e}}$$

where C_m is a factor that relates to end conditions and framing type and can conservatively be taken as 1.0. Remember, f_a is the actual axial stress = P/A

F'_e is a factor that relates to the slenderness ratio of the bending axis of the member:

$$F'_e = \frac{12\pi^2 E}{23\left(\frac{kL_b}{r_b}\right)^2}$$

This seems complex, but the values for F'_e are already calculated for you in a slenderness ratio table (AISC Manual, p. 5-122).

First, don't become overwhelmed by all the numbers and factors. It's all a series of small manageable problems, and with a little thinking, you can usually solve one of these monsters very rapidly.

The most productive way to explain this problem is just to work an example.

Example 13.1

Design the corner column of a structure that has an axial load (compression) of 26.2k and a lateral load applied at the top of 3.5k. The member is 12ft high, is rigidly connected at the base, is pinned at the top, and no lateral support is provided at the top; however, it is not allowed to rotate out of its plane. Use A36 steel.

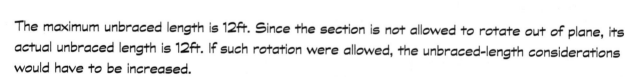

Based on the loading, the member is a column that has the following design criteria:

P = 26.2k

K = 2.1 Fig 10.1, Sec 10.1, condition 5

L = 12ft or 144in

In addition, the member is a beam that cantilevers out of the foundation and has the following loading consideration:

M = 3.5k × 12ft = 42ftk

The maximum unbraced length is 12ft. Since the section is not allowed to rotate out of plane, its actual unbraced length is 12ft. If such rotation were allowed, the unbraced-length considerations would have to be increased.

At this point, we must select a preliminary section: We could either

(a) select a column section that has additional capacity to compensate for the applied moment, or

(b) select a beam section that has additional area to compensate for the axial loading condition.

We would suggest that since we have column tables available that are quick and easy to use, we might start at that point with a preliminary selection.

We'll intentionally make some bad guesses in the beginning to illustrate the ease with which you can evaluate a series of section options:

Looking at the column tables (AISC Manual, p. 3-32):

Select a W8X28: Allowable load capacity at KL = 2.1 x 12ft = 25.2ft

Since both axes of the column are braced at the same points, the Y-Y axis will control the design for this portion:

$$P_{all.} = 36k$$

Percentage capacity utilized as a column: $P_{act.}/P_{all.} = (26.2k)/(361k) = 0.728$

This value is greater than 0.15, so since it is a compression plus bending member, we would have to consider amplification. **However, at this point, we should go on and determine if the member has a chance of working.** We should check the percentage of its capacity utilized at a beam, and if after evaluating both of these criteria, the section still has not exceeded 100% of its capacity, we can check amplification. If, however, we have exceeded 100%, an amplification check would be an exercise in futility.

Checking percentage capacity utilized as a beam:

Required moment capacity = 42.0ftk

Looking at the section modulus table (AISC Manual, p. 2-12), we see that a W8X28 has an L_c of 6.9ft and an L_u of 17.5ft. Since our unbraced L = 12ft, this section would have only partial lateral support and its moment capacity would be based on an allowable stress of $0.60F_y$.

W8X28: $S_x = 24.3in^3$

Allowable moment capacity = $F_b \times S_x = 0.60(36ksi) \times 24.3in^3 = 524.88ink$

$$\frac{524.88ink}{12in/ft} = 43.74ftk$$

We can now determine the percentage capacity utilized as a beam: (required M)/(allowable M)

$$(42.0ftk)/(43.7ftk) = 0.961$$

If we now go back to the base formula with our preliminary numbers, we see:

percentage capacity column + A x percentage capacity beam $\geq$ 1.0

$$0.728 + A(0.961) \gg 1.0$$

Amplification will **always** be a number greater than or equal to 1.0 (otherwise, it would not be "amplifying"), so the effect will be such that the overall result will increase, making it even less acceptable. We won't actually calculate the amplification factor until we see that the selection has a chance of working. This factor will generally be a value of 1.0 to 1.3. If you find a value greater or less than this, double-check your units. Based on this first trial, we can see that the section is only slightly more of a beam than a column, so we might select a new section that has about the same geometry .

Our next selection might be a W10X39. Since its section modulus is almost twice as large and its area is about 50% larger, it will reduce the percentage utilization by roughly these ratios.

Checking the percentage column utilization of a W10X39:

$\qquad$ Allowable load capacity = 75k

By interpolating values between KL= 24 and 26ft,

$$\text{percentage capacity utilized as a column} = P_{act.}/P_{all.} = (26.2k)/(75k) = 0.349$$

Checking the percentage beam utilization of a W10X39:

$\qquad$ $L_c = 8.4ft$; $L_u = 19.9ft$;   use $F_b = 0.60F_y = 21.6ksi$

$\qquad$ Moment capacity = $F_b \times S_x = (21.6ksi)(42.1in^3) = 909.4ink$

$\qquad\qquad$ 909.4ink = 75.8ftk

$$\text{percentage capacity utilized as a beam} = M_{act.}/M_{all.} = (42.0ftk)/(75.8ftk) = 0.554$$

Checking the combined effect (assuming A is approximately 1.0):

$\qquad$ $0.349 + (A = 1.0)0.554 \le 1.0$, so we at least have a chance.

$\qquad\qquad$ Now we must consider **amplification!**

Amplification factor:

$$A = \frac{C_m}{1.0 - \frac{f_a}{F'_e}}$$

W10X39:

$A = 11.5in^2$

$r_x = 4.27in$; this is the r value for the "bending" axis X-X.

$KL_b = 2.1(12.0ft) = 25.2ft$; this is the unbraced L for the "bending" axis X-X.

$C_m = 1.0$ (conservative value)

$f_a = P/A = (26.2k)/(11.5in^2) = 2.28ksi$

F'_e is a function of $KL_b/r_b = 2.1(144in)/(4.27in) = 70.82$

From the AISC Manual (Table 8, p. 5-122), $F'_e = 29.77$ at 71 (conservative value)

Therefore:

$$A = \frac{1.0}{1.0 - \frac{2.28ksi}{29.77ksi}} = 1.083 = 1.08$$

The interaction calculations , including amplification, are:

% capacity utilized as a column + A × % capacity utilized as a beam = ?

0.349 + 1.08(0.554) = 0.947 $\leq$ 1.0   It WORKS!

Long, but not too difficult.

13.2 Sloping "beams"

As we suggested in the beginning of this chapter, some of the typical sloping-beam situations can be found in stairs and roof systems. Some typical techniques for stair construction are illustrated in the following.

The most common type would be to have the stair stringers span only from landing to landing. This would be solved as a simple beam with a pin at one end and a theoretical roller at the other. The roller should be assumed so that any axial loads are compression loads. This would create the worst possible load combination. Remember how critical length is to compression members and that compression members at "their" best are only as strong as tension members at "their" worst. Figure 13.3 illustrates this simple condition.

An alternative technique for stair construction is to allow the stringer to bend and extend onto a wall or beam support and also provide structure for the landing. While the loading seems more complex, again, by using projected surfaces for the live load, it is really fairly simple (Fig. 13.4). The major problem with this system is that the inner stringer that extends above the tread line and turns the corner to connect to the next run of treads (if any) will extend above the surface of the landing and must be trimmed, which will modify its structural characteristics. This is not really a good system for conventional tread-stringer relationships, but it is quite effective for other architectural solutions, such as a single run.

The third technique would be to use an extended stringer to cantilever and form the structure for the landing (Fig. 13. 5). This technique has the same dimensional limitations described in the previous technique. Both of these systems are more applicable to concrete construction since the structural properties can be retained through reinforcing even though dimensions may vary.

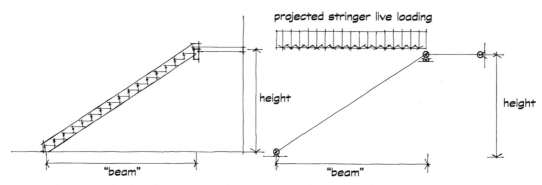

Fig. 13.3 Simple span stair stringer with separate landing system supported at floor and beam and wall.

In all of these cases, you will see that we have a condition in which the axial load varies and accumulates as the force approaches the bottom support. This load would be maximized at the base, whereas the moment is maximized at the center of the span. So what combination of loads do we use? The simplest and most conservative method would be to use the two maximums as though they occur at the same point. This will be conservative, but since the axial loads are usually relatively small, the error is not worth worrying about and it's simple.

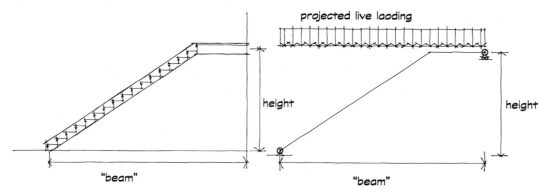

Fig. 13.4 "Bent" beam consisting of stair stringer extended to create landing system supported at floor and wall.

The design methodology for a roof joist is the same; the primary difference is in the determination of the axial component of force. This will be similar to stair-stringer design if a ridge support beam is used and similar to a truss design if a three-hinged arch system is used.

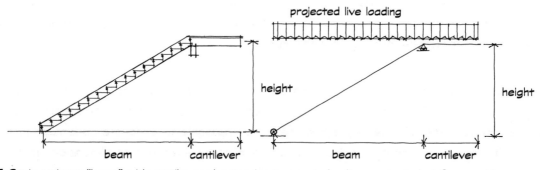

Fig. 13.5 Stair stringer "beam" with cantilevered extension to create landing supported at floor and beam.

Example   **195**

The following example illustrates the design methodology applied to stairs and a roof joist.

Example 13.2

The fire escape stairs are the standard 4ft wide and stringers on each side support the treads. The outside stringers also support the landing, but will be designed as a separate problem. We will assume that the stringers conform to the simplest design, which spans from the bottom to the top of a single run. The live load is 100psf and we will assume a dead load of 30psf, excluding the stringer weight. This would equate to 2-1/2in of concrete. Although the treads are only 1in of concrete fill, the steel tread pans and riser plates would make the total about right. Use A36 steel.

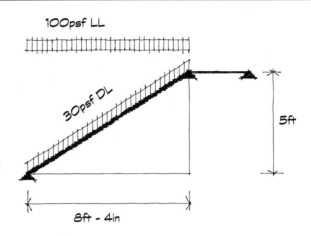

First of all, we must determine the loads on each stringer: Because of the slope, the gravity loads are not perpendicular to the stringer axis (except at the landing), but we can find the perpendicular component that will cause bending and the axial component that will cause compression or tension.

We can determine the actual length of the stringer with the help of Pythagoras:

$$L = \sqrt{(5ft)^2 + (8.3ft)^2} = 9.7ft$$

The DL is uniformly distributed along the stringer; therefore, the DL of the sloping portion is

w_{DL} = (30psf)(2ft) = 60plf + assumed stringer weight of 10plf

w_{DL} = 70plf This load acts along the slope.

This distinction is important since the LL will act on the horizontal projection of the slope.

The LL that is applied to the stringer is most easily calculated by taking the total horizontal load and stretching it out along the stringer.

w_{LL} = (100psf)(2ft) = 200plf acting on 8.3ft = 1660lb

When this load is "stretched out" on the slope, it is equivalent to (1660plf)/(9.7ft) = 171.1plf, so

w_{LL} = 171.1plf

The total gravity load acting on the stringer would be w_{DL} + w_{LL} = w_{TL} = 70plf + 171.1plf = 241.1plf

w_{TL} = 241.1plf

To design the stringer for bending, we must determine the perpendicular components of any forces acting on it. We can now find the two components of w_{total} using the proportional relationships of the stringer slope.

Perpendicular component of w = (8.3ft/9.7ft)155.6plf = 133.1plf

Perpendicular component of w_{LL} = (8.3ft/9.7ft)85.6plf = 73.2plf   (we'll need this for deflection)

Parallel component of w = (5ft/9.7ft)155.6plf = 80.2plf

Both of these components are uniformly distributed along the 9.7ft length of the stringer.

We can now size the stringer for moment, shear, and deflection, keeping in mind that there will be some axial load. We will select a slightly bigger section than we need for bending alone.

$$\text{Maximum Moment} = \frac{(206.3\text{plf})(9.7\text{ft})^2}{8} = 2426.3\text{ftlb}$$

$$\text{Required } S_x = \frac{(2426.3\text{ftlb})(12\text{in/ft})}{0.66(36,000\text{psi})} = 1.225\text{in}^3$$

$$\text{Maximum Shear} = \frac{206.3\text{plf}(9.7\text{ft})}{2} = 1000\text{lb}$$

$$\text{Required } A_w = \frac{1000\text{lb}}{0.4(36,000\text{psi})} = 0.07\text{in}^2$$

We can see at this point that shear is not going to be a consideration.

$$\text{Maximum deflection allowable} = L/360 = (9.7\text{ft})(12\text{in/ft})/360 = 0.32\text{in}$$

$$\text{Required } I = \frac{5\left(\frac{146.4\text{plf}}{12\text{in/ft}}\right)(9.7\text{ft}(12\text{in/ft}))^4}{384(29,000,000\text{psi})(.32\text{in})} = 3.14\text{in}^4$$

The channel section that meets these requirements is a C4X5.4. This is a structural solution, but not an architectural one. This section is simply too small to provide attachment for a standard stairs. If we look a bit further into the channel sections, we'll see a MC10X6.5, only slightly heavier, but more dimensionally compatible with stairs. You should also note that there is a very light MC12X10.6. This section is produced primarily for stair stringers; it is deep enough to provide detailing space for the treads, yet light enough to carry the small loads required for stair stringers.

Although it should be clear that the MC10X6.5 is more than strong enough even for combined axial loading, we'll run through the procedure. Depending on where we assume the stairs has a "roller," the force in the stringer will be either tension (if hung from the top and a roller at the bottom) or compression (if supported at the bottom with a roller at the top). From our understanding of the effect of these two alternative forces, we should clearly assume compression forces to create the worst scenario.

The next issue is at what point do we determine the combined forces? If we assume the point of maximum moment (the center), the maximum amount of axial load will not be developed. At the

Example   **197**

same time, if we use the point of maximum axial load (the bottom), this is a point of 0 moment. A dilemma is solved by taking a conservative approach and assuming both maximums occur at the same point.

Maximum axial load = (124.3plf)(9.7ft) = 1205.7lb

Maximum moment (from previous calculations) = 2426.3ftlb

It's time for the dreaded formula:

$$\frac{f_a}{F_a} + A\frac{f_b}{F_b} \leq 1.0$$

where f_a/F_a = percentage capacity utilized as a column (can be obtained by using $P_{reqd.}/P_{all.}$)

f_b/F_b = percentage capacity utilized as a beam (can be obtained by using $S_{reqd.}/S_{provided}$)

A = amplification factor if the percentage capacity utilized as a column exceeds 15%

(Remember that this factor is usually under 1.3 and never less than 1.0.)

So lets do it!!!

f_a = P/A = (1205.7lb)/(1.9in²) = 634.6psi

F_a is a function of the critical slenderness ratio, which in this case would be KL_x/r_x, since the treads will provide lateral bracing to the weak Y-Y axis.

KL_x/r_x = 1.0(9.7ft)(12in/ft)/3.40in = 34.23

F_a = 19.58ksi at 35   (AISC Manual, Table of Allowable Stresses C-36, p. 3-16)

This means that we are utilizing: (634.6psi)/(19,580psi) = 0.03 or 3% of the column capacity, which means we do not need to consider amplification . . . .YES!!!!

To calculate the percentage capacity utilized as a beam we can take a simple approach and compare the S_x provided to the S_x required:

(1.225in³)/(4.42in³) = 0.277 or 27.7% capacity utilized as a beam

0.03 + 0.277<< 1.0. We are done!

The procedure used in solving this stringer problem would be used for the other structures described at the beginning of the chapter.

METRIC VERSIONS OF ALL EXAMPLES

Example M13.1

Design the corner column of a structure that has an axial load (compression) of 116.54kN and a lateral load applied at the top of 15.57kN. The member is 3.66m high, is rigidly connected at the base, is pinned at the top, and no lateral support is provided at the top; however, it is not allowed to rotate out of its plane. Use 250MPa steel.

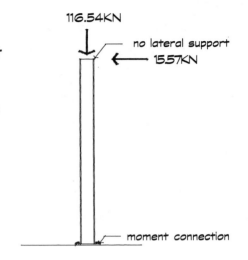

Based on the loading, the member is a column that has the following design criteria:

$$P = 116.54\text{kN}$$
$$K = 2.1 \text{ Fig. 10.1, Section 10.1, condition 5.}$$
$$L = 3.66\text{m}$$

In addition, the member is a beam that cantilevers out of the foundation and has the following loading consideration:

$$M = 15.57\text{kN}(3.66\text{m}) = 57\text{kNm}$$

The maximum unbraced length is 3.66m. Since the section is not allowed to rotate out of plane, its actual unbraced length is 3.66m. If such rotation were allowed, the unbraced-length considerations would have to be increased.

At this point, we must select a preliminary section: We could either select a column section that has additional capacity to compensate for the applied moment or select a beam section that has additional area to compensate for the axial loading condition. We would suggest that since we have column tables available that are quick and easy to use, we might start at that point with a preliminary selection.

We'll intentionally make some bad guesses in the beginning to illustrate the ease with which you can evaluate a series of section options:

Looking at the column tables (AISC Manual, p. 3-32) for U.S. customary units and converting them to metric, a nasty task as best:

  Select a W200X42: Allowable load capacity at KL = 2.1(3.66m) = 7.686m

  Since both axes of the column are braced at the same points, the Y-Y axis will control the design for this portion:

$$P_{all.} = 160\text{kN}$$

Percentage capacity utilized as a column: $P_{act.}/P_{all.} = (116.54\text{kN})/(160\text{kN}) = 0.728$

This value is greater than 0.15, so since it is a compression plus bending member, we would have to consider amplification. However, at this point, we should go on and determine if the member has a chance of working. We should check the percentage of its capacity utilized at a beam, and if after evaluating both of these criteria, the section still has not exceeded 100% of its capacity, we can check amplification. If, however, we have exceeded 100%, an amplification check would be an exercise in futility.

Checking percentage capacity utilized as a beam:

Required moment capacity = 57kNm

Looking at the section modulus table (AISC Manual, p. 2-12), we see that a W200X42 has an L_c of 2.10m and an L_u of 5.34m. Since our unbraced L = 3.66m, this section would have only partial lateral support and its moment capacity would be based on an allowable stress of $0.60F_y$.

W200X42: S_x = 399x10³mm³
Allowable moment capacity = $F_b \times S_x$ = 0.60(250MPa) (399x10³mm³) =
= 59,850x10³Nmm
59,850x10³Nmm/(1000N/kN × 1000mm/m) = 59.85KNm

We can now determine the percentgae capacity utilized as a beam:
(required M)/(allowable M) = (57kNm)/(59.85kNm) = 0.952

If we now go back to the base formula with our preliminary numbers, we see

percentage capacity as a column + A (percentage capacity as a beam) $\leq$ 1.0
0.73 + A(0.952) >> 1.0

Amplification will **always** be a number greater than or equal to 1.0 (otherwise, it would not be "amplifying"), so the effect will be such that the overall result will increase, making it even less acceptable. We won't actually calculate the amplification factor until we see that the selection has a chance of working. This factor will generally be a value of 1.0 to 1.3. If you find a value greater or less than this, double-check your units. Based on this first trial, we can see that the section is only slightly more of a beam than a column, so we might select a new section that has about the same geometry .

Our next selection might be a W250X58. Since its section modulus is almost twice as large and its area is about 50% larger, it will reduce the percentage utilization by roughly these ratios.

Checking the percentage column utilization of a W250X58:

Allowable load capacity = 333.6kN

By interpolating values between KL = 7.32 and 7.93m,

percentage capacity utilized at a column = $P_{act.}/P_{all.}$
$$= (116.54kN)/(333.6kN) = 0.349$$

Checking the percentage beam utilization of a W250X58:

L_c = 2.56m; L_u = 6.07m; use F_b = 0.60F_y = 150MPa
Moment capacity = $F_b \times S_x$ = (150MPa)(693x10³mm³) = 103.95kNm

percentage capacity utilized as a beam = (57kNm)/(103.95kNm) = 0.548

Checking the combined effect (and assuming A is approximately = 1.0):

0.349 + (A = 1.0)0.548 < 1.0,

so we at least have a chance. Now we must consider Ampflication!

Amplification factor:

$$A = \frac{C_m}{1.0 - \frac{f_a}{F'_e}}$$

W250X58:

A = 7400mm²
r_x = 109mm; this is the r value for the "bending" axis X-X.
KL_b =2.1(3.66m) = 7.686m; this is the unbraced L for the "bending" axis X-X.

C_m = 1.0 (conservative value)
f_a = P/A = (116.54kN)/(7400mm²) = 0.01575kN/mm² = 15.57MPa
F'_e is a function of KL_b/r_b = 2.1(3660mm)/109mm = 70.51
From the AISC Manual (Table 8, p. 5-122),

F'_e = 29.77 = 205.23MPa @ 71 (conservative value)

Therefore:

$$A = \frac{1.0}{1.0 - \frac{15.57MPa}{205.23MPa}} = 1.083$$

The utilization calculations, including amplification, are:

% capacity utilized as a column + A x % capacity utilized as a beam = ?

0.349 + 1.08(0.548) = 0.942 < 1.0   It WORKS!

Example M13.2

The fire escape stairs are the standard 1.22m wide and stringers on each side support the treads. The outside stringers also support the landing, but will be designed as a separate problem. We will assume that the stringers conform to the simplest design, which spans from the bottom to the top of a single run. The live load is 4800N/m² and we will assume a dead load of 1440N/m², excluding the stringer weight. This would equate to 63mm of concrete. Although the treads are only 25mm of concrete fill, the steel tread pans and riser plates would make the total about right. Use 250MPa steel.

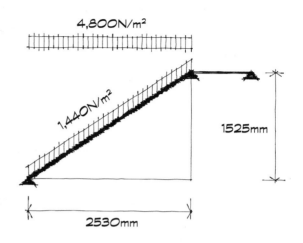

First of all, we must determine the loads on each stringer: Because of the slope, the gravity loads are not perpendicular to the stringer axis (except at the landing), but we can find the perpendicular component that will cause bending and the axial component that will cause compression or tension.

We can determine the actual length of the stringer with the help of Pythagoras:

$$L = \sqrt{(1525mm)^2 + (2530mm)^2} = 2953mm$$

The DL is uniformly distributed along the stringer; therefore the DL of the sloping portion is

$$w_{DL} = (1440N/m^2)(1.22m/2) + (\text{assumed stringer weight of } 147N/m) = 1025N/m$$

The 147N/m corresponds to a C180X15 weight = (15kg/m)(9.81N/kg) = 147N/m.

The w_{DL} of 1025N/m acts along the slope. This distinction is important since the LL will act on the horizontal projection of the slope.

The LL that is applied to the stringer is most easily calculated by taking the total horizontal load and stretching it out along the stringer.

$$w_{LL} = (4800N/m^2)(1.22m/2) = 2928N/m$$

acting on a 2.53m length, so

$$w_{LL} = (2928N/m)(2.53m) = 7408N$$

When this load is "stretched out" on the slope, it is equivalent to (7408N/m)/(2.953m) = 2508N/m, so $w_{LL} = 2508N/m$

The total gravity load acting on the stringer would be

$$w_{DL} + w_{LL} = w_{total} = 1025N/m + 2508N/m = 3533N/m$$

To design the stringer for bending, we must determine the perpendicular components of any forces acting on it. We can now find the two components of w_{total} using the proportional relationships of the stringer slope.

Perpendicular component of w = (2530mm/2953mm)(3533N/m) = 3027N/m

Perpendicular component of w_{LL} = (2530mm/2953mm)(2508N/m) = 2149N/m

(we'll need this for deflection)

Parallel component of w = (1524mm/2953mm)(3533N/m) = 1823N/m

Both of these components are unformly distributed along the 3.31m length of the stringer.

We can now size the stringer for moment, shear, and deflection, keeping in mind that there will be some axial load. We will select a slightly bigger section than we need for bending alone.

$$\text{Maximum Moment} = \frac{3027N/m(2.953m)^2}{8} = 3299Nm$$

$$\text{Required } S_x = \frac{(3,299Nm)(1000mm/m)}{0.66(250MPa)} = 20\times10^3 mm^3$$

$$\text{Maximum Shear} = \frac{3027N/m(2.953N/m)}{2} = 4469N$$

$$\text{Required } A_w = \frac{4469N}{0.4(250MPa)} = 44.7mm^2$$

We can see at this point that shear is not going to be a consideration.

Maximum deflection allowable = L/360 = (2953mm)/360 = 8.2mm

$$\text{Required I} = \frac{5(2149N/m/1000mm/m)(2953mm)^4}{384(200,000MPa)(8.2mm)} = 1.30\times10^6 mm^4$$

The channel section that meets these requirements is a C100X8. This is a structural solution, but not an architectural one. This section is simply too small to provide attachment for a standard stairs. If we look a bit further into the channel sections, we'll see a MC250X9.7, only slightly heavier, but more dimensionally compatible with stairs. You should also note that there is a very light MC310X15.8. This section is produced primarily for stair stringers; it is deep enough to provide detailing space for the treads, yet light enough to meet the small loads required for stair stringers.

Although it should be clear that the MC250X9.7 is more than strong enough even for combined axial loading, we'll run through the procedure. Depending on where we assume the stairs has a "roller," the force in the stringer will be either tension (if hung from the top and with a roller at the bottom) or compression (if supported at the bottom with a roller at the top). From our understanding of the effect of these two alternative forces, we should clearly assume compression forces to create the worst scenario.

The next issue is at what point do we determine the combined forces? If we assume the point of maximum moment (the center), the maximum amount of axial load will not be developed. At the same time, if we use the point of maximum axial load (the bottom), this is a point of 0 moment. A dilemma is solved by taking a conservative approach and assuming both maximums occur at the same point. So . . .

Maximum axial load   P = (1823N/m)(2.953m) = 5383N

Maximum moment (from previous calculations) M = 3299Nm

It's time for the dreaded formula:

$$\frac{f_a}{F_a} + A\frac{f_b}{F_b} \leq 1.0$$

where f_a/F_a = percentage capacity utilized as a column (can be obtained by using $P_{reqd.}/P_{all.}$)

f_b/F_b = % capacity utilized as a beam (can be obtained by using $S_{reqd.}/S_{provided}$)

A = amplification factor if the **percentage capacity utilized as a column exceeds 15%.**
(Remember that this factor is usually under 1.3 and never less than 1.0.)

So let's do it!!!

f_a = P/A = 5383N/1230mm² = 4.38MPa

F_a is a function of the critical slenderness ratio, which in this case would be KL_x/r_x, since the treads will provide lateral bracing to the weak Y-Y axis.

KL_x/r_x = 1.0(2953mm)/(86.5mm) = 34.14

F_a = 19.58ksi = 135MPa at 35 (AISC Manual, Table of Allowable Stresses, p. 3-16)

This means that we are utilizing: (4.38MPa)/(135MPa) = 0.03, or 3% of the column capacity, which means we do not need to consider amplification. YES!!!!

To calculate the percentage capacity utilized as a beam, we can take a simple approach and compare the S_x provided to the S_x required:

(20x10³mm³)/(72.4x10³mm³) = 0.276, or 27.6% capacity utilized as a beam

0.03 + 0.276 << 1.0. We are done!

The procedure used in solving this stringer problem would be used for the other structures described at the beginning of the chapter.

14 TRUSSES

14.1 General comments

The force distribution in a truss is analogous to the force distribution in beams. If the truss is a simple span, the forces that resist the maximum bending of the system will be induced in the top and bottom "chords" of the truss at the center. These forces will be pure compression and pure tension if the loads are applied at the joints or "panel points." The maximum shear forces will be carried by members near the ends: the interior "web" members. These shear forces will be resisted in the truss by the diagonal members also in pure tension and pure compression. This means that we could expect to find the largest chord members in the top and bottom near the middle and the largest web members near the ends. By increasing the distance between the upper and lower chords, the moment of inertia is increased with no additional material, or for the same moment of inertia, the members could be smaller. Trusses are potentially ideal structural solutions for long spans. The definition of the "truss" can be extended to include not only the ordinary planar (primarily two-dimensional) steel trusses, but systems such as folded plates, space frames, and geodesic domes (strongly three-dimensional systems; see Fig. 14.1).

Fig. 14.1 Three-dimensional trusses, using pure compression and tension members: Theme Pavilion at Osaka 70 Expo, Climatron at St. Louis Botanical Gardens

The most common trusses are open-web steel joists. As we mentioned, these are premanufactured steel trusses designed for floor or roof systems in a typically flat configuration. This necessitates the use of a sloping insulation or lightweight concrete fill to create roof drainage, but this is obviously limited to very low slopes in the range of 1/8in per foot. A sloped (upper) chord truss can create a roof slope that can be covered with shingles or metal roofing. These roofing types require a minimum of a 3:12 slope. At the same time, the increase in depth at the interior of the span helps reduce the size of members necessary to carry the load. These two independent criteria are both answered by using the typical triangular truss configuration.

Although some authors attempt to create two different definitions for pitch and slope, we will consistently refer to the "slope" of the truss, since this is the more commonly used architectural expression of a roof "pitch." See how confusing it is. In any case, the slope or pitch of a roof or truss will be defined as the vertical rate of change per foot of sloped member. Figure 14.2 shows how a truss may have two different slopes.

Trusses most efficiently carry loads that are applied to the joints or panel points only. This means the members will not have to use moment and shear to transfer the loads back to the joints where truss action is initiated. Unfortunately, this is seldom the case. Frequently, both in steel and wood, the loads are applied through decking applied continuously along the length of the member. Although this may be constructionally efficient, it causes the truss members to be subjected to combined axial load and moment. In both cases, C + M and T + M, the sizes of the members will be increased over the idealized truss. If "purlins" are used, (small beams that support the roofing load and apply it to the truss at points; see Fig. 14.3), they can increase truss efficiency if they are used to apply loads only at the joints. This again is an idealized system, because frequently the purlins are located at closer intervals due to

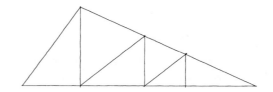

Fig 14.2 Truss with two different slopes

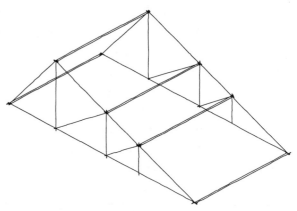

Fig. 14.3 Purlins transferring load to a truss chord

the spanning capability of the decking material. If this is the case, they may also create moment in the truss members. Any load, other than an axial one, applied at points other than the joints will inevitably cause moment in the member. As we have reiterated, the true measure of the cost and efficiency of any system will require the investigation of the entire system, not just an individual component.

A number of different connection techniques are used for manufacturing trusses, depending on their materials, shipping restrictions, and erection considerations. Steel trusses are frequently shop-welded and field-bolted, with some bracing being field-welded. Wood trusses are seldom designed using "bolts" alone. The apparently bolted trusses you have commonly seen are constructed using shear plates or split rings, with the "bolt" being the visible part of the connection. These connection types will be discussed in Chapter 25.

Planar or space trusses are typically designed with pin joints, so that theoretically only axial loads exist in the members. Trusses are used as beam members, horizontal, or as columns, vertical, as shown in (Fig. 2.4, Chapter 2). "Truss" refers not to an orientation, but to any system that conforms to these principles.

The main exception to the pinned joints criterion is the Vierendeel truss with rectangular, nontriangulated, interior geometries. The joints are moment-resisting, making the truss more rigid and the design more complex (statically indeterminate). The Vierendeel truss can be employed whenever large loads are supported on long spans, as shown in Fig. 14.4, or for full-floor-height systems that allow use of the truss volume as functional space. One of the best-known examples of this use is in Boston City Hall. Whereas a typical truss has loads applied whenever possible only to the panel points or joints in an attempt to eliminate any bending stresses in the members, the Vierendeel truss has bending stresses in all members as a result of rigid connections.

In this chapter, we will deal with planar trusses only, although the outline design of space trusses rests on the same basic principles. Trusses are available as preengineered components in many shapes and configurations. Some of the more common are shown in Fig. 14.5.

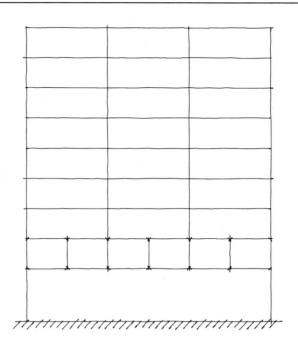

Fig. 14.4 Vierendeel truss as main transfer beam

How great a distance can a truss span? The McCormick Place exhibition hall in Chicago (Fig. 14.6) is covered with a two-way truss system

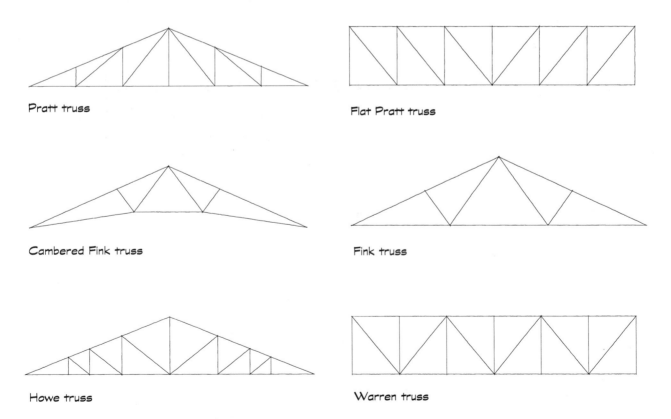

Pratt truss

Flat Pratt truss

Cambered Fink truss

Fink truss

Howe truss

Warren truss

Fig. 14.5 Truss variations based on point of application of loads, support location, bracing of major compression members, and desired form or interior volume profile, i.e., roof slopes or interior clearances

that is 15ft [4.5m] deep supported on columns at 150ft [45m] centers. The use of a space frame, or three-dimensional truss, could reduce the structural depth. How does a truss span? One example of the truss principle in an unusual context would be on the four faces of the John Hancock Building (Chicago) where they are used to resist lateral loads by acting as a trussed cantilevered tube that stands against the wind like a tree (Fig. 14.7).

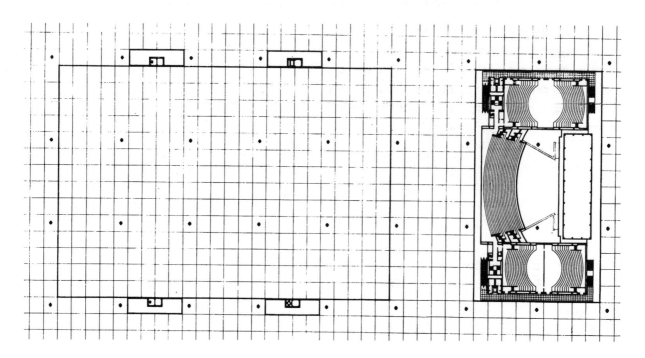

Plan

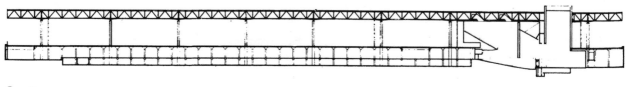

Section

Fig. 14.6 McCormick Place exhibition hall, Chicago, Illinois

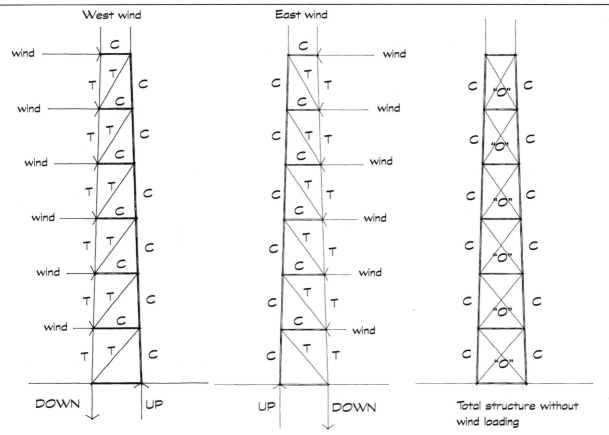

Fig. 14.7 John Handcock Building, Chicago, Illinois. Diagram of wind bracing indicating redundancy depending upon direction of wind forces. In the final version, the diagonal bracing is unloaded without wind loading.

14.2 Design considerations

Frequently, we find that the upper and lower chord members are **one consistent size** over their entire length. This is not a representation of the forces that they resist, but a manufacturing consideration. In fact, trusses, when used in limited number for a specialized situation, may be designed with as few as four members: a maximum compression chord, a maximum tension chord, a maximum compression web, and a maximum tension web. Open-web steel joists are good examples of this simplification. Obviously, this will not be materially efficient, but the engineering time and fabrication time are significantly reduced. If you are creating a large number of a single truss design, it is more likely to have individually designed members, since the cost of the engineering can be divided over several trusses.

Why are there so many different truss configurations? The idea of a truss is to use the material efficiently. A beam is cheap because of manufacturing simplicity, but uses material in a very inefficient manner. In trusses, the material that is not required is simply not provided. The truss configuration not only attempts to use minimum material, but attempts to use the material in the most efficient structural configuration. If you look at truss configurations, you'll see that some of the web members are longer than others. **If the truss is designed efficiently, all of the long web members will be oriented so that they have tensile forces in them.** Remember, the length of a member has no effect on its size when designing in tension. On the other hand, if the member is in compression, its size is significantly affected by unbraced length. The various configurations are simply adaptations for the members to meet this criterion for the unique loading condition that the truss will experience.

If you choose to use trusses, what characteristics will they have? For the purpose of starting your design or simply to establish a general truss size in an architectural solution:

(a) A truss will typically have a depth (at midspan of a simple span) of 8% to 12% of the span. A simple first estimate for most situations would be to use a depth equal to 10% of the design span.

(b) Trusses will be spaced at approximately 25% to 30% of their span. Obviously, this is contingent on the spanning capability of the subsystem that spans between the trusses, but the truss will function efficiently at this spacing.

REMEMBER THESE FIRST TWO APPROXIMATIONS! THEY PROVIDE A SIMPLE ESTIMATE OF THE TRUSS SIZE AND LOCATION FOR "ARCHITECTURAL" DESIGN PURPOSES.

(c) Truss weight is accounted for in the design process by increasing the "applied" loads by 10%. This is not always an accurate representation of the distribution of the truss weight, but it works fairly well and simplifies calculations in truss analysis.

You will remember that we indicated that open-web steel joists were premanufactured trusses. After reading the previous "rules," you'll see that joists are both much shallower and used at a much closer spacing. This makes sense because open-web steel joists are used not only as roof members, but as floors as well. In floors, the 10% depth estimate for widely spaced (25 to 30% of span) would create havoc with the building volume, but if placed at closer intervals, we can reduce the depth. The net result is a depth of 1/2in[0.4m/m] deep per foot of span (4%), instead of 10% of span, but you pay for this depth reduction with much closer spacing, which ultimately consumes more material. The economic key here is life-cycle costing; an initial high one-time cost is offset by continued systems savings over the entire life of the structure.

The configuration of the truss, as mentioned previously, should try to maximize material efficiency and meet the architectural consideration for each design. So, let's do one.

14.3 Truss analysis

Loads in members can be found with a graphic method using a closed, graphical force polygon (equivalent to a geometric summation of forces in equilibrium). This method is potentially not as exact as a mathematical solution (method of sections or method of joints), but it is quick and easy and will allow you to determine which of the members will have the largest forces and then do a mathematical solution of only the most critical members.

This method requires Bow's notation, a labeling system using letters between external forces and numbers in each truss panel. We suggest that you use a letter between **every** load and

every joint. This isn't necessary, but will avoid confusion later, and it doesn't complicate the process.

1.   Find the external loads and support reactions.
2.   Draw a force polygon (forces to scale) to determine values of loads.
3.   Draw force diagrams at each joint to determine T or C or "O" force.

This will result in a "stress diagram," which in reality is not a diagram of stress at all, but a diagram of the **force or load** that the member must resist. Stress is the load per unit area. During the execution of the diagram, member size is not considered. This is one of those accepted "name things" that really confuse the issue but have become the standard over the years. As we begin to design the individual members, we will determine stresses resulting from these forces applied to specific sizes of components.

As usual, the easiest way to explain the technique is to work an example:

Example 14.1

Draw the "stress diagram" for the illustrated truss.

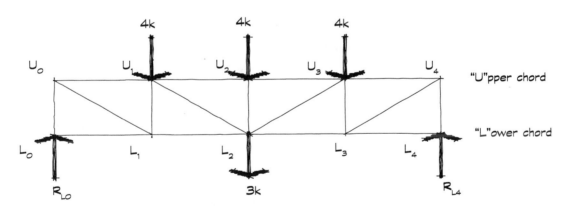

Truss loading and joint nomenclature

1.   Determine the reactions: Since the loads are symmetrical, we can determine the reactions by simply summing the loads and dividing by 2.

$$RL_0 = RL_4 = (4k + 4k + 4k + 3k) / 2 = 7.5k$$

Note that the truss will have several different notations depending on what you're calculating at any particular time. The L_0, L_1, L_2, etc., U_0, U_1, U_2, etc., notation is used to describe the truss for reactions and to label the joints.

2.   Apply all loads and reactions and label the truss using Bow's notation, sequentially lettering "spaces" between joints and/or applied loads around the exterior of the truss and sequentially numbering the "spaces" defined by the members on the interior of the truss.

3. Draw the "stress diagram": This can be accomplished with no possibility for error by "reading" the loads into a simple expression and then drawing **exactly** what the expression says.

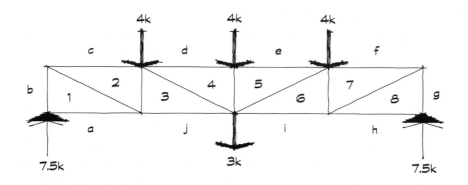

"Between_____ and_____ , I have a _____ line."

This reduces a mathematical procedure to almost a purely mechanical one, but it works and it's easy to apply. Why make it harder?

Some of the inevitabilities of these diagrams are:

1.   No matter where you start a diagram, it will not fit on your paper.
2.   No matter where you put the labels, they will be in the way, graphically

IT MUST CLOSE!!

If you have not returned to your point of origin, you have made an error.

You can now follow the same procedure for the interior "members":

1. "Between b and 1, I have a vertical "line" (draw one). " Between a and 1, I have a horizontal line (draw it). These two lines will intersect at point 1 and will consequently define point no. 1.

No letter or number will exist at more than one point on the diagram.

Complete the diagram in this manner, and it should look like the one illustrated on page 214.

When you get to the end of the diagram, you'll see that you even have a check. The last three lines, g-8, h-8, and 7-8, all have a single point in common (8). If they do not intersect at this single point, you have an error!

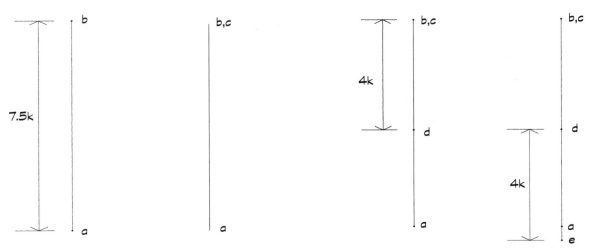

1. Between "a" and "b," I have a 7.5k vertical force (up).

2. Between "b" and "c," I have no change, so just label point "b" as "c" as well.

3: Between "c" and "d," I have a 4k vertical force (down).

4. Between "d" and "e," I have a 4k vertical force (down).

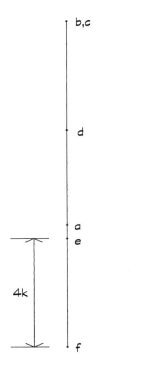

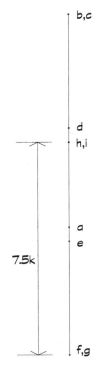

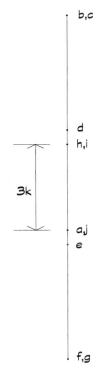

5. Between "e" and "f," I have a 4k vertical force (down). Between "f" and "g," I have no change, so just label "f" as "g" as well.

6. Between "g" and "h," I have a 7.5k vertical force (up). Between "h" and "i," I have no change, so just label "h" as "i" as well.

7. Between "i" and "j," I have a 3k vertical force (down) and no change between"j" and "a," so the system closes. YES!!!

Some ABSOLUTE rules associated with the diagram:

1. This diagram must close: You must return to your point of origin or it's wrong.

2. Slopes on the stress diagram must be **exactly** the same as slopes on the truss.

3. No point can have more than one location definition; i.e., you cannot have two A's or two D's.

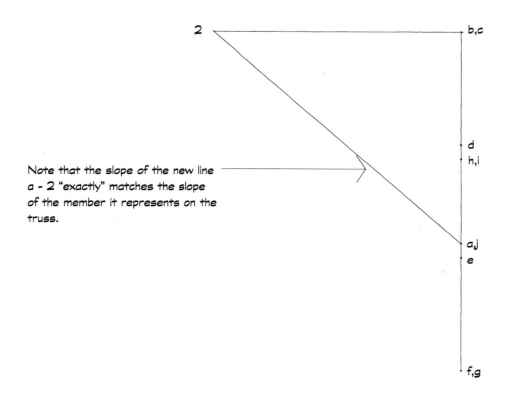

Note that the slope of the new line
a - 2 "exactly" matches the slope
of the member it represents on the
truss.

NO LETTER OR NUMBER WILL EXIST AT MORE THAN ONE POINT ON THE DIAGRAM.
Complete the diagram in this manner and it will look like the one illustrated below:

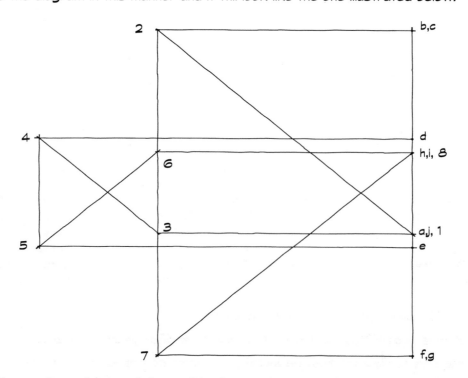

Completed "stress diagram" (not really "stress," but force, yet typically referred to as a stress diagram)

This exercise really only proves that the sum of the vertical forces is "O." If there were horizontal forces applied to the truss, we would simply place those on the diagram as well and simultaneously prove that both the sum of the vertical and the sum of the horizontal forces is "O."

Now that we have the diagram, we can "read" the relative magnitude of the forces in every member of the truss. If we've drawn the diagram to exact scale, the loads will be exact as well. However, even a sketched diagram will identify, in most cases, those members that have the largest forces in them.

In addition, we can isolate individual joints (the diagram is in reality a combination of all the individual joint solutions of the truss) and determine whether the member is in Tension or Compression. As long as we have a known direction of force acting on a joint, either an external force or an internal member load, we can determine the direction of every other force acting on that joint.

Example 14.2

A large exhibition space is organized in plan into 50ft x 50ft bays. The 50ft span can be covered by standard open-web joists, 30K11 series at 6.25ft o.c. These joists are supported on a 50ft trussed girder supported on columns at the ends. We want to give a first shot at the truss size and see what type of shapes could be used for the chords.

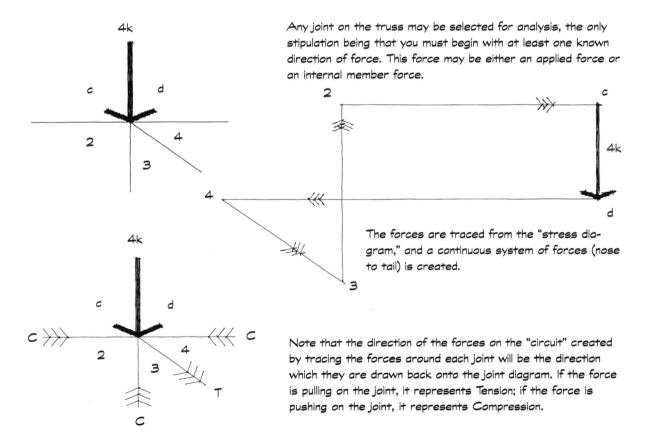

Any joint on the truss may be selected for analysis, the only stipulation being that you must begin with at least one known direction of force. This force may be either an applied force or an internal member force.

The forces are traced from the "stress diagram," and a continuous system of forces (nose to tail) is created.

Note that the direction of the forces on the "circuit" created by tracing the forces around each joint will be the direction which they are drawn back onto the joint diagram. If the force is pulling on the joint, it represents Tension; if the force is pushing on the joint, it represents Compression.

Member analysis using the stress diagram

First of all, it is convenient to have a panel length equal to the joist spacing, so the loads will be at panel points (joints). Assume we have calculated the point loads from the roof, P = 9k each.

By using the rule of thumb of depth equal to 10% of truss span, the depth we assume for a first try is a 50ft span = 5ft depth.

The next step is to find the maximum compression and tension stress in the chords, which is obviously at midspan (by analogy with beams). This can be easily calculated with the method of sections. The equation of equilibrium of moments about joint A is:

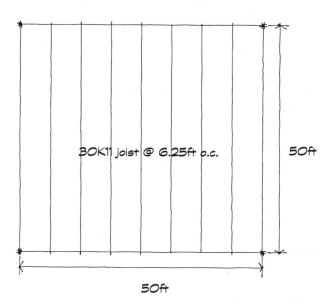

$$F_C (5ft) = 31.5k(21.88ft) - 9k(3.12ft) - 9k(12.5ft) - 9k(18.75ft)$$

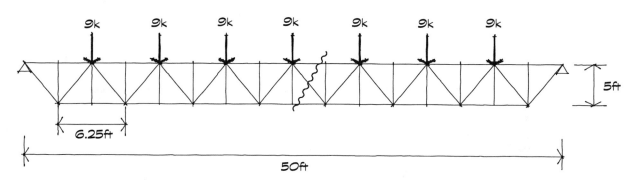

$$F_C = 380ftk/5ft = 76k$$

With this force, we can find a preliminary size for the top chord. In the AISC Manual (Part 3 - Column Design), we can choose 4in x 4in x 3/8in equal-leg double angles, grade 50, an

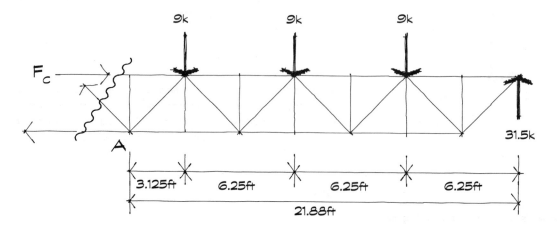

unbraced length 8ft, with allowable load of 87k. The entire top chord could be built with this size for speed of construction.

The usual methods of truss analysis work if the loads are at the panel points. Even in the cases when the loads are applied along the chord members between the panel points, we can obtain a preliminary size. However, always rework the problem with loads applied at their actual locations and include the moments created by this condition prior to any final decision.

The outline procedure for preliminary design might include the following:

1. Calculate the total (roof or floor) load on each member between panel points separating the vertical and horizontal components (if any, due to wind, for instance).

2. Statically resolve the total load between the two ends of the member, and apply the loads coming from all members at the same panel point. This would include both chord loads and any concentrated purlin or beam loads.

3. Find the stresses (axial loads) in the truss members using a stress diagram, method of sections, or method of joints.

4. Size the members for the axial loads. Members loaded between panel points must be designed for axial load and bending using the actual load distribution (step 2) for the bending stresses.

Problem 14.1

This Warren truss is built with equal-leg double angles. Design the tension chord for maximum tension using 36ksi steel.

Solution:

Maximum T = 80k

Required area $A = (80k)/(0.6)(36ksi) = 3.7in^2$

Use two 3-1/2in x 3-1/2in x 5/16in ($A = 4.18in^2$) or two 4in x 4in x 1/4in ($A = 3.88in^2$)

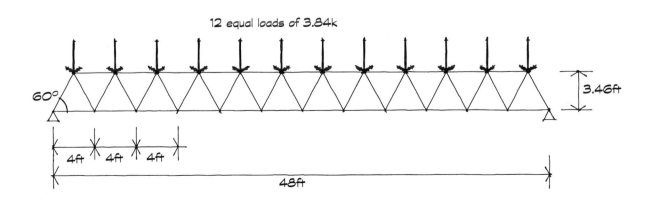

12 equal loads of 3.84k

60°

3.46ft

4ft 4ft 4ft

48ft

Problem 14.2

This truss is supporting structural panels directly on the top chord. Find the equivalent loading configurations with loads at the panel points.

Answer:

1.44k at each joint.

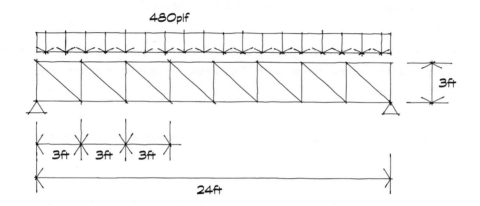

METRIC VERSIONS OF ALL EXAMPLES AND PROBLEMS

Example M14.2

A large exhibition space is organized in plan into 15.2m X 15.2m bays. The 15.2m span can be covered by standard open-web joists, 30K11 series at 1.9m o.c. These joists are supported on a 15.2m trussed girder supported on columns at the ends. We want to give a first shot at the truss size and see what type of shapes could be used for the chords.

First of all, it is convenient to have a panel length equal to the joist spacing, so the loads will be at panel points (joints). Assume we have calculated the point loads from the roof, P = 40kN each.

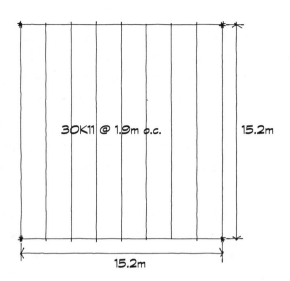

By using the rule of thumb of depth equal to 10% of truss span, the depth we assume for a first try is a 15.2m span = 1.5m depth.

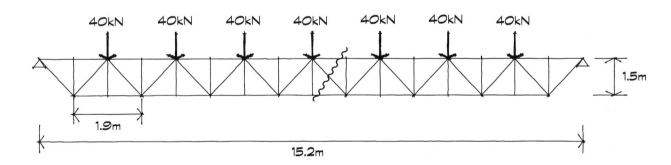

The next step is to find the maximum compression and tension stress in the chords, which is obviously at midspan (by analogy with beams). This can be easily calculated with the method of sections. The equation of equilibrium of moments about joint A is:

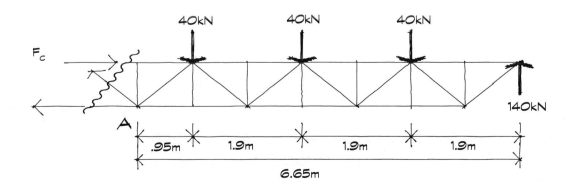

$F_C(1.5m) = (140kN)(6.65m) - (40kN)(.95m) - (40kN)(2.85kN) - (40kN)(4.75m)$

$F_C = (931kNm - 342kNm)/(1.5m) = 392.7kN$

With this we can find a preliminary size for the top chord. In the AISC Manual (Part 3-Column Design), we can choose 102mm x 102mm x 9.5mm equal-leg double angles, 345MPa steel, an unbraced length of 2.4m, with an allowable load of 439kN. The entire top chord could be built with this size for speed of construction.

1. Calculate the total (roof or floor) load on each member between panel points separating the vertical and horizontal components, (if any, due to wind, for instance).

2. Statically resolve the total load between the two ends of the member, and apply the loads coming from all members at the same panel point. This would include both chord loads and any concentrated purlin or beam loads.

3. Find the stresses (axial loads) in the truss members using a stress diagram, method of sections, or method of joints.

4. Size the members for the axial loads. Members loaded between panel points must be designed for axial load and bending using the actual load distribution (step 2) for the bending stresses.

Problem M14.1

This Warren truss is built with equal-leg double angles. Design the tension chord for the maximum tension, using 345MPa steel.

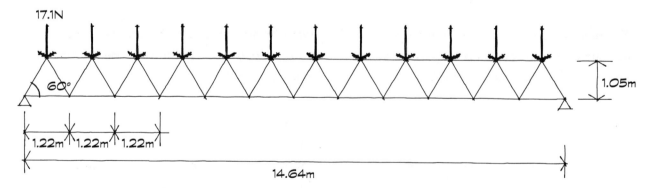

Solution:

Maximum T = 355.84kN, required area A = (355.84kN)/0.6(250MPa) = 2,372mm²

Use two 89mm x 89mm x 7.9mm (A = 2,700mm²) or two 102mm x 102mm x 6.4mm(A = 2,500mm²)

Problem 14.2

This truss is supporting structural panels directly on the top chord. Find the equivalent loading configurations with loads at the panel points.

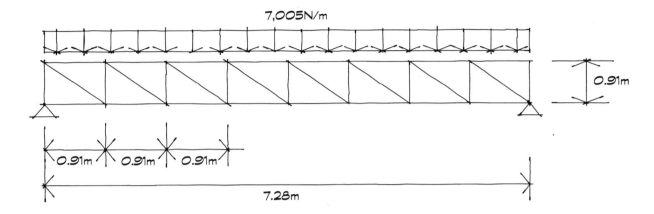

Solution:

6.4kN at each joint.

15 BOLTED CONNECTIONS

15.1 Engineering principles

Connections transfer the loads from one member to another: a beam to a column, a truss chord to a web member, etc. Connections provide continuity in a structural member and in a structural system. We have discussed built-up beams and columns, constructed using plates and other shapes welded together. In the past, prior to the advent of welding, riveted or bolted connections were used to accomplish the same ends. The design of connections is an essential part of structural design and can have a major role in cost and appearance of the finished product. The fundamental design step is to decide the type of connection and the way it transfers the load, consistent with the assumptions made for the structural members that are being connected. The bolted connection shown in Fig. 15.1(a) can be considered pinned in the statics investigation of the system, but if its point of application of load transfer to the column is significantly off center from the centroid of the column, it will create a bending moment in the column of $P \times$ eccentricity = M. Figure 15.1(b) shows a moment-resisting connection that transfers not only shear from the web of the beam, but moment from the flanges.

The design of a connection requires the investigation of two or three components: the members to be connected, the bolts (or welds), and possibly the connecting pieces: gusset plates and angles. We will be discussing bolted connections and for the most part ignoring rivets. Riveted connections are designed in a similar manner, but are seldom used in architectural construction. Their primary benefit is that they will not loosen under conditions of vibration, and their biggest problem is their application process. A railroad or highway bridge that is subjected to severe vibrations would be a potential application for rivets. A typical bolted "framed-beam connection" is shown in Fig. 15.1(c).

15.2 Types of bolted connections

The ASTM types of bolts and fasteners (rivets and threaded parts) are listed in the AISC Manual (p. 4-3) for diameters ranging from 5/8 to 1-1/2in[16 to 36mm]. Larger diameters and different types of steel are also available (Cor-ten, stainless steel, galvanized, etc.).

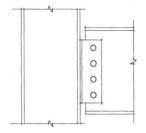

Fig. 15.1a Framed beam connection statically "pinned"

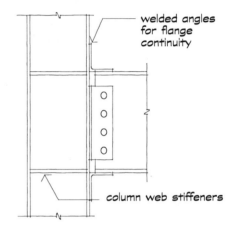

welded angles for flange continuity

column web stiffeners

Fig. 15.1b Framed moment connection

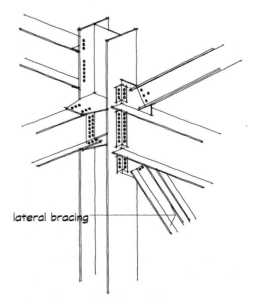

lateral bracing

Fig. 15.1c Typical framed beam connection combination

The ASTM designation is marked on the bolt head (Fig. 15.2). The most common bolt types are listed in Table 15.1.

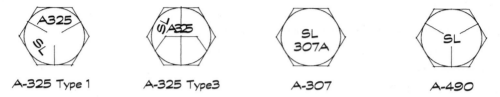

A-325 Type 1   A-325 Type3   A-307   A-490

Fig. 15.2 Registered head markings for structural bolts

Most connections are detailed so that the bolts are subject to a load perpendicular to their shank. In this case, the bolt is loaded in shear and the connection is designated as **shear type**. In some instances (typically, in rigid connections) some of the bolts are in tension, some in shear, and in some cases, both tension and shear. In the design of bearing-type (shear) connections, three criteria must be indicated: **bolt type**, **connection type (SC, N, or X)**, and **hole-type** (AISC Manual, Table 1.D, p.4-5).

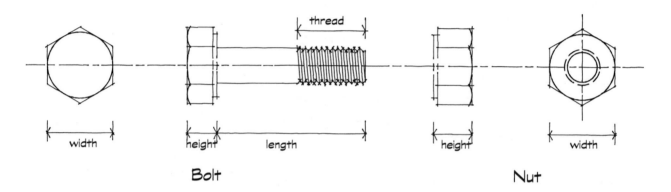

Bolt   Nut

Fig. 15.3 Basic bolt dimensions

A prime consideration in connection design should be the construction or erection process. Although the AISC Manual lists a wide variety of connection types, a few pragmatic considerations will eliminate some of them as options. A coped beam connection (Fig. 15.4) may be possible from an engineering point of view, but has potential field limitations. Precise spacing must be maintained between the angles to allow the flange to be slipped between them; this means the angles must then be bent to bring them into tight contact with the beam web. If the members are not very carefully handled on the job site, the angles can be "bumped" and the critical clearance between them reduced. The best option is to avoid coped flange connections and use those that have the angles welded to the beam in a fabricating shop.

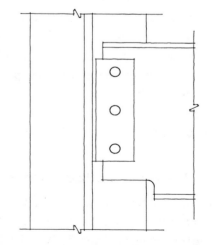

Fig. 15.4 Beams "coped" as shown for field erection should be avoided

Connection eccentricity is also a consideration in two ways (Fig. 15.5). The point of application or transfer of loads should be as symmetrical and as compact as possible. If a one-sided connection is used, torsional forces will be developed in the beam, a small moment will be developed in the girder, and the bolts in connection will be subjected to both shear from a direct vertical load and an additional shear component from the moment that will be induced from the eccentricity. The best option is to make sure the connection is geometrically tight, concentrically loaded, and can be erected on the job site with the least possiblity of error. You should note that the most common use for a coped flange is not to allow the web to slide between the angles, but to provide clearance when a beam frames into another beam or girder having the same top or bottom flange elevation (Fig. 15.6).

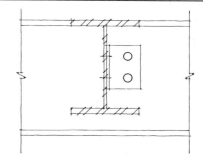

Fig. 15.5 Eccentric framed beam connections should be avoided

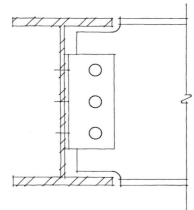

Fig. 15.6 Beam coped for top or bottom flange alignment

15.3 Bolts

The bolts listed in Table 15.1 are the most commonly used bolts in steel construction. The A307 bolts were once the standard of all applications, but are currently used predominately for secondary uses that do not require the high strength of the A325 or A490 bolts.

Table 15.1 Bolt types

A307	Unfinished; not for major connections
A325	High-strength; most common sizes: 3/4in and 7/8in [M20 and M22]
A490	High-strength; most common sizes: 3/4in and 7/8in [M20 and M22]

Source: AISC Manual, p. 4-3.

Connections are specified in basically three types: N, SC, and X:

Type N connections may have threads withiN a shear plane.

Type X connections have all threads eXcluded from shear planes.

Type SC connections are "Slip Critical or friction" types and consequently rely on high tension to create the frictional forces that keep the elements in place.

Type X connections can carry a greater load since the cross-section of the bolt in shear is larger; however, the type N connections carry high loads and it is not necessary for the fabricator to calculate the specific threaded length of every bolt used on the project. This also makes construction much simpler since it is not necessary to maintain such a stringent control over every bolt in every connection. How does one inspect to ensure that the correct bolt has been used based on its threaded portion? It's virtually impossible unless each bolt is coded in some manner - an expensive, time-consuming option. Type SC connections are effective particularly for moment-resisting connections, but in the event that they lose tension at any time during their effective life, they will become bearing-type connections. Consequently, it makes sense that they should also be designed as either a type X or a type N connection. AISC Manual, Table J3.2 (p. 5-73), indicates that the allowable bolt shear stress for SC connections is significantly lower than for the same bolt in a bearing-type connection. In addition, the tension is critical (no pun intended) in the case of SC connections, which is somewhat more expensive to develop than the techniques used for bearing types. The tension in the bolt is obtained by:

1. Using a calibrated torque or impact wrench.
2. Using washers with protrusions that are flattened by the force of the tension in the bolt head or the nut.
3. Using "break-off" bolts, which have an extended shank designed to twist off when proper tension is obtained. This is effectively a "one-time" torque wrench.
4. Using the most elementary method of tightening, when exact tensile force is not critical: the "turn-of-nut" method. In this case, the bolt is tightened until "snug" (not a very precise term) and then turned an additional two-thirds of a turn. This method is similar to that used to change the oil filter in your car.

All this considered, there seems to be little advantage to using SC-type connections; therefore, we'll use type N bolts most of the time for both low expense and simplicity.

Hole type is the next item in the specification for a connection. It can be round or slotted. In order to ensure alignment within construction tolerances, the hole or slot depth needs to be 1/16in [2mm] larger than the bolt diameter; so a 3/4in [M20] bolt will require a 13/16in [M22] hole. Although this isn't terribly significant in the bolt design, it is crucial if the member is in tension, where "net area" is a determinate in the member design. AISC Manual, Table J3.1, gives the maximum dimensions for different hole types.

Two AISC standards will be used in most designs:
1. Standard bolt hole is 1/16in [2mm] larger than the bolt diameter.
2. Standard spacing: 3in [75mm]o.c measured from center to center of hole.

Example   **227**

15.4 Design of connections

Bolted connections must be designed for the following:

1. Shear in the bolt: $F_v = P/A$

2. Bearing on the beam web: $F_p = P/A$ ($F_p = 1.2F_u$)

3. Bearing on the adjacent angle, column, or beam: $F_p = P/A$

4. Edge distance (a bearing problem, but doesn't use $F = P/A$ directly)

5. Shear and/or tension along the bolt line (block shear): $F_v = P/A$, $F_t = P/A$

    $F_v = 0.3F_u$, $F_t = 0.5F_u$

You may have noticed a simple idea behind this multistep design process.

The allowable stress for bearing in bolted connections is not based on the yield-stress capacity, but on the ultimate stress, F_u. This should seem a little strange, using a design value substantially above the yield point of the material. This is done to ensure that the load will be distributed equally among the connectors. As the material yields, a small area of material around (on the bearing side) each bolt will become strain hardened and will act as an internal bearing plate distributing the load out over a larger area until it does not exceed the yield stress of the base material.

Example 15.1

Beam: W18X50, A36 steel

Reaction = 38k; t_w = 0.355in; top flange coped 2in deep

Bolts: 3/4in diameter A325-N (13/16in diameter hole)

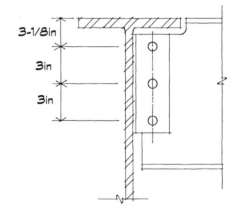

1. Shear in the bolt

    Allowable load tables (AISC Manual, p. 4-5)

    A325-N bolt

    3/4in diameter, D = double shear

    Allowable shear load = 18.6k

    Number of bolts required:

    (38k)/(18.6k/bolt) = 2.04 bolts

    Use two (within 5%) or three bolts.

    Angle legs connected to the columns: each bolt in single shear (S).

    Number of bolts required:

    (38k)/(9.8k/bolt) = 4.09 bolts

Use four bolts, or if you're nervous about the 5% application, you must use six bolts to create a symmetrical pattern for the angle connection.

The symmetrical pattern is not an architectural consideration, but one that ensures that there will be no eccentricity developed in the connection system. Remember, eccentricity means moment that may not have been considered in the design.

YOU SHOULD NOTE THAT THE SECOND CALCULATION WAS REDUNDANT SINCE THE CAPACITY FOR BOLTS IN SINGLE SHEAR IS EXACTLY 50% OF THE CAPACITY OF BOLTS IN DOUBLE SHEAR AND WILL CONSEQUENTLY REQUIRE EXACTLY TWICE THE NUMBER OF BOLTS. THIS IS BECAUSE THE LOAD CAPACITY IN SHEAR IS EQUAL TO F = P/A OR P = FA, AND SINCE THE CROSS-SECTIONAL AREA IS EXACTLY 50%, THEN THE LOAD CAPACITY IS ALSO EXACTLY 50%.

2. Bearing on the beam web:

F_p = P/A or A = P/F_p

where F_p is the allowable bearing pressure.

F_u = 58ksi for A36 steel

   (AISC Manual, p. 5-118)

F_p = 1.2F_u = 1.2(58ksi) = 69.6ksi

   (AISC Manual, p. 4-6)

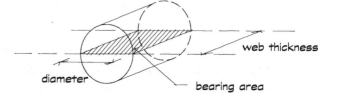

Bearing area A = D(t_w) = (0.75in)(0.355in) = 0.27in^2

This is the **projected area** of the bolt bearing on the web material.

Bearing capacity of each bolt:

P = F_pA = (69.6ksi)(0.27in^2) = 18.8k

Number of bolts required = (38k)/(18.8k/bolt) = 2.02 bolts

Use two bolts (within 5%) or three bolts if you're not an adventurer!

An alternative method would be to use the allowable load table (AISC Manual, p. 4-6).

Allowable load on each bolt per inch of web thickness: 52.2k.

38k = N(52.2k/in)(0.355in web)

N = Number of bolts = (38k)/[(52.2k/in)(0.355in)] = 2.05 bolts

Either method works; just use whichever is easier for you!

Bearing on angles:

The most effective approach at this point is to simply select a standard angle that will work.

A little logic will solve the problem here. If two bolts will transfer the load from 0.355in thick material in the beam web, then the same two bolts will require a total of 0.355in of angle thickness to give that same load back up into the angles. Assume equal-strength materials.

We would suggest you try a 0.355in/2 angle = 0.175in. The closest standard thickness is 3/16in, which is 0.1875in.

Example **229**

If, however, you wish to design the angles:

Required bearing area = P/F_p = (38k)/(1.2)(58ksi) = 0.5459in^2

(0.5459in^2)/2 bolts = 0.273in^2/bolt bearing area

Angle thickness =. (.9273in^2)/(0.75in) bolt diameter = 0.366in total, or

(0.366in)/2 = 0.183in per angle

The reason we get a slightly larger thickness is because we actually needed slightly more than two bolts, but we chose to use two.

3. Edge distance:

From the AISC Manual (p. 4-7), for web: $R = N(P_b)t_w$

where

> R = reaction
>
> N = No. of bolts
>
> P_b = bearing capacity/bolt as a function of edge distance
>
> t_w = web or angle thickness

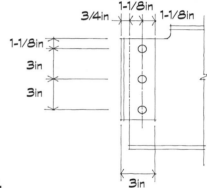

38k = 2(P_b)(0.355in)

P_b = (38k)/(2)(0.355in) = 53.52k/in required per bolt

Enter the table for a value of 53.52k or more under

> F_u = 58ksi

The largest value for 3/4in bolts is 52.2k/bolt; this means that we could use the 1-1/8in edge distance associated with the value of 52.5k/bolt, which is within the 5% margin of error, or use three bolts and recalculate the new pressure requirements.

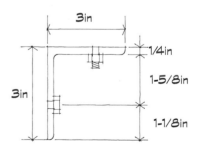

Using these edge distances, we can calculate the angle size:

> 1-1/8in edge distance (beam end)
>
> + 1-1/8in edge distance (angle edge)
>
> + 3/4in standard angle extension beyond end of beam
>
> 3.0in   Use a 3in x 3in x 3/16in angle.

Construction clearances may require additional dimensions for wrench clearance, etc. This explains the different dimensions on the diagrams on the previous page (AISC Manual, pp. 4-12 and 4-137).

The angle length also can be determined by using edge distance and standard bolt spacing.

> 1-1/8in top edge distance (angle)
>
> 3in   spacing between two bolts
>
> <u>1-1/8in</u> bottom edge distance (angle)
>
> 5-1/4in minimum angle length

We would recommend, however, that the angle length be approximately 75% of the beam depth to create more stability in the system during construction. This would suggest that the angle would be 0.75(18in) = 13.5in long.

4. Shear along the bolt line (block shear):

    Block shear in beam web:

    Allowable shear stress (AISC Manual, p. 5-77): $F_v = 0.30F_u = 0.30(58ksi) = 17.4ksi$

    Allowable tensile stress: $F_t = 0.50F_u = 0.50(58ksi) = 29ksi$

These values are slightly higher than the normal allowable shear stress and tensile stress for A36 steel, but could be accounted for by the fact that some of the material around each hole has been strain hardened and will have altered properties.

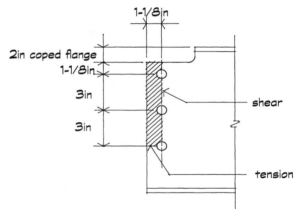

> Allowable P: P (shear) + P (tension)
>
> Since $F = P/A$; $P = AF$,
>
> Allowable $P = A_v F_v + A_t F_t$
>
>   (See AISC Manual, Table p.4-8)
>
> Allowable $P = (L_v t_w)F_v + (L_t t_w)F_t$

where L_v (vertical length of material):

> $L_v = (1\text{-}1/8in + 6in) - 2(13/16in) - (13/16in)/2 = 5.08in$

L_h (horizontal length of material):

> $L_h = 1.125in - (13/16in)/2 = 0.718in$

Therefore the allowable P is

> P= 5.08in × 0.355in × 17.4ksi + 0.718in × 0.355in × 29ksi = 38.77k > 38k   OK!

If we were to make the connection angle 13.5in, as we suggested earlier, the distance between the holes would be increased and the block shear problem disappears.

Block shear in angles: (using the 13-1/2in recommended length)

Minimum net failure length: 13.5in - 2(13/16in) = 11.875in

Allowable P = 11.875in × 2 × 0.1875in × 17.4ksi = 77.48k

Everything works. . . . Finis!

Problem 15.1

A W16X26 beam, whose top flange is coped, is connected to the flange of a W18X60 with a pair of angles. The support reaction at the beam end is 45k, and the steel is A572, grade 50, which has an F_u of 65ksi from the AISC Manual (p. 5-118). Design the connection with A325-N 3/4in bolts.

Solution:

Shear on bolts:

Connection type N, STD hole, D (double shear) load condition, from the AISC Manual, (p. 4-5), P = 18.6k/bolt

Number of bolts = (45k)/(18.6k/bolt) = 2.4 bolts; use three bolts (at this point).

Bearing on web:

Bolt capacity, with t = 1/3in, from the AISC Manual, (p. 4-6), P_b = 14.6k/bolt

Connection capacity with three bolts: P = 3(14.6k/bolt) = 43.9k = 45k within 5%

Edge distance:

Required capacity of each bolt in W16X26 web = (45k)/(3bolts)(0.25in) = 60k/in

(from the AISC Manual, p. 4-7)

Minimum distance = 1-1/8in = 1.125in (the capacity is actually 58.5k, but within our 5% margin)

Angle size:

Length = 3/4 beam depth = 0.75(16in) = 12in

Minimum length = 1.125in + 3.0in + 3.0in + 1.125in = 8.25in < 12in   OK!

Place the top bolt at 1-1/8in from top.

We chose L - 4in × 3-1/2in × 1/4in, so the thickness is not a problem (total t = 1/4in + 1/4in > 1/4in web).

Block shear in web:

L_v = 5.09in, L_h = 0.719in, P = (5.09in)(19.5ksi) + (0.719in)(32.5ksi) = 122.6k > 45k OK!

Use three 3/4in bolts.

Problem 15.2

A tension member consisting of a 3in standard pipe supports an axial load of 66k. The end is adequately welded to a rectangular splice plate 5in wide and 5/8in thick, bolted to a tapered gusset plate 5/8in (.625in) thick. The steel is 36ksi. Design the bolted connection, including the length of the splice plate on the bolts side, using 1in bolts, A325.

Solution:

Shear on bolts:

> Number of bolts = (66k)/(16.5k/bolt) = 4 bolts
>
> Bearing P = 4(48.8k) > 66k   OK!
>
> Edge distance required P = (66k)/(4)(0.625in) = 26.4k/in; L_v = 1in

Block shear:

> L_t = 5in - 17/16in = 3.94in
>
> P = (3.94in × 0.625in)(0.5 × 65ksi) = 80k > 66k   OK!

Length of splice plate = 1.0in + 3.0in + 3.0in + 3.0in + 1.0in = 11.0in + length welded to pipe

---

Problem 15.3

A tension member, made with a pair of L - 3in × 2in × 1/4in angles, short leg outstanding, is spliced with a 3/8in (0.375in) plate and a row of 5/8in (0.625in) bolts. Find the required number of A490 bolts and calculate the appropriate size of the gusset plate with A441 grade-50 steel if the axial load is 59k.

Solution:

Number of bolts = (59k)/(17.2k/bolt)= 3.43 bolts

Use four bolts on each side.

Splice plate:

> A = (59k)/(0.6 × 50ksi) = 1.97in² gross area
>
> A = (59k)/(0.5 × 70ksi) = 1.69in² net area
>
> Bolt hole A = (3/8in × 11/16in) = (0.375in × 0.6875in) = 0.26in²; gross area controls

Note the illusion of supreme accuracy the calculators give us: 0.6875, as though it matters.

> Plate width = (1.97in²)/(0.375in) = 5.25in

Plate length:

> Edge distance for P = (59k)/(4)(0.375in) = 39.3; l_v = 1-1/8in edge distance
>
> Length required for bolts = 2(1.125in + 3.0in + 3.0in + 3.0in + 1.5in) = 23.25in

Use 3/8in thick plate at 6inx24in.

METRIC VERSIONS OF ALL EXAMPLES AND PROBLEMS

The following metric tables are provided for use in the examples and problems in this book. They are conversions of the AISC Manual tables for SI units and are not generallyavailable at this time.

Table M15-1 Bolt properties ASTM A325M-A490M

ASTM designation	Conn. type	Hole type	Loading type	Bolt diameter (mm) and area			
				M16	M20	M22	M24
A325 [a]	N	STD	S	29.1	45.5	55.1	65.5
			D	58.2	91.0	110.2	131.0
A490 (+)N		STD	S	38.8	60.6	73.3	87.2
			D	77.6	121.2	146.6	174.4

[a] $F_v = 145$ MPa (N/mm²)(+) $F_v = 193$ MPa (N/mm²)

Table M15-2 Bolts: shear allowable load in kN

Bolt D (mm)	Hole (mm)	Area (mm²)	US equivalent in. (mm)	
M16	18	201	5/8	15.87
M20	22	314	3/4	19.05
M22	24	380	7/8	22.22
M24	26	452	1	25.40
M27	30		1-1/8	28.60

Table M15-3 Bolts: bearing

Allowable bearing pressure $F_p = 1.2 F_u$

Steel type	F_u(MPa)	F_p (MPa)
250MPa	400	480
345MPa	448	538

Table M15-4 Bolts: edge distance

Allowable load

Bolt diameter	Edge distance	Allowable load (kN)	
		F_u = 400MPa	448MPa
M16	25	5.1	5.7
M20	30	9.1	10.2
M22	33	10.7	12.0
M24	36	12.2	13.7

Table M15-5 Bolts: block shear

Allowable stresses

| Allowable | Steel type | |
stress	250MPa	345MPa
$F_v = 0.30F_u$	75	103
$F_t = 0.50F_u$	125	172

Example M15.1

Beam: W460X74, 250MPa steel

Reaction = 169kN; t_w = 9.02mm; d = 457mm; top flange coped 50mm deep

Bolts: M20 A325-N (22mm diameter hole)

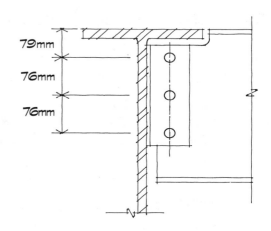

1. Shear in the bolt

       Allowable load tables (Table M15-2)

       A325-N bolt

       M20 diameter, D = double shear

       Allowable shear load = 91.0kN

Number of bolts required:

       (169kN)/(91.0kN/bolt) = 1.86 bolts

       Use two bolts.

Angle legs connected to the columns: each bolt in single shear (S).

       Number of bolts required: (Table M15-3)

       169kN/(45.5kN/bolt) = 3.71 bolts

       Use four bolts.

The symmetrical pattern is not an architectural consideration, but one that ensures that there will be no eccentricity developed in the connection system. Remember, eccentricity means moment that may not have been considered in the design.

YOU SHOULD NOTE THAT THE SECOND CALCULATION WAS REDUNDANT SINCE THE CAPACITY FOR BOLTS IN SINGLE SHEAR IS EXACTLY 50% OF THE CAPACITY OF BOLTS IN DOUBLE SHEAR AND WILL CONSEQUENTLY REQUIRE EXACTLY TWICE THE NUMBER OF BOLTS. THIS IS BECAUSE THE LOAD CAPACITY IN SHEAR IS EQUAL TO F = P/A OR P = FxA, AND SINCE THE CROSS-SECTIONAL AREA IS EXACTLY 50%, THEN THE LOAD CAPACITY IS ALSO EXACTLY 50%.

2. Bearing on the beam web:

$$F_p = P/A \text{ or } A = P/F_p$$

where F_p is the allowable bearing pressure.

F_u = 480MPa for 250MPa steel

$F_p = 1.2F_u = (1.2)(250MPa) = 480MPa$

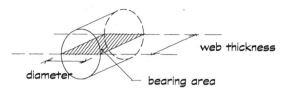

web thickness

diameter — bearing area

Bearing area $A = D(t_w) = (20mm)(9.02mm) = 180.4mm^2$

This is the **projected area** of the bolt bearing on the web material.

Bearing capacity of each bolt:

$P = F_pA = (480MPa)/(1000kN/N)(180.4mm^2) = 86.6kN$

Number of bolts required = (169kN)/(86.6kN) = 1.95 bolts

Use two bolts.

Bearing on angles:

The most effective approach at this point is to simply select a standard angle that will work.

A little logic will solve the problem here. If two bolts will transfer the load from 9.02mm thick material in the beam web, then the same two bolts will require a total of 9.02mm of angle thickness to give that same load back up into the angles. Assume equal-strength materials.

We would suggest you try a (9.02mm)/2 angle = 4.51mm. The closest standard thickness is 4.8mm.

If, however, you wish to design the angles:

Required bearing area = P/F_p = (169kN)/(1.2)(0.250kN/mm²) = 352mm²

(352mm²)/2 bolts = (176mm²)/bolt bearing area

Angle thickness = (176mm²)/(20mm bolt dia.) = 8.8mm total, or

4.4mm per angle < 4.51mm

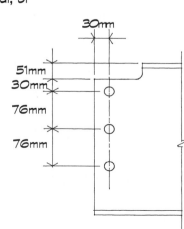

30mm

51mm
30mm

76mm

76mm

3. Edge distance:

From the edge distance table (M15-4), with a M20 bolt, a distance of 30mm gives 9.1kN per bolt per mm of thickness.

Number of bolts = (169kN)/(9.1kN)(9.02mm)

= 2.06 bolts

Two bolts is within 3% of this requirement, so two bolts are acceptable. Use two bolts.

Using these edge distances, we can calculate the angle size:

      30mm web edge distance (beam end)

  + 30mm edge distance (angle edge)

  + <u>20mm</u> standard angle extension beyond end of beam

      80mm

Use a 89mm x 89mm x 6.4mm angle.

Construction clearances may require additional dimensions for wrench clearance, etc.

The angle length can also be determined by using edge distance and standard bolt spacing.

      30mm top edge distance (angle)

      76mm spacing between 2 bolts

      <u>30mm</u> bottom edge distance (angle)

      136mm minimum angle length

We would recommend, however, that the angle length be approximately 75% of the beam depth to create more stability in the system during construction. This would suggest that the angle would be 0.75 (457mm) = 343mm long. Using a 300mm angle will be a good practical option.

4. Shear along the bolt line (block shear):

    Block shear in beam web:

    Allowable shear stress (AISC Manual, p. 5-77):

$$F_v = 0.30F_u = 0.30(400MPa) = 120MPa$$

    Allowable tensile stress: $F_t = 0.50F_u = 0.50(400MPa) = 200MPa$

These values are slightly higher than the normal allowable shear stress and tensile stress for 250MPa steel, but could be accounted for by the fact that some of the material around each hole has been strain hardened and will have altered properties.

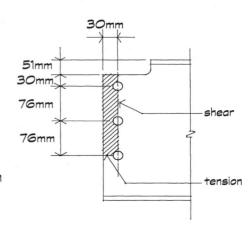

Allowable P: P = P(shear) + P(tension)

Since F = P/A; P = AF

Allowable $P = A_v F_v + A_t F_t$

      $= (L_v t_w) F_v + (L_t t_w) F_t$

where L_v (length of material in shear) =

    (30mm + 240mm - 22mm - 11mm) = 237mm

L_t (length of material in tension) =

    (30mm - 11mm) = 19mm

Therefore the allowable P is

P= (237mm × 9.02mm)(120N/mm²) + (19mmX9.02mm)200N/mm² = 290.8kN

Since 290.8kN > 196kN, it works. OK!

Block shear in angles:

L_v (length of material in shear) = (300-22-22)mm = 256mm

Allowable P = (256mm × 2 × 6.4mm)120N/mm² = 393.2kN > 196kN OK!

Everything works. . . . Finis!

Problem M15.1

A W410X39 beam, whose top flange is coped, is connected to the flange of a W460X89 with a pair of angles. The support reaction at the beam end is 200kN, and the steel is A572 grade-345MPa. Design the connection with A325-N M20 bolts.

Answer:

Use three M20 bolts.

Problem M15.2

A tension member consisting of a 75mm standard pipe supports an axial load of 294kN. The end is adequately welded to a rectangular splice plate 130mm wide and 16mm thick, bolted to a tapered gusset plate 16mm thick. The steel is 250MPa. Design the bolted connection, including the length of the splice plate on the bolts side, using A325 M24 bolts, type N connection.

Answer:

Use four bolts.

Length of splice plate = 280mm + length welded to pipe.

Problem M15.3

A tension member, made with a pair of L 76mm x 51mm x 6.4mm angles, short leg outstanding, is spliced with a 10mm plate and a row of M16 bolts. Find the required number of A490 bolts and calculate the appropriate size of the gusset plate with A441 grade-345MPa steel if the axial load is 262kN.

Answer:

Use four bolts on each side.

Use a 10mm thick plate at 150mm x 600mm.

16 WELDED CONNECTIONS

16.1 Welding and types of welded joints

Welding is a technique of connecting metal parts (called the base material) by simultaneously melting both pieces of base material as well as the electrode itself, using an electrical discharge similar to a short circuit in a wiring system. When properly executed, this will create continuity across the joint or at least adequate continuity to transfer the design load. Although the base metal and filler material can be heated with a mixture of oxygen and acetylene, this is a very slow process. The procedure adopted in structural steel work uses an electric arc to heat the base metal and a rod (electrode) of flux-coated steel matching or stronger than the base material.

The electrode is placed in an insulated clamp held by the operator and supplied with electricity via a flexible cable. The other end of the electrical circuit is connected to the steel to be welded. When the electrode is brought in close proximity to the base material, an "arc" will jump between the two pieces and the circuit is completed. This is a process that requires a skilled craftsman. If any of you have ever tried to arc weld, you soon found that if you brought the electrode too close, it simply welded itself to the base material and began to heat the entire system. Even if you maintain the arc, if proper procedures are not used, you may merely place the weld material on the surface of one of the two parts and, without adequate penetration, it will simply fall apart under a small load.

If the molten steel comes into contact with the air, there will be a chemical reaction that may potentially make the steel brittle and susceptible to corrosion. This is prevented by the flux that coats the electrode. The flux produces an inert gas when heated, shielding the molten metal as well as a slag coating over the weld, further protecting it while it is cooling. The type of welding process is closely related to the type of electrode and flux material used. The most common types of electrodes are listed in Table 16.1.

Table 16.1 Types of electrodes for arc welding

Designation	Yield stress		Allowable stress (ksi) for most applications [a]	MPa
	F_y (ksi)	[MPa]		
E60XX	60	414	$F_a = 0.3(60) = 18$ksi	124MPa
E70XX	70	482	$F_a = 0.3(70) = 21$ksi	145MPa
E80XX	80	551	$F_a = 0.3(80) = 24$ksi	165MPa

[a] The allowable design strength of welds is a function of type of weld, but for most applications of fillet welding, $0.3F_y$ is acceptable.

The allowable stress relates to the strength of the weld metal, which has to be compatible with the base material. E70XX is frequently specified for welding A-36 [250MPa] steel, E80XX is used for A-572 grade 65 [450MPa] steel. You will note that all electrodes are individually marked as to their yield stress, e.g., E70, and with two additional numbers replacing the XX noted before. These values are of little concern to you as a designer, but are significant to the fabricator. They identify other characteristics for which the electrode is designed: DC vs. AC current, vertical or horizontal weld position, etc.

The manual process, known as the shielded-arc process, is used in structural welding, with flux-coated electrodes. In the submerged-arc process, the flux is fed by a separate conduit and covers the weld line. In gas-metal-arc welding, used for aluminum and stainless steel, the flux is replaced by a stream of gas shielding the molten metal. In flux-cored-arc welding, the flux is contained in the core of the electrode. So, for example, an electrode E70T-X indicates flux-cored-arc process. The submerged-arc process is used most commonly in shop fabrication situations.

While cooling, the weld will also shrink, producing residual stresses in the weld and base material; this stress can be reduced if the base material is preheated and expanded, so that it will shrink back with the weld. Preheating is necessary in large steel members to avoid a quick dissipation of heat during welding, which will make fusion difficult. Most importantly, shrinkage takes place across the weld section itself, and the bead of the filler material must be built up to form a convex surface in order to avoid shrinkage cracks. For this reason, welds greater than 3/8in [9.52mm] are generally placed with consecutive passes (multiple layers of weld).

When the weld is created off-site, it is generally referred to as **shop welding**, and when executed at the building site, it is called **field welding**. In order to ensure the best connection, it is preferable to shop weld as much as possible and use bolted connections on the site. This allows the welding to be done under ideal conditions and tested with nondestructive tests (X-rays or magniflux) to ensure their integrity and allows the field connections to be bolted and more easily confirmed. Most of the common welded joints used in steel structures are designated **prequalified** if they conform to the provisions of the Structural Welding Code or the AISC Specifications.

Some other construction materials such as cast iron can be welded, but require preheating to minimize cracking resulting from the differential heating in the area of the weld.

Table 16.2 lists the most common joint and weld types. The most common of all welds is the fillet weld. It is simple and requires little preparation of the materials. You'll note that some joints require grinding to produce enough clearance to allow the electrode to make contact with the material at an appropriate depth. If the arc is not struck in the proper location, the material will not become molten throughout the full weld cross-section and you only have a cosmetic weld.

Table 16.2 Common joint types and weld types

Joint type	Symbol	Section of joint
Butt joint	B	
Corner joint	C	
T - joint	T	
Lap joint		

Weld type	Symbol	Section of weld
Fillet		
Plug or slot		
Groove or butt (square)		
Groove or butt (bevel)		

Since fillet welds are the most common welds used in construction and since the methodology for design of welds is common to all types, we'll use fillet welds in our illustrations.

Fillet weld dimensions:

Weld size = sides of triangular section of weld connected to the base material

Throat dimension = minimum distance from corner to surface of weld

$$= (\text{weld size} \times 0.707 \text{ for a } 90° \text{ corner})$$

Unless stated otherwise, fillet welds are assumed to have equal legs, i.e., 45°angles, and the throat is the least dimension between the two faces.

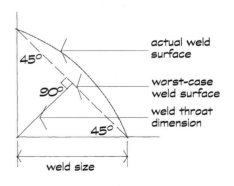

Maximum size of fillet welds:

Material < 1/4in [6mm] thick:

      Not greater than material thickness.

Material = 1/4in [6mm] or more thick:

      1/16in [2mm] less than the material thickness.

16.2 Stresses in welds

The stress in a weld is considered as shear on the throat dimension. This is referred to as "shear on the **effective area**" for complete-penetration groove welds, fillet welds, and plug welds.

Using $F_a = P/A$:   Allowable $P = F_a A$

For fillet welds:

A = (throat X length of weld) = effective area

$F_a = 0.3F_y$ (AISC Manual, p. 5-70)

Electrode E70XX: $F_y = 70$ksi [482MPa] $F_a = 0.3(70$ksi$) = 21$ksi [0.3x482MPa = 145MPa]
Electrode E60XX: $F_y = 60$ksi [414MPa] $F_a = 0.3(60$ksi$) = 18$ksi [0.3x414MPa = 124MPa]

Plug and slot welds (AISC Manual, pp. 5-67, 2a, and 5-68, 3)(the smaller of the following):

    Effective shearing area A = cross-sectional area of the hole or slot, or

    Effective area A of fillet weld in hole or slot = (throat)(length of **centerline** of weld).

Groove welds (AISC Manual, p. 5-65, Art.1):

    Effective area A = (minimum thickness of material)(length of weld)

Example 16.1

A 3/8in x 2in plate, A36 steel, is connected by E70XX fillet welds to a gusset plate. The tension load that can be safely supported by the plate is:

    P = 0.6(36ksi)(0.375in)(2in) = 16.2k

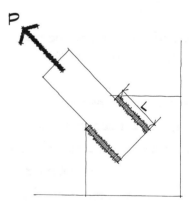

Maximum size of fillet weld:

    D = 3/8in - 1/16in = 5/16in = 0.3125in

Throat dimension:

    T = 0.3125in x 0.707 = 0.221in

Required area of the throat:

    $F_a = P/A$; A = P/F_a = (16.2k)/(0.3)(70ksi) = 0.771in^2

Length of fillet weld required = area/throat dimension, A/T:

    L = A/T = (0.771in^2)/(0.22in) = 3.5in

Example 16.2

A laminated wood beam is supported on a 4in x 4in x 1/4in steel tube column. The support reaction is 19.6k. The connection consists of a 1/4in-thick steel stirrup with blind (hidden) welds, E70XX electrodes.

Maximum size of fillet weld:

D = 1/4in - 1/16in = 3/16in = 0.1875in

Throat dimension:

T = 0.1875in × 0.707 = 0.1325in

Required throat area:

A = P/F = (19.6k)/0.3(70ksi) = 0.93in^2

Length of fillet weld required = area/throat dimension, A/T

L = A/T = (0.93in^2)/(0.1325in) = 7.01in

If we use two circular holes for the blind weld to help provide stability to the connection:

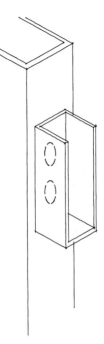

Circumference of center line of each of two welds = (7.01in)/2 = 3.5in

Diameter of center line circumference D = (3.5in)/(3.14) = 1.11in

Diameter of hole = d = 2(size)/2 = 1.11in + 0.187in = 1.297in

Use two 1-5/16in diameter holes.

16.3 Framed beam connections

Column-beam connections can be built using combinations of welds and bolts, or a welded angle. The AISC Manual (Table IV, p. 4.27) gives the load capacity of fillet welds on angles as a function of angle length.

Example 16.3

Using a framed-beam connection, connect a W30X173 beam to a W12X96 column web. Use the standard combination of a welded connection at the beam web and a bolted connection at the column web.

Use E70XX electrodes, 3/4in A-325N bolts.

W30X173: t_w = 0.655in; b_f = 14.98in; T = 26.75in; P = 127k

W12X96: t_w = 0.550in; T = 9.5in

(Since the clearance between the column flanges is less than the flange width of the beam, the sides of the beam flanges will need to be trimmed.)

By using an angle length equal to 75% of the depth of the beam, for stability during construction, an initial selection would look at an angle length of 0.75(30in) = 22.5in (roughly).

From the AISC Manual (tables on p. 4-26 or 4-30):

(These tables are interchangeable, the only difference being that the angle lengths on p. 4-26 are based on the bolted connection requirements and the angle lengths on p. 4-30 are based purely on welded connection requirements.)

A weld type A , 22in long, using a 3/16in weld carries 122k; OK (within 5% of 127k).

Minimum web thickness for this weld at location A = 0.38in < 0.655in; OK!

Minimum angle thickness using a 3/16in weld could be as little as 3/16in, but 1/4in is standard.

Checking the bolted connection at the column web:

Shear (AISC Manual, p.4-5) with bolts in single shear = 9.3k/bolt

Number of bolts = (127k)/(9.3k/bolt) = 13.65 or 14 bolts

> Use seven bolts in each angle leg at 3in spacing:
>
> > 6 spaces × 3.0in = 18in + 2(1in edge distance) = 20in

The welded connection we have selected requires 22in; OK!

Bearing on angles:

> Bearing area = 7bolts × 2 angles × (0..25in × 0.75in) = 2.63 in²
>
> Allowable P = 2.63in² (1.2 × 58ksi) = 183k > 127k. OK!

Edge distance for angles:

Allowable load on each bolt = (127k)/(14 bolts × 1 angle × 0.25in thick) = 36.28k/in

From the AISC Manual (table on p.4-7) for F_u = 58ksi:

This equates to an edge distance = 1-1/4in. This would suggest that the angle length needs to be 18in + 2(1.25in edge) = 20.5in <22in.

Block shear in angles:

Since the vertical and horizontal edge distances are equal and since the allowable stress for material in shear ($0.3F_u$) is less than the allowable stress for material in tension ($0.5F_u$), shear will control! If the distances are not equal, check both possible modes of failure.

> L_v = 22in -7 holes at (13/16in) = 16.31in
>
> Allowable $P_v = F_v A_v$ = 0.3(58ksi) × 2 angles × (16.31in × 0.25in)
>
> > P_v = (17.4ksi)(8.15in²) = 141.8k > 127k

Solution:

Use angle 4in × 3in × 1/4in, 22in long, A-36 steel.

Angle-to-beam connection:

> E-70XX electrodes
>
> 3/16in fillet welds, 22in long each side

Angle-to-column connection:

> Fourteen A-325N bolts, 3/4in diameter, 3in spacing; edge distance of 1-1/4in minimum

Example 16.4

Using the same criteria as Example 16.3, reverse the connection: bolt at the beam web and weld at the column web. To avoid the undesirable construction situation of trying to slide the beam web between the two welded column angles, we will field weld the column connection. When this is done, the column connection will typically use two erection bolts to allow a temporary connection that can then be welded at a more convenient time. If done in this manner, the crane will not need to hold the beam in place while welding is accomplished.

Use 1/4in fillet welds, E70XX electrodes, 3/4in A-325N bolts (threads iNcluded in the shear planes).

W30X173: t_w = 0.655in; b_f = 14.98in; T = 26.75in; P = 127k

W12X96: t_w = 0.550in; T = 9.5in

(Since the clearance between the column flanges is less than the flange width of the beam, the sides of the beam flanges will need to be trimmed.)

By using 1/4in fillet welds, the minimum angle thickness is 1/4in + 1/16in = 5/16in

From the AISC Manual (p. 4-30), weld B capacity 137k with 1/4in weld and L = 22in. You should note that the angle size could also be determined by the bolted-connection requirements.

Shear (AISC Manual, p. 4-5) with bolts in double shear = 18.6k/bolt

Number of bolts = (127k)/(18.6k/bolt) = 6.8 or 7 bolts

Using seven bolts at 3in spacing:

> 6 spaces × 3.0in = 18in + 2×(1in edge distance) = 20in

The welded connection requires 22in; OK!

Bearing on angles:

> Bearing area = 7bolts × 2 angles × (0.3125in × 0.75in) = 3.28in²
>
> Allowable P = (3.28in²)(1.2 × 58ksi) = 228.3k > 127k. OK!

Edge distance for angles:

Allowable load on each bolt = (127k)/(7bolts × 2angles ×0.3125in thick) = 29k/in

From the AISC Manual (table, p.4-7), for F_u = 58 ksi: this equates to an edge distance of 1in. Since the angle thickness is 2(5/16in) < t_w, this is also OK for the beam.

Block shear in angles:

Since the vertical and horizontal edge distances are equal and since the allowable stress for material in shear ($0.3F_u$) is less than the allowable stress for material in tension ($0.5F_u$), shear will control! If the distances are not equal, check both possible modes of failure.

> L_v = 22in -7holes at (13/16in) = 16.31in

Allowable $P_v = F_v A_v$ =).3(58ksi)(16.31in × 0.655in)

P_v = (17.4ksi)(10.68in²) = 185.9k > 127k

Solution: Use angle 4in × 3in × 5/16in, 20in long, A-36 steel.

Angle to column connection:

E-70XX electrodes; 1/4in fillet welds, 22in long each side

Angle-to-beam connection: 7 - A-325N bolts, 3/4in diameter, 3in spacing; edge distance of 1in minimum.

Problem 16.1

A truss is fabricated with 4in pipes, welded at the joints to 1/4in gusset plates, A36 steel, E60XX electrodes. The pipes all converge at a joint. Design the welds of the gusset plate on the top chord as 1/4in fillet welds, one on each side.

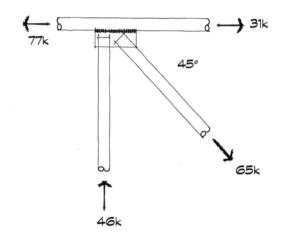

Solution:

The connection of the gusset plate with the chord transfers a load P = 77k - 31k = 46k.

F_a = 0.3(60ksi) = 18ksi

Throat = 0.707(0.25in) = 0.177in

L = (46k)/(0.177in)(18ksi) = 14.44in total

Make L = 8in on each side.

Problem 16.2

A 3/8in end plate for a joist-beam connection is welded to the end of a W8X24 (it will be field-bolted to the girder). If the joist end reaction is 23k, design the welds on the end plate. Use 3/16in fillet welds with E60XX electrodes.

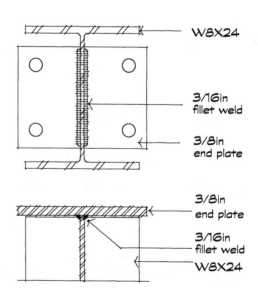

Solution:

Throat = 0.707(3/16in) = 0.132in

L = (23k)/(0.132)(18ksi) = 9.68in total

Use a 5in-long fillet weld on each side of the web.

horizontal section

METRIC VERSIONS OF ALL EXAMPLES AND PROBLEMS

Example M16.1

A 9.5mmX50mm plate, 250MPa steel, is connected by E70XX (482MPa) fillet welds to a gusset plate. The tension load that can be safely supported by the plate is

$$P = 0.6(250\text{MPa})(9.5\text{mm} \times 50\text{mm}) = 71,250\text{N}$$

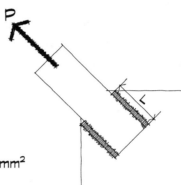

Maximum size of fillet weld:

$$D = 9.5\text{mm} - 2\text{mm} = 7.5\text{mm}$$

Throat dimension:

$$T = 7.5\text{mm}(0.707) = 5.3\text{mm}$$

Required area of the throat:

$$F_a = P/A, \quad A = P/F_a = (71,250\text{N})/(0.3)(482\text{MPa}) = 493\text{mm}^2$$

Length of fillet weld required = area/throat dimension, A/T:

$$L = A/T = (493\text{mm}^2)/(5.3\text{mm}) = 93\text{mm}$$

Example M16.2

A laminated wood beam is supported on a 102mm x 102mm x 6mm steel tube column. The support reaction is 87.2kN. The connection consists of a 6mm thick-steel stirrup with blind (hidden) welds, E70XX (482MPa) electrodes.

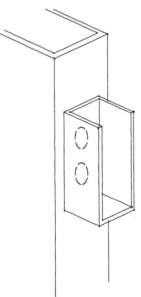

Maximum size of fillet weld:

$$D = 6\text{mm} - 2\text{mm} = 4\text{mm}$$

Throat dimension:

$$T = (4\text{mm})(0.707) = 2.8\text{mm}$$

Required throat area:

$$A = P/F = (87,200\text{N})/(0.3)(482\text{MPa}) = 603\text{mm}^2$$

Length of fillet weld required = area/throat dimension, A/T:

$$L = A/T = (603\text{mm}^2)/(2.8\text{mm}) = 215\text{mm}$$

If we use two circular holes for the blind weld to help provide stability to the connection:

Circumference of center line of weld = (215mm)/2 = 107.5mm

Diameter of center line circumference d_c = (107.5mm)/(3.14) = 34mm

Diameter of hole, d = 2(size)/2 = 34mm +4mm = 38mm

Use two 40mm diameter holes.

Example M16.3

Metric versions of these tables are not available.

Example M16.4

Metric versions of these tables are not available.

Problem M16.1

A truss is fabricated with 162mm pipes, welded at the joints to 6mm gusset plates, 250MPa steel, E60XX (414MPa) electrodes. The pipes all converge at a joint. Design the welds of the gusset plate on the top chord as 6mm fillet welds, one on each side.

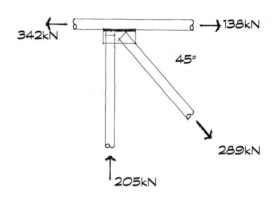

Solution:

The connection of the gusset plate with the chord transfers a load P = 342kN - 138kN = 204kN

F_a = 0.3(414MPa) = 124MPa

Throat = 0.707(6mm) = 4.2mm

L = (204kN)/(4.2mm)(124MPa) = 392mm total

Make L = 200mm on each side.

Problem M16.2

A 10mm end plate for a joist-beam connection is welded to the end of a W200X36 (it will be field-bolted to the girder). If the joist end reaction is 102kN, design the welds on the end plate. Use 5mm fillet welds with E60XX (414MPa) electrodes.

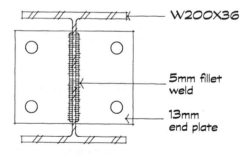

Solution:

Throat = 0.707(5mm) = 3.5mm

L = (102kN)/(3.5mm)(124MPa) = 235mm total

Use a 120mm-long fillet weld on each side of the web.

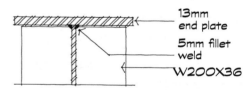

horizontal section

17 LATERAL LOADS

17.1 Wind and earthquakes

Lateral loads are forces acting horizontally or having a component that acts normal (perpendicular) to an inclined surface; they are distinctly different from gravity loads. Lateral loads can be divided into two types: constant and variable. Lateral loads such as soil pressures on a retaining wall are relatively constant, whereas other loads such as wind and earthquakes have variable intensity. The code formulas use static loads in place of these dynamic actions. The magnitude of wind loads depends on the form of the structure or element, and wind direction determines whether the load is positive (pushing) or negative (suction). Wind will exert a pressure on a roof (positive or negative) that is a function of the pitch and orientation of the roof.

Earthquakes result from a rapid release of strain energy built up between tectonic plates that make up the crust of the Earth. This movement is similar to that experienced in buildings when thermal stresses cause movent at expansion joints. These joints were originally "lubricated" with brass plates to reduce friction; now these brass plates have been replaced by Teflon-coated steel or simply Teflon pads. Teflon pads are not yet available for fault lines, so we're forced to deal with the movement and vibration caused by this enormous release of energy.

The energy release (strength of the earthquake) is measured on two types of scales: the Modified Mercalli Intensity Scale, ranging from 1 to 12, which evaluates the earthquake strength based on the degree of destruction observed; and the **Richter Magnitude Scale**, which measures the magnitude of movement observed at a distance of 100km, or 62 miles, from the point of origin (epicenter) of the quake.

The Richter Scale is a logarithmic scale, ranging from 1 to 10, with each step or whole number being **30 times stronger** than the previous number. This means that a magnitude-8 quake is roughly 810,000 times as strong as a magnitude-4 quake: 30x30x30x30 = 810,000. To give you a sense of these abstract numbers, a Richter magnitude-2 earthquake cannot be felt without instrumentation, whereas a magnitude-9 is estimated to be the largest force that can be caused by tectonic plate movement. The Moon striking the Earth would probably be a magnitude 10.

In reinforced-concrete structures, lateral loads can be resisted by a moment-resisting frame, with columns and beams connected rigidly together. In masonry structures, the walls can provide lateral resistance when designed as shear walls; see Fig. 17.1(b). In steel and wood frames, rigid connections are more expensive to build than in reinforced concrete; consequently, a variety of techniques ranging from rigid connections to shear walls to diagonal bracing are typically used (see Chapter 2). The structure must also be designed to resist overturning and uplift due to wind, whereas the same members must be designed for moment, shear, and axial load due to wind or earthquakes. These forces must be combined with the dead loads and a specified percentage of the live loads. Components such as windows and roof finishes must be designed and fastened to resist both wind pressure and wind suction. Lateral loads are assumed to act along the two principal axes of the building. If the structure is safe in these two directions, it is assumed to be adequate for loads acting in any direction. The same wind acting at any angle to a principal axis can be broken into its components acting parallel and perpendicular to the building. These components will produce only a proportion of the full force acting on the principal axis.

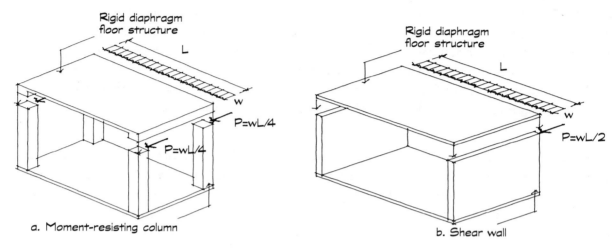

Fig. 17.1 a. Reinforced concrete frame; b. Masonry structure using shear walls

The problem in modern tall buildings built with a relatively light, flexible steel frame is not just strength, but also serviceability; this means, in the case of horizontal loads, drift and vibrations caused by wind, earthquakes, or other dynamic loads such as machinery.

17.2 Structural systems for lateral loads

Wind loads are transferred from the building envelope to the columns, bracing, or shear walls system anchored to the foundations. Earthquake loads similarly are assumed to act at each floor level, and the floor structure and vertical framing must be capable of transferring them to the foundations. Figure 17.2 illustrates, with an exploded view, how wind loads are resisted by vertical walls. Note that this diagram does not show all the loads on the building.

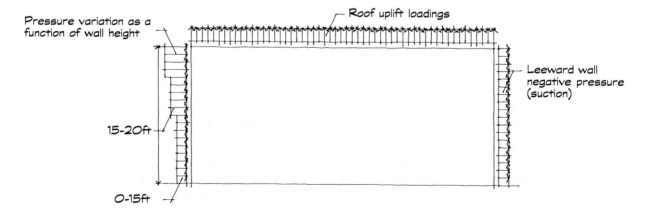

Fig. 17.2 Wind load distribution diaphragm

Two basic types of structural elements used to transfer these loads through the building to the foundations (Fig. 17.3) are as follows:

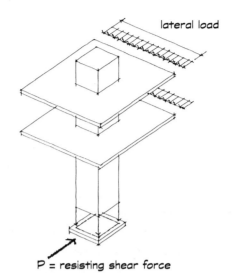

Horizontal elements:

      Beams in a rigid frame

      Trussed floor systems (horizontal cross-bracing)

      Floor diaphragms

Vertical elements:

      Columns in a rigid frame

      Trussed walls (vertical cross-bracing)

      Shear walls

      Rigid cores

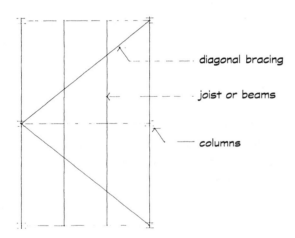

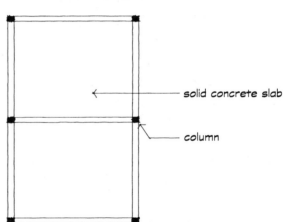

Fig. 17.3 Trussed floor and floor diaphragm systems

A trussed floor is designed for horizontal loads essentially in the same way a truss is designed for gravity loads. As for bracing systems (see Chapter 3), when cross-bracing is used, the diagonal members are designed to work in tension only, which is the most efficient use of material utilizing the smallest possible sizes. Trussed floors are common in steel construction even when concrete-filled decks are specified.

Diaphragms work as deep horizontal beams or as plates spanning between the vertical elements. A diaphragm can be classified as rigid or flexible. Common types of deck construction that can be classified as rigid diaphragms are reinforced-concrete slabs, steel decks with concrete top-pings, and precast concrete tees connected to resist shear. Examples of flexible diaphragms are steel decks without concrete (common on roofs) and plywood. Rigid diaphragms can resist tor-sion, and some eccentricity in the design of vertical elements is possible. Flexible diaphragms, however, require vertical shear supports at both sides in each direction (Fig. 17.4).

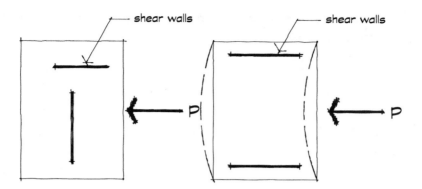

Fig. 17.4 a. Rigid diaphragm; b. Flexible diaphragm

Some steel decks can be considered rigid diaphragms, depending on the gage (thickness of the steel) and depth. Manufacturers publish data on the use of their deck as a diaphragm, particularly shear strength and degree of flexibility. The connection between beams and decking is critical. This is accomplished with welds on each panel placed where the deck rib rests on the beam flange. In addition, deck panels should be attached to each other by side-seam welds, or screws placed at intervals not exceeding 36in [900mm]. In any event, the specific manufacturer must be consulted and its recommendations adhered to.

These horizontal diaphragms transfer their loads to braced vertical structural elements. In steel systems, such an element might be a cantilevered column, a solid cantilevered wall, a trussed cantilevered structure, a rigid frame with moment connections, or even a simple diagonal bracing system. In concrete frames, moment connections are easier to create than to avoid. Consequently, they may or may not utilize additional visually observable bracing systems. In order to resist lateral loads, we have basically four techniques or combinations of techniques.

1. Short cantilever, or shear wall, working primarily in (surprise) shear.
2. Long cantilever: rigid three-dimensional core, tall shear walls, or a pierced tube (a core with holes) acting as a beam or truss working in bending as well as shear.
3. Trussed cantilever, as the name implies, a vertical truss acting in bending and shear, but developing primarily axial loads in tension or compression (a truss).
4. Rigid frame, a "standard" structural system of columns and beams with moment connections to provide lateral stability. The moment connections cause all **members** to be subject to combined axial load and bending. These are statically indeterminate structures and are typically analyzed on a computer.

The combination of a frame and a wall or core element is called a "dual" or "hybrid" system. You'll see this referred to in the UBC 2711 discussions of framing systems. The types of framing systems that can be designed for lateral loads are endless, and decisions depend on architectural considerations.

The code provisions for reinforced-concrete buildings in seismic areas assume the use of rigid "moment-resisting" frames with or without shear walls, and the prescriptions are mainly concerning reinforcement details and maximum stress values. Some of the earthquake issuesthat are easily considered in preliminary design are minimum sizes of beams and columns. In seismic zones 3 and 4, beams must be at least 10in [254mm] wide and the width-to-depth ratio must be at least O.3; i.e., the minimum beam would be 10in X 33.3in [254mm X 846mm]. For columns, the smallest dimension in cross-section must be 12in [305mm]. The column requirement is mainly to ensure that there is enough space provided for the necessary reinforcing bars and ties.

Although we have discussed primarily steel and concrete systems, most buildings are a combination of systems. Whereas concrete and steel systems can be used in almost any combination, the code does not allow the use of concrete or masonry systems on wood support structures, although horizontal wood structures may be used on concrete or masonry supports. The notable exception is that steel joist systems may be supported on wood columns or walls.

17.3 Wind loads

Wind exerts a positive pressure or a negative pressure (suction) on the exposed surfaces of buildings. The magnitude of the forces depends on a number of factors: The most important are wind velocity, exposure, and building shape.

Wind velocity changes continuously with time, but for our purposes, it is defined with an average value. The effects of dynamic gust velocity depend on the stiffness of the structure. The UBC quantifies wind pressure using the *fastest-mile wind speed*, defined as the highest sustained average wind speed based on the time required for a 1-mile-long [1.6km] sample of air to pass a fixed point. The wind speed is measured by the National Oceanographic and Atmospheric Administration and recorded on wind-velocity maps. The value of the wind speed used in the calculations is the fastest-mile wind speed associated with an annual probability of O.O2 measured at a point 33ft [10m] above the ground.

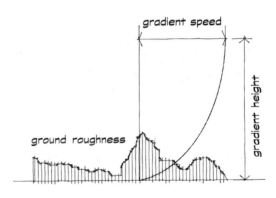

Fig. 17.5 A wind velocity gradient

Wind velocity is reduced by friction resulting from the irregularities of the terrain. Above certain heights, the air can move without being affected by ground conditions. This height is called the "gradient height" and the velocity is called "q," the "wind speed."

The code defines three degrees of exposure:

Exposure B has terrain with buildings, forest, or surface irregularities 2Oft [6m] or more in height covering at least 2O% of the area extending 1 mile [1.6km] or more from the site.

Exposure C has terrain that is flat and generally open, extending 1/2 mile [O.8km] or more from the site in any full quadrant.

Exposure D represents the most severe exposure in areas with basic wind speeds of 80mph
[130km/h] or greater and that have terrain that is flat and unobstructed facing large
bodies of water over 1 mile [1.6km] or more in width relative to any quadrant of the building
site. Exposure D extends inland from the shoreline 1/4 mile [400m] or 10 times the
building height, whichever is greater.

The minimum basic wind speed for determining wind pressure can be found in Fig. 17.6.

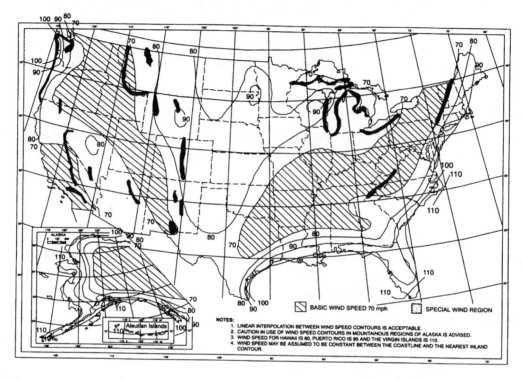

Fig. 17.6 Wind speed map for the United States
Source: Adopted from the Uniform Building Code, 1997

Design wind loads are calculated with the formula from the UBC.

$$P = C_e C_q q_s I_w$$

where C_e = combined height, gust factor, and exposure coefficient (Table 17.1)

   C_q = pressure coefficient depending on the type of structure (Table 17.3)

   q_s = wind stagnation pressure (Table 17.2)

   I_w = Importance factor depending on occupancy category (Table 17.4)

   P = pressure measured in psf

NOTE:

When you investigate the code, you will see that it refers to two ways of determining loads: the
normal-force method and the projected-area method. These two methods will ultimately yield

the same answers for any condition, but can be used with some forethought to get what you really want to know in the easiest manner. The normal-force method will give you directly the force that acts normal or "perpendicular" to any surface either inward or outward. This would be a desirable technique if you were designing a sloped roof joist since you would directly obtain the force acting perpendicular to the joist that creates moment, shear, and deflection in the joist. You would need to calculate the component of force that acts parallel to the joist if the slope was steep enough (12:12) to create significant axial loads. On the other hand, if you were trying to determine the total force that was attempting to overturn or slide the structure, the projected-area method would be the most direct. The projected-area method simply applies all coefficients to both positive and negative pressure on the vertical or horizontal projection of the entire building. This technique would yield a force that acts on the vertical (or horizontal) plane of the basic elevation (or plan) dimensions of the building.

Table 17.1 C_e Combined height, exposure, and gust factor coefficient

Height above ave. level of adjoining ground (ft)[1]	Exposure D	Exposure C	Exposure B
0-15ft	1.39	1.06	0.62
20ft	1.45	1.13	0.67
25ft	1.50	1.19	0.72
30ft	1.54	1.23	0.76
40ft	1.62	1.31	0.84
60ft	1.73	1.43	0.95
80ft	1.81	1.53	1.04
100ft	1.88	1.61	1.13
120ft	1.93	1.67	1.20
160ft	2.02	1.79	1.31
200ft	2.10	1.87	1.42
300ft	2.23	2.05	1.63
400ft	2.34	2.19	1.80

Source: Adopted from the Uniform Building Code, 1997 [multiply values × 304.8 for mm]

1. Values for intermediate heights above 15ft [4.5m] may be interpolated; i.e., for height = 22.5ft [7m], Exposure D, C_e = 1.45 + (1.50-1.45)/2 = 1.475. That was obviously an easy one, but you get the idea.

Table 17.2 q_s Wind stagnation pressure at standard height of 33ft [10m]

Basic wind speed (mph)[1] [× 1.61 for km/h]	70	80	90	100	110	120	130
Pressure q_s (psf) [× 0.0479 for kN/m²]	12.6	16.4	20.8	25.6	31.0	36.9	43.3

Source: Adopted from the Uniform Building Code, 1997

1. Wind speed from Basic Wind Speed Map

Table 17.3 C_q Pressure coefficient

Structure (or part thereof)	Description	C_q factor
Primary frames and systems	Method 1 (normal force method)	
	Walls:	
	Windward wall	0.8 inward
	Leeward wall	0.5 outward
	Roofs[1]:	
	Wind perpendicular to ridge	
	Leeward roof or flat roof	0.7 outward
	Windward side:	
	Less than 2:12 slope	0.7 outward
	Slope 2:12 < 9:12	0.9 outward or 0.3 inward
	Slope 9:12 $\leq$ 12:12	0.4 inward
	Slope > 12:12	0.7 inward
	Wind parallel to ridge/flat roofs	0.7 outward
	Method 2 (projected area method)	
	On vertical projected area	
	Structures 40ft or less in height	1.3 horizontal in any direction
	Structures over 40ft in height	1.4 horizontal in any direction
	On horizontal projected area	0.7 upward
Elements and components not in areas of discontinuity[2]	Wall elements	
	All structures	1.2 inward
	Enclosed and unenclosed structures	1.2 outward
	Partially enclosed structures	1.6 outward
	Parapet walls	1.3 inward or outward
	Roof elements	
	Enclosed and unenclosed structures	
	Slope < 7:12	1.3 outward
	Slope 7:12	1.3 outward or inward
	Partially enclosed structures	
	Slope <2:12	1.7 outward
	Slope 2:12 to 7:12	1.6 outward or 0.8 inward
	Slope > 7:12 to 12:12	1.7 outward or inward
Elements and components in areas of discontinuities[2,4,5]	Wall Corners[6]	1.5 outward or 1.2 inward
	Roof eaves, rakes, or ridges without overhangs[6]	
	Slope < 2:12	
	Slope 2:12 to 7:12	2.3 upward
	Slope >7:12 to 12:12	2.6 outward
	For slopes less than 2:12	1.6 outward
	Overhangs at roof eaves, rakes, or ridges and canopies	0.5 added to vaules above
Chimneys, tanks, and solid towers	Square or rectangular	1.4 in any direction
	Hexagonal or octagonal	1.1 in any dir4ection
	Round or elliptical	0.8 in any direction
Open-frame towers[7,8]	Square or recttangular	
	Diagonal	4.0
	Normal	3.6
	Triangular	3.2
Signs, flagopoles lightpoles, minor structures[8]		1.4 in any direction

Source: Adopted from the Uniform Building Code, 1997

1. For one story or the top story of multistory partially enclosed structures, an additional value of 0.5 shall be added to the outward C_q. The most critical combination shall be used for design. For definition of partially enclosed structures see 1997 UBC Sec 1616.

2. C_q values listed are for 10sqft [0.93m²] areas. For tributary areas of 100sqft [9.29m²], the value of 0.3 may be subtracted from C_q, except for areas at discontinuities with slopes less than 7:12 where the value of 0.8 may be subtracted from C_q. Interpolation may be used for tributary areas between 10 and 100sqft [0.93 and 9.29m²]. For tributary areas greater than 1,000sqft [92.9m²], use primary frame values.

3. For slopes greater than 12:12, use wall element values.

4. Local pressures shall apply over a distance from the discontinuity of 10ft [3m] or 0.1 times the least width of the structure, whichever is smaller.

5. Discontinuities at wall corners or roof ridges are defined as discontinuous breaks in the surface where the included interior angle measures 170 degrees or less.

6. Load is to be applied on either side of discontinuity but not simultaneously on both sides.

7. Wind pressures shall be applied to the total normal projected area of all elements on one face. The forces shall be assumed to act parallel to the wind direction.

8. Factors for cylindrical elements are two thirds of those for flat or angular elements.

Table 17.4 I, I_p Seismic Importance factor; I_w Wind Importance factor; Importance factor

Category	Occupancy or functions of structure	Seismic I	Seismic I_p[1]	Wind I_w
Essential facilities[2]	Group I, Division 1 occupancies having surgery and emergency-treatment areas Fire and police stations Garages and shelters for emergency vehicles and emergency aircraft Structures and shelters in emergency-preparedness centers. Aviation control towers Structures and equipment in government communication centers and other facilities required for emergency response Standby power-generating equipment for Category 1 facilities Tanks or other structures containing housing or supporting water or other fire-suppression material or equipment required for the protection of Category 1,2, or 3 structures	1.25	1.50	1.15
Hazardous facilities	Group H, Divisions 1,2, 6, and 7 occupancies and structures therein housing or supporting toxic or explosive chemicals or substances Nonbuilding structures housing, supporting, or containing quantities of toxic or explosive substances that, if contained within a building, would cause that building to be classified as a Group H, Division 1,2, or 7 occupancy	1.25	1.50	1.15
Special occupancy structures[3]	Group A, Divisions 1, 2, and 2.1 occupancies Buildings housing Group E, Divisions 1 and 3 occupancies with a capacity greater than 300 students Buildings housing Group B occupancies used for college or adult education with a capacity greater than 500 students Group I, Divisions 1 and 2 occupancies with 50 or more resident incapacitated patients, but not included in Category 1 Group I, Division 3 occupancies All structures with an occupancy greater than 5,000 persons Structures and equipment in power-generating stations, and other public utility faculties not included in preceeding Category 1 or 2, and required for continued operation	1.00	1.00	1.00
Standard occupancy structures[3]	All structures housing occupancies or having functions not listed in Category 1,2, or 3 and Group U occupancy towers	1.00	1.00	1.00
Miscellaneous structures	Group U occupancies except for towers	1.00	1.00	1.00

Source: Adopted from the Uniform Building Code, 1997

1. The limitation of I_p for panel connections in Section 1633.2.4 shall be 1.0 for the entire connector.

2. Structural observation requirements are given in Section 1702.

3. For anchorage of machinery and equipment required for life-safety systems, the value of I_p shall be taken as 1.5.

While it would appear that Categories 1 and 2 and 3, 4 and 5 could be combined into two simple groups, a distinction is made for future changes in the code.

Example 17.1

A school in a flat urban area, Indianapolis, has a wall height of 24ft, a flat roof, and exposure B. Determine the total horizontal and vertical forces acting on this building.

(a) Positive pressure (inward):

$$C_e = 0.62 \quad \text{0-15ft (Table 17.1)}$$
$$= 0.67 \quad \text{15-20ft}$$
$$= 0.72 \quad \text{20-24ft}$$
$$C_q = 0.8 \text{ (Table 17.3)}$$

Basic wind speed from map 23-1 is 70mph; therefore, $q_s = 12.6$psf

$I_w = 1.0$ (special occupancy structure, Table 17.4)

0-15ft	$P_1 = 0.62 \times 0.8 \times 12.6\text{psf} \times 1.0 = 6.25\text{psf}$
15-20ft	$P_2 = 0.67 \times 0.8 \times 12.6\text{psf} \times 1.0 = 6.75\text{psf}$
20-24ft	$P_3 = 0.72 \times 0.8 \times 12.6\text{psf} \times 1.0 = 7.26\text{psf}$

(b) Negative pressure (outward):

$$C_e = 0.62 \quad \text{0-15ft (Table 17.1)}$$
$$= 0.67 \quad \text{15-20ft}$$
$$= 0.72 \quad \text{20-24ft}$$
$$C_q = 0.5 \text{ (Table 17.3)}$$

0-15ft	$P_1 = 0.62 \times 0.5 \times 12.6\text{psf} \times 1.0 = 3.91\text{psf}$
15-20ft	$P_2 = 0.67 \times 0.5 \times 12.6\text{psf} \times 1.0 = 4.22\text{psf}$
20-24ft	$P_3 = 0.72 \times 0.5 \times 12.6\text{psf} \times 1.0 = 4.54\text{psf}$

(c) Roof uplift (outward):

$C_q = 0.7$ (Table 17.3)

$C_e = 0.72$ (based on the mean roof height $= 24$ft) (Table 17.1)

$P_4 = 0.72 \times 0.7 \times 12.6\text{psf} \times 1.0 = 6.35\text{psf}$

These pressures must be applied simultaneously perpendicular to the walls and to the roof surfaces. The vertical loads can be considered to resist overturning and suction.

Example 17.2

We'll now work Example 17.1 using the projected-area method.

(a) Total horizontal pressure: $C_e = 0.62 \quad \text{0-15ft (Table 17.1)}$
$$= 0.67 \quad \text{15-20ft}$$
$$= 0.72 \quad \text{20-24ft}$$
$$C_q = 1.3 \text{ (Table 17.3)}$$

(b) Total gross pressure (inward + outward):

Basic wind speed from map 23-1 is 70mph; therefore, q_s = 12.6psf

$\quad$ I_w = 1.0 Special occupancy structure   (Table 17.4)

$\quad\quad$ 0-15ft $\quad$ P_1 = 0.62 × 1.3 × 12.6psf × 1.0 = 10.15psf

$\quad\quad$ 15-20ft $\quad$ P_2 = 0.67 × 1.3 × 12.6psf × 1.0 = 10.97psf

$\quad\quad$ 20-24ft $\quad$ P_3 = 0.72 × 1.3 × 12.6psf × 1.0 = 11.79psf

If you take the values for positive and negative pressure from Example 17.1 you get:

$\quad\quad$ 0-15ft $\quad$ P_1 = 6.25psf + 3.91psf = 10.16psf

$\quad\quad\quad\quad\quad\quad$ (compared to 10.15psf in this solution)

$\quad\quad$ 15-20ft $\quad$ P_2 = 6.75psf + 4.22psf = 10.97psf

$\quad\quad\quad\quad\quad\quad$ (compared to 10.97psf in this solution)

$\quad\quad$ 20-24ft $\quad$ P_3 = 7.26psf + 4.54psf = 11.8psf

$\quad\quad\quad\quad\quad\quad$ (compared to 11.79psf in this solution)

These answers show that if you want only the horizontal forces, the projected area method gives you the same answers with 50% of the work. Take your choice.

(c) Roof uplift (outward):

Since the roof is flat, this calculation will remain the same.

c) Roof uplift (outward):

$\quad\quad$ C_q $\quad$ = 0.7 (Table 17.3)

$\quad\quad$ C_e $\quad$ = 0.72 (based on the mean roof height = 24ft) (Table 17.1)

$\quad\quad$ P_4 $\quad$ = 0.72 × 0.7 ×12.6psf × 1.0 = 6.35psf

17.4 Seismic loads

Earthquake forces on structures result from movement of the ground on which a building is supported. The ground moves both vertically and horizontally in a wavelike motion, but the vertical components are typically neglected since structures are designed for code-specified vertical live loads and construction dead loads. These code-defined loads will generally compensate, along with material factors of safety, for the vertical loads generated by vertical components of earthquake movements. The horizontal dynamic forces are of more concern, because the oscillation of the building masses can amplify the effect of these seismic forces much as a child causing its swing to "swing" by alternately shifting its weight to reinforce the movement of the swing. This is particularly true of very flexible structures subjected to several cycles of lateral movement of the ground and their subsequent rebound.

In order to understand how earthquake forces act on buildings, we can start by studying a simplified model of a mass M connected to the ground. If the acceleration of the ground due to the earthquake is "a," the corresponding inertial force acting on the mass is F = Ma (Newton's law). The base of the building would have to resist this horizontal shear force and the other effects due to the point of application of this force, as shown in Fig. 17.8. Rarely do we find a building that reacts as a perfectly rigid mass; all structures and materials are elastic. The effect of an earthquake on a building is very similar to what a person standing on a bus experiences when the bus starts with a sudden acceleration. Since the center of the body mass is located well above the floor, the inertial force will push the body backward, whereas the feet tend to remain in place. A person will instinctively step back to maintain balance (stepping on your foot). For a building connected to the ground, this is not possible; it would be like a person with feet glued to the floor of the bus, and the consequences of a sudden acceleration or stop would be very painful.

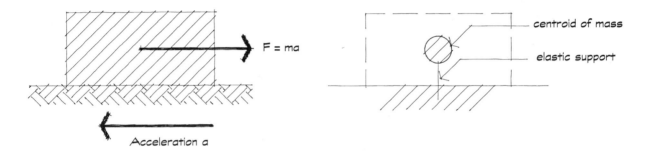

Fig. 17.8 Point of application of seismic force   Fig. 17.9 Analogy of mass on an elastic support

Since most buildings, somewhat like the human body, are flexible and have no straps to hang on to, we can model this type of effect with a mass connected to the ground by a spring or even a person riding a horse sitting lightly in the saddle. This represents the effect of dead loads supported on elastic columns (Fig. 17.9). If you pull the mass, extending the spring, or bending an elastic stick, and release it, it will start oscillating with a type of movement mathematically similar to that of a pendulum subject to the force of gravity. The oscillation of the model is independent from the initial pull and depends on the elastic characteristics of the spring. The time it takes to complete a single oscillation, back and forth, is called **period of oscillation**, and is described mathematically as

$$T = 2\pi\sqrt{M/k}$$

where M is the mass, and k is the rigidity of the cantilever or the force constant of the spring.

In the base shear formula, the appropriate period T for each structure affects the intensity of the inertial force due to the quake through the coefficient C_v or C_a, called the **base shear coefficient**. The peak acceleration of the ground (corresponding to the worst earthquake expected) is basically included in coefficient C_v or C_a, which are a function of the **seismic zone factor, Z**.

If a series of pendulums is moved through the same ground motion that occurred in any given earthquake recorded by a seismograph, the maximum response of each pendulum can be recorded. The response may be deflection, velocity, acceleration, or shear, whichever is chosen, since they are all related. The curve of the response has very similar shapes for different earthquakes, although the magnitude of the response varies with the magnitude of the earthquake.

The actual magnitude of the earthquake is expressed as a percentage of the acceleration of gravity g = 32.2ft/sec/sec [9.81m/sec²] of the ground.

It is very important, therefore, to know the seismic history of an area in order to determine the maximum acceleration to be expected. The equivalent static force to be applied in the design of a building can then be calculated using Newton's law, F = Ma, or

$$\text{or} \quad F = (a/g)W$$

where F is the equivalent static lateral load of a building or element having a weight W, and a/g is the seismic factor.

In arriving at the seismic factor, the code takes into account the degree of probability of a large earthquake, the type of building occupancy, the flexibility of the structure, and the physical site characteristics. By using symbols to represent these considerations, the seismic factor is given by a product of coefficients without units:

$$a/g = \frac{CI}{RT}$$

and the equivalent static force becomes:

$$V = \frac{C_v I}{RT} W$$

where $T = C_t(h)^{3/4}$

$$C_t = 0.035 \ [0.0853] \ \text{steel moment resisting frames}$$
$$C_t = 0.030 \ [0.0731] \ \text{reinforced concrete frames}$$
$$C_t = 0.020 \ [0.0488] \ \text{all other systems}$$

with maximum value of

$$V = \frac{2.5 C_a I}{R} W \quad \text{and a minimum value of} \quad V = [0.11 C_a I] W$$

Also for structures located in a seismic zone 4, the total base shear cannot be less than:

$$V = \frac{0.8 Z N_v I}{R} W$$

where V is called **base shear**. The coefficients are found accordingly in Tables 17.6 through 17.11. There is an alternative method for calculation of base shear, but this seems to be the simpliest and most direct.

Table 17.5   Seismic zone map of the United States

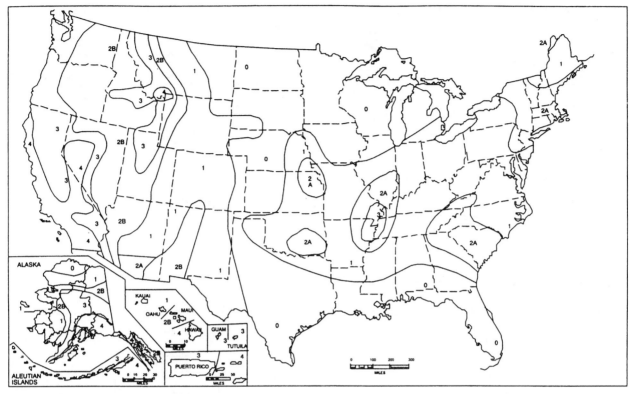

Source: Adopted from UBC, 1997

Table 17.6 Seismic zone factor Z

Zone	1	2A	2B	3	4
Z	0.075	0.15	0.20	0.30	0.40

Source: Adopted from the Uniform Building Code, 1997

Zone designation shall be determined from the seismic zone map Table 17.5

Table 17.7 Soil profile types

Soil profile	Soil generic description	Average soil properties for top 100ft of profile		
		Shear wave velocity V_s feet/ second	Penetration test N blows per foot	Undrained shear strength
S_A	Hard rock	>5,000f/s		
S_B	Rock	2,500 to 5,000fps		
S_C	Very dense soil and soft rock	1,200 to 2,500fps	>50	>2,000psf
S_D	Stiff soil profile	600 to 1,200fps	15 to 50	1,000 to 2,000psf
S_E[1]	Soft soil profile	<600fps	<15	<1,000psf
S_F	note 1			

Source: Adopted from the Uniform Building Code, 1997

1. Any soil with more than 10ft of soft clay with a plasticity index of PI>20, $w_{mc} \geq 40\%$, s_u <500psf.

Table 17.8 Seismic coefficient C_v

Soil profile type	Z = 0.075	Z = 0.15	Z = 0.2	Z = 0.3	Z = 0.4
S_A	0.06	0.12	0.16	0.24	$0.32N_a$
S_B	0.08	0.15	0.20	0.30	$0.40N_a$
S_C	0.13	0.25	0.32	0.45	$0.56N_a$
S_D	0.18	0.32	0.40	0.54	$0.64N_a$
S_E	0.26	0.50	0.64	0.84	$0.96N_a$
S_F	note 1	note 1	note 1	note 1	note 1

Source: Adopted from the Uniform Building Code, 1997

1. Site-specific geotechnical investigation and dynamic site response analysis shall be performed to determine seismic coefficients for Soil Profile Type S_F.

Table 17.9 Seismic coefficient C_a

Soil profile type	Z = 0.075	Z = 0.15	Z = 0.2	Z = 0.3	Z = 0.4
S_A	0.06	0.12	0.16	0.24	$0.32N_a$
S_B	0.08	0.15	0.20	0.30	$0.40N_a$
S_C	0.09	0.18	0.24	0.33	$0.40N_a$
S_D	0.12	0.22	0.28	0.36	$0.44N_a$
S_E	0.19	0.30	0.34	0.36	$0.36N_a$
S_F	note 1	note 1	note 1	note 1	note 1

Source: Adopted from the Uniform Building Code, 1997

1. Site-specific geotechnical investigation and dynamic site response analysis shall be performed to determine seismic coefficients for Soil Profile Type S_F.

Table 17.10 Structural system coefficient R

Basic structural system[1]	Lateral-force-resisting system	R	Max. height[2]
Bearing wall system	1. Light-framed walls with shear panels		
	(a) Wood structural panel walls for structures three stories or less	5.5	65ft
	(b) All other light-framed walls	4.5	65ft
	2. Shear walls		
	(a) Concrete or masonry	4.5	160ft
	3. Light steel-framed bearing walls with tension-only bracing	2.8	65ft
	4. Braced frames where bracing carries gravity load		
	(a) Steel	4.4	60ft
	(b) Concrete[3]	2.8	-------
	(c) Heavy timber	2.8	65ft
Building frame system	1. Steel eccentrically braced frame (EBF)	7.0	240ft
	2. Light-framed walls with shear panels		
	(a) Wood structural panel walls for structures three stories or less	6.5	65ft
	(b) All other light-framed walls	5.0	65ft
	3. Shear walls		
	(a) Concrete	5.5	240ft
	(b) Masonry	5.5	160ft
	4. Ordinary braced frames		
	(a) Steel	5.6	160ft
	(b) Concrete[3]	5.6	-------
	(c) Heavy timber	5.6	65ft
	5. Special concentrically braced frames		
	(a) Steel	6.4	240ft
Moment-resisting frame system	1. Special moment-resisting frame (SMRF)		
	(a) Steel	8.5	No limit
	(b) Concrete[4]	8.5	No limit
	2. Masonry moment-resisting wall frame (MMRWF)	6.5	160ft
	3. Concrete intermediate moment-resisting frame (IMRF)[5]	5.5	-------
	4. Ordinary moment-resisting frame (OMRF)		
	(a) Steel[6]	4.5	160ft
	(b) Concrete[7]	3.5	-------
	5. Special truss moment frames of steel (STMF)	6.5	240ft
Dual systems	1. Shear walls		
	(a) Concrete with SMRF	8.5	No limit
	(b) Concrete with steel OMRF	4.2	160ft
	(c) Concrete with concrete IMRF[5]	6.5	160ft
	(d) Masonry with SMRF	5.5	160ft
	(e) Masonry with steel OMRF	4.2	160ft
	(f) Masonry with concrete IMRF[3]	4.2	-------
	(g) Masonry with masonry MMRWF	6.0	160ft
	2. Steel EBF		
	(a) Steel with steel SMRF	8.5	No limit
	(b) Steel with steel OMRF	4.2	160ft
	3. Ordinary braced frames		
	(a) Steel with steel SMRF	6.5	No limit
	(b) Steel with steel OMRF	4.2	160ft
	(c) Concrete with concrete SMRF[3]	6.5	--------
	(d) Concrete with concrete IMRF[3]	4.2	--------

(handwritten note next to "Moment-resisting frame system": if connections are all fixed to not move)

Table 17.10 Structural system coefficient R (continued)

Dual systems (continued)	14. Special concentrically braced frames		
	(a) Steel with steel SMRF	7.5	No limit
	(b) Steel with steeel OMRF	4.2	160ft
Cantilevered column building systems	1. Cantilevered column elements	2.2	35ft[6]
Shear wall-frame interaction systems	1. Concrete[8]	5.5	160ft

Source: Adopted from the Uniform Building Code, 1997

1. Basic structural systems are defined in UBC Section 1629.6

2. Height limit for Seismic Zones 3 and 4 [X304.8 for mm].

3. Prohibited in Seismic Zones 3 and 4.

4. Includes precast concrete conforming to UBC Section 1921.2.7.

5. Prohibited in Seismic Zones 3 and 4, except as permitted in UBC Section 1634.2.

6. Ordinary moment-resisting frames in Seismic Zone 1 meeting the requirements of UBC Section 2211.6 may use an R value of 8.

7. Total height of the building including cantilevered columns.

8. Prohibited in Seismic Zones 2A, 2B, 3, and 4. See UBC Section 1633.2.7.

Table 17.11 Near-source factor N_a[1]

	Closest distance to known seismic source[2,3]		
Seismic source type	$\leq$ 1.2 miles [2 Km]	3 miles [5km]	$\geq$ 6 miles [10km]
A	1.5	1.2	1
B	1.3	1	1
C	1	1	1

Source: Adopted from the Uniform Building Code, 1997

1. The near-source factor may be based on the linear interpolation of values for distances other than those shown in the table.

2. The location and type of seismic sources to be used for design shall be established based on approved geotechnical data. i.e., US Geological Survey or California Division of Mines and Geology.

3. The closest distance to seismic source shall be taken as the minimum distance between the site and the area described by the vertical projection of the source on the surface of the earth. The largest value of the near-source factor considering all sources shall be used for design.

Seismic Source Type Definitions:

A. "Faults that are capable of producing large magnitude events and that have a high rate of seismic activity".

B. "All faults other than Types A and C".

C. "Faults that are not capable of producing large magnitude earthquakes and that have a relatively low rate of seismic activity".

Table 17.12 Near-source factor N_v[1]

Seismic source type	Closest distance to known seismic source[2,3]			
	≤ 1.2 miles [2 Km]	3 miles [5km]	6 miles [10km]	9miles [15km]
A	2	1.6	1.2	1
B	1.6	1.2	1	1
C	1	1	1	1

Source: Adopted from the Uniform Building Code, 1997

1. The near-source factor may be based on the linear interpolation of values for distances other than those shown in the table.

2. The location and type of seismic sources to be used for design shall be established based on approved geotechnical data. i.e., US Geological Survey or California Division of Mines and Geology.

3. The closest distance to seismic source shall be taken as the minimum distance between the site and the area described by the vertical projection of the source on the surface of the earth. The largest value of the near-source factor considering all sources shall be used for design.

In reality, structures do not behave like perfect elastic bodies, and the response is affected by factors that decrease the magnitude of the response. This effect is called **damping** and can be caused by structural elements that absorb the accumulated energy of the structure (deformation of partitions and nonbearing walls, and other forms of bracing with dampers or energy dissipaters). The coefficients used take damping into consideration with the C and R factors.

Example 17.3

A one-story steel-frame commercial building, in Indianapolis, Indiana, is 14ft high, and the soil is soft clay. The total load of the roof structure is:

DL = 48psf + LL = 20psf = 68psf

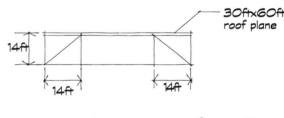

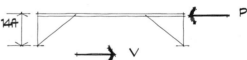

From the UBC Seismic Zone map:

Indianapolis, Indiana, is a Zone 1

W = 68psf (30ft)(60ft) = 122,400lb = 22.4k

Z = 0.075 seismic zone factor for Zone 1, Table 17.7

I = 1.0 standard occupancy, Table 17.4

R_w = 5.6 for a concentrically braced frame (members designed for axial loads and not as moment-resisting frame), Table 17.10

$T = C_t (h_n)^{3/4}$ period of vibration

$C_t = 0.02$ not a moment-resisting frame

$T = 0.02 (14ft)^{0.75} = 0.14sec$

Substituting these values in the base shear formula:

$$V = [0.26(1.0)(122.4k)]/[(5.6)(0.14sec)] = 40.59k$$

$$\text{Maximum } V = 2.5(I)(C_a)(W)/(R) = 2.5(1.0)(0.19)(122.4k)/5.6 = 10.38k$$

In our case, the force (V) is resisted by two diagonal braces on each side of the building, therefore each brace takes 10.38k/2 = 5.2k, and the axial load on the brace is $P = 5.2k \times 1.414 = 7.25k$ tension in one brace and 7.25k compression (not a good idea) in the other brace.

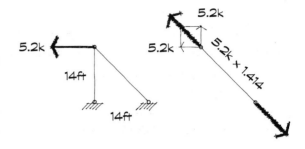

17.5 Shear-force distribution

The base shear, as the name suggests, must be resisted at the base and at each level as it accumulates down the building. At each floor, the shear is proportional to the height and to the weight (dead load) as follows:

$$V_x = V_{total}\left[\frac{W_x(h_x)}{\sum W_i(h_i)}\right]$$

where x represents the number of a specific floor, as shown in Fig. 17.10, W is the dead load, and h is the height from the ground. The weight is calculated by adding the floor weight, half of the weight of the wall above, and half of the weight of the wall below.

For buildings with low slenderness, with low elasticity (represented by the coefficient k in the equation for T), the fundamental mode is more important ("pendulum" motion), and the distribution of the forces at each floor for design purposes is corresponding to this mode. In more slender buildings with higher elasticity, T is longer, and the secondary mode could become significant, due to the higher proportion of inertia at the top of the building ("whipping" motion). In this case, the codes require the application of portion F_t of the total force (base shear) at the top of the building, in addition to the force V_n (where n is the top floor) calculated with the previous equation:

$$F_t = 0.07TV \leq 0.25V, \text{ for } T > 0.7 \text{ sec}$$
$$F_t = 0 \qquad\qquad\qquad \text{for } T \leq 0.7 \text{ sec}$$

where V is the base shear, and T is the fundamental period of vibration.

The equivalent static force to be applied at the generic floor x is

$$V_x = \frac{(V - F_t)W_x h_x}{\sum W_i h_i}$$

and the total shear at the base must be

$$V = F_t + \sum V_i$$

Example 17.4

A multistory reinforced-concrete building in Zone 3 has a height of 36ft and has a moment-resisting frame. The foundation rests on rock bed (S_B). The occupancy type is residential.

$W = (84psf + 100psf + 100psf)(30ft \times 60ft) = 511,200lb = 511.2k$

Soil profile S_B for Zone 3, $Z = 0.03$

$I = 1.0$

$R_w = 3.5$ (steel OMRF)

$C_V = 0.3$

$T = C_t (h)^{3/4}$

$T = 0.03(36ft)^{0.75} = 0.44sec$

$V = C_V IW/RT$

$V = 0.3(1.0)(511.2k)/(3.5)(0.44) = 99.6k$

$C_a = 0.36$

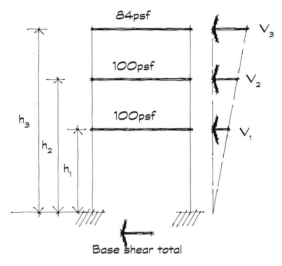

Fig. 17.10 Distribution of base shear within a structure

Maximum $V = 2.5(0.36)(1.0))(511.2k)/3.5 = 131.4k > 99.6k$ OK! use 99.6k

Vertical distribution:

$\Sigma W_i h_i = (84k \times 36ft) + (100k \times 24ft) + (100k \times 12ft) = 6624ftk$

$V_1 = \dfrac{99.6k (100k \times 12ft)}{6624ftk} = 18.04k$

$V_2 = \dfrac{99.6k (100k \times 24ft)}{6624ftk} = 36.09k$

$V_3 = \dfrac{99.6k (84k \times 36ft)}{6624ftk} = 47.47k$

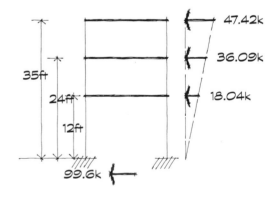

Total V   99.6k

The base shear is distributed at the base of the columns.

In addition to the primary structural elements, parts of the structures and attachments must be checked. Examples are nonbearing walls, floor connections to vertical structures, chimneys, tanks, etc. Attachments include anchorage of shear walls and bracing. These parts must resist a force of:

$$F_p = 4.0 C_a I_p W_p$$

where C_a and I are the same as for buildings, but for life-safety systems, such as a water tank supplying sprinklers, I is 1.5. The coefficient C_a is given (Table 17.7) for elements, components, and rigid equipment. Attachments for floor or roof-mounted equipment weighing less than 400lbs [181kg] need not be designed for seismic forces.

The UBC defines the **basic structural systems**, for the purpose of earthquake design, for application of R factor.

- (a) Bearing-wall system
- (b) Building-frame system
- (c) Moment-resisting frame
- (d) Dual system

Table 17.13 Alternative framing systems for seismic resistance

Braced systems
Concentrically braced frame (CBF)

Diagonal or cross-bracing connections are considered to be pinned in the plane of bracing; therefore, the loads induced by lateral forces in the frame members are axial (concentric). This is comparable to truss design.

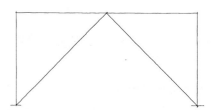

Eccentrically braced frame (EBF)

Bracing members do not converge on one point of the beam axis; therefore, bending moments are induced by lateral loads. The design of the frame takes into consideration the ductility of the beam.

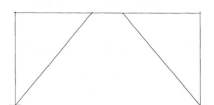

Moment-resisting frame (MRF)

The frame provides resistance to lateral loads by flexural action of the members. The connections are all or partially rigid.

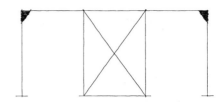

Dual system

Special moment-resisting frames (SMRF)

Ordinary moment-resisting frames (OMRF)

These are moment-resisting frames combined with bracing or shear walls. SMRF must comply with special strength and ductility requirements.

17.6 Earthquake-resistant construction

We have suggested that the vertical components of the structure will resist the loads in bending and/or shear. In small steel buildings, you could assume that the entire base shear will be taken by the column bases, and in small masonry buildings, the shear will be taken by all walls. This is a very simplified view, and a more accurate analysis should be made by a structural engineer. This approach essentially requires that we design the structural members to provide bending and shear resistance and ensure that the connections will hold the pieces together (exterior wall panels must be adequately attached to the main building frame even if they have no structural function).

As mentioned in Section 17.4, an important consideration is damping. From the base-shear coefficient formula:

$$T = C_t \, (h_n)^{3/4}$$

it is clear that a short T (a building with short oscillations and high-frequency vibrations) will give higher values of base shear V, compared to a building with a large T (a building swinging slowly with long oscillations). If the structure is very strong and rigid, it will probably have a high frequency of vibration, so by building it too "strong," we risk increasing the inertial forces that try to destroy it. Damping can be introduced in the form of building elements that crack in a ductile way, absorbing large deformations, and dissipating energy during an earthquake by frictional forces. These could be masonry wall infill between steel columns. Damping devices can also be introduced in the bracing system that work like shock absorbers in a car or like the collapsible steering wheel columns that crumble instead of transferring the impact to the driver. Damper elements can be installed as a retrofit in place of or in addition to normal diagonal bracing, a technique used commonly in California and Japan.

A different strategy that has implications on the overall building design is "seismic isolation." This technique consists of separating the building superstructure from the foundations to allow vertical loads to be transferred to the ground, whereas ground movements cannot be transmitted to the superstructure. Conceptually, this is very simple, since you can imagine the column supported on sliding pads; if the foundations are shaken suddenly, the building remains still, held by its inertia (imagine a waiter, pulling the tablecloth from under the dishes with a sudden jolt). Don't forget that an earthquake does not move the building, but only the ground where the building is attached. If the building can slide on the ground, it will not experience, at least theoretically, any earthquake.

Of course, this floating movement must be restricted; consequently, the isolators are also damped, allowing only a limited movement (10in)[250mm] and some transfer of lateral forces. An example of seismic isolators used in California consists of alternating layers of copolymers bonded to steel plates, separating the base of each column from the foundation.

In steel construction, the systems employed are braced-frame systems or moment-resisting frames. These systems are described in Table 17.10. Shear walls may be part of the system, often in the form of reinforced concrete elevator and stair shafts. The systems also can be used in combination, for example, a moment-resisting frame with braced core or braced bents to "resist" quake forces.

An alternative system, a Tuned Pendulum System, is being experimented with to compensate for the forces and reduce their effect on the structure. This is accomplished by "suspending" a large mass of material, typically water stored for firefighting or mechanical systems. When a lateral force is applied to the structure, it moves sideways, but the pendulum swings out of phase with the building movement. The pendulum is laterally braced by massive springs or shock absorbers, which reduce its motion. An example of this principle is the child in the swing, which we previously discussed. If the child leans or shifts its weight properly, it will increase the amplitude of the "swing," whereas if the "timing" is off, this same shifting of mass will cause no additional acceleration and it may even stop the swing.

Several major concerns immediately come to mind when you put these issues all together and the code identifies them as well. Buildings that have either vertical or horizontal discontinuities are subject to serious problems during an earthquake. Some of the most easily understood discontinuities would include the following:

A nonuniform distribution of mass

A clear example would be to use a long vertical metal rod and place an apple on it. As you subject the rod to a horizontal force and move the apple to different locations, you'll immediately note that the movement is reduced or amplified, depending on the location of the apple relative to the base.

Sections of a building having different degrees of stiffness

Using the steel rod from the previous example along with two others of different stiffnesses (diameter or cross-sections) held together at the base, subject them to a sharp horizontal force. In this case, you'll see or can imagine that they would move at different rates. Imagine what would happen if there were parts of the building connected together. This differential in movement will immediately tear them apart.

Irregular geometries

As an example, you might look at an L-shaped plan and imagine a horizontal force applied parallel to either leg.

Fig. 17.10 Transamerica building

One portion of the building would have the length of the leg resisting the horizontal force and the other would have the width of the leg resisting the force. Obviously, one is much stronger (stiffer) than the other; consequently, the point of connection between the legs would be trying to reconcile two different magnitudes of movement. Result? Crash!

The UBC and the NEHRP elaborate on these basic ideas, but, in general, irregularities in form or mass distribution are not good ideas. Think about the Transamerica building (Fig. 17.10) in San Francisco. Is there a simple earthquake principle at work there?

Problem 17.1

A five-story steel-frame office building (OMRF type) in seismic Zone 3 is 62ft high. The soil is medium stiff clay (S_D). The total weight of the building is 1,120k, and the weight of the top floor (roof structure) and lower floors is shown in the illustration. Calculate (a) the base shear, and (b) the seismic force V_5 at the top floor.

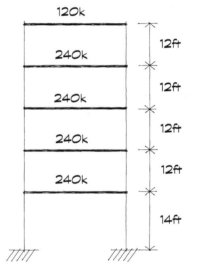

Solution:

Soil profile S_D for Zone 3 with $Z = 0.30$

$C_V = 0.54$

$R = 4.5$ (moment-resisting system, Table 17.10, case 4-a)

$I = 1.0$

$C_t = 0.035$

$T = C_t(h)^{3/4} = 0.035(62ft)^{3/4} = 0.77$

$T = 0.77 > 0.7$, therefore $F_t = 0.07TV$

$$V = \frac{CI}{RT}(W) = \frac{0.54(1.0)}{4.5(0.77)}(1,120k) = 174.5k$$

Base shear $V = 174.5k$

Maximum $V = 2.5(1.0)(0.36)(1,120k)/4.5 = 224.0k > 174.5k$

Minimum $V = 0.11(0.36)(1.0)(1,120k) = 44.3k < 174.5k$

Use $V = 174.5k$

with $T = 0.77 > 0.7$ sec., add $F_t = 0.07TV \leq 0.25V$

$F_t = 0.07(0.77)(174.5k) = 9.41k < 0.25(174.5k) = 43.6k$ OK!

$\Sigma W_i h_i = 38,160k$

$W_5 h_5 = 120k(62ft) = 7,440ftk$

$V_5 = (174.5k)(7,440ftk/38,160ftk) = 34.02k$ (without force increment F_t)

V_5 (total) $= 34.02k + 9.41k = 43.43k$ (total shear force at the top floor)

Problem 17.2

A two-story house, 18ft high, is built in Los Angeles. Calculate the fundamental period of vibration if the structure is made of (a) load-bearing concrete walls, and (b) a steel moment-resisting frame with masonry shear walls. Which is better for earthquake resistance? What is the ratio of concrete/steel base shear V values?

Solution:

Concrete $T = 0.020(18ft)^{3/4} = 0.17 sec., R = 4.5$

Steel $T = 0.035(18ft)^{3/4} = 0.31 sec., R = 4.2$

The steel system has a larger larger T, resulting in a smaller V, so the steel system is better.

$V_{concrete} = 1.7 V_{steel}$

METRIC VERSIONS OF ALL EXAMPLES AND PROBLEMS

Example M17.1

A school in a flat urban area, Indianapolis, has a wall height of 7.2m, a flat roof, and exposure B. Determine the total horizontal and vertical forces acting on this building.

(a) Positive pressure (inward):

$$C_e = 0.62 \quad 0\text{-}5m \quad (\text{table } 17.1)$$
$$= 0.67 \quad 5\text{-}6m$$
$$= 0.72 \quad 6\text{-}7.2m$$
$$C_q = 0.8 \ (\text{Table M17.3})$$

Basic wind speed from map 23-1 is 70mph, therefore $q_s = 87MPa$

$$I = 1.0 \quad \text{Special occupancy structure} \quad (\text{Table } 17.4)$$

0-5m	$P_1 = 0.62 \times 0.8 \times 87MPa \times 1.0 = 43.1MPa$
15-20ft	$P_2 = 0.67 \times 0.8 \times 87MPa \times 1.0 = 46.6MPa$
20-24ft	$P_1 = 0.72 \times 0.8 \times 87MPa \times 1.0 = 50.1MPa$

(b) Negative pressure (outward):

$$C_e = 0.62 \quad 0\text{-}5m \quad (\text{Table } 17.1)$$
$$= 0.67 \quad 6m$$
$$= 0.72 \quad 7.2m$$
$$C_q = 0.5 \ (\text{Table } 17.3)$$

0-15ft	$P_1 = 0.62 \times 0.5 \times 87MPa \times 1.0 = 27.0MPa$
15-20ft	$P_2 = 0.67 \times 0.5 \times 87MPa \times 1.0 = 29.1MPa$
20-24ft	$P_3 = 0.72 \times 0.5 \times 87MPa \times 1.0 = 31.3MPa$

(c) Roof uplift (outward):

$$C_q = 0.7 \ (\text{Table M17.3})$$
$$C_e = 0.72 \ (\text{based on the mean roof height} = 7.2m) \ (\text{Table } 17.1)$$
$$P_3 = 0.72 \times 0.7 \times 87MPa \times 1.0 = 43.8MPa$$

These pressures must be applied simultaneously perpendicular to the walls and to the roof surfaces. The vertical loads can be considered to resist overturning and suction.

Example M17.2

We'll now work Example M17.1 using the projected-area method.

(a) Total horizontal pressure:

$$C_e = 0.62 \quad 0\text{-}5m \quad (\text{Table } 17.1)$$
$$= 0.67 \quad 6m$$
$$= 0.72 \quad 7.2m$$
$$C_q = 1.3 \ (\text{Table } 17.\ 3)$$

Basic wind speed from map 23-1 is70mph; therefore, $q_s = 87MPa$

(b) Total gross pressure (inward + outward):

$I = 1.0$ (special occupancy structure, Table 17.4)

0-5m	$P_1 = 0.62 \times 1.3 \times 87\text{MPa} \times 1.0 = 70.1\text{MPa}$
6m	$P_2 = 0.67 \times 1.3 \times 87\text{MPa} \times 1.0 = 75.8\text{MPa}$
7.2m	$P_3 = 0.72 \times 1.3 \times 87\text{MPa} \times 1.0 = 81.4\text{MPa}$

If you take the values for positive and negative pressure from Example M17.1, you get:

0-5m $P_1 = 43.1\text{MPa} + 27\text{MPa} = 70.1\text{MPa}$ (same as in this solution)

6m $P_2 = 46.6\text{MPa} + 29.1\text{MPa} = 75.7\text{MPa}$

               (compared to 75.8MPa in this solution)

7.2m $P_3 = 50.1\text{MPa} + 31.3\text{MPa} = 81.4\text{MPa}$

               (same as in this solution)

These answers show that if you want only the horizontal forces, the projected-area method gives you the same answers with 50% of the work. Take your choice.

(c) Roof uplift (outward): Since the roof is flat, this calculation will remain the same.

$C_q = 0.7$ (Table M17.3)

$C_e = 0.72$ (based on the mean roof height $= 7.2$m) (Table 17.1)

$P_3 = 0.72 \times 0.7 \times 87\text{MPa} \times 1.0 = 43.8\text{MPa}$

Example M17.3

One-story steel-frame commercial building in Indianapolis, Indiana, is 4.27m high, and the soil is soft clay. The total load of the roof structure is:

$DL = 2300\text{N/m}^2 + LL = 1000\text{N/m}^2 = 3300\text{N/m}^2$

From the UBC Seismic Zone map:

Indianapolis, Indiana is a Zone 1

$W = 3300\text{N/m}^2(9.1\text{m})(18.2\text{m}) = 546.5\text{kN}$

$Z = 0.075$ seismic zone factor for Zone 1, Table 17.6

$I = 1.0$ standard occupancy, Table 17.4

$R_w = 5.6$ for a concentrically braced frame (members designed for axial loads and not as moment-resisting frame), Table 17.10

$T = C_t (h_n)^{3/4}$ period of vibration

$C_t = 0.0488$ not a moment-resisting frame

$T = 0.0488 (4.27m)^{0.75} = 0.14sec$

Substituting these values in the base shear formula:

$$V = [0.26(1.0)(546.4kN)]/[(5.6)(0.14sec)] = 397.3kN$$

$$\text{Maximum } V = 2.5(C_a)(I)(W)/R = 2.5(0.19)(1.0)(546.4kN)/5.6 = 46.34kN$$

In our case, the force (V) is resisted by two diagonal braces on each side of the building, therefore each brace takes $46.3kN/2 = 23.1kN$, and the axial load on the brace is $P = 23.1kNk \times 1.414 = 32.7kN$ tension in one brace and 32.7kN compression (not a good idea) in the other brace.

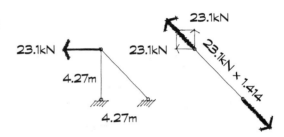

Example M17.4

A multistory reinforced-concrete building in Zone 3 has a height of 11.1m and has a moment-resisting frame. The foundation rests on rock bed. The occupancy type is residential.

$$W = (4021N/m^2 + 4788N/m^2 + 4788N/m^2)(9.1m \times 18.2m) = 2,252kN$$

> Soil profile S_B for Zone 3, $Z = 0.03$
>
> $I = 1.0$
>
> $R_w = 3.5$ (steel OMRF)
>
> $C_V = 0.3$
>
> $T = C_t(h)^{3/4}$
>
> $T = 0.0731(11.1m)^{3/4} = 0.44sec$
>
> $C_a = 0.36$
>
> $V = C_v IW/RT$
>
> $V = 0.3(2,252kN)/(3.5)(0.44) = 438.7kN$

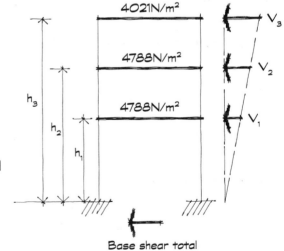

Base shear total

$\text{Maximum } V = 2.5C_a IW/R$

$V = 2.5(0.36)(1.0)(2252kN)/3.5 = 579kN > 438.7kN$ OK! use 438.7kN

Vertical distribution:

$$W_i h_i = (793kN \times 3.7m) + (793kN \times 7.4m) + (666kN \times 11.1m) = 16,194.9Nm$$

$$V_2 = \frac{438.7kN(793kN \times 3.7m)}{16,194.9Nm} = 79.48kN$$

$$V_3 = \frac{438.7kN(793kN \times 7.4m)}{16,194.9Nm} = 158.96kN$$

$$V_4 = \frac{438.7kN(666kN \times 11.1m)}{16,194.9Nm} = 200.26kN$$

Total V   438.7kN

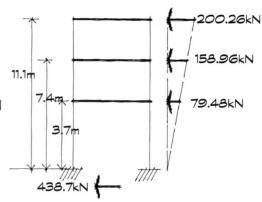

The base shear is distributed at the base of the columns.

Problem M17.1

A five-story steel-frame office building (OMRF type) in seismic Zone 3 is 18.7m high. The soil is medium stiff clay. The total weight of the building is 4,985kN, and the weight of the top floor (roof structure) and lower floors is shown in the illustration. Calculate (a) the base shear, and (b) the seismic force V_5 at the top floor.

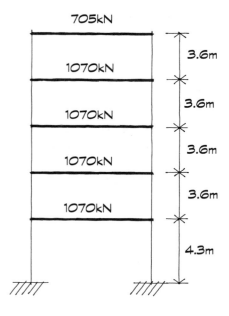

Solution:

Soil profile S_D for Zone 3 with Z = 0.30

$C_V = 0.54$

R = 4.5 (moment-resisting system Table 17.10, case 4-a)

I = 1.0

$C_t = 0.0853$

$T = C_t(h)^{3/4}$

$T = 0.0853(18.7)^{3/4}$

T = 0.77 > 0.7

Base shear V = 776.8kN

Maximum V = 2.5(0.36)(1.0)(4,985kN)/4.5 = 997kN

Minimum V = 0.11(0.36)(4,985kN) = 197kN

Use V = 776.88kN

$F_t = 0.07(0.77)(776.88kN) = 41.87kN$

$\Sigma W_i h_i = 52,251kNm$

$W_5 h_5 = 10,004.5kNm$

$V_5 = (776.88kN - 41.87kN)(10,004.5kNm/55,251kNm) = 140.73kN) = 4.88k$

Problem M17.2

A two-story house, 18ft high, is built in Los Angeles. Calculate the fundamental period of vibration if the structure is made of (a) load-bearing concrete walls, and (b) a steel moment-resisting frame with masonry shear walls. Which is better for earthquake resistance? What is the ratio of concrete/steel base shear V values?

Solution:

Concrete $\quad T = 0.0488(5.5m)^{3/4} = 0.17$ sec., $R = 4.5$

Steel $\quad\quad T = 0.0853(5.5m)^{3/4} = 0.31$ sec., $R = 4.2$

The steel system has a larger larger T, resulting in a smaller V, so the steel system is better.

$V_{concrete} = 1.7 V_{steel}$

PART TWO

WOOD

Ch. Angre' and E. Dexheimer; ASTEC Engineering B.E.T. Exhibition Hall of the Avignon Fair, France (courtesy of Techniques & Architecture)

18 MATERIALS AND PROPERTIES

18.1 Physical properties

Wood is a cellular organic material from which lumber is cut for construction purposes. There are a number of tree species that produce two broad categories of wood: softwood and hardwood, the first from conifers and the latter from deciduous trees. Most of the lumber used in construction in North America is softwood (spruce, pine, and fir), but hardwood species (poplar, aspen, cottonwood, etc.) are available as commercial species. Structural strength has nothing to do with this terminology: Contrary to what the name suggests, the density of most hardwoods is lower than some of the best softwood (southern pine or douglas fir). The National Design Specification Supplement lists 43 species or species combinations (species with comparable structural properties). For instance, the combination spruce-pine-fir includes eight species of pine, fir, and spruce. The choice of wood species for structural use will depend as much as on availability in the local lumberyards as on required properties.

Because wood is an organic material, lumber cut from the same tree will not usually exhibit constant or homogeneous properties. As the tree grows, the cells at the center become inactive and assume a darker color: This is called the heartwood. The outer layer, or sapwood, contains living cells and is lighter in color. Heartwood and sapwood may have the same structural properties, but growth characteristics do affect wood strength as well as appearance. The annual growth rings show, in many species, a lighter layer corresponding to rapid spring growth (earlywood) and a darker ring of slow summer growth (latewood). Latewood contains more dense wood material than earlywood and it is consequently stronger.

Other growth characteristics may decrease the strength of a structural member, depending on their size, number, and location. Some of the most important are:

Density, associated with the rate of growth

Knots, formed by the branches as they become embedded in the growth of the trunk.

Deviations of the grain from the axis of the tree

Reaction wood, growing in a compression zone (softwood) or in a tension zone (hardwood) of the tree, e.g., in the case of a branch

The wood cells have an orientation that gives different wood properties, depending on the direction of the grain. For example, shrinkage is maximum tangential to the grain, but is an average of 50% less in the radial direction. This is the main reason for the tendency of wood to warp once it is sawn tangentially, since it is cut across more rings (Fig. 18.1). There is little shrinkage parallel to the direction of the grain. Strength and elastic properties are designated according to the two principal axes of wood: parallel to the grain and perpendicular to the grain.

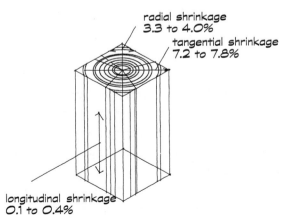

radial shrinkage
3.3 to 4.0%

tangential shrinkage
7.2 to 7.8%

longitudinal shrinkage
0.1 to 0.4%

Fig. 18.1 Differential shrinkage of lumber

The mechanical properties depend on the wood species and grade (quality) as well as on the way the timber is sawn and used. These properties are affected by the moisture retained in the grain. both in the voids of the grain and in the cell walls themselves. This is especially critical since some of the structural properties are affected as much as 5% for every 1% change in moisture content. Standing timber may have as much as 35-40% moisture content. Air drying will reduce this value down to around 19%; this value is accounted for by the removal of the moisture within the open cells and can be further reduced by kiln drying, which drives moisture out of the cell walls and reduces the moisture content to 15%. Moisture driven out of the cell walls will not be reabsorbed; however, the open cells can regain moisture, and, consequently, it is essential that the wood be maintained in a dry condition both during construction and after installation. The swelling and shrinkage associated with changes in moisture can produce checking and distortion, as well as dramatic changes in strength. The cracks resulting from this distortion can affect the horizontal shear strength of lumber.

Shrinkage may influence decisions on the configuration of the structural system. Although it is possible to reduce the absolute rate of shrinkage, the main objective should be to reduce differential shrinkage, e.g., between the wood frame and the interior partitions, or wood combined with other materials, such as steel beams or facing material, that could cause damage to the finishes. One of the most common shrinkage problems occurs in the platform frame construction at floor level, where vertical framing is interrupted by joists and headers. Since shrinkage is negligible in the longitudinal (grain) direction, the studs will not shorten appreciably.

The joists and plates, on the other hand, will shrink across the grain 4% or more, and there may be up to 1 inch of shrinkage in the full height of a two-story building. The technique of using a single top plate, by placing the joists over the studs, can be part of the solution. This will reduce shrinkage in surfaces that have interior finishes applied. Good architectural detailing can also provide "space" for shrinkage on both the interior and the exterior. Other solutions are shown in Fig. 18.2, and include using trussed joists and jacks or blocking between plates. An alternative or additional measure is the use of lumber of moisture content below 19%, in equilibrium with the expected service conditions (say, 12%). In fact, lumber will often be dryer than 19% when delivered to the site, and it is likely to dry further during construction. The differential shrinkage for large, multistory wood buildings (three stories or more) has to be calculated in relation to the other materials and tolerable deformation of the finishes and plumbing.

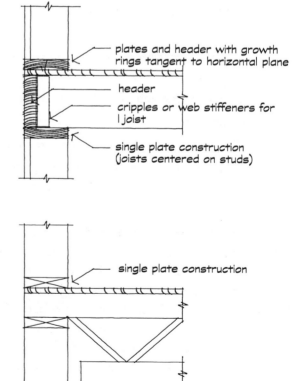

Fig. 18.2 Methods of providing space for shrinkage

18.2 Density and weight of wood

The weight of wood depends on the density of the fibers and on water content. Density is defined as the weight of the wood and its held water in a unit volume (usually in a cubic foot or in a cubic meter). As a piece of wood dries, it also shrinks, i.e., its volume decreases. So 1 cubic foot of green wood will became both lighter and smaller as it dries. The weight of 1 cubic foot (or 1 cubic meter) of wood with a given moisture content to the specific weight of water ($62.4 lb/ft^3$) [$100 Kg/m^3$] is called **specific gravity**, or G = weight of wood (and held moisture)(lb/ft^3)/(62.4 lb/ft^3) [$G = (Kg/m^3)/(1000 kg/m^3)$].

Therefore, if a species of wood weighs $30 lb/ft^3$ [$480 Kg/m^3$] when dry (moisture content, m.c. = 19%), the specific gravity will be $G = 30 lb/ft^3/(62.4 lb/ft^3) = 0.48$ [$G = 480 Kg/m^3/(1000 Kg/m^3) = 0.48$]. The specific gravity of dry construction lumber can vary between 0.70 (oak, beech) and 0.30 (cedar, poplar). The density of the wood substance represented by the compact fiber or grain without moisture and with no air voids is about 50% higher than water, giving an average specific gravity of $G = 1.5$. When wood is green or saturated with water, its unit weight may be equal to or greater than that of water. This is why green wood does not float well.

Table 18.1 gives G and density for some common species. The density of the wood species can be used to calculate the weight of structural components. For instance, red pine 7/8in solid T&G boarding weighs ($7/8$in × 31.5pcf)/12in/ft) = 2.3psf [(22mm×504Kg/m³)/(1000mm/m) = 11.1Kg/m²].

Density is also related to wood strength and capacity of holding nails and screws. A nail driven in a very light wood, such as balsa, can be pulled out by hand; if the wood is oak, it will be hard to pull the nail out even with a crowbar (or to drive it in the first place). NDS relates the load-bearing capacity of connectors to the specific gravity of wood species.

Table 18.1

Specific gravity and density of common species of wood at 12% m.c.

Species	G	Density (lb/ft³) (including 12% held water)	[Kg/m³]
Western cedars	0.35	24.5	392
California redwood	0.36	25.2	403
Eastern white pine	0.37	25.9	414
Hem-fir	0.42	29.4	470
Ponderosa pine	0.42	29.4	470
Red pine	0.45	31.5	509
Northern pine	0.46	32.2	515
Douglas fir-larch	0.49	34.3	545
Southern pine	0.52	36.4	582
Red oak	0.67	40.8	653

18.3 Protection from decay and fire

Wood is subject to decay and insect attack. Decay or rot is caused by fungi grown from microscopic spores present wherever wood is used. Their growth is encouraged by moisture, temperature, and air: Fungi will not grow in water, with moisture below 20%, or temperatures below freezing. Fungi stain penetrates into the wood, and unlike surface molds cannot be removed. During decay, the wood changes in color and texture as well as strength.

The first rule to prevent decay is to eliminate the causes for such conditions to occur by providing moisture barriers and membranes, and allowing ventilation on the **cold side of** insulated surfaces of roof attic spaces and exterior walls. Some details on wood protection are shown in Fig. 18.3.

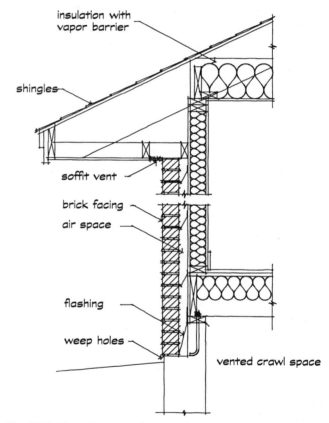

It is possible, however, to design wood members to work in wet conditions if the wood is adequately treated or protected. Protection techniques include painting and treating with wood preservatives. Pressure-treated wood should be used for wall plates anchored to concrete or masonry foundations, and for permanent wood foundations. The liquids used to impregnate the wood are EPA-registered pesticides containing inorganic arsenic or pentachlorophenol. Exposure to these pesticides may be dangerous, and precautions should be taken when handling the treated wood and when disposing of the waste. Treated wood should not be used when it may come into direct contact with drinking water or food. For residential, commercial, and marine applications, waterborne preservatives, such as chromated copper arsenate, are preferred.

Fig. 18.3 Typical details for humidity and moisture control

The codes place restrictions on the use of timber structure because of its combustibility. Even when a wood structure is allowed, fire insurance rates are higher. However, when heavy timber exposed is to fires reaching temperatures around 1000°F to 2000°F [538° to 1076°C], the surface chars. This creates a protective layer that retards combustion and helps protect the wood core, which retains its structural capabilities.

Wood species can be assigned char rates or thickness of charred section formed per minute at a given temperature. The char rate for high-grade solid timber and other wood products such as glulam is about 1/40in [0.635mm] per minute. It is therefore possible to calculate the additional thickness of a structural member for the required fire resistance.

Example 18.1

The required structural dimension of a Douglas fir timber beam is 6inx12in nominal. If the beam is left exposed and unprotected in the building, what is the additional thickness required for a 1-hour fire resistance?

Char rate 1/40in = 0.025in/minute.

Charred thickness after 1 hour = (0.025in) × (60 minutes) = 1.5in each face.

The depth and breadth of the beam have to be increased by 2 × 1.5in, or 3in. The closest nominal size is 10in × 16in, or (147.3in)/(63.25sqin), which is 2.3 times the original size.

This example shows that charring is not an economical technique for achieving fire resistance in small timber members, because it virtually requires doubling their size. However, if charring is only meant to prevent complete collapse of the structure during the fire, design loads and deflection limitations can be reduced to evaluate the size of the uncharred core.

It is possible to make wood highly resistant to the spread of fire by pressure impregnating or coating with approved chemicals; these will reduce the rate of destruction and transmission of heat. The first method of injecting fire retardants, such as chromated zinc chloride, is more effective, but many of these chemicals are very hygroscopic and therefore are not recommended when the relative humidity reached in the building is high (over 80%). In this situation, fire-retardant salt treatments may affect the wood strength (remember our discussion of the effect of moisture content on structural properties).

18.4 Design values

The sawn lumber from each species is ranked according to its structural strength from Select Structural (best) to Utility (worst). The rules associated with this grading system are established by seven agencies representing producers (e.g., WWPA, Western Wood Products Association), based on provisions of ASTM standards. This involves an adjustment of strength properties, obtained from the test of a specimen for the effect of knots, slope of grain, and several influencing factors by means of visual inspection. These sawn lumber members therefore are called **visually graded**, and they are assigned the design values for that grade. The design values are allowable stresses for bending, shear, tension, compression (parallel and perpendicular to the grain), and modulus of elasticity, and are published in the NDS Supplement.

Individual members of lumber also can be machine-stress rated (MSR) or machine evaluated (MEL) with nondestructive tests (Fig. 18.4), and in these cases species are not indicated. The design values for **mechanically graded** lumber are in the NDS Supplement, Table 4C.

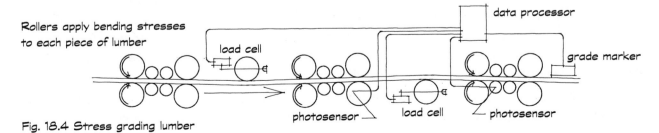

Fig. 18.4 Stress grading lumber

The design values for each species (visually graded) are listed in order of

1. decreasing strength, the best being normally **Select Structural**, then **No. 1, No. 2, No. 3, Stud, Construction,** and **Utility** (special-use products, such as decking, are listed as separate categories, with decking graded as **Select and Commercial**)
2. increasing width, in the case of Southern pine only

Different design values may be assigned by different grading agencies. For this reason, it is probably best to use the most conservative values, since you may not be guaranteed access to wood from specific grading agencies.

18.5 Size classifications

Actual sizes of sawn lumber are smaller than the nominal sizes due to shrinkage during the drying and surfacing processes. You will notice that the nominal and actual sizes do not form a consistent system. Whereas the percentage of shrinkage remains constant, a large piece of lumber will experience a greater total change in dimension than a small piece, So, as you might expect, larger lumber sizes have a smaller relative actual dimension. However, a 2X10 is 1.5in x 9.25in [38mmx235mm] and a 10X10 is 9.5in x 9.5in [241mmx241mm].

For complete definitions of sizing, see NDS; but for most situations, Table 18.2 will be helpful.

Table 18.2 Classification of lumber according to size

Definition	Nominal thickness	Nominal width	Typical use
	[actual mm]	[actual mm]	
Dimension lumber	2 to 4in	2in or more	Light framing
	[38 to 89mm]	[38mm or more]	
Timbers			
Beams and stringers	5in or more	Thickness + 2in or more	Beams
	[114mm or more]	[114mm or more]	
Posts and timbers	5in or more	5in or more	Columns
	[114mm or more]	[114mm or more]	
Decking	2 to 4in	4 to 12in (T&G)	Roof/floors
	[38 to 89mm]	[89 to 292mm]	
Note: Metric lumber sizes are always actual.			

Source: Adopted from the NDS.

The actual sizes of dry lumber corresponding to the nominal sizes are called **standard dressed** or **S4S** (Surfaced on 4 Sides), and they are given in the NDS (Tables 1A and 1B, pp. 9,10, and 11). "Dry" indicates a maximum moisture content (m.c.) of 19% at the time of manufacture and use. For a higher moisture, the lumber is called "green," and its dressed size is 1/16in to 1/4in [2mm to 6mm] greater than the corresponding dry size (NDS).

The different size groups have different design values. These depend on the criteria used to saw the log (Figs. 18.5 and 18.6) and on grading procedures. For example, a 2X6 [38X140] dimension lumber Douglas fir-larch select structural has a bending design value F_b = 1450psi [10MPa]; a 6X8 [140X191] timber piece of the same wood has F_b = 1600psi [11MPa]; the No. 2 grades of each have the same F_b (875psi)[5.9MPa] and a 2X6 [38X140] of hem-fir has and F_b = 850psi [5.6MPa] and a 6X8 [140X191] hem-fir has an F_b = 575psi [4MPa]. Is bigger better, the same, or worse? Yes! It is hard to clearly understand these apparent discrepancies; however, any time you change size classifications in a design process, you should verify the design values.

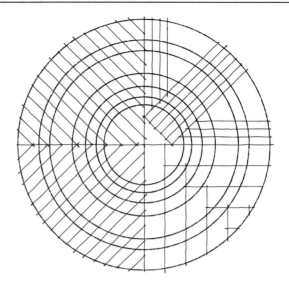

Fig. 18.5 Methods of sawing boards from a log

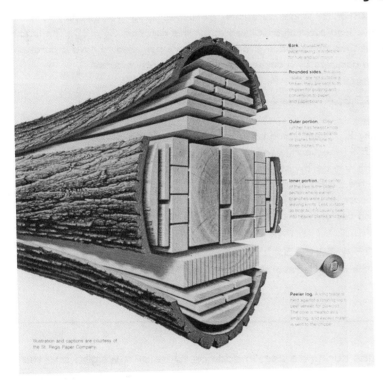

Fig. 18.6 Contemporary lumber cuts

18.6 Adjustment factors

The characteristics of wood growth affecting strength are taken into account in the grading process. Wood members are also stronger or weaker according the manner and conditions under which they are used, i.e., time, moisture, temperature, and even orientation. One of the most important considerations is that most members resist loads of short duration better than loads applied continuously for a period of years. The organic material of wood responds much like the material of your body. You can hold a "small" weight at the end or your arm extended horizontally for a short period of time, but after a while, your arm begins to tire and you must re-move the load (lower the weight). Wood acts in a similar manner, but the time interval for "tiring" is much longer. Another common cause of change of structural properties during use is the exposure of wood for a prolonged time to a moist or very high-temperature environment. The conditions of use are represented by adjustment factors listed in the NDS (Table 2.3.1, p. 5). This is one of the tables most frequently used, since it is important to check in every situation which design values are modified by the applicable adjustment factors. In this part, Wood, you will find factors to use for various structural members; the factors listed in Table 18.3 apply to most design problems. Finally, the orientation of a bending member is a consideration. You'll note that bending components have correction factors for "flat use" conditions. This might be a short construction site ramp or a child's teeter-totter.

Table 18.3 Adjustment factors for load, wet service, and temperature

C_D Load duration   (NDS, Sec. 2.3.3, Table 2.3.2, Appendix B)

Wood can carry greater loads for short durations than for longer durations. Tabulated design values apply to normal load durations. "Normal" means a cumulative load duration of 10 years; for this duration, $C_D = 1.0$. The load duration assigned to types of loads is conventional: Occupancy loads given in codes are assumed to have a dura-tion of 10 years, even if an escape corridor will be occupied by a dense crowd for only a few minutes.

C_M Wet service   (NDS, Sec. 2.3.3, All Tables 4 and 5 of the NDS Supplement)

High moisture content dramatically reduces strength. Green wood has a m.c. generally in the 30%+ range, dry lumber m.c. is 19%, kiln dried lumber is generally 15% (frequently used by truss manufacturing companies), and some specialty wood products may use lumber with m.c. as low as 4% (i.e., laminated timber). Service conditions are important because wood can absorb some of the moisture previously removed. A m.c. of 19% is in equilibrium with an 85% relative humidity (summer months, indoor conditions). Fire-retardant treatments can be hygroscopic (absorbing or attracting moisture from the air), and use of the wet service factor is recommended in those cases.

C_t Temperature   (NDS, Sec. 2.3.4, Table 2.3.4)
Heating decreases wood strength; up to 150° F, the effect is reversible. Permanent loss of strength can be caused by prolonged heating.

A structure will be subject to different loads during its use, in addition to its own weight, and the problem arises of selecting the correct load-duration factor. For instance, a roof rafter will be supporting construction loads, wind loads, and snow loads. First of all, the building codes will tell which loads should be considered acting simultaneously (e.g., earthquake loads are not usually considered together with wind or snow loads). Once these loads have been determined, you have to find the critical load combination, according to the NDS criteria (NDS, Appendix B). The critical load is the largest found of those calculated as:

$$\frac{\text{Dead loads}}{0.9} \quad \text{and} \quad \frac{\text{Dead loads + any possible combination of live loads}}{C_D}$$

where C_D is the load-duration factor for the **shortest-time-period** live load in the combination.

Example 18.2

A roof beam carries a dead load (weight of structure and finishes) of 220plf, a construction load (materials and people during construction) of 200plf, and a snow load of 200plf. The critical load is the largest of the following combinations (refer to NDS, Table 2.3.2):

$DL/0.9 = (220plf)/0.9 = 244plf$

$(DL + CL)/1.25 = (220plf + 200plf)/1.25 = 336plf$

$(DL + SL)/1.15 = (220plf + 200plf)/1.15 = 365plf$

$(DL + CL + SL)/1.25 = (220plf + 200plf + 200plf)/1.25 = 496plf$

Since the critical load combination is DL + CL + SL, the load-duration factor to be used for the design of the beam is 1.25. This is because we have assumed the possibility of construction loads with a maximum snow load. If we exclude this possibility, the critical load is given by DL + SL, and $C_D = 1.15$.

This method seems somewhat cumbersome and arcane, but it has some rational foundations. When the combination of loads includes a short-duration load such as wind, it may at first seem optimistic to attribute a greater strength to the material just because the wind will blow sometimes. However, the structure in fact will be sized according to the total load, and it will be oversized when there is no wind if the critical-load method is followed. So this procedure is realistic and is directed toward an efficient use of the material. In many practical cases, the critical load combination is the one that corresponds to the addition of all applicable loads. This is virtually always the case for wood floor structures, since DL + occupancy load is always substantially larger than DL/0.9 (the live load is rarely as small as 10% of the dead load), and the C_D is therefore taken equal to 1.0.

C_D does not modify the modulus of elasticity, but the effect of load duration is felt also on deflections. Bending members are subject to a slow increase in deflection, causing a permanent set under long-term loads. This progressive deformation is called **creep** and it is more pronounced in unseasoned wood. Creep is accounted for with a factor that increases deflection for long-term loads, but this is not part of the adjustment factors listed in NDS. Table 18.4 lists the criteria for deflection of wood members.

Table 18.4 Maximum allowable deflection for structural members

Type of member	Allowable LL deflection	Allowable LL + KDL deflection
Roof supporting plaster or floor member	L/360	L/240

The value for K varies for each material discussed in this book. In this part, Wood, the value for K is 0.5. This means that the maximum allowable deflection is a function of either LL at L/360 or LL + 0.5DL at L/240. Since L/240 allows 50% more deflection than L/360, **the total DL would need to be at least equal to the LL for this criterion to control. This rarely happens in wood systems.**

18.7 Engineered wood products

Lumber is gradually being replaced by wood products that combine the advantages of wood with the qualities of an industrially produced material: Plywood, oriented strand board, and laminated timber in different forms are the most common (Fig. 18.7). Alternatives to solid timber components are also available in the form of I joists and different types of trussed beams. These components and materials are essentially composites, manufactured with wood products of different properties. New systems of fastening have been added to the traditional mechanical connections of staples and metal plates, in addition to a wide range of metal connectors (hangers, anchors, etc.). Glues are especially important in manufactured components; also on-site gluing of sheathing to joists is becoming common, since it helps reduce squeaking.

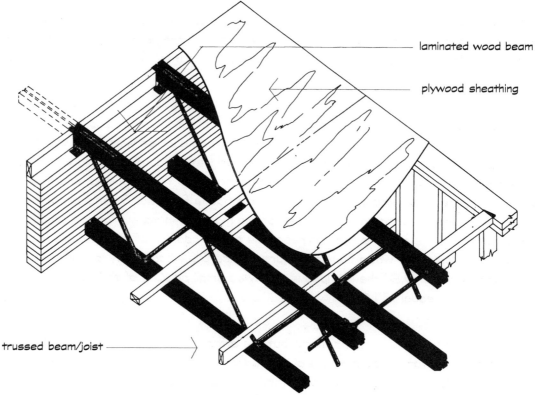

Fig. 18.7 Framing with engineered component made of composite wood systems

---

Problem 18.1

Using the NDS tables, find the allowable tension stress F_t' for a visually graded timber subject to a wind load. The timber is a 6X6 Eastern spruce, select structural (timbers). Assume normal conditions of use.

Answer:

$F_t' = 1.6(675\text{psi}) = 1080\text{psi}$

Problem 18.2

Find the bending design value in the NDS tables for a 2X8 joist of visually graded Southern pine dimension lumber, No. 2 dense.

Answer:

F_b = 1400psi

Problem 18.3

Using the NDS tables, find the approximate weight of a 2X12 joist of Ponderosa pine, 10ft long. Can a single person lift it?

Answer:

W = (3.516plf)(10ft) = 35lb

This is the weight of a suitcase full of books.

Problem 18.4

An open pier on the shore of Lake Erie is built with Douglas fir-larch visually graded decking. Will the modulus of elasticity E be affected by the exposure to the water and, if yes, by how much?

Answer:

E' = 0.9E, or 10% decrease

Problem 18.5

You have two options for the design of a bench seat for a park: a slotted arrangement using five nominal 2X3 pieces or a solid 2X14 board. In either case, the material is red oak.

Find the best solution purely in terms of (a) allowable bending stress, (b) moment of inertia, (c) quantity of material.

Solution:

2X3s on edge:

(a) $F'_b = C_F C_r F_b$ = (1.5)(1.15)(1150psi) = 1984psi

(b) I_x = (5)(1.953in^4) = 9.765in^4

(c) A = 18.75in^2

Flat 2X14 board:

(a) $F'_b = C_{fu} F_b$ = (1.2)(1150psi) = 1380psi

(b) I_x = 3.727in^4

(c) A = 19.88in^2

Problem 18.6

An exterior deck of a mountain lodge is used as a dancing area, with an occupancy load of 100psf. The local code requires a 40psf snow load. The weight of the structure is 18psf. Find the applicable C_D.

Answer:

Assume the deck is cleared when occupied, and not used with snow. The load combinations are

(a) (18psf)/0.9 = 20psf

(b) (18psf + 100psf)/1.0 = 118psf

(c) (18psf + 40psf)/1.15 = 50psf

$$C_D = 1.0$$

METRIC VERSIONS OF ALL EXAMPLES AND PROBLEMS

Example M18.1

The required structural dimension of a Douglas fir timber beam is 140mmX292mm. If the beam is left exposed and unprotected in the building, what is the additional thickness required for a 1-hour fire resistance?

Char rate 0.635mm per minute.

Charred thickness after 1-hour = (0..635mm/min) × (60min) = 38mm each face.

The depth and breadth of the beam have to be increased by 2 × 38mm or 76mm. The closest size is 241mmX394mm = 94.95mm^2, which is 2.3 times the original size.

Example M18.2

A roof beam carries a dead load (weight of structure and finishes) of 10,500N/m^2, a construction load (materials and people during construction) of 9,500N/m^2, and a snow load of 9,500N/m^2. The critical load is the largest of the following combinations (refer to NDS, Table 2.3.2):

DL/0.9 = (10,500N/m^2)/0.9 = 11,666N/m^2

(DL + CL)/1.25 = (10,500N/m^2 + 9,500N/m^2)/1.25 = 16,000N/m^2

(DL + SL)/1.15 = (10,500N/m^2 + 9,500N/m^2)/1.15 = 17,391N/m^2

(DL + CL + SL)/1.25 = (10,500N/m^2 + 9,500N/m^2 + 9,500N/m^2)/1.25 = 23,600N/m^2

Since the critical load combination is DL + CL + SL, the load-duration factor to be used for the design of the beam is 1.25. This is because we have assumed the possibility of construction loads with a maximum snow load. If we exclude this possibility, the critical load is given by DL + SL, and C_D = 1.15.

Problem M18.1

Using the NDS tables, find the allowable tension stress F_t' for a visually graded timber subject to a wind load. The timber is a 140mmX140mm Eastern spruce, select structural (timbers). Assume normal conditions of use.

Answer:

F_t' = 1.6(4.65MPa) = 7.4MPa

Problem M18.2

Find the bending design value in the NDS tables for a 38mmX184mm joist of visually graded Southern pine dimension lumber, No. 2 dense.

Answer:

F_b =9.65MPa

Problem M18.3

Using the NDS tables, find the approximate weight of a 38mmX286mm joist of ponderosa pine, 3m long. Can a single person lift it?

Answer:

W = (5.22Kg/m)(3m) = 15.7Kg

This is the weight of a suitcase full of books.

Problem M18.4

An open pier on the shore of Lake Erie is built with Douglas fir-larch visually graded decking. Will the modulus of elasticity E be affected by the exposure to the water and, if yes, by how much?

Answer:

E' = 0.9E or 10% decrease

Problem M18.5

You have two options for the design of a bench seat for a park: a slotted arrangement using five nominal 38mmX63mm pieces, or a solid 38mmX286mm board. In either case, the material is red oak. Find the best solution purely in terms of (a) allowable bending stress, (b) moment of inertia, and (c) quantity of material.

Solution:

38X63s on edge:

(a) $F_b = C_F C_r F_b$ = (1.5)(1.15)7.9MPa = 13.7MPa

(b) I_x = 5(0.813x10^{6}mm^4) = 4.065x10^6 mm^4

(c) A = 5(38mmx63mm) = 11,970mm^2

Flat 38X286 board:

(a) $F'_b = C_{fu} F_b$ = (1.2)(7.9MPa) = 9.5MPa

(b) I_x = 1.551x10^{6}mm^4

(c) A = 10,868mm^2

Problem M18.6

An exterior deck of a mountain lodge is used as a dancing area, with an occupancy load of 4800N/m^2. The local code requires a 1920N/m^2 snow load. The weight of the structure is 862N/m^2. Find the applicable C_D.

Answer:

Assume the deck is cleared when occupied, and not used with snow. The load combinations are

(a) $(862 N/m^2)/0.9 = 958 N/m^2$

(b) $(862 N/m^2 + 4,788 N/m^2)/1.0 = 5,650 N/m^2$

(c) $(862 N/m^2 + 1,915 N/m^2)/1.15 = 2,415 N/m^2$

$C_D = 1.0$

19 WOOD STRUCTURES IN ARCHITECTURE

19.1 Construction and architectural philosophies

Wood is a traditional building material, probably the oldest in the world together with stone. The combination of masonry walls with wood horizontal members for floors and roofs and structures completely of wood are still used today for houses and small buildings in many parts of the world, although wood products and construction technologies have changed radically during the last 50 years. Colonial structures in the United States were originally derived from European models, with heavy timber frames (Fig. 19.1), still found in barn buildings and some residential structures (Fig. 19.2).

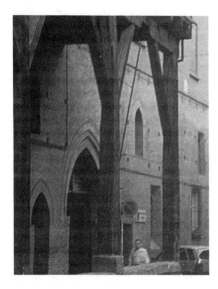 

Fig. 19.1 Heavy timber frame in Bologna, Italy   Fig. 19.2 Barn frame in central Indiana

Wood construction in North America was revolutionized by the invention of the "balloon frame" around 1833 in Chicago, which made use of small-dimension lumber produced economically in saw-mills. By the 1830s, machine-made nails were also available at a fraction of the former price of hand-made nails. The next most important invention was probably plywood, which replaced lumber board subfloors and wall sheathing. Now even plywood is being replaced by materials that use even smaller pieces of lumber: particle board, flake board, and oriented strand board. These products create less waste in production and use material properties even more efficiently.

Many Native American Indian buildings made use of wood in a variety of structural shapes. Most of these were constructed of a very lightweight framing, using straight or bent saplings (Fig. 19.3) connected together in a space grid, a structural concept common to modern lamella fram-ing. In many cultures, the primitive, archetypal wood construction system was canonized into a form that persisted over a long time, even when the wood was replaced by stone or masonry, as in the case of the classical Greek temple. Sometime in prehistory, the use of heavy, straight logs led to the development of wood post-and-beam structures and bent saplings created the primor-dial form of domes and vaults. These different systems are important since they correspond to the early discovery of architectural systems, each with its scale of components, forms, and spaces.

The idea that classical orders derived from prehistoric or primitive wood construction techniques has been discussed throughout the history of architectural theory, from Vitruvius to Laugier, a French theorist of the seventeenth century. In order to demonstrate the superiority of lineage of the column in architectural composition, Laugier argued that the column, a direct derivation of the trunk of the tree, was the oldest form of construction associated with civilization. This suggests how wood structural systems have played a role in shaping architectural forms as we see them today, and how the development of structural systems is interrelated with architectural aesthetics and architectural philosophy.

Fig. 19.3 Wigwam structure

In our times, architectural philosophies have been greatly influenced by the question of availability and use of resources. Architects like Walter Segal have designed houses with minimal wood framing, by carefully calculating each component, an approach similar to the research on minimum-weight structures by Buckminster Fuller and Frei Otto.

Figure 19.4 illustrates a contemporary version of historic post-and-beam construction used in residential construction. This is particularily successful in areas that have difficult sloping sites or soil conditions which do not allow significant penetration (rock) or, in general, require minimum site disruption.

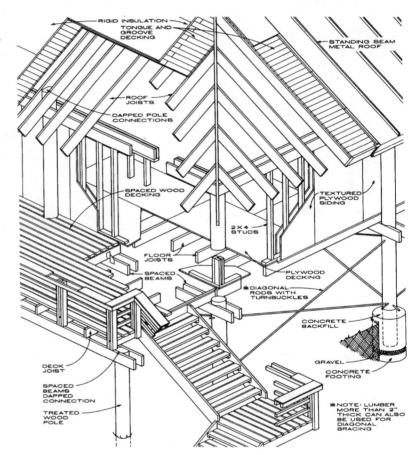

Fig. 19.4 Contemporary pole framing

19.2 The architect's responsibility in structural wood design

Wood is a renewable material using the Sun to transform chemicals extracted by the roots from the soil into cellulose fiber and lignin. Deciduous trees (like maple and oak) can grow from the old roots if they are cut, whereas conifers, producing softwood, have to be replanted. Until late in the nineteenth century, the United States considered its forest inexhaustible, and only in 1891 was forest management introduced to ensure a continuous supply for shipbuilding and construction.

In 1970, the total area of commercial forest in the United States was nearly 500 million acres [*200 million hectares*], including the national forests. This area has been shrinking due to the expansion of urban areas, infrastructure (airports, interstates, etc.), and agriculture. Between 1962 and 1970, the commercial forest lost 8.5 million acres [*3.4 million hectares*], and a loss of 5 million acres [*2 million hectares*] is expected to continue to the year 2000 despite the fact that the nation ceased to be self-sufficient in timber in 1941. Twenty percent of U.S. timber needs must be supplied from Canada and Southeast Asia. Ecological concerns have led to the need for preserving a variety of ecosystems, some of which are threatened by logging. Implications are wider than loss of ecosystems, and include hydrogeological problems, climate, and landscape quality.

New wood products are making more efficient use of wood, although this means the ability to face an increased consumption more than decreasing logging activities. In fact, the production of industrial wood grew by 65% between 1942 and 1972, and logging production to supply that demand grew by only 56%. Types of laminated timber, reconstituted wood boards, and composite components have been replacing solid lumber in construction. These use resins and glues that have to be checked for chemical emissions; for example, formaldhyde used in particle board some years ago is toxic. Chemically sensitive persons are affected by virtually all emissions from preservatives, glues, plastics, and asphalt.

Architects have the responsibility to evaluate the best constructional solution, in terms of cost, health, energy consumption for the building, as well as for the community at large. Speed of construction and light weight make wood framing an economic system that, if properly built, has sufficient resistance to the natural forces of hurricanes and earthquakes. This requires careful supervision to ensure proper construction practices that will ensure safety and durability. Desirable wood properties include good thermal insulation, which helps avoiding cold bridges often created by steel or concrete structures (at perimeter columns, spandrel beams, window lintels); on the other hand, wood structures have low thermal-storage capacity. A holistic design approach balances human comfort and energy requirements with structural design, functional and aesthetic considerations.

19.3 Selection and configuration of wood systems

We told you not to be surprised when you see this section, as a reference, included in each material description. The issues are similar and a reiteration will probably help you remember what you're looking for when making system selection decisions.

Selection of a wood system involves a complex set of interrelated issues, so there is no single answer. You need to at least consider the following:

Spatial requirements

What are the volumetric requirements of the function that the space will house?

Is there a planning module or a leasing module that should be column-free or could be easily isolated? Parking garages, office structures, schools, hospitals, and retail establishments all have identifiable modules that are more desirable. Unfortunately these are not absolute, perhaps with the exception of parking facilities, and must be identified during programming and schematic design stages.

In residential construction, there are modules that may be determined by materials (carpet, drywall, etc.), as well as repetitive space modules such as bedrooms.

Expansion

Is the spatial system likely to require frequent remodeling?

What is the likelihood of expansion of the system?

Integration of systems

Will the environmental/lighting systems require frequent change?

Does the spatial system require isolated volumes of space: for temperature, sound, security, utilities management, etc.?

Ease of erection/construction

What is the availability of skilled labor?

Are material or component production facilities easily accessible?

Fire resistance

What are the fire-rating requirements of the building as a whole?

What are the isolation characteristics of zones of the building?

What are the patterns of access required for emergency egress?

Soil conditions

What is the nature of the soil on the site?

Chemical reactions with wood products?

Bearing capacities?

Wood structures are light structures: This may be an advantage for foundation design and a disadvantage for thermal balancing. Wood products do have a good inherent insulation quality, but you will still have some thermal transfer through structural components. You might look at the dirt collected on painted surfaces over joists or studs. This is a result of convection currents at points of heat transfer. A wood structure will generally weigh slightly less than its equivalent in steel and significantly less than its concrete counterpart. This could be a major consideration if the building is supported by weak soils, which would require an excessive investment in foundations.

Spans and loadings

Some common combinations of span/loading/mechanical integration characteristics for simple systems would include the following:

S4S lumber is available today in relatively short dimensions, 20 to 24ft [6 to 7.2m], and is a common material for small buildings, typically single-family residences and other low-rise structures. However, small lumber pieces can be combined in structural systems such as trusses or lamella vaults to cover large spans, and manufactured components such as laminated wood members allow the production of beams and arches capable of spanning long distances. This ability is achieved by carefully utilizing the strength characteristics of the wood in their most advantageous way and by eliminating the "flaws" that are naturally found in solid wood members.

Knots, splits, shakes, etc., all reduce wood capacities below their ideal values. The use of short, thin pieces that are selected and edited for flaws is a double benefit in the design of laminated members.

There are now several construction systems competitive with traditional wood framing for small residential construction as well as for larger buildings, and there are no absolute criteria to allow us to select one material or system over another. The choice, as we discussed in the selection of open-web steel joists, will depend on unique situations, and on an analysis of the several factors that may be more or less influential in wood structures.

Total system cost. The same system can be more expensive in one material than in another. The optimal construction cost may correspond to different configurations in different materials (e.g., certain structural spans that are economical for steel may not be economical for glulam). The economy has to be considered in relation to the total cost, since one should add to the cost of the structure considerations of finishes, building volume, gross floor area, and fire protection. Wood components can require fire or insect treatment or painting. Solid glulam beams do not allow ductwork to be run through the web, so additional building height may add to the cost. Heavy timber systems have an intrinsic fire resistance, but sprinkler systems could be required by the fire code.

Local factors, such as cultural issues and availability. Wood is not readily available in all parts of the world. In areas of the world where wood is scarce or where durability and fire resistance are the main requisites, wood is not considered a desirable material (e.g., for housing in Mexico), although laminated wood has gained acceptance where its structural advantages are combined with the aesthetic quality of the exposed material. Wood species (including the ones used in engineered wood) have very different structural properties, so it is important to make sure that the lumber with the required properties is available locally.

Environmental aspects. These are both global and local. Wood is a renewable resource, but the current rate of consumption is creating supply shortages and is leading to environmental protection measures worldwide. A lot of waste can be generated by design solutions that do not consider accurately timber properties and sizes. The construction of a new family house can generate two to three tons of waste, part of which is wood cuttings; these go into landfills and are not recycled.

Human comfort. The light weight of wood, with its workability, makes it ideal for manual handling on-site and for semiskilled labor; its disadvantage is the low degree of acoustic insulation intrinsic in common floor assemblies and partitions. Its thermal insulation is good compared to that of steel and concrete, but, on the other hand, wood has less thermal-storage capacity, making it potentially less desirable in areas of extreme daily temperature fluctuations. These thermal characteristics should be considered early on in the design of the building.

The codes impose limits to the size of a wood building due to the combustibility of wood structures, as shown in Table 19.1.

Table 19.1 Maximum residential building size based on fire resistance of the structure

Type of construction	Number of stories (Height in feet)	Floor area (ft²) [m²]	
Non-combustible frame (fire-resistive), Types I and II	Unlimited	Unlimited	
Combustible frame (treated), Type III	4	13,500	[1,254]
Heavy timber	4	13,500	[1,254]
Other, Type V	3	6,500	[604]

Source: Adapted from UBC, Chapter 17, Group R occupancy, Division I (hotels and apartment houses)

In many cases, a system will consist of a combination of components made of different materials: the lumber or timber may constitute the bending members, whereas the vertical supports could be wood, steel, or concrete columns, or wood or masonry walls. The selection of a wood system for a large building requires the evaluation of alternative solutions in different materials that depend on the type and functional requirements of the building. Whereas structural strength and fire regulations do not allow high-rise construction in wood, it is often a desirable material for long spans. It is useful to understand which are the possible systems and which are the main selection and design criteria.

As we have seen for steel, most systems include horizontal bending members, or floor and roof assemblies, that transfer floor loads to the vertical supports. The horizontal assemblies often work as diaphragms, or basically rigid horizontal platforms that transfer the lateral loads (acting on walls and the floors themselves in the case of earthquakes) to the bracing system. These systems are made of frames or panels. The following section illustrates the most common types of systems using wood members.

19.4 Frame systems

19.4.1 Traditional light framing

Eastern braced frame or barn frame

The main load-bearing members are timber posts (typically, 4×4 or 6×6) [89×89mm or 140×140mm] continuous for the building height supporting second-floor girths. The knee braces are for lateral bracing and also reduce girth span. The joints are mortise and tenon, with wood pegs or metal connections. The main function of the studs is supporting the wall sheathing, commonly horizontal 7/8in [22mm] boarding, and locally resisting wind loads. The spans and structural bays are limited by the length of the available lumber, typically about 20ft [6m].

Balloon frame

The innovation of this system over the previous is the replacement of heavy timber with built-up posts using dimension lumber (with a nominal thickness of 2in [*38mm*]) and with ribbands of dimension lumber for the girths. The thin lumber eliminated the labor-consuming mortise and tenon connections and could be put together with nails only. The knee braces are also eliminated and replaced with diagonal solid wood sheathing, plywood sheathing, or combinations of insulation and plywood sheathing at the corners nailed to the studs. The studs take a more important structural role and create a bearing wall to support the floor loads.

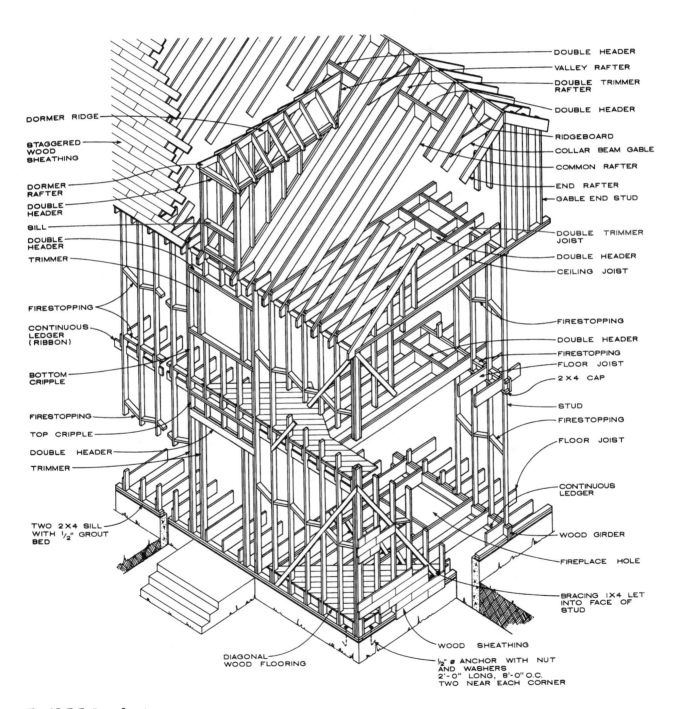

Fig. 19.5 Balloon framing

Platform frame

The name comes from the floor system built as a rigid platform. Each story is built as a structurally stable unit, connected to the platform of the floor below. The studs therefore are one-story high, requiring shorter pieces than the balloon frame. Rafters can be used for the roof, as in the previous systems, but prefabricated lumber trusses are now common. Bracing was provided originally by diagonal board sheathing or diagonal 1inX4in [19mmX89mm] let-in braces (nailed into notches in the studs), extending from the top to the bottom plate; but today structural panels (plywood or other composite material) or, once again, combinations of insulation panels and plywood sheathing at the corners are almost exclusively used. The connections are all nailed; however, metal connectors are sometimes required to resist large lateral loads such as earthquakes. This is currently the most common system for low-rise residential construction.

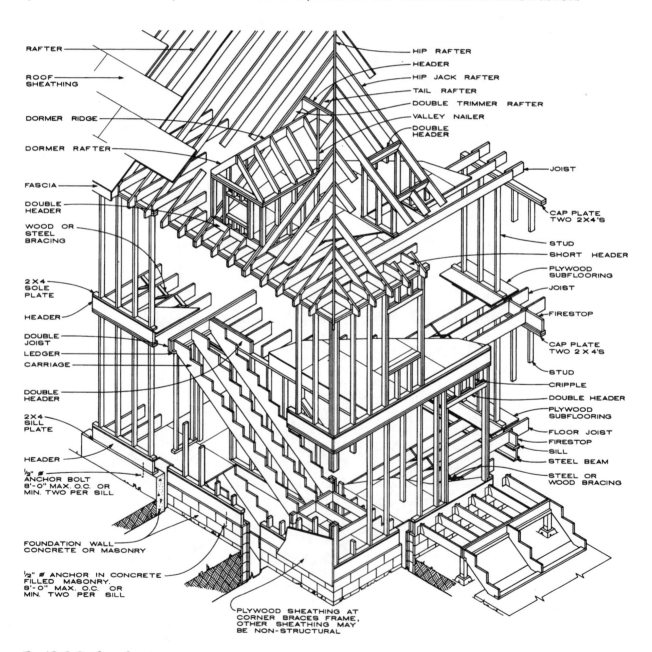

Fig. 19.6 Platform framing

19.4.2 Post-and-beam systems
Single-story, simply supported beams

Currently, laminated wood products allow spans of 100ft [30m] and more, depending on loads and spacing of beams.

System (a) is planned on a rectangular grid, with short beams and long engineered joists. The cost of the beams is minimized by placing them on the short span. The economy of this system depends on the cost of the joists (typically, members with laminated timber flanges and plywood or metal webs). The joists can be run on top of the beams (e.g., to create an overhang or to run ductwork) or can be dropped flush with the beams.

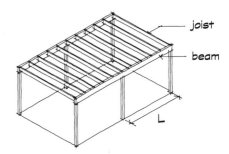

(a) Short beam and long joist system

System (b) is a typical rectangular grid used for very large spans (over 100ft [30m]), when long engineered joists are not available. In this case, the beam spacing is likely to be around 20ft [6m], and a system of purlins or joists is supported directly on the main beams and covered with a structural roofing system. Various types of structural stressed skin panels are sometimes used in place of the purlins.

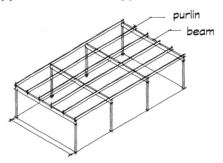

(b) Long beam and short joist system

System (c) is planned on a square grid and uses secondary beams to break the span of the joists or purlins. Secondary beams can be usefully introduced also in system (a), since they will reduce the size of the joists substantially, particularly if the joists can be run over the beams as continuous members.

The primary beams or girders will be deep and will constitute an expensive item if large unobstructed spans are required.

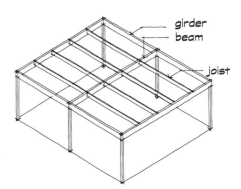

(c) Girders and secondary beams

Any of the preceeding systems, when they are multiple-span, can be built with the cantilever system(see Gerber beams in Chapter 5).

System (d) is common when using glulam beams for large-span, one-story buildings. The system consists of a beam cantilevering on the two supporting columns every other bay. Shorter beams are then simply supported on the ends of the cantilevers. This system is economical with medium- and long-span roofs (about 30 to 80ft [9 to 24m]).

(d) Cantilevered (Gerber) beams

Two-story, continuous girder

System (e) is conceptually similar to the platform frame, but since the columns carry heavier loads than the studs, they cannot rest directly on the girders, since the bearing stresses are usually too large. The columns are spliced and the vertical loads are transferred through steel plates. The continuous girder can be used to create cantilevers at each floor.

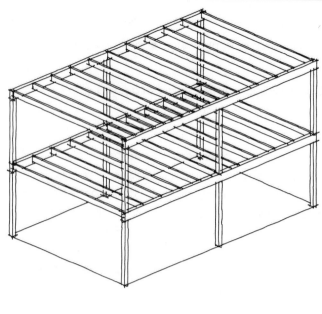

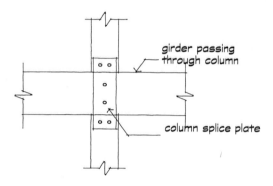

(e) Single girder and column system

System (f) is a twin-girder system that allows the continuity of the column; the column can be spliced for constructional reasons, and this is done often above the beam-column connection, similar to steel and precast concrete systems.

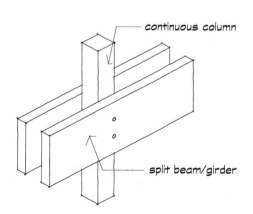

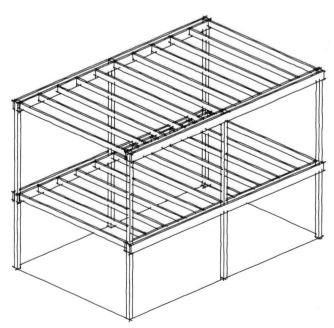

(f) Twin girder and solid column

System (g) illustrates the alternative use of a
double column to allow the continuity of the
girder. The choice between systems (f) and (g)
may depend on the design of the connection. A
variation of this system is the quadruple column, with the four members connected with
wood spacers or specially made steel connectors. This latter system allows for continuous
girders in two directions. This solution is
suitable to create large spaces with few
columns, but it will require a considerable
structural depth of the floors and roof.

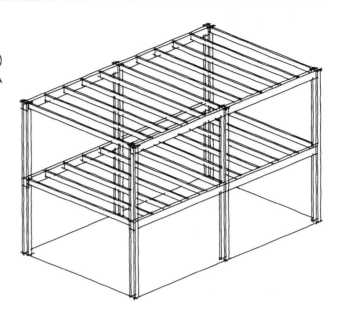

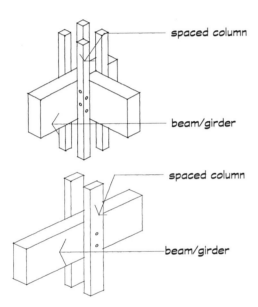

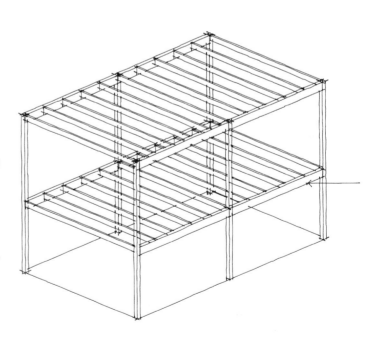

(h) Tie beam system

Tie beam system

This system is based on a continuous column
and on beams connecting the columns at each
floor, usually bearing on steel hangers. It is
likely to have steel columns continuous for
two or three floors, with rigid connections
forming a moment-resisting frame; otherwise,
splices can be used for the columns, in combination with a bracing system. The main
girders cannot be cantilevered.

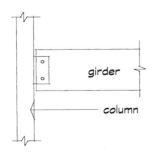

(g) Single girder and spaced column

19.5 Long-span systems

Note: A more detailed description of these systems can be found in Chapter 22, Timber and Laminated Beams; in Chapter 23, Timber Truss Design; and in Chapter 25, Arches, Vaults, and Domes.

19.5.1 Axial load and bending roof systems

Stayed or propped girders

These systems are simply supported on vertical supports and are not shown here.

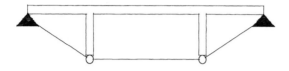 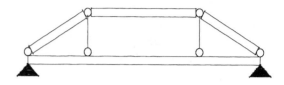

Suspension systems

These systems require stayed or cantilevered vertical supports.

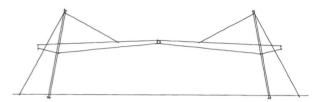

19.5.2 Frames and arches

Three-hinged frames and arches

The construction can be

(a) Composite plywood-lumber

(b) Trussed frame

(c) Laminated timber

In all cases, the construction must achieve a rigid connection or continuity between the "beam" portion and the "column" portion of the frame. This continuity is implicit in the geometry of the circular arch.

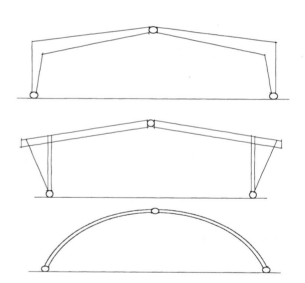

Two-hinged frames and arches

These systems are similar to the three-hinged frames, but are more rigid for roof and lateral loads and are better suited for very long spans. The problems are associated with the transportation of very long members and/or the assembly of rigid joints. It is usually necessary to manufacture the arch in sections that are spliced and connected on-site with mechanical fasteners.

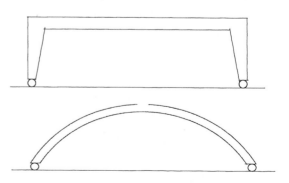

Cantilevered, three-hinged arches

These systems are typical of grandstands for open-air sports facilities. The cantilever can reach 100ft [30m]. Two cantilever systems connected with a hinge form a five-hinge arch also called a propped arch.

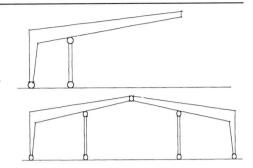

19.6 Bracing

Diaphragms and shear walls

Roof sheathing is built with a structural panel material (plywood, oriented strand board, or similar material). This is nailed or stapled on the roof members. Structural insulated panels can double as structural sheathing and insulation system.

Shear walls can be built with the same concepts used in roof construction to form vertical diaphragms called racking panels.

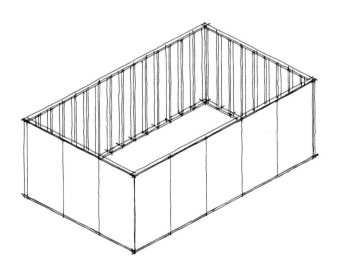

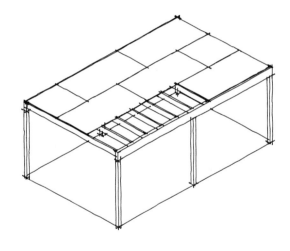

Diagonal bracing

Knee or eccentric bracing

Knee bracing introduces bending moments in the columns and beams, as we have seen in steel systems. Other types of eccentric wall bracing are K bracing and chevron bracing.

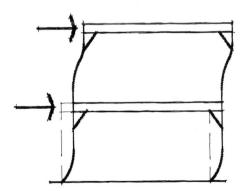

Diagonal or cross-bracing

Used for roofs and walls, this system utilizes lumber or steel, and it requires the use of steel plates bolted or screwed to the wood.

Rigid-frame system

This is not a "bracing" system, but a system able to resist lateral loads. Unlike steel and concrete systems, this is not common in wood construction, except in the traditional pole barn system and in large laminated-timber systems. In the pole barn system, the post (pressure-treated with preservative) is embedded in a concrete foundation to create a moment connection. With large laminated beams, moment-resisting frames can be built by increasing the "beam" section at the connection.

20 ROOF AND FLOOR SYSTEMS

20.1 Floor framing

Floor framing consists of a system of sills, beams, joists, or floor trusses and subflooring. Although bending members can be in any orientation, the simplest expression of such a member is a joist supporting floor gravity loads. The live loads are carried first of all by the subfloor, usually a structural panel material such as T&G plywood or OSB (oriented strand board). This panel is essentially a bending member spanning between the joists; therefore, joist spacing affects the thickness of the subfloor. Generally, we do not need to calculate the thickness of the subfloor since the span rating given by the manufacturers (see Sec. 20.4, Subfloors). In this chapter, we will focus on the design of joists as bending members.

The spacing of joists will determine the tributary area or portion of the floor load carried by each joist. Typical spacing of floor joists are 12, 16, 19.2, and 24in [305mm, 406mm, 488mm, and 610mm] o.c. At 12in [305mm] o.c. spacing, the introduction of mechanical systems and plumbing is cumbersome. Typically, 16in [406mm] o.c. spacing is the standard of the building industry for floors with 1/2in [12mm] gypsum board ceilings and 3/4in [19mm] T&G plywood floor sheathing. At 19.2in [488mm] o.c., a savings could be realized: The same ceiling and floor sheathing materials will work, and over an 8ft [2.44m] section, the system will require one less joist. This can be considerably more cost-effective if the same size joist will work at both 16in [406mm] and 19.2in [488mm] o.c.; however, something as simple as the absence of a standardized premarked tape measure keeps this from becoming a popular option. Spacing of 24in [610mm] o.c. is becoming common, particularly with the introduction of engineered joists. This spacing saves labor, and when coordinated (aligned vertically) with 24in [610mm] o.c. studs allows the use of a single top plate, giving additional savings and reducing thermal bridges. With S4S lumber, however, the use of 16in [406mm] o.c. spacing remains the most popular construction system for small buildings.

The most common type of joist is S4S sawn lumber 2in [38mm] thick and from 6 to 14in deep (nominal dimensions) [140mm to 343mm (actual)], classified as dimension lumber. Although 2X14 dimension lumber is available, its use is generally limited both by cost and local availability. Typically, although tables list a wide range of grades, No. 2 is the most common yard grade lumber used in construction. With No. 2 lumber, spans exceeding 18ft [5.5m] are likely to be beyond the capability of most species for floor construction; however, some species such as Southern pine may span over 20ft [6m]. Although some premanufactured systems will exceed this limitation, designers should be aware of the potential cost impact of a 20ft [6m] span on a wood floor system. All bending member design must also consider deflection limitations, as described in Table 20.1. Stricter limits may be placed on deflections (L/480) to prevent damage of ceramic floors or reduce the sensation of vibrations. Since this S4S wood construction is so common in North America, maximum-span tables for lumber joists are frequently used for joist selection. An example is Table 20.1 for Southern pine. These tables are valid for the loads and deflection indicated. If the safe span for a different maximum deflection is required, it is possible to multiply the span in the table by the ratio between the given and desired deflection values.

Table 20.1 Southern pine spans for flat roof and floor joists

Design Criteria:
Deflection of L/360 (based on live load only)
Loading of 30psf (roof live load) or 40psf (residential floor live load) + 10psf (dead load)

| | | Size (inches) and spacing (inches on center) | | | | | |
| | | 2X8 | | 2X10 | | 2X12 | |
Grade	Live Load	12in o.c.	16in o.c.	12in o.c.	16in o.c.	12in o.c.	16in o.c.
No. 1	30psf	15ft-10in	14ft-5in	20ft-3in	18ft-5in	24ft-8in	22ft-5in
	40psf	14ft-5in	13ft-1in	18ft-5in	16ft-9in	22ft-5in	20ft-4in
No. 2	30psf	15ft-7in	14ft-2in	19ft-10in	18ft-0in	24ft-2in	21ft-1in
	40psf	14ft-2in	12ft-10in	18ft-0in	16ft-1in	21ft-9in	18ft-10in

Example 20.1

Find the size of floor joists for a live load of 40psf and a span of 12ft-6in. Use Southern pine.

From Table 20.1 (LL = 40 psf, deflection = L/360), we can write down some options, according to different lumber grades:

Grade size	Spacing	Safe span
No. 1 2X8	16in	13ft-1in
No. 2 2X8	16in	12ft-10in
2X10	24in	13ft-2in

The most economical choice corresponds to the least joist area per foot width of floor.
With 2X8 at 16in o.c.:   A = (10.88sqin)/(1.33ft)per ft = 8.18sqin/ft
With 2X10 at 24in o.c.:   A = ([13.88sqin)/(2ft)per ft = 6.94sqin/ft

The 2X10 uses the least material, but used at 24in o.c., it will require a thicker subfloor and ceiling drywall. Consequently, the best choice from the point of view of "system" cost may be a 2X8. As in an architectural studio, there never seems to be an "absolute" answer.

For spans beyond 18ft [5.5m], an alternative would be the use of a composite action of the joist and the floor sheathing. This alternative, although requiring both more engineering time and more constructional effort (nailing), may reduce the joist by two full sizes or can significantly increase the spanning capabilities of the existing size. This is also an excellent technique for reinforcing underdesigned structures or those undergoing a change in function requiring additional load-carrying capacity.

Early information on the dimensions and optimum spans of joists is a key to cost savings, since it enables the architect to design the building with minimum waste of material and labor. In the United States, where most of the dimension lumber is used in the highly organized home building industry, standard sizes of materials and components have made construction extremely rapid and comparatively inexpensive. The basis for economic design is the 4ft [1.22m] modular dimension corresponding to 4ft × 8ft [1.22m × 2.44m] and 4ft × 12ft [2.44m × 3.66m]sheathing and drywall materials. Using lumber framing and panel products on a designed modular basis in floors and roofs eliminates unnecessary waste of materials.

The NFPA adopts a modular design for floors based on a planning grid with a 4in module and overall dimensions that are all multiples of 4in [102mm] (Fig. 20.1). Basic modular dimensions are on the outside face of the stud walls. Floor joist lengths for modular house construction are based on multiples of 2 or 4ft [610 or 1220mm] which are used for house exterior measurements.

Span tables for joists and rafters published by the NFPA and other wood industry associations permit selection of the most economical grades and sizes of locally available lumber. Floor joists are spaced 12, 16, or 24in [305,406, or 610mm] o.c., depending on the design floor loads. Joist spacings coordinated with an 8ft [2.44m] module are listed in Table 20.2.

Table 20.2 Dimensionally coordinated joist spacing

Spacing	4ft module	8ft module
12in o.c.	12in × 4 = 48in	12in × 8 = 96in
13.7in o.c.		13.7in × 7 = 96in
16in o.c.	16in × 3 = 48in	16in × 6 = 96in
19.2in o.c.		19.2in × 5 = 96in
24in	24in × 2 = 48in	24in × 4 = 96in

Use of the 48in [1220mm] module on house dimensions permits greater use of full 4ft × 8ft [1.22m × 2.44m] plywood subflooring and minimizes cutting and waste. In many new homes, 6% to 17% of the cost of floor framing is wasted. If the front-to-back dimension of the floor plan measured between the outside surfaces of the exterior wall studs is not evenly divisible by 4, usable lengths of floor joists are being lost. Lumber joists are produced in length increments of 2ft [610mm] with a tolerance of + 3in [76mm] and - 0in [0mm, obvious conversion]. Joists 12ft [3.66m] and longer typically have a plus 1/2in [12mm] or more to allow for cutting into shorter standard lengths. (The additional 1/2in [12mm] compensates for material removed by the saw cuts.) Joists are normally lapped 3in minimum on a center support, which means they extend 1-1/2in [38mm] beyond the center support. By adding the 1-1/2in [38mm] header thickness on each side, the house dimension is equal to the modular dimension of the joist (Fig. 20.1). Even in this case, the lap length is kept to a minimum and a modular joist length corresponds to a modular house dimension.

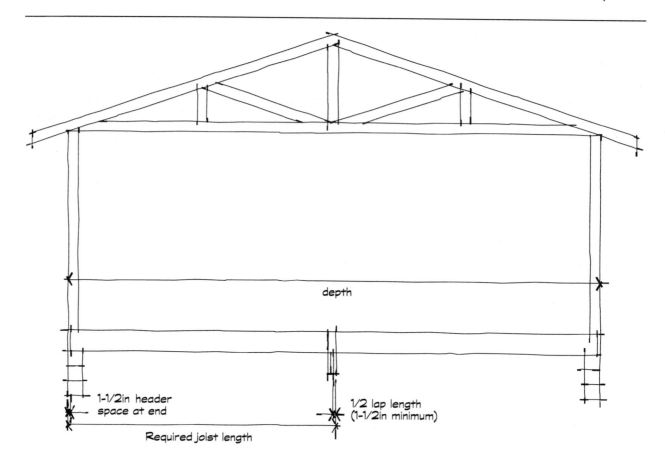

depth

1-1/2in header
space at end

1/2 lap length
(1-1/2in minimum)

Required joist length

Fig. 20.1 Lapped joists and house dimensions in typical floor framing

20.2 Joist design

The procedure for joist design normally requires the designer to make assumptions on span, spacing (based on wall framing and sheathing type), wood species, and conditions of use. The basic steps of the procedure consist of (a) solving the statics; (b) finding the allowable stresses F'_b, F'_v, and E'; (c) calculating the required section properties (S, A, I); (d) selecting the adequate size. Step (b) requires the application of the adjustment factors.

Table 20.3 Adjustment factors for joists

Factor		Adjusts
C_M	Wet service (see Chapter 2)	F_b, F_v, E
C_t	Temperature (see Chapter 2)	F_b, F_v, E
C_D	Load duration (NDS, Table 2.3.2, p. 6; see Chapter 2)	F_b, F_v
C_r	Repetitive member (NDS, p.19; and design values tables, Supplement)	F_b
	Applies to three or more parallel members of the same size spaced at 24in o.c. maximum.	
C_F	Size (NDS, p.19; and design values tables, Supplement)	F_b
C_L	Lateral stability	F_b

20.2.1 Adjustment factors

In order to take into account the effect of the conditions of use mentioned in Chapter 19, the tabulated design values are multiplied by adjustment factors (NDS, p. 5). The factors that have to be considered in joist design are listed in Table 20.3. Three new factors are introduced here: the repetitive member, form, and beam-stability factors.

Load duration factor (C_D) This factor is identified in the tables as one that can be used as a multiplier for all wood properties except E. This is a bit misleading. If you have already developed the critical-loading condition using the worst combination of loads with their respective C_D's, it would be inappropriate to include this factor in the stress calculations as well. The three potential ways to use C_D are as follows:

1. To adjust the load conditions using the critical-load-duration factor, use these loads for moment and shear calculations (but never for deflection) and DO NOT use C_D as a properties multiplier.
2. To calculate the critical moment and shear values, use all possible loading conditions and then divide these values by their respective C_D values to determine the design values. This is a cumbersome technique and we do not recommend it, although it will produce the correct answers.
3. To determine the critical moment and shear values for all possible loading conditions, calculate the required area and section modulus using design values that have the alternative C_D values applied. At this point you, would use the set of values that requires the largest properties of all combinations. Again, this is not recommended, simply because it is clumsy and requires more work.

Techniques 2 and 3 are mentioned only because they are sometimes referenced and it is important that you understand the differences and implications of the three techniques.

The critical issue is DO NOT APPLY C_D MORE THAN ONCE TO A SOLUTION.

Repetitive member factor (C_r) This factor recognizes the distribution of loads over more joists (at least three) connected together by a subfloor and bridging. It increases the design values by 15%.

Size factor (C_F) This factor reflects the relative decrease of the strength of wood as the depth of the joist (width of lumber) varies. This factor does not apply to mechanically graded lumber.

Beam stability factor (C_L) This factor applies to joists and timber beams, but in practice its value is 1.0 for ordinary floor joists. This is because (according to the NDS considerations in Art. 4.4.1 that will be discussed further for beam design) the subfloor is nailed to the compression edge of joists, thus preventing buckling or sidesway; the end supports are also usually restrained against rotation, being nailed to the wall plates and headers. With this type of construction, stability is ensured for joists as deep as a 2X12 [38mmx286mm]. Keep in mind that it is important to restrain the compression edge; thus, in a cantilever, where the compression edge is not the floor surface, it may be necessary to attach a plywood soffit to provide support. Another case could be a long-span 2X2 [38mmx286mm] ceiling joist (with no occupied floor above), where a subfloor should be nailed to the joists, although it may not be required for functional reasons. You should realize that when you do this, the owner will probably attempt to utilize this new "floor" space at least for storage. Consequently, it is worth considering a design live load greater than the code-required 10psf [480N/m²]for ceilings. For deeper joists (in practice, 2X14 [38mmx336mm] is the maximum size used), the tension edge also has to be held in place; a nailed drywall ceiling can serve the purpose.

Bridging, in the form of diagonal cross-struts or solid blocking between joists, in the past (until 1973) required a maximum spacing of 8ft [2.44m] for 2X12 [38mmx286mm] joists. Today, the role of bridging is recognized in helping the distribution of point loads on adjacent joists, and in keeping the joists straight as they dry and shrink.

Required section properties. The horizontal shear stress is calculated with the general shear formula (NDS):

$$f_v = \frac{VQ}{Ib} = \frac{VA\tilde{Y}}{Ib} = \frac{V\Sigma a\tilde{y}}{Ib}$$

which becomes (for a **SOLID RECTANGULAR SECTION ONLY** of dimensions b × d):

$$f_v = \frac{V\left[b\left(\frac{d}{2}\right)\left(\frac{1}{2}\right)\left(\frac{d}{2}\right)\right]}{\left(\frac{bd^3}{12}\right)\left(\frac{b}{1}\right)} = Vb\left[\frac{d}{2}\left(\frac{d}{4}\right)\right]\left(\frac{12}{bd^3}\right)\left(\frac{1}{b}\right) = \frac{3}{2}\frac{V}{bd} = 1.5\frac{V}{bd} = 1.5\frac{V}{A}$$

This is really no more than a variation of the general horizontal shear formula solved for a simple rectangle b inches wide and d inches high. Note that this is similar to, but not the same as, the approximated $f_v = V/A$ form that we previously used for steel web design. When calculating the shear force, V, the loads within a distance d from the support can be ignored, since it is assumed that they bear on the support, as shown in Fig. 20.2. This is a commonly ignored practice

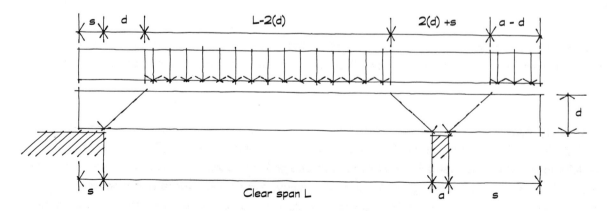

Fig. 20.2 Actual loads creating shear stress in a beam

in both wood and steel construction, since the beam depths are quite small compared to the span and no significant savings are made, with the exception of short lintels and cantilevers. However, this is relevant in deep wood sections and concrete beam design.

The bending stress is given by the flexure formula:

$$f_b = \frac{M}{S_x}$$

which, solved for the section modulus, becomes

$$\text{Required } S_x = \frac{M}{F_b}$$

I can be calculated from deflection formulas. Due to creep, the deflection caused by the long-term loads (DL) is multiplied by 1.5 for dry lumber and laminated timber. Thus, the total deflection is

$$\Delta = 1.5\Delta(\text{long-term DL}) + \Delta \text{ (short-term LL)}$$

The required section can be selected in tables of section properties of standard, dressed sawn lumber (NDS-S) listing nominal sizes of 2X(d) and 4X(d).

If no satisfactory section can be found in the tables, the following options are possible:

1. Change spacing: Standard spacing values, based on 8ft sheathing length, are 12, 16, 19.2, and 24in [305, 406, 488, and 610mm].
2. Change lumber type: species or grade. This may be a "mathematical" solution, but not a practical one based on availability.
3. Create a composite design solution.

Example 20.2

Design of a Flat Roof Joist

Roof joists 16in o.c., 18ft span

Snow load	30psf
Dead load (ballast, roofing, sheathing, joists, ceiling)	18.9psf
Total load	48.9psf

(a) Adjustment factors

Adjustment factors: Snow load, $C_D = 1.15$; permanent DL, $C_D = 0.9$

Critical load: Divide each combination of loads including DL by the highest applicable adjustment factor.

DL alone = (18.9psf)/0.9 = 21psf

Snow + D_L = (48.9psf)/(1.15) = 42.5 psf > 21psf critical

Therefore, $C_D = 1.15$ for all values except E, F_c (NDS).

Repetitive member factor (more than three joists): $C_r = 1.15$ for bending only (NDS-S).
Distributed load on a single joist:

Snow + DL: w = (48.9psf)[(16in o.c.)/(12in/ft)] = 65.2plf

Maximum moment: $M = \dfrac{(65.2plf)18ft)^2}{8} = 2640.6ftlb$

Maximum shear: $V = \dfrac{(65.2plf)(18ft)}{2} = 586.8lb$

Allowable deflection: $\Delta = \dfrac{(18ft)(12in/ft)}{360} = 0.6in$ for LL

$\Delta = 0.6in(360)/(240) = 0.9in$ for TL

Since the DL is small, the deflection of the floor is controlled by the live loads inclusive of creep. (Based on these relationships L/240 and L/360, if the DL is less than 100% of the TL, it will NOT control.) Although a number of different grades are available, No.2 is the most common "lumber yard" variety. It is usually referred to in the yard as No.2 and better. The particular species you should specify might be logically a consideration of those species commonly carried by lumber suppliers in the area of your project.

Use Douglas fir-larch No. 2 (NDS-S).

Allowable shear stress: $F'_v = C_D F_v = 1.15(95psi) = 109.3psi$
Allowable bending stress: $F'_b = C_D C_r F_b = 1.15(1.15)(875psi) = 1157.2psi$
Modulus of elasticity: $E = 1,600,000psi$

(b) Design of the joist

Moment: required $S_x = \dfrac{M}{F_b} = \dfrac{(2640.6ftlb)(12in/ft)}{1157.2psi} = 27.38in^3$

Shear: required $A = \dfrac{1.5V}{F_v} = \dfrac{1.5(586.8lb)}{109.3psi} = 8.95in^2$

Deflection: snow load w_S = (30psf)(16in o.c.)/(12in/ft) = 40plf
          dead load w_D = (18.9psf)(16in o.c.)/(12in/ft) = 25.2plf

Required $I_{LL} = \dfrac{5(40plf/12in/ft)[(18ft)(12in/ft)]^4}{384(1,600,000psi)(0.6in)} = 98.4in^4$ for snow (short duration loads)

The creep factor increases the deflection due to long-term loads by 1.5; therefore, the additional deflection developed over time is assumed = $0.5\Delta DL$. The additional I can be found by proportion:

Required $\quad I_{DL} = 98.4\text{in}^4(0.5)\dfrac{25.2\text{plf}}{40\text{plf}} = 30.99\text{in}^4$

Total required $I = 98.4\text{in}^4 + 30.99\text{in}4 = 129.4\text{in}^4$

From NDS, Table 1B, 2X12 has $A = 16.875\text{in}^2$, $S_x = 31.64\text{in}^3$, and $I_x = 178\text{in}^4$

A 2X12 joist meets all criteria, BUT the size factor for 2X12 lumber is $C_F = 1.0$. It would be theoretically possible to use a smaller joist size if moment is the controlling factor, such as 2X10, since $C_F = 1.1$ and $F'_b = 1.1(1157.2\text{psi})$.

Required $S'_x = M/F'_b = S_x/1.1 = (27.38\text{in}^3)/1.1 = 24.89\text{in}^3$

A 2X10 has $S_x = 21.39\text{in}^3 < S'_x$ and $I_x = 98.93\text{in}^4 <$ total required I.

Therefore, use 2X12 Douglas fir-larch No. 2 at 16in o.c.

Lateral stability: $d/b = 12/2 = 6:1$. No bridging required since the compression edge is held in place by sheathing and the ends are fastened to the header and wall plate. However, bridging at midspan is **recommended**. Use V-shaped 16-gage galvanized steel bridging or 2X4 [*38×89*] lumber cross-bridging.

Note: A 2X12 is a critical shape, since a 2X14 has $d/b = 7:1$, requiring both edges held in line for the entire length.

20.3 Engineered joists

A variety of composite structural wood joists are available today, and they can be classified into three types: (a) I joists, (b) trussed joists, and (c) structural composite lumber joists. Each of these types can be manufactured with different wood materials. These joists have the following advantages over sawn lumber:

1. They can be produced in greater lengths (40ft [*12.2m*] lengths are common),
2. They have shallower depths when comparing equal spans and spacing,
3. They are lighter, which means that they can frequently be erected without using cranes or other heavy construction equipment,
4. They are more dimensionally accurate and stable,
5. They provide a wider nailing surface (in some configurations),
6. They use less old-growth logs, since they can use smaller components. This consideration is becoming more important in an era when both professionals and clients are concerned with the impact on the environment.

Engineered joists become cost-effective on medium or large spans, over 16 to18ft [4.8 to 5.5m]. It is important to design the span of the building for maximum efficiency of the framing, and therefore to be aware, from the inception, of the most economical use of each product. I joists can be easily used in systems with a variety of spans since they can be cut to length in the field. Many "trussed joists," however, are premanufactured to length, and using a variety of lengths on a project increases their manufacturing cost as well as coordination problems at the job site.

20.3.1 I joists

I joists use the concept of steel I-beams, where the flanges resist most of the bending stresses and the web resists most of the shear. The flanges are made of lumber or LVL; the web can be plywood or oriented strand board to take advantage of their superior shear capabilities. Chases for ductwork and pipes can be cut in the web according to the manufacturer's recom-mendations. The headers for platform framing can be made with 3/4in [19mm] plywood or equivalent; in this case, stiffeners are nailed at the ends of the joist, serving as nailers for the header as well as supports for the wall framing.

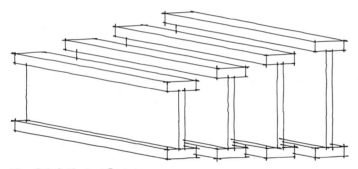

Sizes range from 9-1/2 to 16in with spanning capabilities of up to 32ft.

Fig. 20.3 Typical I joists

Flanges are typically 3-1/2in wide.

20.3.2 Trussed joists

Two types are common: lumber members with metal plate connectors at the joints (Fig. 20.4) and lumber flanges and metal webs. Like sawn lumber, they are manufactured in increments of 2ft [610mm]. The open web allows installation of ductwork and plumbing. The flanges, made with 2X4s [38mmx89mm], provide a 3.5in [89mm] wide nailing surface. The metal webs are nailed to the sides of the flanges or pin jointed at the axis of the flange. Depths are usually greater than those of sawn lumber, ranging from 11.25 to 22in [286 to 559mm] or more. These products are mainly for use on medium spans of 16 to 30ft [4.8 to 9.0m].

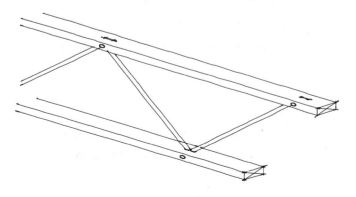

Sizes range from 10 to 63in with spanning capabilities of up to 70ft.

Flanges are typically 3-1/2 or 5-1/2in wide.

Fig. 20.4 "Trussed" joist with metal or wood web members

20.3.3 Structural composite lumber

This category includes LVL, PSL, and OSB combined with LVL (Fig. 20.5). LVL and PSL are reconstituted lumber produced in rectangular sections, similar to conventional joists, but with higher design values. The last type, made of LVL flanges and OSB webs, is an I section, from an engineering point of view, because of the different design values of web and flange materials; however, the construction use is the same as for rectangular section lumber joists (no stiffeners required for the web).

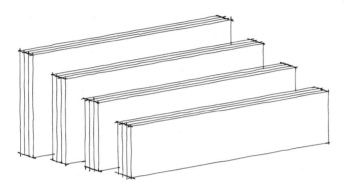

Sizes range from 7-1/4 to 18in with spanning capabilities of up to 20ft.

Member widths are generally 1-3/4 to 5-1/4in.

Fig. 20.5 Structural composite lumber (LVL)

The design of engineered joists is carried out following the manufacturer's data. The technical specifications literature for each product will contain section properties and tables of safe loads (psf) for spans at 2ft [610mm] increments.

It is important to check that the loads do not exceed a maximum deflection of L/360 or less for live loads, and L/240 for the total load. Tables for roof joists will usually indicate the use of a C_D factor of 1.15, or an increase of 15% of the base design values (due to snow loads), and floor joists tables will have a 0% increase (occupancy loads are long-term, according to the NDS). Table 20.4 is an example of design tables for I joists of the type shown in Fig. 20.3. The table indicates maximum spans, and one may look for the optimum combination of joist size and spacing for a given load per square foot.

Table 20.4 Engineered joist span table (40psf LL + 10psf DL)

Joist depth	Series	Spacing 12in o.c.		Spacing 16in o.c.		Spacing 24in o.c.	
		L/360	L/480	L/360	L/480	L/360	L/480
9-1/2in	LPI26	17ft-0in	16ft-10in	17ft-0in	15ft-4in	14ft-10in	13ft-4in
9-1/2in	LPI30	17ft-0in	17ft-0in	17ft-0in	16ft-2in	15ft-8in	14ft-1in
11-7/8in	LPI30	21ft-3in	21ft-1in	21ft-3in	19ft-3in	18ft-7in	16ft-9in
11-7/8in	LPI36	23ft-9in	22ft-10in	23ft-0in	20ft-9in	20ft-0in	19ft-1in
14in	LPI30	25ft-1in	23ft-11in	24ft-1in	21ft-9in	19ft-10in	19ft-0in
14in	LPI36	28ft-0in	25ft-10in	25ft-11in	23ft-5in	22ft-8in	20ft-6in
16in	LPI30	28ft-7in	26ft-4in	26ft-7in	24ft-0in	19ft-10in	19ft-10in
16in	LPI36	31ft-6in	28ft-6in	28ft-8in	25ft-11i	23ft-10in	22ft-7in

Example 20.3

Select an engineered joist for a span of 22ft and a maximum deflection of L/360.

Live load = 40psf

Total load = 49psf

Table 20.4 lists three alternative spacings 12, 16, and 24in o.c. Some alternatives are: LPI36 (11-7/8in deep) joists at 16in o.c. (maximum span 23ft-0in), or LPI30 (14in deep) joists (maximum span 19ft-0in) at 24in o.c. (within 5%). A LPI30 (16in deep) also works at 24in o.c. (masimum span 19ft-10in), but there is no advantage in load carrying capacity and it is 2in deeper. The 11-7/8in deep joist are the shallowest and at a 16in o.c. spacing will require a thinner decking than the 14in joists at 24in o.c. While the 24in o.c. spacing will require less labor (fewer pieces) than the 16in o.c. spacing, the overall economy of the system will probably be determined by the thinner sheathing. In a case like this, where engineered systems are used, the manufacturer should be consulted about the most appropriate use of the product. The tables are used for preliminary design only. It is important to remember that engineered joists are NOT standardized the same way as open-web steel joists.

20.4 Subfloors

At the beginning of this chapter, we have described the subfloor as a bending member spanning between joists. Some indications are given here to allow a selection of the appropriate material for the subfloor. In Chapter 21, we are going to look at the way the subfloor works with the other floor members to ensure the structural integrity of the building for lateral loads.

Traditional floor systems consist of the subfloor, sometimes an underlayment, and a finish floor. The ceiling finish and fire protection are provided by gypsum board. The dead load should include an allowance for ductwork and plumbing. A conventional system built this way weighs between 9 and 12psf [431 and 575N/m²]. In early framing, the subfloor was made of 7/8in [22mm] square-edged wood boards laid at 45° with the joists. This allowed the underlay or finished floor (typically, wood boards) to run perpendicular to a wall in any direction without matching the joints between the boards of the subfloor. Currently, subfloors are built with structural wood panels made of a type of reconstituted wood: plywood, oriented strand board (OSB), oriented wafer board (OWB), or composite plywood (often known as Comply) produced in a variety of thicknesses mostly between 3/8 to 1-1/8in [9.5 to 28.5mm] (see Table 20.5), with a square or tongue-and-groove edge.

Table 20.5 Oriented strand board thickness

Thickness (in)	[mm]	Weight (psf)	[N/m²]
7/16in	11	1.5	71.9
15/32 and 1/2in	12.5	1.7	81.4
19/32 and 5/8in	15.5	2.1	100.5
23/32 and 3/4in	19.0	2.5	119.7
1-1/8in	28.5	3.8	181.9

The surface is usually unsanded, and for floor sheathing, it can even be mill scuffed or roughened for improved worker safety. Some panels are sanded in the manufacturing process, so that an underlayment is not needed for some types of floor finishes. A 23/32in [19mm] T&G subfloor can be suitable for combined subfloor-underlayment (APA rated Sturd-I-Floor). Subfloors do not need to be calculated if the product is code-accepted for sheathing; the type of structural wood panel is specified for a specific maximum span between joists. The major code-accepted quality assurance agencies for structural wood panels are the American Plywood Association (APA), Timberco, Inc. (TECO), Professional Service Industries, Inc., and Pittsburgh Testing Laboratory Division. Structural wood panels may be manufactured to either prescriptive or performance standards. The prescriptive standard, U.S. Product Standard PS 1-83 for Construction and Industrial Plywood, is for plywood only. One code-recognized performance standard is APA PRP-108, Performance Standards and Policies for Structural-Use Panels. Both plywood and mat-formed panels (like OWB) are currently manufactured to performance standards. Sheathing grades are used as decks in wood roof systems and floors. Each sheathing panel has a span rating included in the panel trademark. Span ratings denote the maximum recommended support spacing over which the panel should be placed in construction applications. Sheathing span ratings have two numbers separated by a slash, e.g., 32/16. The first number is the maximum spacing in inches for roof supports for average loading conditions (30psf snow load)[1440N/m²]. The second is the maximum span between joists under average residential floor load (40psf)[1920N/m²]. The panels are sufficiently strong to carry concentrated loads such as pianos, kitchen appliances, and water heaters. A panel rate Sturd-I-Floor is intended for floor use and thus only the floor joist spacing (20 or 24in o.c.)[508 or 610mm o.c.] is marked. This is a performance specification; therefore, no thickness or wood species is indicated. The panel must be installed across three or more supports, and it must be at least 24in [610mm] wide. Table 20.6 indicates thickness, spans, and nailing requirements for typical subflooring.

Table 20.6 Subflooring (APA-rated sheathing)

Panel span rating	Minimum panel thickness	Maximum span	Nail size	Nail spacing supported edge	interior support
24in/16in	7/16in	16in	6d	6in	12in
32in/16in	15/32in	16in[a]	8d	6in	12in
40in/20in	19/32in	20in[a]	8d	6in	12in
48in/24in	23/32in	24in	8d	6in	12in

[a] Sheathing span may be 24in if 3/4in wood strip flooring is installed at right angles to joist span.

Other structural wood panel grades such as sanded siding or underlayment can be used as roof decking, although they do not have sheathing span ratings. In this case, it is necessary to obtain the structural properties or design values of the panel from the manufacturer. These materials should not be confused with particleboard or chipboard, manufactured with small wood particles without sufficient fiber length to provide structural strength for sheathing, but used in cabinetry for shelves and laminated countertops. Structural particleboard is used for underlayment.

Structural panels are produced with a low moisture content (well below 16%) and will absorb moisture during construction. It is often specified that a 1/8in [3mm] space be left at the joints to accommodate moisture expansion. This is not considered necessary by most manufacturers, since the panels are glued and fastened on the framing, and 1/8in [3mm] expansion and contraction would be impossible without failure of the connections and panel damage. Moreover, in order to avoid thermal leakage from roof and floors, it is good practice to tighten the panel joints as much as possible, using tongue-and-groove edges glued on the framing and caulked with mastic.

20.5 Fire protection and sound insulation

Dimension lumber has a fire-endurance performance that the code allows us to calculate. Times are assigned for the contribution of wood frame construction in fire assembly calculation sections and Tables 43-9-V and 3106.2B of the UBC and Standard Building Code, respectively. Wood floor and roof joists 16in o.c. [406mm] have a time of 10 minutes assigned to them. Fire-endurance assemblies also have been developed for metal plate connected trusses.

To properly design a building for fire safety means, according to code procedures, breaking it into fire-resistant compartments. To prevent the fire from spreading from one compartment to the next, the code requires finished assemblies to withstand full fire exposure without major damage and, at the same time, act as barriers to heat transfer.

Standard fire tests measure the fire-endurance performance of a variety of structural assemblies and boundary conditions that make up compartments. In the United States, ASTM Standard E119 sets terms of the assembly's ability to withstand a severe fire for a period of time. Performance times are measured in hours: 1-hour rated, 2-hour rated, etc. Different building construction types and occupancies are referenced to these hourly requirements. The two major source documents for dimension lumber and truss fire-endurance assemblies are the Fire Resistance, Sound Control Design Manual, published by the Gypsum Association, and the Fire Resistance Directory, published by the Underwriters Laboratory, Inc. (UL).

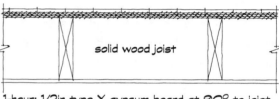

1-hour; 1/2in type X gypsum board at 90° to joist

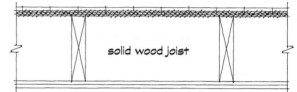

1-hour; 1/2in type X gypsum board on furring channels at 90° to joist

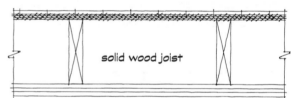

2-hour; 2 layers of 5/8in type X gypsum board with furring channels between the layers

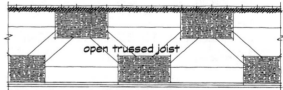

1-hour; 2 layers of 1/2in type X gypsum board applied directly at 90° to the joist

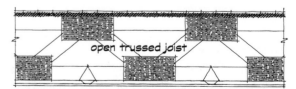

1-hour; 1 layer of 5.8in type X gypsum board applied to furring channels at 90° to the joist

Fig. 20.6 Fire-rated floor assemblies

The assemblies in these documents range in performance from a 1-hour rating to a 2-hour rating, providing flexibility for any project need. The most common dimension lumber and truss assembly types are detailed in Fig. 20.6, is excerpted from the *Gypsum Association* manual. Included are assemblies having direct gypsum application, resilient channels, insulation, and a 2-hour assembly. For suspended ceiling assemblies, see the UL directory.

Sound-transmission ratings are mentioned here since they are closely aligned with fire-endurance ratings for assemblies. This is because flame penetration and sound penetration follow similar, least-resistance paths. Methods to increase sound insulation may affect the construction and loads of a floor.

20.6 Roof Construction

Roof construction can consist of joists for a flat roof, rafters or trusses for a pitched roof, or stressed skin panels. Joists, rafters, and trusses require a deck or sheathing. There are two ways of designing rafters:

1. As simply supported beams spanning from a ridge "beam" (not a ridge board) to exterior walls or other beams. In this case, since the members span between the ridge "beam" and a wall or beam, no horizontal thrust is developed by the system, and open ceiling volumes are possible.

2. As three-hinged arches tied by a collar beam ceiling joist or some other lower-tension member. The ceiling joists are designed to transfer the tension from the rafter heel.

At the ridge, the rafters are nailed to a ridge board. (Note that ridge "boards" are not spanning members.)

Trusses are usually a lightweight engineered type and may be designed by the fabricator based on your specific loading conditions.

In all cases, the rafters or trusses must be braced for horizontal loads in the direction of the ridge, by diagonal braces or wood panels attached to form a continuous sheathing. Bracing is necessary also during construction before the sheathing is installed. Performance requirements for roof sheathing typically require a deflection limitation of 1/180 of the span between rafters or beams for a total load and 1/240 for a live load only. It is important to connect the edges of wood panels to avoid differential deflection. This is done by connecting the edges with a tongue and groove or special roof clips or by blocking the edges with a wood member.

Example 20.4

The system shown in the illustration consists of hem-fir No. 2 rafters at 24in o.c. on purlins. The ABC rafter is made of two pieces, since the actual length of the pitch is L = (22ft)(12/9) = 29.33ft. The design is controlled by the BC rafter, since the overhang is helping to reduce the bending in the AB span. The BC distance is L = (10ft)(12/9) = 13.33ft.

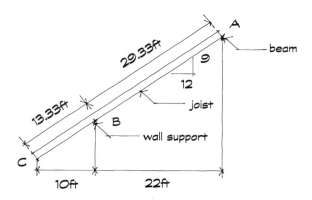

Local codes require roofs to support a snow load of 40psf with no wind, and a wind pressure of 7.5psf combined with 50% of the maximum snow load. It is clear that the load combination causing the maximum shear and bending stresses is snow plus the dead load. Let's solve the statics for this load combination:

Vertical load on each rafter:

Snow:	(40psf)(2ft)(10ft)	=	800lb
DL:	(12psf)(2ft)(13.33ft)	=	320lb
	Total	=	1,120lb

Vertical reactions R = 1,120lb/2 = 560lb

Component of a distributed load perpendicular to the rafter axis:

W = (1,120lb)(12/15) = 896lb (perpendicular to the span)

w = (896lb)/(13.33ft) = 67.2plf

Maximum shear V = (896lb)/2 = 448lb

Maximum moment M = $\dfrac{(67.2\text{lb/ft})(13.33\text{ft})^2}{8}$ = 1493ftlb

There is a component of the load parallel to the rafter that causes an axial load (tension on the upper half, and compression in the lower half). Since the axial load is low where the bending is high, we will neglect the combination of axial and bending stresses.

Adjustment factors: For normal conditions, C_M = 1.0 and C_t = 1.0. We have to determine C_D with the critical-load criteria, since the strength of the wood depends on the duration of the loads.

The components of the loads perpendicular to the rafters are

DL = (320lb)(12/15) = 256lb

SL = (800lb)(12/15) = 640lb

WL = (7.5psf)(2ft)(13.33ft) = 200lb

Load combinations:

1. DL/0.9 = (256lb)/0.9 = 284lb
2. (DL + SL)/1.15 = (256lb + 640lb)/1.15 = 779lb
3. (DL + 0.5SL + WL)/1.6 = (256lb + 320lb + 200lb)/1.6 = 485lb

The critical load combination is DL + SL; therefore, C_D = 1.15.

Notice, **as we previously discussed**, that the considerations on the load-duration factor, which adjusts the design values of the material, are quite independent from the working stress calculations (which depend on the statics of the system). In other words, you use the alternative load combinations and their respective LDF ONLY to determine which factor creates the critical-loading condition. You then use actual loads for statics calculations and apply the critical LDF previously determined to the design stresses, OR use the adjusted loads for statics and DO NOT apply any further LDF to the design stresses. Either method will work; however, since we sometimes carry static loads through a structure and design a series of related members, it is probably safer, in practice, to use actual loads and apply LDF to the respective stresses for each independent calculation. **REMEMBER ALSO THAT LDF DOES NOT APPLY TO DEFLECTION.**

The applicable adjustment factors are

C_D = 1.15   (snow load)

C_R = 1.15   (repetitive member)

C_F = 1.1   (2X10 assumed)

C_L = 1.0   (compression edge held in line)

Allowable stresses:

F'_v = (1.15)(1.15)(75psi) = 99.2psi

F'_b = (1.15)(1.15)(1.1)(850psi) = 1,236psi

TL = DL + SL = 256lb + 640lb = 892lb

Maximum V = (892lb)/2 = 448lb

Required area for shear:

$$A = \frac{1.5(448lb)}{99.2psi} = 6.77in^2$$

Required section modulus for bending:

$$S = \frac{(1,493ftlb)(12in/ft)}{1,236psi} = 14.5in^3$$

The deflection limitation is L/180 for total loads, since the slope is > 3/12 (see Table 20.1) and we assume there is no drywall attached to the ceiling.

Allowable deflection = L/180 = (13.33ft×12in/ft)/180 = 0.89in including creep.

Required I for total load = $\dfrac{5wL^4}{E\Delta}$ = $\dfrac{5(67.2plf/12in/ft)[(13.33ft)(12in.ft)]^4}{384(1,300,000psi)0.89in}$ = 41.27in^4

Required I for creep = 0.5(12 psf/52 psf)(41.27in^4) = 4.76in^4

        Total required I = 46.03in^4

        A 2X8 has: A = 10.88 > 6.77in^2
                            S = 13.14 < 14.5in^3
                            I = 47.63in^4 > 46.03in^4

The S$_x$ of the 2X8 is not adequate, but the C$_F$ is 1.2 versus the assumed 1.1 for a 2X10; there-fore, the adjusted S$_x$ = 14.5in^3 (1.1/1.2) = 13.29in^3 is sufficiently close to that provided (within our tolerance of 5%).

Use 2X8 Hem-Fir No. 2 rafters at 24in o.c.

20.7 Bearing and stress concentration

The design of roof and joist systems cannot be complete without some consideration about localized stresses such as bearing points. When a joist or rafter is supported on a wall plate, the gravity loads are transferred through the area of contact. On a horizontal support (such as a plate or a beam), if R is the support reaction, the bearing stress will be f = R/A, where A is the contact area. This type of bearing stress is perpendicular to the grain, one of the worst ways of loading wood. Wood strength perpendicular to the grain is considerably lower than parallel to the grain (which is the stress in wood columns); for instance, Douglas fir-larch select structural has a design value for compression parallel to the grain of 1700psi [11.7MPa], whereas its design value perpendicular to the grain is only 625psi [4.3MPa] (the same for all grades). If you think of wood as a "bundle of straws," you can easily understand that these hollow tubes are very strong parallel to their longitudinal axis and very weak perpendicular to that axis. Thus, bearing stresses can be a critical factor in the design of a member.

We'll come back to this question in Chapter 22 on columns, since compression members (such as studs) often bear on a wood plate.

The only adjustment factors applicable to bearing are C$_M$ and C$_T$ (the value is independent of C$_D$), and thus the allowable stress is

$F_c' = F_c(C_M)(C_T)$

The UBC requires a minimum 1-1/2in [*38mm*] bearing length for a joist on wood and steel, and 3in [*76mm*] on masonry. An exception is made for the balloon frame, where joists can bear on a 1 × 4 ribband if they are also nailed to the studs. When joists are framed into the side of a wood beam, the framing anchor or ledger strip should not be less than 2in [*51mm*]. Finally, when joists come from opposite sides of a beam or partition, they must be lapped at least 3in [*76mm*], or otherwise connected together with metal straps, to ensure continuity of the floor construction and to avoid bumps on the flooring. It is obvious that the bearing stress must be checked on the joist as well as on the plate or ledger supporting the joist.

Example 20.5

A floor joist made of hem-fir No. 1 has a vertical support reaction at the end of 598lb. The joist rests on a 2x4 nomimal wall plate made of Western Woods No. 3 and it is nailed to the header, so its actual bearing length is 3in, which is greater than the 1.5in required by the code. However, we want to check the bearing pressure. The allowable stress for hem-fir is 405psi and for Western Woods, it is 335psi; therefore, the plate controls.

$$f = P/A = (598lb)/(3in)(1.5in) = 133psi < 335psi \quad OK!$$

Rafter supports are somehow more complicated because the members meet the wall plates at an angle, and they are usually shaped or notched to create a good contact surface and allow toe nailing. This is called a bird's-mouth joint. In this case, the support reaction is at an angle with the grain, and the design value can be found by interpolating between the values of the compression parallel to the grain and of the compression perpendicular to the grain. This interpolation can be done very quickly with the help of the "Solution of the Hankinson Formula" diagram, explained in Chapter 25.

Example 20.6

A 2X12 Eastern softwood No.1 rafter has a slope of 5:12 and bears 2in on a steel girder. The vertical support reaction is 950lb. Check the bearing pressure.

The design values are:

F_c = 335psi,

$F_c' = F_c(C_D) = (1200psi)(1.15) = 1380psi$

(snow load, rafter in compression)

The tangent of the angle of the slope is: $\tan \theta = 5/12 = 0.417$; by using the inverse tangent function of your calculator, which tells you the angle that has a tangent of 0.417, you find $\theta = 22.6°$. The angle of the reaction with the grain is $90° - \theta = 67.4°$

From the NDS (Appendix J), we find the approximate value of F_θ = 350psi (the interpolation is very approximate for small loads values and the formula gives more accurately F_θ= 377psi). The actual bearing pressure is

F_p = (950psi)/(1.5in)(2in) = 317psi < 350psi   OK!

20.8 Notched bending members

Notches are not usually necessary in modern joining practices, with some exceptions such as the bird's mouth in rafters. Plumbers and electricians have the habit of notching and drilling through joists to run pipes and wires. A simple rule of thumb applies in these situations: **Never notch;** drill only in the middle third at half the depth (neutral axis). Chases can be cut in the web of engineered joists following the manufacturer's instructions, but the flanges can never be cut. If it is absolutely necessary to notch solid lumber, the NDS does not allow you to cut more than one-fourth of the depth at the support. The reduced section must be checked for shear with the following formula (see Fig. 20.7):

$$f_v = \frac{3V}{2bd_n}\frac{d}{d_n}$$

where the d/d_n factor increases the stress to take into account the stress concentration at the sudden change of section.

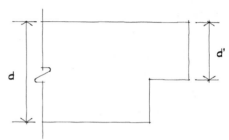

Fig. 20.7 Notched beam sections

Example 20.7

A 2X10 redwood clear structural joist is part of an exposed floor system and, to create a clean connection with a beam, the joists are "dropped" and supported on a redwood ledge. The support reaction is 460lb. How much can the joist be safely notched at the support?

We can solve the notched-end formula for d_1 (depth of material remaining at the notch) as follows:

$$f'_v = \frac{3V}{2(bd_1)}\left(\frac{d}{d_1}\right) \quad \text{or this yields:} \quad d_1^2 = \frac{3Vd}{2bf'_v}$$

$$d_n = \sqrt{\frac{3(460\text{lb})(9.5\text{in})}{2(1.5\text{in})(145\text{psi})}} = 5.49\text{in}$$

The minimum code d_1 is $3(d)/4 = 3(9.5\text{in})/4 = 7.12\text{in} > 5.49\text{in}$;
therefore, the code controls. Cut 2-3/8in and use a 2in x 2-3/8in redwood ledge.

There are lots of small but sophisticated wood structures around us, such as furniture, garden structures, and playground equipment. The spindles of a Windsor chair are an example of repetitive bending members working together as a system, connected transversely by the curved armrests and by the top "header." The connections, like bridging in joists, distribute the loads on all spindles. The spindles are thicker at the fixed end, where the bending moment is greatest. The connections, of course, are more complex than anything built into building structures.

A park bench is a good example of horizontal repetitive bending members. The "joists" are subject to "wet service conditions," and they are connected to the frame with screws or bolts. The compression edge is not always held in line, so lateral stability is a consideration.

Wood decks and trellises are designed with the same principles as floor and roof structures. A trellis is in fact, also in practice, similar to a joist system that carries live loads (plants growing on it, snow accumulated on the plants and the trellis members). The joists overhang on both ends and are notched at the supports to make them more stable. The size of these members can be calculated using the same procedure we illustrated for roof joists or rafters.

A garden deck is a very simple structure to build (many of you have built or will build one). However, if you attempt to size the decking according to the procedures for floor and roof systems, you will run into two slight complications of statics. First, you have to think of the worst loading situation. A 200lb [90kg] person could step on a single board of the decking, or someone could even carry a grand piano onto the deck! So the design criteria for the decking could be that the boards should support a 200lb [90kg] point load at midspan. Second, the boards are not simple span members, but are continuous over more joists, placed at 12 or 16in [305 or 610mm] o.c. In other words, these boards (like the sheathing on a floor) are multiple-span, statically indeterminate members. The equations to solve for shear, moment, and deflection can be easily found in a structure manual (e.g., the Manual of Steel Construction). Once the stresses are known, it is possible to calculate the required section properties for the decking. The commercial size for decking is different from that shown in the NDS tables for decking, the thickness being, for example, 1-1/4in × 6in nominal, or 1in × 5-1/2in actual [25mm×140mm actual], obtained from a 2x6 [38mm×140mm]. Two edges are also rounded, and for structural calculations, an actual thickness of 7/8in [22mm] may be taken. Although the boards are identical, the conditions for the repetitive member factor C_R do not apply: The boards are not collaborating to support point loads (unlike decking for floor sheathing that is connected on each side with T&G) and there is no sheathing. The flat-use factor C_{fu} should be applied when using the dimension lumber tables from the NDS. The other factors are C_M and C_D (in theory, 1.0 for occupancy load).

Problem 20.1
Using Table 20.2, determine the maximum span of No. 2 grade 2X12 floor joists 16in o.c., with a live load of 40psf.

Answer:

18ft-10in

Problem 20.2
Using Southern pine No. 2, find the size of joists, with 16in spacing and a span of 16ft, for a floor that carries a live load of 50psf and a dead load of 20psf including the joists.

Answer:

2X12

Problem 20.3

Design the joists for a floor span of 14ft using Western Woods No. 1. LL = 40psf, DL = 8psf, and allowable deflection = L/360 for live load.

Answer:

2X12 at 16in o.c.

Problem 20.4

The floor structure of an attic is built with 2x6 spruce-pine-fir, select structural grade, 12in o.c., spanning 10ft. The dead load of the floor assembly is 8psf. Find the maximum occupancy load per square foot the joists can carry with a maximum deflection of L/360.

Problem 20.5

An outdoor pedestrian bridge is designed with douglas fir Select Dex heavy decking, nominal thickness of 2in, spanning 4ft between two supporting beams. The code live load is 100psf (occupancy). Check if the thickness is adequate for shear, moment, and maximum deflection = L/240 for total load.

METRIC VERSIONS OF ALL EXAMPLES AND PROBLEMS

The following tables are provided for use in the metric versions of examples and problems in this chapter. Metrification is in a constant state of evolution in the United States; consequently, we would recommend that you consult the most current version of all manufacturers' and producers' catalogues of the materials you are using for the most current information.

Table M20.1 Design values for visually graded dimension lumber[a]

Species and grade	F_b [MPa]	F_+ [MPa]	F_v [MPa]	$F_{c\_}$ [MPa]	F_c [MPa]	E [MPa]
Douglas F-larch						
Select structural	10.0	6.9	0.6	4.3	11.7	13,000
No. 1	6.9	4.6	0.6	4.3	10	12,000
No. 2	6.0	4.0	0.6	4.3	9.0	11,000
Stud	4.6	3.1	0.6	4.3	5.7	9,600
Eastern softwoods						
Select structural	18.6	4.0	0.5	2.3	8.3	8,300
No. 1	5.3	2.4	0.5	2.3	6.9	7,600
No. 2	4.0	1.9	0.5	2.3	5.7	7,600
Stud	3.1	1.4	0.5	2.3	3.6	6,200
Red oak						
Select structural	7.9	4.6	0.6	5.6	6.9	9,700
No. 1	5.7	3.4	0.6	5.6	5.7	9,000
No. 2	5.5	3.3	0.6	5.6	4.9	8,300
Stud	4.9	2.6	0.6	5.6	4.4	7,600
Redwood						
Clear structural	12.1	6.9	1.0	4.5	12.7	9,700
Select structural	9.3	5.5	0.5	4.5	10.3	9,700
No. 1	6.7	4.0	0.5	4.5	8.3	9,000
No. 2	5.3	3.6	0.5	4.5	6.5	8,300

[a] 38 or 114mm thick (MPa)

Table M20.2 Southern pine safe span (m)

Deflection L/360 (based on live loads only)

Grade	Live load	Size 38 × 184mm Spacing [mm]			Size 38 × 235mm Spacing [mm]			Size 38 × 286mm Spacing [mm]		
		305	406	610	305	406	610	305	406	610
No. 1										
	1440N/m²	4.8	4.4	3.8	6.2	5.6	4.9	7.5	6.8	5.9
	1920N/m²	4.4	4.0	3.5	5.6	5.1	4.4	6.8	6.2	5.3
No. 2										
	1440N/m²	4.7	4.3	3.8	6.0	5.5	4.5	7.4	6.4	5.2
	1920N/m²	4.3	4.7	3.3	5.5	4.9	4.0	6.6	5.7	4.7

For roof live loads 1440N/m² or 1920N/m² + 480N/m² dead load

Table M20.3 Design values: Visually graded Southern pine bending members[a]

Grade	Width inches [mm]	F_b [MPa]	F_v [MPa]	E [MPa]
Select structural	8in	15.8	0.6	12,400
	10in	14.1	0.6	
	12in	13.1	0.6	
No. 1	8in	10.3	0.6	11,700
	10in	9.0	0.6	
	12in	8.6	0.6	
No. 2	8in	8.3	0.6	11,000
	10in	7.2	0.6	
	12in	6.7	0.6	
No. 3	8in	4.8	0.6	9,600
	10in	4.1	0.6	

[a]Dimension lumber 38mm thick, wet service (MPa)

Table M20.4 Design values for visually graded beams and stringers[a]

Species and grade	F_b [MPa]	F_t [MPa]	F_v [MPa]	$F_{c\perp}$ [MPa]	F_c [MPa]	E [MPa]
Douglas F-larch						
Select structural	11.0	6.5	0.6	4.3	7.6	11,000
No. 1	9.3	4.6	0.6	4.3	6.4	11,000
No. 2	6.0	2.9	0.6	4.3	4.1	9,000
Eastern spruce						
Select structural	7.2	5.0	0.4	2.7	5.2	9,600
No. 1	6.2	4.1	0.4	2.7	4.3	9,600
No. 2	4.0	1.9	0.4	2.7	2.6	7,000
Red oak						
Select structural	9.3	5.5	0.5	5.6	5.7	8,300
No. 1	7.9	3.8	0.5	5.6	4.8	8,300
No. 2	5.0	2.6	0.5	5.6	3.1	7,000
Southern pine						
Select structural	10.3	6.9	0.8	2.6	6.5	10,300
No. 1	9.3	6.2	0.8	2.6	5.7	10,300

[a] Timbers 114mm×114mm or larger (MPa)

Example M20.1

Find the size of floor joists for a live load of 1920N/m² and a span of 3.8m. Use Southern pine.

From Table M20.2 (LL = 1920N/m², deflection = L/360), we can write down some options, according to different lumber grades:

Grade	Size	Spacing (mm)	Safe span (m)
No. 1	38X184	406	4.0
No. 2	38X184	406	4.7
	38X235	601	4.0

The most economical choice corresponds to the least joist area per foot width of floor.

With 38X184 at 406mm o.c.:

$$A = [(6992mm^2)/(406mm)] \text{ per meter} = 17.22mm^2/m$$

With 38X235 at 601mm o.c.:

$$A = [(8930mm^2)/(601mm)] \text{ per meter} = 14.86mm^2/m$$

The 38X235 uses the least material, but used at 601mm o.c., it will require a thicker subfloor and ceiling drywall. Consequently, the best choice from the point of view of "system" cost may be a 38X184. As in an architectural studio, there never seems to be an "absolute" answer.

Example M20.2

Design of a Flat Roof Joist

Roof joists 406mm o.c., 5.5m span

Snow load	1,440N/m²
Dead load (ballast, roofing, sheathing, joists, ceiling)	905N/m²
Total load	2,345N/m²

(a) Adjustment factors

Adjustment factors: Snow load, $C_D = 1.15$; permanent DL, $C_D = 0.9$

Critical load: Divide each combination of loads including DL by the highest applicable adjustment factor.

DL alone = (905N/m²)/0.9 = 1,005N/m²

Snow + DL = (2,345N/m²)/1.15 = 2,039N/m² >1,005N/m² critical

Therefore, $C_D = 1.15$ for all values except E, F_c (NDS).

Repetitive member factor (more than three joists): $C_r = 1.15$ for bending only (NDS-S).

Distributed load on single joist:

Snow + DL:

$$w = (2,345N/m^2)(0.406m) = 952N/m$$

Maximum moment:

$$M = \frac{952N/m(5.5m)^2}{8} = 3597Nm$$

Maximum shear:

$$V = \frac{952N/m(5.5)}{2} = 2618N$$

Allowable deflection:

$$\Delta = (5500mm)/360 = 15mm \text{ for LL}$$
$$\Delta = (5500mm)/240 = 23mm \text{ for TL}$$

(Based on these relationships L/240 and L/360, if the DL is less than 50% of the LL, it will NOT control.) Although a number of different grades are available, No. 2 is the most common "lumber yard" variety. It is usually referred to in the yard as No. 2 and better. Your specification of a particular species may logically be based on a consideration of those species commonly carried by lumber suppliers in the area of your project.

Use Douglas fir-larch No. 2 (NDS-S).

Allowable shear stress: $F_v' = C_D F_v = 1.15(0.6 \text{MPa}) = 0.69 \text{MPa (or N/mm}^2)$

Allowable bending stress: $F_b' = C_D C_r F_b = 1.15(1.15)(6.0 \text{MPa}) = 7.93 \text{MPa}$

Modulus of elasticity: $E = 11,000 \text{MPa}$

(b) Design of the joist

Moment: Required $S = \dfrac{M}{F_b} = \dfrac{3597 \text{Nm}(1000 \text{mm/m})}{7.93 \text{N/mm}^2} = 453.6 \times 10^3 \text{mm}^3$

Shear: Required $A = \dfrac{1.5V}{F_v} = \dfrac{1.5(2618 \text{N})}{0.69 \text{MPa}} = 5691 \text{mm}^2$

Deflection: Snow load $w_S = (1,440 \text{N/m}^2)(0.406 \text{m}) = 585 \text{N/m}$

Total load $w_T = 952 \text{N/m}$ previously calculated

Required $I_{LL} = \dfrac{5(585 \text{N/m}/1000 \text{mm/m})(5500 \text{mm})}{384(11,000 \text{MPa})(15 \text{mm})} = 42.24 \times 10^6 \text{mm}^4$ for snow

The I for total load can be found by proportion:

Required $I_{TL} = 42.24 \times 10^6 \text{mm}^4 \dfrac{0.5(905 \text{N/m}^2)}{(1440 \text{N/m}^2)} = 13.27 \times 10^6 \text{mm}^4$

The I for total load controls, as expected since DL > 50% LL.

From Table M20.1, 38X286 has A= 10,868mm^2, $S_x = 518.04 \times 10^3$mm^3, and $I_x = 178$in^4
A 38X86 joist meets all criteria, BUT the size factor for 38X286 lumber is $C_F = 1.0$. It would be theoretically possible to use a smaller joist size if moment is the controlling factor, such as 38X235, since $C_F = 1.1$ and $F_b' = 1.1(6.0 \text{MPa}) = 6.6 \text{MPa}$.

Required $S_x' = M/F_b' = S_x/1.1 = (453.6 \times 10^3 \text{mm}^3)/1.1 = 412.36 \times 10^3 \text{mm}^3$
38X235 has $S_x = 349.76 \times 10^3$mm$^3 < S_x'$ and $I_x = 41.09 \times 10^6$mm$^4 <$ total required I

Therefore, use **38X286 Douglas fir-larch No. 2 at 406mm o.c.**

Lateral stability: d/b = 286/38 = 6:1 No bridging required since the compression edge is held in place by sheathing and the ends are fastened to the header and wall plate. However, bridging at midspan is **recommended.** Use V-shaped 16-gage galvanized steel bridging or 38X89 lumber cross-bridging.

Example M20.3 Metric version of the table required for this problem is not available.

Example M20.4

The system shown in the illustrastion consists of hem-fir No. 2 rafters at 610mm o.c. on purlins. The ABC rafter is made of two pieces, since the actual length of the pitch is L = (6.71m)(10/7.5) = 8.95m. The design is controlled by the BC rafter, since the overhang is helping to reduce the bending in the AB span. The BC distance is L = (3.05m)(10/7.5) = 4.07m.

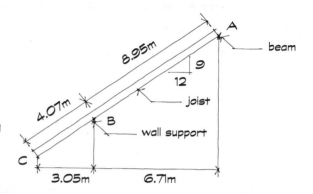

Local codes require roofs to support a snow load of 1920N/m² with no wind, and a wind pressure of 359N/m² combined with 50% of the maximum snow load. It is clear that the load combination causing the maximum shear and bending stresses is snow plus the dead load. Let's solve the statics for this load combination:

Vertical load on each rafter:

Snow: $(1920N/m^2)(0.61m)(3.05m) = 3,572.2N$
DL: $(575N/m^2)(0.61m)(4.07m)$ $= \underline{1,427.5N}$
                          Total $= 4,999.7N$, say 5,000N

Component of a distributed load perpendicular to the rafter axis:

$W = (5,000N)(10/12.5) = 4,000N$ (perpendicular to the span)

$w = (4,000N)/(4.07m) = 982.8N/m$

Maximum shear $V = (4,000N)/2 = 2,000N$

Maximum moment $M = \dfrac{982.8N/m(4.07m)^2}{8} = 2035Nm$

There is a component of the load parallel to the rafter that causes an axial load (tension on the upper half, and compression in the lower half). Since the axial load is low where the bending is high, we will neglect the combination of axial and bending stresses.

Adjustment factors: For normal conditions, $C_M = 1.0$, and $C_t = 1.0$. We have to determine C_D with the critical-load criteria, since the strength of the wood depends on the duration of the loads.

The components of the loads perpendicular to the rafters are

DL = (1,427.5N)(10/12.5) = 1,142N

SL = (3,572.2N)(10/12.5) = 2,857.8N

WL = (359N)(0.61m)(4.07m) = 891.3N

Load combinations:

1. DL/0.9 = (1,142N)/0.9 = 1,269N
2. (DL + SL)/1.15 = (1,142N + 2,857.8N)/1.15 = 3,478N
3. (DL + 0.5SL + WL)/1.6 = (1,142N + 0.5 × 2,857.8N + 891.3N)/1.6 = 2,164N

The critical load combination is DL + SL = 1,142N + 2,857.8N = 4,000N; therefore, C_D = 1.15. The applicable adjustment factors are

C_D = 1.15   (snow load)

C_R = 1.15   (repetitive member)

C_F = 1.1   (38X235 assumed)

C_L = 1.0   (compression edge held in line)

Allowable stresses:

F_v' = 1.15(1.15)(0.5MPa) = 0.66MPa

F_b' = 1.15(1.15)(1.1)(5.0MPa) = 8.53MPa

Max V = (4,000N)/2 = 2,000N

Required area for shear:

$$A = \frac{1.5(2000N)}{0.66MPa} = 4545mm^2$$

Required section modulus for moment:

$$S_x = \frac{2035Nm(1000mm/m)}{8.53MPa} = 238.6 \times 1000mm^3$$

The deflection limitation is L/180 for total loads, since the slope is > 3/12 (see Table 20.1) and we assume there is no drywall attached to the ceiling.

Allowable deflection = L/180 = (4,070mm)/180 = 22.6mm

Required I for total load = $I_x = \dfrac{5(982.8N/m(1000mm/m)(4070mm)^4}{384(9000MPa)(22.6mm)} = 17,263 \times 10^3 mm^4$

A 38X184mm has: A = 6,992mm² > 4,545mm²

S_x = 214.42x10³ mm³ > 238x10³mm³

I_x = 19,726x10³mm⁴ > 17,263x103mm⁴

The S_x of the 38X184 is not adequate, but the C_F is 1.2 for a 38X184 versus the assumed 1.1 for a 38X235; therefore, the required S_x = 238.6x10³mm³(1.1/1.2) = 219x10³mm³, which is sufficiently close to that required (within 2%).

Use 38x184 hem-fir No. 2 rafters @ 610mm o.c.

Example M20.5

A floor joist made of hem-fir No. 1 has a vertical support reaction at the end of 2,660N. The joist rests on a 38x89mm wall plate made of Western Woods No.3 and it is nailed to the header, so its actual bearing length is 76mm > 38mm, required by the code. However, we want to check the bearing pressure. The allowable stress for hem-fir is 2.8MPa, and for Western Woods it is 2.3MPa; therefore, the plate controls.

f = R/A = 2660N/(76mmX38mm) = 0.92MPa < 2.3MPa OK!

Example M20.6 A metric version of the Hankinson formula graphs is not available.

Example M20.7

A 38X235 redwood clear structural joist is part of an exposed floor system and, to create a clean connection with a beam, the joists are "dropped" and supported on a redwood ledge. The support reaction is 2046N. How much can the joist be safely notched at the support?

We can solve the notched-end formula for d_1 (depth of material remaining at the notch) as follows:

$$F'_v = \frac{3V}{2bd_1}\frac{d}{d_1} \qquad \text{thus} \qquad d_1^2 = \frac{3Vd}{2bF'_v}$$

$$d_1 = \sqrt{\frac{3(2046N)(235mm)}{2(38mm)(1.0MPa)}} = 138mm$$

The minimum code d_1 is 3(d)/4 = 3(235mm)/4 = 176mm > 138mm;
therefore, the code controls. Cut 235mm - 176mm = 59mm and use a 38X59mm redwood ledge.

Problem M20.1

Using Table 20.2, determine the maximum span of No. 2 grade 38X286 floor joists 406mm o.c., with a live load of 1,920N/m².

Answer:

5.7m.

Problem M20.2

Using Southern pine No. 2, find the size of joists, with 16in spacing and a span of 16ft, for a floor that carries a live load of 50psf and a dead load of 20psf including the joists.

Answer:

2X12

Problem M20.3

Design the joists for a floor span of 4.3m using Western Woods No. 1. LL =1,920N/m², DL = 384N/m², and allowable deflection = L/360 for live loads.

Answer:

38X286 at 406mm o.c.

Problem M20.4

The floor structure of an attic is built with 38X140 spruce-pine-fir, select structural grade, 305mm o.c., spanning 3m. The dead load of the floor assembly is 384N/m². Find the maximum occupancy load per square foot the joists can carry with a maximum deflection of L/360.

Problem M20.5

An outdoor pedestrian bridge is designed with douglas fir Select Dex heavy decking, nominal thickness of 38mm, spanning 1.2m between two supporting beams. The code live load is 4,800N/m² (occupancy). Check if the thickness is adequate for shear, moment, and maximum deflection = L/240 for total load.

21 SHEATHING AND DIAPHRAGM DESIGN

21.1 Diaphragm construction

Floor and roof structural systems are important not only to carry vertical loads, but also to transfer horizontal (lateral) loads to the bracing system, columns, and the foundations. A floor or roof that is designed to carry in-plane forces is referred to as a **diaphragm**. In order to perform this function, diaphragms have to be built as rigid elements, virtually as horizontal beams loaded on their plane by wind and earthquake forces and resisting bending and shear. The subfloor or sheathing is essential for this function, and it must be connected to headers and joists or rafters. The flange, often called the chord, of a diaphragm consists of the double top plate of a wood frame wall or the plates and the joist header between them. The web of the diaphragm is the sheathing material, typically, plywood or OSB. The web 4ft x 8ft [1.22m x 2.44m] panels are spliced over framing members, and nails and steel connectors provide the stress transfer between the web and the flange.

Testing of plywood-sheathed assemblies as diaphragms has shown that panel thickness, panel layout, and the relationship to framing member orientation and size, fastener size, and spacing, all influence the shear carrying capacity of the diaphragm. These findings have been incorporated into tables developed by the APA and adopted by building codes.

The in-plane loads parallel to joints between panel edges cause one panel to displace relative to the adjacent panel. The displacement can be almost eliminated by nailing the two edges to a common framing member along the joint. Tests have shown that these connections greatly increase the shear capacity of a diaphragm. Diaphragms are thus referred to as **blocked** and **unblocked**, depending on the presence or absence of attachment between panels. The difference between blocking for diaphragm loads and for loads perpendicular to the panel must be kept in mind.

There are restrictions that limit the depth-to-span ratio of wood diaphragms. These limits are given in Table 21.1 and are usually applied not only to diaphragms as a whole, but also to individual segments of a diaphragm, for example, the solid areas adjacent to openings. The depth is taken parallel to the load, and the span is the distance between supports providing the diaphragm reactions.

Table 21.1   Maximum diaphragm dimension ratios

Material	Maximum span/depth ratio
Diagonal sheathing, conventional	3:1
Diagonal sheathing, special	4:1
Plywood, nailed all edges	4:1
Plywood, blocking omitted at intermediate joints	4:1

Reproduced with permission of the McGraw-Hill Companies. Adapted from: Keith Faherty and Thomas Williamson. *Wood Engineering and Construction Handbook*. McGraw-Hill, 1995.

Diaphragm design is based on shear resistance of the connection between the floor and the wall structure. Although the detailed design of connections can be found in Chapter 25, here we are introducing some principles of statics and giving some practical information on the construction of this system.

If F_i is the force acting at the floor and above, calculated according to the criteria explained in the chapters referring to lateral loads, this force will be resisted by the vertical supports below that floor. We assume here, in order to simplify the argument, that these supports are walls located at the ends of the diaphragm. If the diaphragm is treated like a beam that is loaded, not vertically as we might normally expect, but on the (horizontal) plane of the floor, the support reaction provided by each wall is $V = F_i/2$ (Fig. 21.1), equal to the maximum shear between the top of each wall and the edge of the floor. At the top of the wall, we can assume the shear is uniformly distributed; thus

$$f_v = V/b \text{ (plf) } [N/m]$$

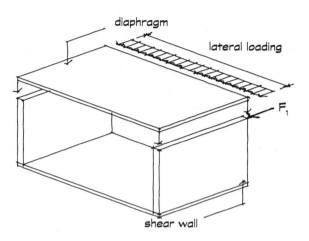

Fig. 21.1 Diaphragm action of a floor system

Table 21.2 allows specifying the nailing schedule (the spacing and type of nails), which depends on the magnitude of the shear stress f_v. The following criteria apply:

- Use minimum 8d wire nails at 4in [102mm] o.c. at all panel edges to nail the sheathing to the floor framing.

- When V < 270plf [3940N/m], nails can be spaced at 6in [152mm].

- Use 8d nails at 12in [305mm] o.c. to nail the sheathing on intermediate supports (joists and other members).

The design of stressed-skin panels as bending members is discussed in the following section, where plywood properties are presented. Stressed-skin panel technology was developed in aircraft design during the 1930s and was employed for civil uses after World War II.

Figure 21.2 shows an example of diaphragm action in an early design of an airplane wing. While materials have advanced significantly since we built wooden wings, the principle remains the same in contemporary airplane design. Here, where weight is at a premium, the use of a stressed skin structural system provides a logical alternative.

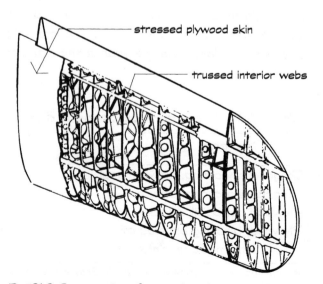

Fig. 21.2 Construction of a wooden wing

Table 21.2 Maximum shear for horizontal diaphragms for wind and earthquake loads

| Nail size | Minimum sheathing thickness (inches) | Maximum shear stress v (plf) | | | |
| | | Unblocked diaphragm Nail spacing = 6in [a] | | Blocked diaphragm Nail spacing [b] | |
		Case 1	Case 2	6in	4in
6d	5/16in	150plf	110plf	170plf	225plf
	3/8in	165plf	125plf	185plf	250plf
8d	3/8in	215plf	160plf	240plf	320plf
	7/16in	230plf	170plf	255plf	340plf
	15/32in	240plf	180plf	270plf	360plf

Source: Adapted from the APA.

[a] Nailing at supported edges.

[b] Nail spacing at all continuous panel edges and diaphragm boundaries. At staggered joints, the space is 6in [152mm], and at intermediate supports the space is 12in [305mm].

Note: Assumed 2in [38mm actual] thickness (nominal) of framing and maximum sheathing thickness of 3/4in [19mm]; nails larger than 8d require thicker framing lumber.

Example 21.1

A two-story house is subject to an earthquake load below the second-floor level, W = 45k. The diaphragm works like a beam lying on its side, with the flanges represented by the wall plates and headers.

The bending moment is given by a couple acting on the flanges, where C is the maximum compression force and T the tension force acting alternatively on each flange at each oscillation:

$$M = WL^2/8 = (45,000lb)(40ft)^2/8 = 225,000ftlb$$

$$T = M/b = (225,000ftlb)/(24in) = 9,375lb$$

Assuming the use of hem-fir No. 2, the allowable tensile stress is
$$F_t' = (500psi)C_D = (500psi)(1.6) = 800psi$$

The header by itself has A = 16.88in²
Tension capacity $T_{all.} = F_t'A = (800psi)(16.88in^2) = 13,504lb$ OK!
The header has to be continuous for 40ft, and splices will be necessary with connections able to carry that tension load at midspan.

21.2 Shear walls

Diaphragm action provides in-plane stiffness to a floor or roof that enables the floor or roof to carry forces imposed on it to the vertical bracing elements, which in turn carry the forces down to the foundations. One type of vertical bracing element, and the type found in most wood frame buildings, although not always as a designed element, is a wall with in-plane rigidity referred to as a shear wall (Fig. 21.1).

You've seen lateral load action and bracing design in Part One, "Steel." Wall sheathing with plywood or OSB has virtually replaced diagonal bracing in wood framing. Whereas floor diaphragms are essentially horizontal beams loaded on their edges and supported on walls usually at the ends, shear walls are loaded at the top by the floor diaphragms with lateral loads. These loads are the shear reactions at the connections with the floor diaphragm. The walls work as cantilever beams fixed on the foundations, and are designed accordingly for shear and bending moment. The design is simplified by the assumptions that

(a) The shear is uniformly distributed along the top of the wall.

(b) The bending moment is resisted by a couple applied to members at the boundary of the wall.

One of the forces of the couple will always be in tension, and, consequently, anchorage must be provided at both ends. This tension anchorage is referred to as the tie down, and consists of anchor bolts through the sill plates or other metal connectors for posts and bracing (see Chapter 25). The connections between the top of the shear wall and the floor diaphragm are also important.

Table 21.3 (adapted from the APA) gives the recommended maximum shear for APA panel shear walls.

Table 21.3 Maximum shear stress on shear walls (plf)

Nail size	Panel thickness (in)	Nail spacing at the edges		
		6in	4in	3in
6d	5/16	200	300	390
8d	3/8	230	360	460
	7/16	255	395	505
	15/32	280	430	550

Example 21.2

The total wind load to be resisted by the first-floor shear walls of a house is 11,370lb. Assuming the two end walls, acting as shear walls, are identical, each has to resist a shear load

$$V = (11,370lb)/2 = 5,685lb$$

The shear stress in the panel wall is:

$$f_v = V/b = (5{,}685lb)/(20ft) = 284plf$$

From Table 21.3, we find 6d nails with 5/16in plywood at 4in o.c. have an allowable load of 300plf > 284plf　OK!

Construction systems consisting of stressed skin panels, including foam core panels, are inherently rigid to lateral loads, since the panel skin is securely fastened to the web or core material, and forms a continuous floor, roof, or wall diaphragm.

21.3 Composite bending members

As we discussed earlier in the chapter, lumber and timber members are gradually being replaced by more efficient engineered components such as I-beams and joists, variable-section laminated wood beams, and box beams. These components can be called composites, since they are built by combining different wood-derived materials (such as plywood and laminated wood) or by combining wood with different design values, for example, at the web and flange portion of a laminated timber beam (Fig. 21.3).

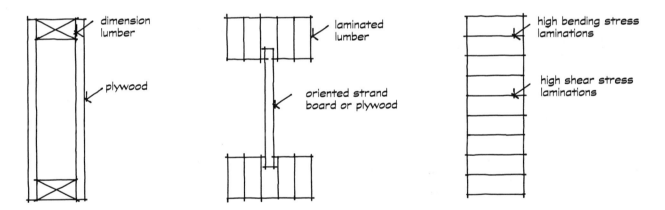

Fig. 21.3 Alternative composite bending member designs

The concept of a composite wood beam is parallel to that of the design of a steel I-beam: The flanges resist bending with the couple C (the resultant of the compression stresses) and T (the resultant of the tension stresses), and the web is calculated to resist all shear. One of the critical aspects of the design for wood members is the connection between the web and the flanges, which has to resist the horizontal shear force P_v.

Another type of composite member includes stressed skin panels. These generally industrially produced panels are conceptually box beams with plywood or OSB flanges (skin) and lumber webs. In foam core panels, the skin is glued directly to a polystyrene or urethane "core" without lumber webs, and in this case, the shear stress must be taken by the rigid foam (Fig. 21.4).

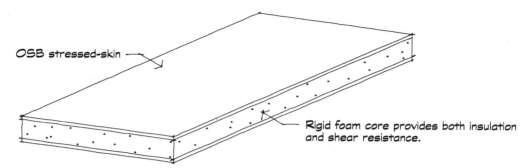

OSB stressed-skin

Rigid foam core provides both insulation and shear resistance.

Fig. 21.4 Premanufactured insulated stressed-skin panel

Box beams may be built in a fabricating plant, but are frequently site-built (Fig. 21.5). Dimension lumber or laminated wood is used for the flanges and plywood or oriented strand board (OSB) for the web (side panels). The structural concept and the design procedure are essentially the same for other types of engineered beams or joists shown in Figs. 20.4 and 20.5.

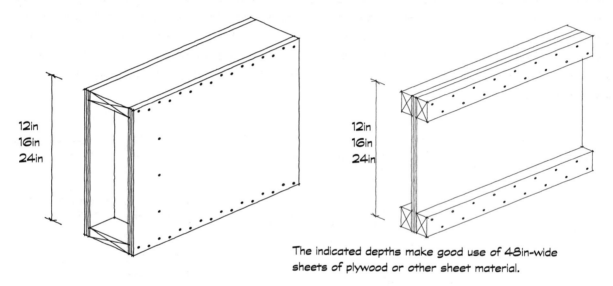

12in
16in
24in

12in
16in
24in

The indicated depths make good use of 48in-wide sheets of plywood or other sheet material.

Fig. 21.5 Site-built composite plywood and wood beams

21.4 Plywood structural properties

Plywood can be manufactured from over 70 species of wood, but it is classified into five groups (see Table 21.4). The allowable stresses depend on

(a) species group, 1 to 5

(b) grade stress level (Table 21.6)

(c) moisture condition

(d) adjustment factors; the values listed in Table 21.5 are for "normal" load duration ($C_D = 1.0$). These values must be adjusted for all applicable factors.

The moisture condition can be dry (m.c. $\leq$ 16%) for indoor use or wet (m.c. > 16%) for exterior use. Different stress values are associated that correspond to wet or dry use (Table 21.5).

Table 21.4 Classification of species

Group 1	Group 2	Group 3	Group 4	Group 5
Apitong	Cedar, Port Orford	Alder, red	Aspen, bigtooth	Basswood
Beech, American	Cypress	Birch, paper	Cativo	Poplar, balsam
Birch, sweet	Douglas fir #2	Cedar, Alaska	Cedar, incense	
Birch, yellow	Fir, balsam	Fir, subalpine	Cedar, Western red	
Douglas fir #1	Fir, California red	Hemlock, Eastern	Cottonwood, Eastern	
Kapur	Fir, grand	Maple, bigleaf	Cottonwood, black	
Keruing	Fir, noble	Pine, jack	Pine, Eastern white	
Larch, Western	Fir, Pacific silver	Pine, lodgepole	Pine, sugar	
Maple, sugar	Fir, white	Pine, ponderosa		
Pine, Caribbean	Hemlock, Western	Pine, spruce		
Pine, ocote	Lauan, Almon	Redwood		
Pine, Southern	Lauan, Bagtikan	Spruce, Engelmann		
Pine, loblolly	Lauan, Mayapis	Spruce, white		
Pine, longleaf	Lauan, red			
Pine, shortleaf	Lauan, tangile			
Pine, slash	Lauan, white			
Tanoak	Maple, black			
	Mengkulang			
	Maranti, red			
	Mersawa			
	Pine, pond			
	Pine, red			
	Pine, Virginia			
	Pine, Western white			
	Poplar, yellow			
	Spruce, black			
	Spruce, red			
	Spruce, sitka			
	Sweetgum			
	Tamarack			

Table 21.5 Allowable stresses for plywood (Dry use only)

Stress type	Species group of face ply	Stress grade level		
		S-1	S-2	S-3
Bending: F_b	1	2000psi	1650psi	1650psi
Tension F_t in plane of plies				
Face grain parallel or	2, 3	1400psi	1200psi	1200psi
perpendicular to span	4	1330psi	1110psi	1110psi
Load applied at 45° to face grain: Use $(1/6)F_t$				
Compression: F_c	1	1640psi	1540psi	1540psi
In plane of plies,	2	1200psi	1100psi	1100psi
parallel or perpendicular	3	1060psi	990psi	990psi
to face grain	4	1000psi	950psi	950psi
Load applied at 45° to face grain: Use $(1/3)F_c$				
Shear: F_v	1	190psi	190psi	160psi
(through thickness)	2, 3	140psi	140psi	120psi
parallel or perpendicular	4	130psi	130psi	115psi
to face grain				
Load applied at 45° to face grain: Use $2F_v$				
Rolling Shear: F_s	1	75psi	75psi	do not use
parallel or perpendicular				
to face grain	2, 3, 4	53psi	53psi	48psi
Load applied at 45° to face grain: Use $(4/3)F_s$				
Bearing: F_c (perpendicular)	1	340psi	340psi	340psi
perpendicular to plane	2, 3	210psi	210psi	210psi
of plies	4	160psi	160psi	160psi
Modulus of Elasticity: E	1	1,800,000psi for all grades		
Face grain parallel or	2	1,500,000psi for all grades		
perpendicular to span	3	1,200,000psi for all grades		
	4	1,000,000psi for all grades		

For wet usage, consult American Plywood Association recommendations.

You'll notice a new stress value listed in the left-hand column of Table 21.5, "Rolling Shear." This is a failure mode that is unique to plywood, but has a simple explanation. If you were to glue a piece of plywood to a joist, at some point (layer) in the plywood, a ply would be perpendicular to the attachment surface. If the plywood is stressed at the surface and that force is transferred internally through the plies, at some point it would be similar to placing a layer of pencils between your hands and trying to slide your hands past one another. The pencils would roll and actually make the sliding action very easy. (This is one of the ways it is speculated that the great stones at Stonehenge were moved the 300 miles from their point of origin.) In the case of plywood, the glue bond between the plies resists this "rolling" action, but there is a limit to its ability to transfer load; therefore, the "rolling shear" stress value must not be exceeded. You'll also notice from Table 21.5 that the value is 33% higher if the load is applied at 45° to the face. This makes sense, since the "pencils" would be trying to roll at an angle to their axis. If you're bored, try it!

Table 21.6 Grade stress levels per UBC and APA

| | Stress Grade Level | |
S-1	S-2	S-3
APA Rated sheathing, exterior	Sturd-I-Floor, exterior	All exposure 2 plywood panels
Structural I, exterior	C-C plugged, exterior	(All interior plywood panels)
Structural II, exterior	Underlayment, exterior	
	All exposure 1 plywood panels	
Exterior A-A	Exterior A-B	All other grades, including C-D
EXterior A-C	Exterior B-B	
Exterior C-C	Exterior B-C	
Structural I A-C (group 1)	Exterior C-C plugged	
Structural I C-C (group 1)	Structural I C-D (group 1)	
	Structural II C-D (group 2)	
	C-D sheathing exterior glue	
	All interior grades with exterior glue	

21.5 Box beam design

In a box beam, the flanges are designed to carry the bending moment, and the web to carry the shear load. The design procedure is as follows:

1. Prepare V and M diagrams.
2. Design the web (preliminary) to carry the shear load.
3. Find the moment capacity of the plywood web.
4. Design the flanges to provide the additional required moment capacity.
5. Verify the actual web shear stresses.
6. Find the moment of inertia of the flanges required for maximum deflection.
7. Specify web stiffeners.
8. Check the lateral stability criteria.
9. Calculate the connection between the flanges and web.

Creep is not taken into account in the same way as in the design of solid lumber beams. Box beams have deflections higher than calculated with deflection formulas, due to the combined effect of creep and nailed connections. The calculated value of the deflection, or the required value of I for maximum deflection, can be multiplied by the correction factors listed in Table 21.7. The design procedure is illustrated in Example 21.3.

Table 21.7 Correction factors for box beam deflection

Span/depth ratio	Correction factor
10	1.5
15	1.2
20	1.0

Example 21.3

Design a site-built beam to support the floor joist system shown in the illustration. The beam will span 20ft and will be designed using white fir plywood, Group 2, APA-rated exposure 1, stress grade level S-2, dry condition.

F_b = 1,200psi

F_v = 140psi

E = 1,500,000psi

Live load = 40psf

Dead load = 10psf

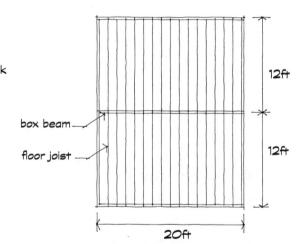

Adjustment factors:

Load duration C_D = 1.0 (occupancy live loads)

The tributary area of the beam is 12ft wide; therefore, the distributed load is

w = (50psf)(12ft) = 600plf

R = (600plf)(20ft)/2 = 6,000lb

V can be reduced by taking off the load at each end for a length equal to the depth of the beam. In this case, where the depth is a significant proportion, it is probably worth considering. Assuming the beam will be 24in deep:

V = (600plf)(20ft - 4ft)/2 = 4800lb

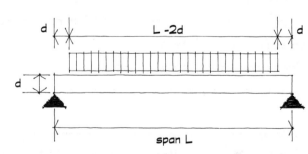

Step 1: Design web for shear

Plywood F_v = 140psi

Required area of plywood:

A = (1.5)(4,800lb)/(140psi) = 51.4in²

Try a double layer of 5/8in plywood on each side, which would equal 4x0.625in = 2.5in.

$$d = \frac{51.4\text{in}^2}{4(0.625\text{in}^2)} = 20.57\text{in}^2$$

Make d = 24in to efficiently use a 4ft plywood sheet.

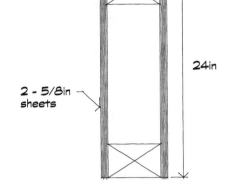

2 - 5/8in sheets

24in

Step 2: Determine the moment capacity of plywood web

$$\text{Web moment} = \frac{F_b I_{web}}{c} = \frac{1200\text{psi}\frac{2.5\text{in}(24\text{in})^3}{12}}{12\text{in}} = 288{,}000\text{inlb}$$

Total required M = (30,000ftlb)(12in/ft) = 360,000inlb

Moment resistance to be provided by flanges:

Required moment capacity	360,000inlb
Web moment capacity	- 288,000inlb
Flange required capacity	72,000 inlb

Step 3: Flange design

As a first try, hem-fir North No. 2 is selected. C_D = 1.0

F_b' = 1.0(1,000psi) = 1,000psi.

Required I flanges =

$$\frac{M(\text{flanges})c}{F_b} = \frac{72{,}000\text{inlb}(12\text{in})}{1000\text{psi}} = 864\text{in}^4$$

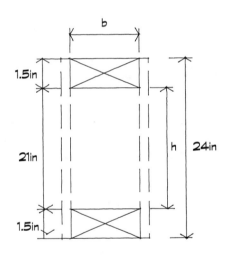

b

1.5in

21in

1.5in

h   24in

I flanges =

$$\frac{b(H^3 - h^3)}{12} = \frac{b(24^3 - 21^3)}{12} = 864\text{in}^4$$

which, solved for b, gives the minimum width of the lumber flange:

$$b = \frac{864\text{in}^4(12)}{24^3 - 21^3} = 2.27\text{in}$$

Since this is smaller than a 2X4, a lower-strength lumber will work. Using hem-fir Stud grade, F_b' = 675psi, and E = 1,200,000psi:

$$\text{Required I flanges} = \frac{M_{r(\text{flanges})}c}{F_b} = \frac{72{,}000\text{inlb}(12\text{in})}{675\text{psi}} = 1280\text{in}^4$$

$$b = \frac{1280\text{in}^4(12\text{in/ft})}{24^3 - 21^3} = 3.36\text{in}$$

Use nominal 2X4 hem-fir Stud grade for the flanges.

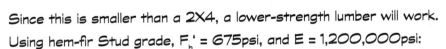

Step 4: Deflection

$$\Delta = \frac{5wL^4}{384E_{ply}I_{ply}}$$

$$\Delta_{max} = \frac{20ft(12in/ft)}{360} = 0.67in \text{ for live load}$$

Live load = (40psf)(12ft) = 480plf

Dead load = (10psf)(12ft) = 120plf

At this point, we must compensate for the effects of "creep," as noted in Table 21.7. In our case, the span/depth ratio is (20ft)(12in/ft)/24in = 10; therefore, the correction factor is 1.5.

Since the box beam is a composite of two materials with two different values of E, we will start considering the required I for a homogeneous beam made of plywood:

$$I_{required} = 1.5\frac{5wL^4}{E_{ply}\Delta} = 1.5\frac{5\left(\frac{480lb/ft}{12in/ft}\right)(20ft \times 12in/ft)^4}{384(1,500,000psi)0.67} = 2579in^4$$

Using the web material calculated in Step 1: A total of four layers of 5/8in plywood, 24in deep,

$$I_{web} = \frac{(2.5in)(24in)^3}{12in} = 2880in^4$$

Therefore, the web alone is sufficient for deflection; moment controls.

In order to understand the procedure for the deflection design of a composite member, we will calculate the deflection of this box beam, taking into account the contribution of web and flanges of different materials with different E values.

$$E_{ply} = 1,500,000psi$$
$$E_{wood} = 1,200,000psi$$

Since we have two different values of E, we will find the I equivalent to a beam section made entirely of plywood. $E_{ply} > E_{wood}$; therefore, the wood flanges will be equivalent to plywood flanges with a smaller I value:

Moment of inertia of the equivalent plywood flange:

$$I_{ply} = \frac{I_{wood(E_{wood})}}{E_{ply}}$$

therefore:

$$I_{wood} = \frac{3.5in\left[(24in)^3 - (21in)^3\right]}{12} = 1332in^4$$

Plywood webs I = 2880in⁴

$$\text{plywood flanges } I = \frac{1,200,000\text{psi}(1332\text{in}^4)}{1,500,000\text{psi}} = 1066\text{in}^4$$

Equivalent plywood beam $\quad I = 2880\text{in}^4 + 1066\text{in}^4 = 3946\text{ in}^4$

$$\Delta = 1.5 \frac{5\left(\frac{480\text{lb/ft}}{12\text{in/ft}}\right)(20\text{ft} \times 12\text{in/ft})^4}{384(1,500,000\text{psi})3946\text{in}^4} = 0.44\text{in} \leq 0.67\text{in} \quad \text{determined in deflection calculation}$$

If Δ is the deflection of the composite beam with $I = I_c$, a beam made entirely of plywood will have the same deflection value with a value of $I = I_{ply}$ different from I_c. The same can be said for a beam made entirely of wood:

$$\Delta = \frac{5wL^4}{384E_{wood}I_{wood}}$$

Since the two deflections are the same, it must be

$$\Delta = \frac{5wL^4}{384E_{ply}I_{ly}} = \frac{5wL^4}{384E_{wood}I_{wood}}$$

which can be also written as

$$\frac{1}{E_{wood}I_{wood}} = \frac{1}{E_{ply}I_{ply}}$$

Therefore:

$$I_{ply} = \frac{E_{wood}}{E_{ply}}I_{wood}$$

which is the I of a plywood flange equivalent to the wood flange. Therefore, to transform a wood part into a plywood part, equivalent for deflection purposes, we have to multiply its moment of inertia by the ratio between the E values.

Step 5: Web stiffeners

The web is subject to buckling due to compression stress in bending, or yielding due to concentrated loads on the flanges such as at the supports. Stiffeners should be provided in the form of vertical 2X2 [38X38] lumber ribs; vertical stiffeners as wide as the flange pieces, nailed between the flanges, are also a convenient solution. These stiffeners provide the same function as those in a built-up steel plate girder of a railroad bridge or in some of Mies's exposed spandrel beams. The web is relatively thin and when subjected to vertical load, or even the forces of horizonal shear, it has a tendency to buckle. The stiffeners provide lateral support and reduce this likelihood.

The minimum spacing of stiffeners can be derived from Table 21.8. Enter in the table the clear distance between flanges on the horizontal scale: $d_c = 21$ in [533mm].

From this value, draw a vertical line until you meet the curve corresponding to the plywood thickness. In our case, the total thickness is 2.5in [63/5mm], but the plywood sheets will buckle individually; therefore, the smallest plywood sheet thickness is the critical value, or 5/8in [16mm].

Table 21.8   Spacing of stiffeners for plywood beams

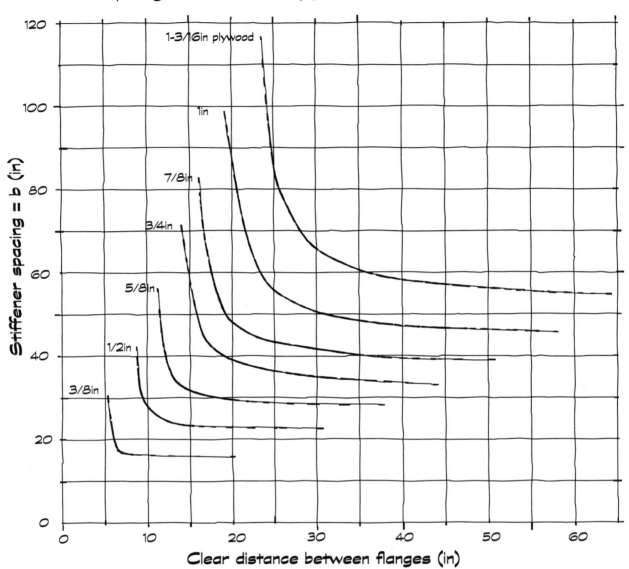

On the scale of the clear distance between stiffeners b, we read 29in [737mm]. If the stiffeners are 2X2 [38X38], the spacing can be 30in or 2.5ft [762mm].

Stiffener spacing can be increased up to 3b when the shear decreases to one half the maximum value. In our beam, $V = V_{max}/2$ at 6ft [1.83m] from the ends. Therefore, four stiffeners can be placed at each end 30in [762mm] o.c. and no other stiffeners are needed at midspan since 3(30in) = 90in = 7.5ft [3(72mm) = 2286mm].

Step 6: Lateral stability

Refer to Table 21.9. In our case, the compression flange is fully restrained by the joists supported on the beam, particularly if joist hangers nailed on the flanges are used. In all cases where the compression flange is not fully supported laterally, the I_x/I_y ratio should be checked and the given criteria should be followed.

Table 21.9 Lateral stability criteria for built-up beams

For adequate lateral stability in built-up plywood beams, the following criteria apply:

1. If the ratio of I_x to I_y does not exceed 5:1, no lateral support is required.

2. If the ratio of I_x to I_y is between 5:1 and 10:1, the ends of the beam should be held in position at the bottom flange at the support.

3. If the ratio of I_x to I_y is between 10:1, and 20:1, the ends of the beam should be held in position at both the top and bottom flange at the support.

4. If the ratio of I_x to I_y is between 20:1 and 30:1, one edge should be held in line.

5. If the ratio of I_x to I_y is between 30:1 and 40:1, the beam should have bridging or other bracing at intervals of not more than 8 feet.

6. If the ratio of I_x to I_y exceeds 40:1, the compression flange should be fully restrained and forced to deflect in a plane.

Step 7: Web-flanges connection

The design of glued and nailed connections is explained in Chapter 26.

Problem 21.1

Determine the moment capacity of this engineered beam that is simply supported, in dry conditions, and occupancy loads. Assume the web
and flanges are adequately connected.

Answer:
30.3ftk

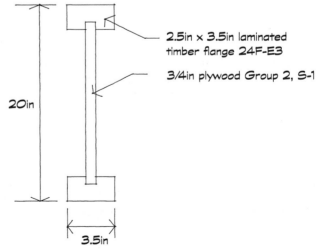

2.5in x 3.5in laminated
timber flange 24F-E3

3/4in plywood Group 2, S-1

20in

3.5in

Problem 21.2

The stressed-skin panel in the figure is a roof system component. Determine the maximum span based on V, M, and maximum deflection of L/360. The snow load = 30psf and the dead load = 14psf, including panel weight. Neglect the structural contribution of the polystyrene.

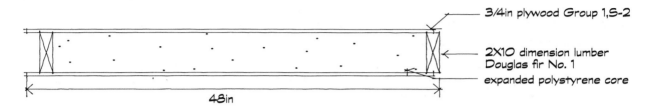

3/4in plywood Group 1,S-2

2X10 dimension lumber Douglas fir No. 1

expanded polystyrene core

48in

Answer:

23ft

Problem 21.3

Check if this beam has adequate shear capacity. If not, design a solution. $C_D = 1.0$.

Solution:

The required web area is 34.6in², whereas the existing is 27in², so the shear capacity is not adequate. By using plywood of the same grade, an additional 0.42in web, or 1/4in on each side, is to be added.

Problem 21.4

This composite joist is going to be used to form an overhang with V_{max} = 3,000lb at the support B. Check the shear capacity, and if not adequate, design the reinforcement of the joist with plywood of the same species.

Answer:

An additional web area of 26.7in² is required. Add 3/4in plywood on each side.

Problem 21.5

Compare the deflection value of a 2X8 floor joist alone and the deflection of the same 2X8 floor joist acting in composite action with the subfloor. The 2X8 is MSR (machine stress rated) dimension lumber 1200f-1.5E. The occupancy load is 40psf.

Answer:

Deflection of joist = 0.26in

The composite joist has I = 310.4in⁴

Deflection of composite = (0.26in)(47.63in⁴/310.4in⁴) = 0.04in

METRIC VERSIONS OF ALL EXAMPLES AND PROBLEMS

Table M21.2 Maximum shear for horizontal diaphragms for wind and earthquake loads

Nail size	Minimum sheathing thickness (mm)	Allowable shear stress v (N/m)			
		Unblocked diaphragm Nail spacing = 150mm		Blocked diaphragm Nail spacing	
		Case 1	Case 2	150mm	100mm
6d	8mm	2188N/m	1605N/m	2480N/m	3283N/m
	9.5mm	2407N/m	1824N/m	2699N/m	3647N/m
8d	9.5mm	3137N/m	2334N/m	3502N/m	4669N/m
	11mm	3356N/m	2480N/m	3720N/m	4961N/m
	12mm	3502N/m	2626N/m	3939N/m	5252N/m

Source: Adapted from the APA

Table M21.3 Maximum shear stress on shear walls (N/m²)

Nail size	Panel thickness (mm)	Nail spacing at the edges (mm)		
		150	100	75
6d	8mm	2918mm	4377mm	5690mm
8d	9.5mm	3356mm	5252mm	6711mm
	11mm	3720mm	5763mm	7368mm
	12mm	4085mm	6274mm	8024mm

Example M21.1

A two-story house is subject to an earthquake load below the second-floor level, W = 200kN. The diaphragm works like a beam lying on its side, with the flanges represented by the wall plates and headers.

The bending moment is given by a couple acting on the flanges, where C is the maximum compression force and T the tension force acting alternatively on each flange at each oscillation:

$$M = WL^2/8 = (200kN)(12.2m)^2/8 = 305kNm$$

$$T = M/b = (305kNm)/(7.3m) = 41.8kN$$

Assuming the use of hem-fir No. 2, the allowable tensile stress is

$$F_t' = (3.4MPa)C_D = (3.4MPa)(1.6) = 5.5MPa$$

The header by itself has A = 10,862mm²

Tension capacity $T_{all.}$ = $F_t'A$ = (10,862mm²)(5.5MPa) = 59,739N = 59.7kN > 41.8kN   OK!

The header has to be continuous for 12.2m, and splices will be necessary with connections able to carry that tension load at midspan.

Example M21.2

The total wind load to be resisted by the first-floor shear walls of a house is 50,574N. Assuming the two end walls, acting as shear walls, are identical, each has to resist a shear load V = (50,574N)/2 = 25,287N.

The shear stress in the panel wall is:

f_v = V/b = (25,287N)/(6.1m) = 4,145N/m

From Table M21.3, we find 6d nails with 8mm plywood at 100mm o.c. have an allowable load of 4,377N/m > 4,145N/m   OK!

Example M21.3

Design a site-built beam to support the floor joist system shown in the illustration. The beam will span 6.1m and will be designed using white fir plywood, Group 2, APA-rated exposure 1, stress grade level S-2, dry condition.

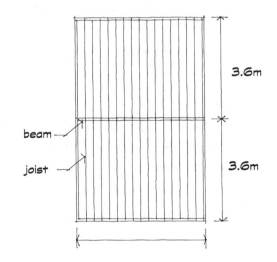

F_b = 8.3Mpa
F_v = 1.0MPa
E = 10,300MPa
Live Load = 1920N/m²
Dead Load = 480N/m²
Adjustment factors:
Load duration C_D = 1.0 (occupancy live loads)

The tributary area of the beam is 3.6m wide, therefore the distributed load is :

w = 2400N/m²(3.6m) = 8640N/m
R = 8640N/m²(6.1m)/2 = 26,352N

V can be reduced by taking off the load at each end for a length equal to the depth of the beam. In this case, where the depth is a significant proportion, it is probably worth considering.

Assuming the beam will be 600mm deep:

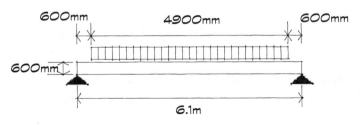

$$V = \frac{8640N/m(6.1m - 1.2m)}{2} = 21,168N$$

$$M = \frac{8640N/m(6.1m)^2}{8} = 40,187Nm$$

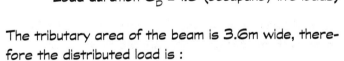

Design web for shear:

Plywood $F_v = 1.0$MPa
Required area of plywood:

$$A = \frac{1.5(21,168N)}{1.0MPa} = 31,752mm^2$$

Use double layer of 16mm plywood.

$$d = \frac{31,752mm^2}{4(16mm)} = 496mm$$

Make d = 600mm to efficiently use a 1.2m plywood sheet.

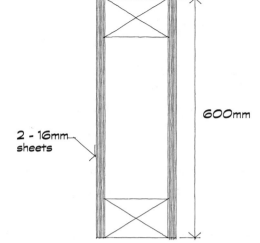

Determine moment capacity of plywood web:

$$\text{Web moment} = \frac{F_b I_{web}}{c} = \frac{8.3MPa(1,152\times10^6mm^4)}{300mm} = 31,872\times10^3Nmm$$

$$\text{where } I_{web} = \frac{64mm(600mm)^3}{12} = 1,152\times10^6mm^4$$

Total required M = $40,187\times10^3$Nmm
Moment resistance to be provided by flanges
= $40,187\times10^3$Nmm - $31,872\times10^3$Nmm = $8,315\times10^3$Nmm

Flange design:

As a first try, hem-fir North No. 2 is selected. $C_D = 1.0$
$F_b' = 1.0(6.9MPa) = 6.9$MPa

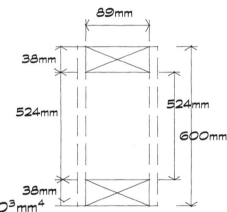

$$I_{flanges} = \frac{M_{flanges}c}{F_b} = \frac{8,315\times10^3Nmm(300mm)}{6.9MPa} = 361,522\times10^3mm^4$$

$$I_{flanges} = \frac{b[(600mm)^3 - (524mm)^3]}{12} = 361,522\times10^3mm^4$$

which, solved for b, gives the minimum width of the lumber flange:

$$b = \frac{(361,522\times10^3mm^4)12}{72,122\times10^3mm^3} = 60mm$$

Since this is smaller than a 38X89, a lower-strength lumber will work.
Using hem-fir Stud grade, $F_b' = 4.6$MPa, E = 8,300MPa

$$I_{flange} = \frac{M_{r(flanges)}}{F_b} = \frac{8,315\times10^3Nmm(300mm)}{4.6MPa} = 542,283\times10^3mm^4$$

$$b = \frac{(542,283\times10^3mm^4)12}{72,122\times10^3mm^3} = 90mm$$

Use a 38X89 hem-fir Stud grade for flanges.

Deflection:

$$\Delta = \frac{5wL^4}{384E_{ply}I_{ply}}$$

$$\Delta_{all} = \frac{6,100mm}{360} = 17mm \text{ for live loads.}$$

Live load = $1920N/m^2(3.6m) = 6912N/m$

Dead load = $480N/m^2(3.6m) = 1728N/m$

At this point, we must compensate for the effects of "creep," as noted in Table 21.7. In our case, the span/depth ratio is (6100mm)/(600mm) = 10; therefore, the correction factor is 1.5.

Since the box beam is a composite of two materials with two different values of E, we will start considering the required I for a homogeneous beam made of plywood:

$$I_{required} = 1.5\frac{5wL^4}{E_{ply}\Delta} = 1.5\frac{5(6912N/m)/1000mm/m(6100mm)^4}{384(10,300MPa)(17mm)} = 1067.5X10^6mm^4$$

As previously calculated

$$I_{web} = \frac{64mm(600mm)^3}{12} = 1,152X10^6mm^4$$

Therefore, the web alone is sufficient for deflection; moment controls.

In order to understand the procedure for deflection design of a composite member, we will calculate the deflection of this box beam, taking into account the contribution of web and flanges of different materials with different E values.

$$E_{ply} = 10,300MPa$$
$$E_{wood} = 8,300MPa$$
$$I_{ply} = \frac{I_{wood}E_{wood}}{E_{ply}}$$

$$I_{wood} = \frac{89mm[(600mm)^3 - (524mm)^3]}{12} = 535X10^6mm^4$$

plywood webs I = $1,152X106mm^4$

plywood flanges I = $\frac{8,300MPa(535X10^6mm^4)}{10,300MPa} = 431X10^6mm^4$

Equivalent plywood beam I = $1,152X10^6mm^4 + 431X10^6mm^4 = 1583X106mm^4$

$$\Delta = 1.5\frac{5(6912N/m)/1000mm/m(6100mm)^4}{384(10,300MPa)1583X106mm^4} = 11.5mm < 17mm$$

22 TIMBER AND LAMINATED TIMBER BEAMS

22.1 Timber beams

Beams are primary structural elements of floors and roofs, supporting secondary elements like joists, trusses, and panels. Historically, beams were made with unsawn logs for primarily practical reasons (Fig. 22.1). This practice was originally merely a labor-saving technique, but also used the material efficiently, since the continuous surface fiber gives the log a strength higher than the equivalent square timber beam cut from a larger tree (however, this is not recognized in the current codes).

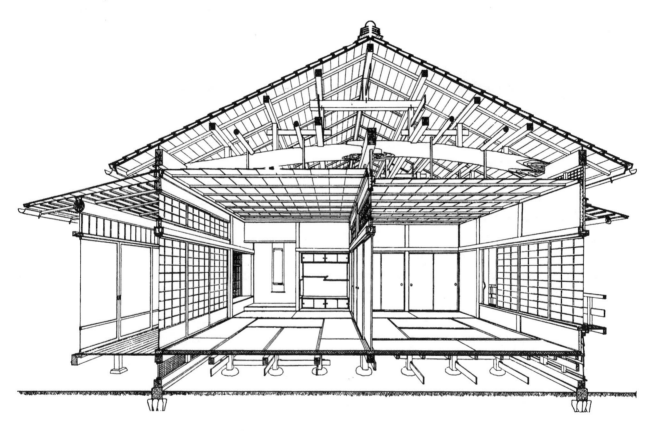

Fig. 22.1 Japanese farmhouse, Nineteenth century

When lumber was abundant, even industrial and commercial buildings had timber floor structures., A common system was known as the "Mill Construction" or "Slow-burning Mill Construction," developed for use in factory or mill buildings in New England. All columns and beams were made of large sections; joists were eliminated where possible and replaced by a heavy, thick floor spanning some feet. The name "slow-burning" refers to the fact that during fire, the timber tends to char rather than burn, and when properly sprinkled, allowed comparatively low insurance rates. In this system, the columns take heavy loads, and they can not bear directly on the beams, due to the low strength perpendicular to the fiber. Today, large solid lumber members are more difficult to find, and timber beams are made by laminating wood in various ways.

Timber beams are designed like all other bending members. The distinction between timber and dimension lumber is essentially one of size. Timber beams have a 5inx5in [114X114mm] section or larger, and the design values are presented in specific tables of the NDS Supplement (Table 4D) under the size classifications of beams and stringers.

These design values have to be adjusted according to Table 22.1.

Table 22.1 Adjustment factors applicable to timber beams for moment, shear, and deflection

Symbol	Definition and applicability	Adjusts:
C_D	Load duration	F_b, F_v
C_M	Wet service: Moisture content > 19%, except Southern pine.	F_b, F_v, E
C_T	Temperature	F_b, F_v, E
C_L	Lateral stability: see text discussion.	F_b
C_F	Size: Applies only to visually graded sawn lumber and round timber members. See text discussion.	F_b
C_f	Form: Applies to beams with a circular section or with a square section loaded in the plane of the diagonal (diamond section) (NDS, Sec. 2.3.8). For a round section, $C_f = 1.18$.	F_b

Size factor C_F applies when depth d exceeds 12in [286mm]. Size factor C_F (see also the NDS, Sec. 4.3) is calculated by

$$C_F = (12/d)^{1/9} \leq 1.0$$

Since joists are usually not deeper than 12in [286mm], this is one of the differences between dimension lumber and timber beam design. For beams of circular cross-section with a diameter greater than 13.5in [322mm] or for 12in [286mm] or larger square beams loaded in the plane of the diagonal, the size factor can be determined on the basis of an equivalent conventionally loaded square beam of the same cross-sectional area. **The size factor applies only to visually graded sawn lumber or round timber members.**

An additional complication arises when the beam does not have full lateral support. In this case, the bending stress should be adjusted also with the beam stability factor C_L, using the same formulas discussed in Example 22.1 for a glulam beam. This factor is determined with formulas and criteria explained in the NDS, Sec. 3.3.3. However, the NDS allows the use of some simplified assumptions for lateral stability, based on the ratio of depth to breadth (NDS, Sec.

4.4.1). Up to a ratio of 4:1, corresponding in practice to a 6X24 [140X597mm] for timbers (since no deeper sizes are tabulated for sawn lumber in the NDS), only the ends need be restrained. Consequently, it can be inferred that in practice there are no lateral stability concerns for solid timber beams, and the C_L factor becomes important for glulam and other composite beams available with higher depth/breadth ratios.

22.2 Built-up beams

When no adequate sawn lumber sections are readily available for a given condition, it is possible to combine pieces on-site with glue, nails, or other mechanical fasteners. The simplest case is when two pieces are connected side by side. A doubling or even tripling joist under a wall or heavy load is a common example of this situation. In these cases the beam capacity will obviously be the sum of the two or three separate lumber members. This system is not the most efficient use of the material, since flexural strength can be better achieved by increasing depth. The problem with fabricating deeper beams with lumber (Figs. 22.2 and 22.3) is that horizontal joints that result from "stacking" sections have no horizontal shear capacity, and these joints must either be located in areas of low horizontal shear load or be reinforced to transfer the load to an adjacent component. In some orientations, a nailed joint may ensure adequate shear resistance, but the nailed connection should be designed with a procedure similar to that explained for plywood box beams (Chapter 21).

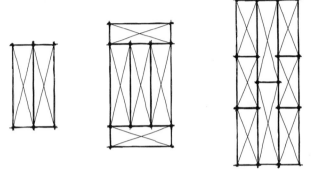

Fig. 22.2 Nailed and glued built-up beam sections

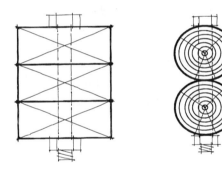

Fig. 22.3 Bolted beam sections

22.3 Laminated timber beams

Laminated timber beams are horizontally glued built-up sections (Fig. 22.4) that use the wood structural properties very efficiently by removing any major defects in the natural lumber. There are three grades that refer to the appearance of glue-laminated members: industrial, architectural, and premium. These appearance grades refer to the surface characteristics of lumber and not to the structural

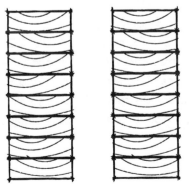

Fig. 22.4 Built-up laminated beam sections

qualities. Members produced to ANSI specifications bear the APA EWS (Engineered Wood System) trademark. Members can be bent or "cambered" during lamination to produce arches or beams. Even straight beams will be cambered to compensate for dead load deflection. The sides can also be sawed and planed to produce tapered and curved components (Fig. 22.5).

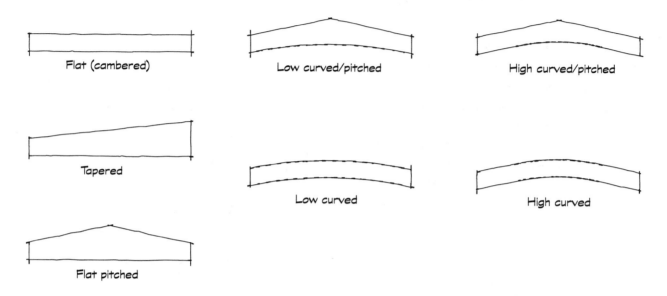

Fig. 22.5 Common types of laminated beams

The size of laminated timber is not nominal, but finished. Laminations are 1-1/2in [38mm] or 1-3/8in [35mm] thick; lumber for laminations can be spliced with scarf joints or finger joints to reach any desired length. The weight of laminated timber varies between 27pcf [4240N/m³](hem-fir) and 36pcf [5650N/m³](Southern pine), with a standard value of 35pcf [5500N/m³] for Western species.

Although most of the laminated timber uses softwood, members are also produced using laminated hardwood timbers of maple or ash. The design values are in the NDS, Table 5C.

22.4 Design of laminated timber beams

The design of laminated timber beams differs from that of sawn timber beams primarily in the use of adjustment factors (see Table 22.2). The calculation of some of these factors (C_v and C_L) is cumbersome, and the design can be speeded up with the help of tables supplied by manufacturers associations. The American Wood System (related to the American Plywood Association) has published design tables including volume factors for a range of sizes and spans of Western species, which is reproduced in Table 22.3.

Table 22.2 Adjustment factors for laminated timber beams (moment, shear, and deflection)

Symbol	Definition and applicability	Adjusts:
C_D	Load duration	F_b, F_v
C_M	Wet service: It applies for m.c. > 16% (NDS Supplement).	F_b, F_v, E
C_T	Temperature	F_b, F_v, E
C_V	Volume: It does not apply simultaneously with the beam-stability factor C_L (NDS Supplement, Sec. 5.3.2). Its function is effectively the same as the size factor C_F for timber.	F_b
C_L	Beam stability (NDS Part III, Sec. 3.3.3): Modifies F_b. It applies to beams without full lateral support. It is calculated with the formula in the NDS (Eq. 3.3-6). Note that in this calculation, the modulus of elasticity E_{yy} must be used (Sec. 5.4.2). For beams with full lateral support, $C_L = 1$.	F_b
C_{fu}	Flat use: Applicable when the beam is loaded parallel to the wide face of the laminations, and the width is less than 12in (NDS, Sec. 5.3.3). The values are given in the NDS Supplement, Table 5A.	F_b

C_L takes into account the possibility of buckling in the compression edge of the beam. The formula giving this factor contains Euler's critical buckling design value $F_{bE} = K_{bE}E'/R_B^2$, where R_B is the slenderness ratio given as

$$R_b = \sqrt{\frac{\frac{L_e}{d}}{b^2}}$$

where

L_e = effective length (NDS, Table 3.3.3) based on the unbraced length of the compression edge.

K_{bE} = coefficient depending on the coefficient of variation of E of COV_E (NDS Table F1).

Glulam with six or more laminations have a $COV_E \leq 0.11$.

Table 22.3 C_v
Volume factors for simple span laminated beams (Western species)

2-1/2in width   Span in feet

Depth (in)	16	18	20	22	24	26	28	30	32	34	36	38	40
13.5	1	1	1	1	1	1	1	1	1	1	1	1	0.99
15	1	1	1	1	1	1	1	1	1	1	0.99	0.99	0.99
16.5	1	1	1	1	1	1	1	1	0.99	0.99	0.99	0.98	0.98
18	1	1	1	1	1	1	1	1	0.99	0.98	0.98	0.97	0.97
19.5	1	1	1	1	1	1	0.99	0.99	0.98	0.98	0.97	0.96	0.96
21	1	1	1	1	1	0.99	0.99	0.98	0.97	0.97	0.96	0.96	0.95
22.5	1	1	1	1	1	0.99	0.98	0.97	0.97	0.96	0.96	0.95	0.95
24	1	1	1	1	0.99	0.98	0.97	0.97	0.96	0.96	0.95	0.95	0.94
25.5	1	1	1	0.99	0.98	0.98	0.97	0.96	0.96	0.95	0.94	0.94	0.93
27	1	1	1	0.99	0.98	0.97	0.96	0.96	0.95	0.94	0.94	0.93	0.93

3-1/8in width   Span in feet

Depth (in)	18	20	22	24	26	28	30	32	34	36	38	40	42
13.5	1	1	1	1	1	1	1	1	0.99	0.98	0.98	0.97	0.97
15	1	1	1	1	1	1	0.99	0.99	0.99	0.98	0.97	0.97	0.96
16.5	1	1	1	1	1	0.99	0.98	0.98	0.97	0.96	0.96	0.95	0.95
18	1	1	1	1	0.99	0.98	0.97	0.97	0.96	0.96	0.95	0.95	0.94
19.5	1	1	1	0.99	0.98	0.97	0.97	0.96	0.95	0.95	0.94	0.94	0.93
21	1	1	0.99	0.98	0.97	0.97	0.96	0.95	0.95	0.94	0.94	0.93	0.93
22.5	1	0.99	0.98	0.97	0.97	0.96	0.95	0.95	0.94	0.93	0.93	0.93	0.92
24	1	0.99	0.98	0.97	0.96	0.95	0.95	0.94	0.93	0.93	0.92	0.91	0.91
25.5	0.99	0.98	0.97	0.96	0.95	0.95	0.94	0.93	0.93	0.92	0.92	0.91	0.91
27	0.98	0.97	0.96	0.96	0.95	0.94	0.93	0.93	0.92	0.92	0.91	0.91	0.9

5-1/8in width   Span in feet

Depth (in)	18	20	22	24	26	28	30	32	34	36	38	40	42
13.5	1	0.99	0.98	0.98	0.97	0.96	0.95	0.95	0.94	0.94	0.93	0.93	0.92
15	0.99	0.98	0.97	0.97	0.96	0.95	0.94	0.94	0.93	0.93	0.92	0.92	0.91
16.5	0.98	0.97	0.96	0.96	0.95	0.94	0.93	0.93	0.92	0.92	0.91	0.91	0.9
18	0.98	0.97	0.96	0.95	0.94	0.93	0.93	0.92	0.92	0.91	0.9	0.9	0.9
19.5	0.97	0.96	0.95	0.94	0.93	0.93	0.92	0.91	0.91	0.9	0.9	0.89	0.89
21	0.96	0.95	0.94	0.93	0.93	0.92	0.91	0.91	0.9	0.9	0.89	0.89	0.88
22.5	0.95	0.94	0.93	0.93	0.92	0.91	0.91	0.9	0.9	0.89	0.88	0.88	0.88

Table 22.3 C$_v$
Volume factors for simple span laminated beams (Western species)

Depth (in)	18	20	22	24	26	28	30	32	34	36	38	40	42
24	0.95	0.94	0.93	0.92	91	0.91	0.9	0.89	0.89	0.88	0.88	0.87	0.87
25.5	0.94	0.93	0.92	0.91	0.91	0.9	0.89	0.89	0.88	0.88	0.87	0.87	0.86
27	0.94	0.93	0.92	0.91	0.9	0.9	0.89	0.88	0.88	0.87	0.87	0.86	0.86
28.5	0.93	0.92	0.91	0.9	0.9	0.89	0.88	0.88	0.87	0.87	0.86	0.86	0.86

6-3/4in width   Span in feet

Depth (in)	18	20	22	24	26	28	30	32	34	36	38	40	42
18	0.95	0.94	0.93	0.92	0.91	0.91	0.9	0.9	0.89	0.89	0.88	0.88	0.87
19.5	0.94	0.93	0.92	0.91	0.91	0.9	0.89	0.89	0.88	0.88	0.87	0.87	0.86
21	0.93	0.92	0.92	0.91	0.9	0.89	0.89	0.88	0.88	0.87	0.87	0.86	0.86
22.5	0.93	0.92	0.91	0.9	0.89	0.89	0.88	0.88	0.87	0.87	0.86	0.86	0.85
24	0.92	0.91	0.9	0.9	0.89	0.88	0.88	0.87	0.87	0.86	0.86	0.85	0.85
25.5	0.92	0.91	0.9	0.89	0.88	0.88	0.87	0.87	0.86	0.85	0.85	0.85	0.84
27	0.91	0.9	0.89	0.89	0.88	0.87	0.87	0.86	0.85	0.85	0.85	0.84	0.84
28.5	0.91	0.9	0.89	0.88	0.87	0.87	0.86	0.86	0.85	0.84	0.84	0.84	0.83
30	0.9	0.89	0.88	0.88	0.87	0.86	0.86	0.85	0.85	0.84	0.84	0.83	0.83
31.5	0.9	0.89	0.88	0.87	0.86	0.86	0.85	0.85	0.84	0.84	0.83	0.83	0.82
33	0.89	0.88	0.88	0.87	0.86	0.85	0.85	0.84	0.84	0.83	0.83	0.82	0.82
34.5	0.89	0.88	0.87	0.86	0.86	0.85	0.84	0.84	0.83	0.83	0.82	0.82	0.82

8-3/4in width   Span in feet

Depth (in)	18	20	22	24	26	28	30	32	34	36	38	40	42
24	0.9	0.89	0.88	0.87	0.87	0.86	0.85	0.85	0.84	0.84	0.84	0.83	0.83
25.5	0.89	0.88	0.88	0.87	0.86	0.85	0.85	0.84	0.84	0.83	0.83	0.82	0.82
27	0.89	0.88	0.87	0.86	0.86	0.85	0.84	0.84	0.83	0.83	0.82	0.82	0.82
28.5	0.88	0.87	0.87	0.86	0.85	0.84	0.84	0.83	0.83	0.82	0.82	0.81	0.81
30	0.88	0.87	0.86	0.85	0.85	0.84	0.83	0.83	0.82	0.82	0.82	0.81	0.81
31.5	0.87	0.86	0.86	0.85	0.84	0.84	0.83	0.83	0.82	0.82	0.81	0.81	0.8
33	0.87	0.86	0.85	0.85	0.84	0.83	0.83	0.82	0.82	0.81	0.81	0.8	0.8
34.5	0.87	0.86	0.85	0.84	0.83	0.83	0.82	0.82	0.81	0.81	0.8	0.8	0.8
36	0.86	0.85	0.85	0.84	0.83	0.83	0.82	0.81	0.81	80	0.8	0.8	0.79
37.5	0.86	0.85	0.84	0.83	0.83	0.82	0.82	0.81	0.81	0.8	0.8	0.79	0.79
39	0.86	0.85	0.84	0.83	0.82	0.82	0.81	0.81	0.8	0.8	0.79	0.79	0.79
40.5	0.85	0.84	0.84	0.83	0.82	0.82	0.81	0.8	0.8	0.8	0.79	0.79	0.78

All values are based on the following formula.

$$C_v = [12/d]^{1/10} \times [5.125/b]^{1/10} \times [21/L]^{1/10} \leq 1.0 \quad \text{Use the lesser value.}$$

The Southern Pine Marketing Council, with the American Institute of Timber Construction has published design tables for Southern pine laminated beams. Parts are reproduced here as Tables 22.4 to 22.7, and will be used in design examples.

Table 22.4 Southern pine laminated timber beams

F_b = 2,400psi F_v = 200psi   E = 1.7x10⁶psi C_D = 1.15

Roof loadings (with deflection limited to L/180)   1-month controlling loading combination

Total load capacity, uniform loading in plf

Width	Depth	Wt.	Span	Span	Span	Span	Span	Span	Span	Span	Span	Span	Span
inches	inches	plf	10ft	12ft	14ft	16ft	18ft	20ft	24ft	28ft	32ft	34ft	40ft
3	5-1/2	4.1	210	121	76	51	36	26	15	10	6	4	3
3	6-7/8	5.2	409	237	149	100	70	51	30	19	12	9	6
3	8-1/4	6.2	626	409	258	173	121	88	51	32	22	15	11
3	9-5/8	7.2	852	592	409	274	193	140	81	51	34	24	18
3	11-0	8.3	1,113	773	568	409	287	210	121	76	51	36	26
3	12-3/8	9.3	1,409	978	719	550	409	298	173	109	73	51	37
3	13-3/4	10.3	1,641	1,208	887	679	537	409	237	149	100	70	51
3	15-1/8	11.3	1,861	1,462	1,074	822	650	526	315	198	133	93	68
3	16-1/2	12.4	2,094	1,641	1,278	978	773	626	409	258	173	121	88
3	17-7/8	13.4	2,342	1,823	1,492	1,148	907	735	510	328	219	154	112
5	5-1/2	6.9	349	202	127	85	60	44	25	16	11	7	5
5	6-7/8	8.6	682	395	249	167	117	85	49	31	21	15	11
5	8-1/4	10.3	1,044	682	429	288	202	147	85	54	36	25	18
5	9-5/8	12	1,420	986	682	457	321	234	135	85	57	40	29
5	11-0	13.8	1,855	1,288	947	682	479	349	202	127	85	60	44
5	12-3/8	15.5	2,348	1,631	1,198	917	682	497	288	181	121	85	62
5	13-3/4	17.2	2,735	2,013	1,479	1,132	895	682	395	249	167	117	85
5	15-1/8	18.9	3,101	2,436	1,790	1,370	1,077	863	525	331	222	156	113
5	16-1/2	20.6	3,490	2,735	2,130	1,628	1,271	1,019	682	429	288	202	147
5	17-7/8	22.3	3,904	3,038	2,487	1,895	1,480	1,186	809	546	366	257	187
5	19-1/4	24.1	4,346	3,357	2,735	2,181	1,703	1,365	931	674	457	321	234
5	20-5/8	25.8	4,819	3,693	2,994	2,487	1,942	1,557	1,061	768	562	395	288
5	22-0	27.5	5,326	4,048	3,265	2,735	2,195	1,760	1,200	868	656	479	349
5	23-3/8	29.2	5,872	4,423	3,547	2,961	2,463	1,974	1,346	974	736	575	419
6-3/4	6-7/8	11.6	921	533	336	225	158	115	67	42	28	20	14
6-3/4	8-1/4	13.9	1,409	921	580	388	273	199	115	72	49	34	25

width	depth	wt.	10ft	12ft	14ft	16ft	18ft	20ft	24ft	28ft	32ft	36ft	40ft
6-3/4	9-5/8	16.2	1,918	1,332	921	617	433	316	183	115	77	54	39
6-3/4	11-0	18.6	2,505	1,739	1,278	921	647	471	273	172	115	81	59
6-3/4	12-3/8	20.9	3,170	2,201	1,617	1,230	921	671	388	245	164	115	84
6-3/4	15-1/8	25.5	4,550	3,574	2,592	1,958	1,529	1,225	709	447	299	210	153
6-3/4	16-1/2	27.8	5,121	4,014	3,058	2,310	1,804	1,446	921	580	388	273	199
6-3/4	17-7/8	30.2	5,728	4,458	3,560	2,689	2,100	1,683	1,148	737	494	347	253
6-3/4	19-1/4	32.5	6,377	4,927	4,014	3,096	2,418	1,938	1,321	921	617	433	316
6-3/4	20-5/8	34.8	7,071	5,420	4,393	3,530	2,756	2,209	1,506	1,090	759	533	388
6-3/4	22-0	37.1	7,816	5,940	4,790	3,990	3,116	2,497	1,703	1,232	921	647	471
6-3/4	23-3/8	39.4	8,616	6,490	5,205	4,345	3,496	2,808	1,911	1,382	1,044	776	565
6-3/4	24-3/4	41.8	9,479	7,071	5,639	4,689	3,897	3,124	2,130	1,541	1,164	909	671
6-3/4	26-1/8	44.1	10,411	7,688	6,094	5,047	4,308	3,462	2,360	1,708	1,290	1,007	789
6-3/4	27-1/2	46.4	11,423	8,343	6,571	5,420	4,612	3,816	2,602	1,882	1,422	1,111	890

TABLE SPECIFICATIONS:

This table applies to straight, simply supported, glued laminated timber beams under dry conditions of use. Beams must be laterally supported at the top along the entire length of the beam and at the top and bottom at the end bearing points. The load-carrying capacities listed are for TOTAL LOAD, including the weight of the member. A unit weight of 36pcf was assumed to determine the plf beam weights.

DESIGN VALUE MODIFICATIONS:

The allowable stress in bending, F_b, has been modified by the volume factor, C_v. For determination of load-carrying capacities governed by shear, loads within a distance d (the depth of the beam) from the ends have been disregarded.

DEFLECTION LIMITS:

For roof beams, a deflection limit of 1/180 for total load has been used; consequently, they must be used ONLY FOR SLOPED ROOF SYSTEMS WITHOUT A RIGID CEILING SURFACE ATTACHED. For flat roofs or rigid ceiling finishes, 1/360 of the span is the maximum allowable deflection.

IN ALL CASES, tables should be used for PRELIMINARY DESIGN ONLY! You will seldom encounter a pure, simple, and uniform loading condition; however, the information will provide a reasonable size approximation for preliminary architectural design purposes.

Table 22.5 Southern pine laminated timber beams

$F_b = 2,400\,psi$ $\qquad$ $F_v = 200\,psi$ $\quad$ $E = 1.7 \times 10^6\,psi$ $\qquad$ $C_D = 1.00$

Floor loading (with deflection limited to L/360) "permanent" controlling loading combination

Total load capacity, uniform loading in plf

Width	Depth	Wt.	Span	Span	Span	Span	Span	Span	Span	Span	Span	Span	Span
inches	inches	plf	10ft	12ft	14ft	16ft	18ft	20ft	24ft	28ft	32ft	34ft	40ft
3	5-1/2	4.1	131	76	48	32	22	16	9	6	4	3	2
3	6-7/8	5.2	256	148	93	62	44	32	19	12	8	5	4
3	8-1/4	6.2	422	256	161	108	76	55	32	20	13	9	7
3	9-5/8	7.2	702	406	256	171	120	88	51	32	21	15	11
3	11-0	8.3	968	606	382	256	180	131	76	48	32	22	16
3	12-3/8	9.3	1,225	851	544	364	256	186	108	68	46	32	23
3	13-3/4	10.3	1,427	1,050	746	500	351	256	148	93	62	44	32
3	15-1/8	11.3	1,618	1,271	934	665	467	340	197	124	83	58	43
3	16-1/2	12.4	1,821	1,427	1,111	851	606	442	256	161	108	76	55
3	17-7/8	13.4	2,037	1,585	1,298	998	771	562	325	205	137	96	70
5	5-1/2	6.9	218	126	80	53	37	27	16	10	7	5	3
5	6-7/8	8.6	426	247	155	104	73	53	31	19	13	9	7
5	8-1/4	10.3	737	426	268	180	126	92	53	34	22	16	12
5	9-5/8	12	1,170	677	426	286	201	146	85	53	36	25	18
5	11-0	13.8	1,613	1,010	636	426	299	218	126	80	53	37	27
5	12-3/8	15.5	2,042	1,418	906	607	426	311	180	113	76	53	39
5	13-3/4	17.2	2,378	1,715	1,243	833	585	426	247	155	104	73	53
5	15-1/8	18.9	2,696	2,118	1,556	1,108	778	567	328	207	139	97	71
5	16-1/2	20.6	3,034	2,378	1,852	1,415	1,010	737	426	268	180	126	92
5	17-7/8	22.3	3,395	2,642	2,163	1,648	1,285	936	542	341	229	161	117
5	19-1/4	24.1	3,779	2,919	2,378	1,897	1,481	1,170	677	426	286	201	146
5	20-5/8	25.8	4,190	3,212	2,604	2,163	1,689	1,353	833	524	351	247	180
5	22-0	27.5	4,632	3,520	2,839	2,378	1,909	1,530	1,010	636	426	299	218
5	23-3/8	29.2	5,106	3,846	3,085	2,575	2,142	1,717	1,171	763	511	359	262
6-3/4	6-7/8	11.6	575	333	210	140	99	72	42	26	18	12	9
6-3/4	8-1/4	13.9	994	575	362	243	170	124	72	45	30	21	16
6-3/4	9-5/8	16.2	1,579	914	575	385	271	197	114	72	48	34	25
6-3/4	11-0	18.6	2,178	1,364	859	575	404	295	170	107	72	51	37
6-3/4	12-3/8	20.9	2,757	1,914	1,223	819	575	419	243	153	102	72	52

width	depth	wt.	10ft	12ft	14ft	16ft	18ft	20ft	24ft	28ft	32ft	36ft	40ft
6-3/4	13-3/4	23.2	3,211	2,363	1,678	1,124	789	575	333	210	140	99	72
6-3/4	15-1/8	25.5	3,640	2,860	2,073	1,496	1,051	766	443	279	187	131	96
6-3/4	16-1/2	27.8	4,097	3,211	2,446	1,848	1,364	994	575	362	243	170	124
6-3/4	17-7/8	30.2	4,583	3,567	2,848	2,151	1,680	1,264	732	461	309	217	158
6-3/4	19-1/4	32.5	5,102	3,941	3,211	2,477	1,934	1,550	914	575	385	271	197
6-3/4	20-5/8	34.8	5,657	4,336	3,515	2,824	2,205	1,767	1,124	708	474	333	243
6-3/4	22-0	37.1	6,253	4,752	3,832	3,192	2,493	1,998	1,362	859	575	404	295
6-3/4	23-3/8	39.4	6,893	5,192	4,164	3,476	2,797	2,242	1,529	1,030	690	485	353
6-3/4	24-3/4	41.8	7,583	5,657	4,511	3,752	3,118	2,499	1,704	1,223	819	575	419
6-3/4	26-1/8	44.1	8,329	6,150	4,875	4,038	3,446	2,769	1,888	1,366	964	677	493
6-3/4	27-1/2	46.4	9,138	6,674	5,257	4,336	3,689	3,053	2,082	1,506	1,124	789	575

TABLE SPECIFICATIONS:

This table applies to straight, simply supported, glued laminated timber beams under dry conditions of use. Beams must be laterally supported at the top along the entire length of the beam and at the top and bottom at the end bearing points. The load-carrying capacities listed are for TOTAL LOAD, including the weight of the member. A unit weight of 36pcf was assumed to determine the plf beam weights.

DESIGN VALUE MODIFICATIONS:

The allowable stress in bending, F_b, has been modified by the volume factor, C_v. For determination of load-carrying capacities governed by shear, loads within a distance d (the depth of the beam) from the ends have been disregarded.

DEFLECTION LIMITS:

For roof beams, a deflection limit of 1/180 for total load has been used; consequently, they must be used ONLY FOR SLOPED ROOF SYSTEMS WITHOUT A RIGID CEILING SURFACE ATTACHED. For flat roofs or rigid ceiling finishes, 1/360 of the span is the maximum allowable deflection.

IN ALL CASES, tables should be used for PRELIMINARY DESIGN ONLY! You will seldom encounter a pure, simple, and uniform loading condition; however, the information will provide a reasonable size approximation for preliminary architectural design purposes.

Table 22.6   Western species laminated timber beams

$F_b = 2,400$psi $F_v = 165$psi $E = 1.8 \times 10^6$psi $C_D = 1.15$

Roof loadings (with deflection limited to L/180)   1-month controlling loading combination

Total load capacity, uniform loading in plf

Width	Depth	Wt.	Span	Span	Span	Span	Span	Span	Span	Span	Span	Span	Span
inches	inches	plf	10ft	12ft	14ft	16ft	18ft	20ft	24ft	28ft	32ft	36ft	40ft
2-1/2	6	3.6	240	139	87	59							
2-1/2	7-1/2	4.6	431	271	171	114	80	59					
2-1/2	9	5.5	621	431	295	198	139	101	59				
2-1/2	10-1/2	6.4	805	587	431	314	221	161	93	59			
2-1/2	12	7.3	949	759	563	431	329	240	139	87	59		
2-1/2	13-1/2	8.2	1,102	876	7,131	546	431	342	198	125	83	59	
2-1/2	15	9.1	1,265	999	825	674	532	431	271	171	114	80	59
2-1/2	16-1/2	10	1,439	1,128	928	788	644	522	361	227	152	107	78
2-1/2	18	10.9	1,626	1,265	1,035	876	759	621	431	295	198	139	101
2-1/2	19-1/2	11.8	1,827	1,410	1,147	967	836	729	506	370	251	177	129
3-1/8	6	4.6	300	174	109	73	51						
3-1/8	7-1/2	5.7	539	339	214	143	100	73					
3-1/8	9	6.8	776	539	369	247	174	127	73				
3-1/8	10-1/2	8	1,006	734	539	393	276	201	116	73			
3-1/8	12	9.1	1,186	949	704	539	412	300	174	109	73	51	
3-1/8	13-1/2	10.3	1,377	1,095	891	682	539	427	247	156	104	73	53
3-1/8	15	11.4	1,581	1,248	1,031	842	666	539	339	214	143	100	73
3-1/8	16-1/2	12.5	1,799	1,410	1,160	985	805	652	451	284	190	134	97
3-1/8	18	13.7	2,033	1,581	1,294	1,095	949	776	537	369	247	174	127
3-1/8	19-1/2	14.8	2,284	1,762	1,434	1,209	1,045	911	625	452	314	221	161
3-1/8	21	16	2,554	1,953	1,581	1,328	1,145	1,006	719	520	393	276	201
3-1/8	22-1/2	17.1	2,846	2,156	1,736	1,452	1,248	1,095	820	593	448	339	247
3-1/8	24	18.2	3,162	2,372	1,897	1,581	1,355	1,186	927	671	507	396	300
3-1/8	25-1/2	19.4	3,506	2,601	2,068	1,716	1,466	1,280	1,021	753	569	444	356
5-1/8	12	14.9	1,945	1,556	1,155	884	675	492	285	179	120	84	61
5-1/8	13-1/2	16.8	2,259	1,795	1,461	1,119	884	701	405	255	171	120	88
5-1/8	15	18.7	2,593	2,047	1,691	1,381	1,084	869	556	350	235	165	120
5-1/8	16-1/2	20.6	2,951	2,313	1,902	1,615	1,299	1,041	710	466	312	219	160
5-1/8	18	22.4	3,334	2,593	2,122	1,795	1,533	1,228	838	605	405	285	208

width	depth	wt.	10ft	12ft	14ft	16ft	18ft	20ft	24ft	28ft	32ft	36ft	40ft
5-1/8	19-1/2	24.3	3,746	2,890	2,352	1,983	1,714	1,430	975	706	515	362	264
5-1/8	21	26.2	4,189	3,203	2,593	2,178	1,878	1,646	1,123	812	614	452	330
5-1/8	22-1/2	28	4,668	3,536	2,846	2,382	2,047	1,795	1,280	926	700	546	405
5-1/8	24	29.9	5,186	3,890	3,112	2,593	2,223	1,945	1,447	1,047	791	618	492
5-1/8	25-1/2	31.8	5,750	4,266	3,391	2,814	2,405	2,099	1,624	1,175	887	693	555
5-1/8	27	33.6	6,365	4,668	3,685	3,044	2,593	2,259	1,795	1,309	989	772	619
5-1/8	28-1/2	35.5	7,039	5,097	3,995	3,285	2,789	2,423	1,920	1,451	1,096	856	686
6-3/4	13-1/2	22.1	2,975	2,365	1,925	1,456	1,137	911	534	336	225	158	115
6-3/4	15	24.5	3,415	2,696	2,227	1,779	1,389	1,113	732	461	309	217	158
6-3/4	16-1/2	27	3,887	3,046	2,505	2,127	1,665	1,334	910	614	411	289	211
6-3/4	18	29.5	4,391	3,415	2,794	2,365	1,964	1,574	1,073	776	534	375	273
6-3/4	19-1/2	32	4,933	3,806	3,098	2,612	2,258	1,833	1,250	904	679	477	348
6-3/4	21	34.5	5,517	4,219	3,415	2,869	2,473	2,110	1,439	1,041	786	595	434
6-3/4	22-1/2	36.9	6,148	4,657	3,749	3,137	2,696	2,365	1,640	1,186	896	700	534
6-3/4	24	39.4	6,831	5,123	4,099	3,415	2,928	2,562	1,854	1,341	1,013	791	634
6-3/4	25-1/2	41.8	7,573	5,619	4,466	3,706	3,167	2,765	2,080	1,505	1,137	888	712
6-3/4	27	44.3	8,383	6,148	4,854	4,009	3,415	2,975	2,319	1,678	1,267	990	793
6-3/4	28-1/2	46.8	9,271	6,713	5,262	4,326	3,673	3,192	2,528	1,859	1,405	1,097	879
6-3/4	30	49.2	10,246	7,319	5,692	4,657	3,941	3,415	2,696	2,049	1,548	1,209	969
8-3/4	15	31.9	4,427	3,495	2,887	2,246	1,754	1,406	949	598	401	281	205
8-3/4	16-1/2	35.1	5,038	3,949	3,247	2,692	2,102	1,685	1,149	796	533	374	273
8-3/4	18	38.3	5,692	4,427	3,622	3,065	2,480	1,988	1,356	981	692	486	354
8-3/4	19-1/2	41.1	6,395	4,933	4,016	3,386	2,888	2,315	1,578	1,142	863	618	451
8-3/4	21	44.6	7,152	5,469	4,427	3,719	3,206	2,665	1,817	1,315	993	772	563
8-3/4	22-1/2	47.8	7,969	6,037	4,859	4,066	3,495	3,038	2,072	1,499	1,132	884	692
8-3/4	24	51	8,855	6,641	5,313	4,427	3,795	3,321	2,342	1,694	1,280	999	801
8-3/4	25-1/2	54.2	9,817	7,284	5,790	4,804	4,105	3,584	2,628	1,901	1,436	1,121	899
8-3/4	27	57.4	10,867	7,969	6,292	5,197	4,427	3,856	2,929	2,119	1,601	1,250	1,002
8-3/4	28-1/2	60.6	12,017	8,702	6,821	5,608	4,762	4,137	3,246	2,348	1,774	1,385	1,110
8-3/4	30	63.8	13,282	9,487	7,379	6,037	5,109	4,427	3,495	2,589	1,956	1,527	1,224
8-3/4	31-1/2	67	14,681	10,331	7,969	6,487	5,469	4,728	3,719	2,840	2,146	1,675	1,343
8-3/4	33	70.2	16,234	11,239	8,595	6,957	5,844	5,038	3,949	3,103	2,344	1,830	1,467

TABLE SPECIFICATIONS:

This table applies to straight, simply supported, glued laminated timber beams under dry conditions of use. Beams must be laterally supported at the top along the entire length of the beam and at the top and bottom at the end bearing points. The load-carrying capacities listed are for TOTAL LOAD, including the weight of the member. A unit weight of 36pcf was assumed to determine the plf beam weights.

DESIGN VALUE MODIFICATIONS:

The allowable stress in bending, F_b, has been modified by the volume factor, C_v. For determination of load-carrying capacities governed by shear, loads within a distance d (the depth of the beam) from the ends have been disregarded.

DEFLECTION LIMITS:

For roof beams, a deflection limit of 1/180 for total load has been used; consequently, they must be used ONLY FOR SLOPED ROOF SYSTEMS WITHOUT A RIGID CEILING SURFACE ATTACHED. For flat roofs or rigid ceiling finishes, 1/360 of the span is the maximum allowable deflection.

IN ALL CASES, tables should be used for PRELIMINARY DESIGN ONLY! You will seldom encounter a pure, simple, and uniform loading condition; however, the information will provide a reasonable size approximation for preliminary architectural design purposes.

Table 22.7 Western species laminated timber beams

$F_b = 2,400\,psi$ $\qquad F_v = 165\,psi$ $\qquad E = 1.8\times10^6\,psi$ $\qquad C_D = 1.00$

Floor loading (with deflection limited to L/360) "permanent" controlling loading combination

Total load capacity, uniform loading in plf

Width	Depth	Wt.	Span	Span	Span	Span	Span	Span	Span	Span	Span	Span	Span
inches	inches	plf	10ft	12ft	14ft	16ft	18ft	20ft	24ft	28ft	32ft	36ft	40ft
2-1/2	6	3.6	150	87	55								
2-1/2	7-1/2	4.6	293	170	107	72	50						
2-1/2	9	5.5	506	293	184	124	87	63					
2-1/2	10-1/2	6.4	700	465	293	196	138	100	58				
2-1/2	12	7.3	825	660	437	293	206	150	87	55			
2-1/2	13-1/2	8.2	958	762	620	417	293	214	124	78	52		
2-1/2	15	9.1	1,100	868	717	572	402	293	170	107	72	50	
2-1/2	16-1/2	10	1,252	981	807	685	535	390	226	142	95	67	
2-1/2	18	10.9	1,414	1,100	900	762	660	506	293	184	124	87	63
2-1/2	19-1/2	11.8	1,589	1,226	998	841	727	634	372	235	157	110	80
3-1/8	6	4.6	187	109	68								
3-1/8	7-1/2	5.7	366	212	133	89	63						
3-1/8	9	6.8	633	366	231	154	109	79					
3-1/8	10-1/2	8	875	582	366	245	172	126	73				
3-1/8	12	9.1	1,031	825	547	366	257	187	109	68			
3-1/8	13-1/2	10.3	1,198	952	775	521	366	267	154	97	65		
3-1/8	15	11.4	1,375	1,086	897	715	502	366	212	133	89	63	
3-1/8	16-1/2	12.5	1,565	1,226	1,008	856	669	487	282	178	119	84	61
3-1/8	18	13.7	1,768	1,375	1,125	952	825	633	366	231	154	109	79
3-1/8	19-1/2	14.8	1,986	1,532	1,247	1,051	909	792	466	293	196	138	101
3-1/8	21	16	2,221	1,699	1,375	1,155	996	875	582	366	245	172	126
3-1/8	22-1/2	17.1	2,475	1,875	1,509	1,263	1,086	952	713	450	302	212	154
3-1/8	24	18.2	2,750	2,062	1,650	1,375	1,179	1,031	806	547	366	257	187
3-1/8	25-1/2	19.4	3,049	2,262	1,798	1,492	1,275	1,113	888	654	439	309	225
5-1/8	12	14.9	1,691	1,353	897	601	422	307	178	112	75	53	
5-1/8	13-1/2	16.8	1,964	1,561	1,271	855	601	438	253	160	107	75	55
5-1/8	15	18.7	2,255	1,780	1,471	1,173	824	601	348	219	147	103	75
5-1/8	16-1/2	20.6	2,566	2,011	1,654	1,404	1,097	799	463	291	195	137	100
5-1/8	18	22.4	2,899	2,255	1,845	1,561	1,333	1,038	601	378	253	178	130
5-1/8	19-1/2	24.3	3,257	2,513	2,045	1,724	1,491	1,244	764	481	322	226	165
5-1/8	21	26.2	3,643	2,786	2,255	1,894	1,633	1,432	954	601	402	283	206
5-1/8	22-1/2	28	4,059	3,075	2,475	2,071	1,780	1,561	1,113	739	495	348	253
5-1/8	24	29.9	4,510	3,382	2,706	2,255	1,933	1,691	1,258	897	601	422	307

width	depth	wt.	10ft	12ft	14ft	16ft	18ft	20ft	24ft	28ft	32ft	36ft	40ft
5-1/8	25-1/2	31.8	5,000	3,710	2,949	2,447	2,091	1,825	1,412	1,021	720	506	369
5-1/8	27	33.6	5,535	4,059	3,204	2,647	2,255	1,964	1,561	1,139	855	601	438
5-1/8	28-1/2	35.5	6,121	4,432	3,474	2,856	2,425	2,107	1,669	1,262	953	706	515
6-3/4	13-1/2	22.1	2,587	2,056	1,674	1,126	791	577	334	210	141		
6-3/4	15	24.5	2,970	2,345	1,937	1,545	1,085	791	458	288	193	136	
6-3/4	16-1/2	27	3,380	2,649	2,178	1,849	1,444	1,053	609	384	257	181	132
6-3/4	18	29.5	3,819	2,970	2,430	2,056	1,708	1,367	791	498	334	234	171
6-3/4	19-1/2	32	4,290	3,309	2,694	2,271	1,963	1,594	1,006	633	424	298	217
6-3/4	21	34.5	4,798	3,669	2,970	2,495	2,151	1,834	1,251	791	530	372	271
6-3/4	22-1/2	36.9	5,346	4,050	3,260	2,728	2,345	2,056	1,426	973	652	458	334
6-3/4	24	39.4	5,940	4,455	3,564	2,970	2,546	2,227	1,612	1,166	791	556	405
6-3/4	25-1/2	41.8	6,586	4,886	3,884	3,223	2,754	2,404	1,809	1,309	949	666	486
6-3/4	27	44.3	7,290	5,346	4,221	3,487	2,970	2,587	2,017	1,459	1,102	791	577
6-3/4	28-1/2	46.8	8,061	5,838	4,575	3,762	3,194	2,775	2,199	1,617	1,221	930	678
6-3/4	30	49.2	8,910	6,364	4,950	4,050	3,427	2,970	2,345	1,782	1,346	1,051	791
8-3/4	15	31.9	3,850	3,039	2,511	1,953	1,407	1,025	593	374	250	176	128
8-3/4	16-1/2	35.1	4,381	3,434	2,823	2,341	1,828	1,365	790	497	333	234	171
8-3/4	18	38.3	4,950	3,850	3,150	2,665	2,157	1,729	1,025	646	433	304	221
8-3/4	19-1/2	41.1	5,561	4,290	3,492	2,944	2,511	2,013	1,304	821	550	386	282
8-3/4	21	44.6	6,219	4,756	3,850	3,234	2,788	2,317	1,580	1,025	687	482	352
8-3/4	22-1/2	47.8	6,930	5,250	4,226	3,536	3,039	2,642	1,801	1,261	845	593	433
8-3/4	24	51	7,700	5,775	4,620	3,850	3,300	2,887	2,036	1,473	1,025	720	525
8-3/4	25-1/2	54.2	8,537	6,334	5,035	4,178	3,570	3,117	2,285	1,653	1,230	864	630
8-3/4	27	57.4	9,450	6,930	5,471	4,520	3,850	3,353	2,547	1,843	1,392	1,025	748
8-3/4	28-1/2	60.6	10,450	7,567	5,931	4,877	4,141	3,598	2,823	2,042	1,543	1,205	879
8-3/4	30	63.8	11,550	8,250	6,417	5,250	4,442	3,850	3,039	2,251	1,701	1,328	1,025
8-3/4	31-1/2	67	12,766	8,983	6,930	5,641	4,756	4,111	3,234	2,470	1,866	1,457	1,168
8-3/4	33	70.2	14,117	9,773	7,474	6,050	5,082	4,381	3,434	2,698	2,038	1,592	1,276

TABLE SPECIFICATIONS:

This table applies to straight, simply supported, glued laminated timber beams under dry conditions of use. Beams must be laterally supported at the top along the entire length of the beam and at the top and bottom at the end bearing points. The load-carrying capacities listed are for TOTAL LOAD, including the weight of the member. A unit weight of 36pcf was assumed to determine the plf beam weights.

DESIGN VALUE MODIFICATIONS:

The allowable stress in bending, F_b, has been modified by the volume factor, C_v. For determination of load-carrying capacities governed by shear, loads within a distance d (the depth of the beam) from the ends have been disregarded.

DEFLECTION LIMITS:

For roof beams, a deflection limit of 1/180 for total load has been used; consequently, they must be used ONLY FOR SLOPED ROOF SYSTEMS WITHOUT A RIGID CEILING SURFACE ATTACHED. For flat roofs or rigid ceiling finishes, 1/360 of the span is the maximum allowable deflection.

IN ALL CASES, tables should be used for PRELIMINARY DESIGN ONLY! You will seldom encounter a pure, simple, and uniform loading condition; however, the information will provide a reasonable size approximation for preliminary architectural design purposes.

22.5 Types of structures using glulam

Since the manufacturing process does not put limitations on the size of members, glulam is employed in long-span buildings in the form of rectangular beams, three-hinged arches, and lamella vaults and domes. Rectilinear beams are stressed in bending, whereas curved or pitched beams (as shown in Fig. 22.5) are effectively arches and therefore subject to a combination of bending and compression stresses. In a system such as the one shown in Fig. 22.6, beams (in this case, curved) arranged in a nonorthagonal geometry can create a wide range of spatial volumes. These types of systems are similar to three-hinged arches, with members loaded along their length, and they are presented in Chapter 24. Hip beams, like rafters and other sloping beams, also have a component of axial load, and are discussed in Chapter 23. Beams or arches are spaced typically

Fig. 22.6 Northern Arizona University stadium, Flagstaff, Arizona

16 to 20ft [5 to 6m] or more, depending on the secondary structure used for roofing (Fig. 22.7). The most common system is to support purlins at 4 to 8ft [1.2 to 2.4m] on the primary beams, and on the roof with metal decking or stressed-skin panels fastened to the purlins. In this way, the purlins (single or double-span) are loaded uniformly and are continuously supported along most

of the compression zone fibers. The purlins generate point loads on the primary beams or arches. An example will illustrate the design procedure for a simple system.

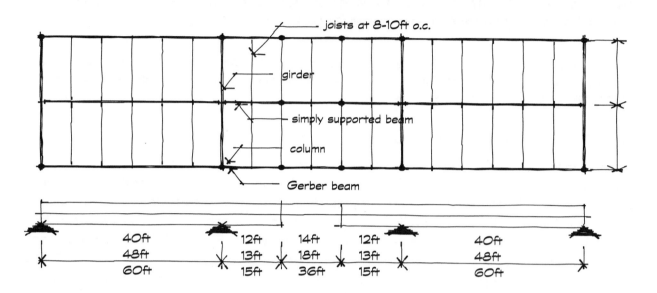

Fig. 22.7 Framing system using Gerber beams, simple beams, and joists

Example 22.1

A floor timber beam with a 24ft span, simply supported at the ends, must carry a load of 975plf. The beam is fully laterally supported.

Use Southern pine glued laminated timber with F_b = 2400psi.

From Table 22.5, a 5X22 carries 1,010plf; a 6-3/4X23-3/8 carries 1,030plf. The respective weights are 27.5plf and 39.4plf, so we choose the lighter, 5X22.

Example 22.2

For the same beam of Example 22.1, use Western species with F_b = 2400 psi.

From Table 22.7, a 5-1/8X22-1/2 carries 1,113plf; a 6-3/4X19-1/2 carries 1,006plf. The respective weights are 28plf and 32plf, so we choose the lighter, 5-1/8X22-1/2.

Tables 22.1 to 22.7 are not usable when the load configuration is different from a uniformly distributed load and when the compression edge is not fully supported. This case is common for large-span beam systems, where the main beams support secondary beams; the beam length between secondary beams may be unbraced. This affects the value of C_L with a consequent reduction of F_b'. The following example shows the design procedure of a glulam beam using the normal procedure for timber beams; the tables are used only to determine the C_v value.

Example 22.3

A double-span structure is made of 40ft beams at 40ft o.c. and supports 40ft purlins 8ft o.c. The beam is unbraced between purlins. Foam core roofing panels,16ft long, are glued and nailed to the purlins. The roof slope is negligible for structural calculation purposes and the beams are considered horizontal.

Purlin design

Single-span beam 40ft, 8ft o.c., and full lateral support from roofing panels

Roof loads:

 Snow load = 30psf

 Dead load = 12psf (excluding beam weight)

Use Western species laminated timber.

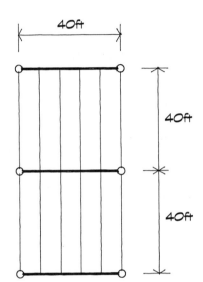

(a) Loads

 Uniformly distributed load:

 Snow: 30psf x 8ft o.c = 240plf

 Dead load: 12psf x 8ft o.c. = 96plf

 Estimate of purlin weight:

 d = (1/2in/ft)(40ft) = 20in

 b = (20in)/4 = 5in

 The closest section is 5-1/8inX21in.

$$w_{beam} = (5.125in)(21in)(35pcf)/(144in^2/ft^2) = 26plf$$

The total load supported by the beam is 362plf.

Critical load combination using LDF:

Dead load only = (26plf + 96plf)/0.9 = 135.5plf

Snow load + dead load = (362plf)/1.15 = 314.8plf   Critical: use C_D = 1.15

(b) Statics

 R = (362plf)(40ft)/2 = 7240lb

 V = 7,240lb

 M = 72,400ftlb

 Maximum Δ = (40ft)(12in/ft)/360 = 1.33in

(c) Design values

Adjustment factors:

$C_D = 1.15$

C_V to be determined after the first trial section

$C_L = 1.0$ (full lateral support)

Visually graded Western species 24F-V1:

$F_b = 2,400$psi $F_b' = 1.15(2,400$psi$) = 2,760$psi

$F_v = 140$psi $F_v' = 1.15(140$psi$) = 161$psi

(d) Section properties required

Required $A = 1.5V/F_v = 1.5(7,240$lb$)/(161$psi$) = 67.4$in^2

Required $S = M/F_b = (72,400$ftlb$)(12$in/ft$)/(2,760$psi$) = 314.8$in^3

Required $I = I_{snow} + 0.5(I_{DL}) = \dfrac{5wL^4}{384E\Delta}$

where $w = w_{snow} + 0.5w_{DL} = 240$lb/ft $+ 0.5(96$plf$+26$plf$) = 301$plf

Required $I = \dfrac{5\left(\dfrac{301\text{plf}}{12\text{in/ft}}\right)(40\text{ft}\times12\text{in/ft})^4}{384(1,700,000\text{psi})1.33\text{in}} = 7668$in^4

Deflection controls. 5-1/8X27 has:

$A = 138.4$in^2

$S = 623$in^3

$C_V = 0.8646$ for a span of 40ft

(from Table 22.3, C_V for Western species); therefore:

$S_x = (623$in$^3)(0.8646) = 539$in$^3 > 315$in^3 required

$I_x = 8,406$in$^4 > 7,668$in^4 required

Since the roof panels provide continuous lateral support to the compressed fibers, $C_L = 1.0$.

The purlin weighs 33.6plf, which is more than the initial estimate, but this would cause only a slight increase (about 1%) of the required I calculated, so no adjustments are necessary. Use 5-1/8X27 with 1-1/2in camber to compensate for the deflection caused by $0.5\ w_{DL}$ (creep), which is calculated with the deflection formula.

Example 22.4

For the same system in Example 22.3, design the primary beam supporting the purlins. Loads directly over the supports are not considered in M, V, or deflection.

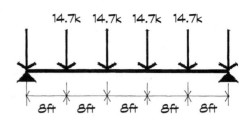

Laminated timber beams are 40ft o.c. There is a span of 40ft with four purlins 8ft o.c. supporting the roof panels.

Roof loads:

> Snow load = 30psf
>
> Dead load = 12psf
>
> Purlin weight = 33.6plf

(a) Loads

Concentrated loads P from purlins with tributary area of 40ft x 8ft:

Snow load = (30psf)(40ft x 8ft) =	9,600lb
Dead load = (12psf)(40ft x 8ft) =	3,840lb
Purlin weight = (20ft + 20ft)(33.6lb/ft) =	1,344lb
Dead load subtotal = 3840lb + 1344lb =	<u>5,184lb</u>
P (live load + dead load)	14,784lb

Distributed load: weight of beam, Western species, 35pcf.

Estimated beam size:

d = (3/4in)(40ft) = 30in

b = (30in)/4 = 7.5in

The closest size is 8-3/4X30; therefore,

Beam weight (dead load)

$\quad$ w = (8.75in)(30in)(35lb/ft³)/(144in²/ft²) = 64plf

Critical load combination:

DL = [(5,184lb)(4) + (63.8plf)(40ft)]/0.9 = (23,296lb)/0.9 = 25,884lb

$\quad$ (total dead load on each beam)

Snow + DL = [(9,600lb)(4) + (25,884lb)]/1.15 = 55,899lb $\qquad$ Critical: Use C_D = 1.15.

(b) Statics

R = 4P/2 + wL/2 = 4(14,784lb)/2 + (64plf)(40ft)/2 =

$\qquad$ 2,9568lb + 1,280lb = 30,848lb

V = 30,848lb

M = aPL = 0.6(14784lb)40ft = 354,816ftlb

(AISC tables, p. 2-295)

Maximum Δ = L/240 = 40ft(12in/ft)/240 = 2in for the total load.

This is pushing the limits of deflection for an essentially flat roof, but if we assume some slight slope to eliminate any ponding, we can probably get away with it.

(c) Design values

Adjustment factors:

$C_D = 1.15$

C_V and C_L are to be determined after the first trial section. Assume they are 1.0.

Visually graded Western species 22F-V1 design values:

$F_b = 2,200$psi	$F_b' = 1.15(2,200$psi$) = 2,530$psi
$F_v = 140$psi	$F_v' = 1.15(140$psi$) = 161$psi
$E_x = 1,600,000$psi	
$E_y = 1,300,000$psi	

(d) Section properties required

Required A = 1.5 V/F_v = 1.5(30,848lb)/(161psi) = 287 in^2

Required S = M/F_b = (354,816ftlb)(12in/ft)/2,530psi = 1,683 in^3

Required I for snow load + 0.5 DL = 9,600lb +.5(5,184lb) = 12,192lb

Since this is a symmetrical series of point loads, we can use a simplified technique to calculate deflection. According to the AISC Manual (p. 2-295) this condition would have a maximum deflection of

$$I_x = e\left(\frac{PL^3}{EI}\right) = .063\left(\frac{12192lb(40ft \times 12in/ft)^3}{1,600,000psi(2.0in)}\right) = 26,545in^4$$

Required I for w = 64plf (purlin weight); the load is increased by 0.5 to account for creep.

$$I_w = 0.5\frac{5\left(\frac{64plf}{12in/ft}\right)(40ft \times 12in/ft)^4}{384(1,600,000psi)2in} = 576in^4$$

Total required I = 27,121in^4

(e) Section selection

Remembering that the required A = 287in^2 and the required S = 1683in^3, from the NDS Supplement, Table 1C and C_v Table 22.3:

Size	A	S	C_v	I
8-3/4X36	15.0in²	1890in³	0.7963	34,020in⁴
8-3/4X37.5	328.1in²	2051in³	0.7930	38,452in⁴
8-3/4X39	341.3in²	2218in³	0.7899	43,253in⁴

Of the three sections, the first has adequate S, but the required S must be adjusted with the volume factor C_v and then recalculated with the stability factor C_L (NDS).

First trial: 8.75X36, and C_v = 0.7963   Required S = (1683in³)/0.7963 = 2113 in³ NO!

Second trial: 8.75X37.5, and C_v = 0.730   Required S = (1683in³)/0.7930 = 2122 in³ OK!

The 8.75inX37.5in is sufficiently close, so we calculate the stability factor C_L:

L_u = 8ft

L_e = 1.68(8ft) = 13.44ft, (NDS, Table 3.3.3)

Slenderness ratio $R_B = \sqrt{\dfrac{l_e d}{b^2}} = \sqrt{\dfrac{13.44ft(12in/ft)(37.5in)}{(8.75in)^2}} = 8.9 \leq 50$ OK

$F_b^* = F_b' = 2{,}530psi$

K_{bE} = 0.610 for laminated timber (NDS, Table F1).

$F_{bE} = [(0.610)(1{,}300{,}000psi)]/(8.9)^2 = 9{,}995psi$

(Note: E_y used in this formula for lateral stability.)

$F_{bE}/F_b^* = (9{,}995psi)/(2{,}530psi) = 4.25$

$$C_L = \frac{1+4.25}{1.9} - \sqrt{\left(\frac{(1+4.25)}{1.9}\right)^2 - \frac{4.25}{0.95}} = 2.76 - \sqrt{7.63 - 4.47} = 0.98$$

New design value for bending:

$F_b' = C_L(C_D F_b) = 0.98(2{,}530psi) = 2{,}479psi$

C_V is not applied simultaneously with C_L; therefore, the required S value should be compared with S_x and not S_v.

Required S_x = (354,816ftlb)(12in/ft)/(2,507psi) = 1,698in³

A 8-3/4X37-1/2 has S_x = 2,051in³ > required S. OK!

Note that both requirements for S (with F_b' adjusted with C_v and then C_L independently) must be satisfied.

A 2in camber can be built in the beam for dead loads and creep deflection.

Problem 22.1

A 6X12 Eastern hemlock timber beam, 10ft long, extends 2ft over a wall. Assuming deflection does not control, which wood grade is required to support a uniform load of 100lb/ft with a point load of 2,800lb applied at the free end?

Problem 22.2

A simple 3-1/2inX18in laminated beam of Douglas fir, grade 16F-V8, spans 24ft. Find the allowable bending stress F_b' and the corresponding moment capacity M_r. Assume the compression edge and the ends of the beam have full lateral support.

Problem 22.3

A Western species laminated beam, 22F-V1, has a span of 20ft, LL = 480plf, and DL = 160plf, excluding beam weight. Calculate the minimum size required and determine the recommended camber.

Problem 22.4

Design a 40ft beam with a roof LL = 40psf and DL = 17psf excluding the beam weight. Find the required size based on M, V, and deflection with an allowable deflection of L/240 for a live load only. What camber would you recommend?

Problem 22.5

Design a 32ft beam, with load being applied by purlins at 8ft o.c. The purlins have an end reaction of a 12,000lb snow load and 4,800lb dead load (excluding beam weight). Design the most economical beam using Douglas fir, grade 16F-V8.

METRIC VERSIONS OF EXAMPLES AND PROBLEMS

CRITERIA FOR METRIC VERSIONS OF LAMINATED BEAMS EXAMPLES AND PROBLEMS

The tables used for wood design (NDS, Southern Pine, Western species) do not exist in metric. Therefore, in this case two criteria are followed:

a. HARD METRIC. Some metric tables have been created, e.g., for lumber sizes and glulam properties based on Canadian standards (Canadian Wood Council). These tables are inserted at the end of each chapter with the examples and problems associated with that chapter. The design procedure depends on how the tables are built, and it can be somehow different in metric. Quantities, coefficients, and formulas for the metric procedure are explained before the examples (this happens also for concrete and masonry).

b. SOFT METRIC. Some tables cannot be recreated in metric, e.g., design values for connectors. In this case it is assumed that the design will have to be in lb-in and then converted to metric if necessary. So the initial data and the results of calculations are inserted in metric in the text, and no full metric version exists.

c. NO METRIC. Problems that are solved directly with a table (e.g., Southern pine joists safe loads) do not make sense in metric and there is no metrication.

USE OF THE GLULAM TABLES

The following glulam beam tables are based on metric sizes of Western wood products for 20f-E Spruce-pine and 24f-E Douglas fir stress grades. The tables give the moment and shear capacity for each size, and EI assuming all adjustment factors = 1.0. If different factors apply, the values in the table must be multiplied by the appropriate factors. The lateral stablity factor C_L for metric problems is calculated as follows.

1. Calculate the slenderness ratio R_B with the metric dimensions.
2. Determine C_L from table of beam stability factors.
3. Multiply M_r from the glulam table by C_L.

Table M22.8 Laminated beam metric properties

Size bxd (mm)	20f-E M_r kNm	SP V_r kN	80mm $EI \times 10^9$ Nmm²	24f-E M_r kNm	DF V_r kN	80mm $EI \times 10^9$ Nmm²
80x114	3.9	9.58	102	4.8	10.9	129
80x152	7.1	12.8	241	8.5	14.6	307
80x190	11.1	16.1	471	13.3	18.2	599
80x228	16.1	19.2	814	19.1	21.9	1,040
80x266	21.7	22.3	1,290	26.1	25.5	1,640
80x304	28.4	25.5	1,930	33.9	29.2	2,450
80x342	35.9	28.7	2,750	42.9	32.8	3,490
80x380	44.4	31.9	3,770	53.1	36.5	4,790
80x418	53.7	35.1	5,020	64.2	40.1	6,380
80x456	63.9	38.3	6,510	76.4	43.8	8,280
80x494	75.1	41.5	8,280	89.6	47.4	10,500
80x532	86.9	44.7	10,300	104.1	51.1	13,100
80x570	99.8	47.9	12,700	119.1	54.7	16,200

Size bxd (mm)	20f-E M_r kNm	SP V_r kN	130mm $EI \times 10^9$ Nmm²	24f-E M_r kNm	DF V_r kN	130mm $EI \times 10^9$ Nmm²
130x152	11.5	20.7	392	13.8	23.7	498
130x190	18.1	25.9	765	21.5	29.6	973
130x228	26.1	31.1	1,320	31.1	35.6	1,680
130x266	35.3	36.3	2,100	42.2	41.5	2,670
130x304	46.7	41.5	3,130	55.1	47.4	3,990
130x342	58.4	46.7	4,460	69.8	53.4	5,680
130x380	72.1	51.9	6,120	86.2	59.3	7,790
130x418	87.2	57.1	8,150	104	65.2	10,400
130x456	104	62.2	10,600	124	71.1	13,500
130x494	122	67.4	13,500	146	77.1	17,100
130x532	141	72.6	16,800	169	83.1	21,400
130x570	162	77.8	20,700	194	88.9	26,300
130x608	185	83.1	25,100	221	94.8	31,900

Table M22.8 Laminated beam metric properties (continued)

130x646	208	88.2	30,100	249	101	38,300
130x684	234	93.4	35,700	279	107	45,400
130x722	260	98.6	42,000	311	113	53,400
130x760	288	104	49,000	345	119	62,300
130x798	318	109	56,700	380	124	72,100
130x836	349	114	65,200	417	130	82,900
130x874	381	119	74,500	456	136	94,700
130x912	415	124	84,600	496	142	108,000
130x950	451	130	95,700	539	148	122,000

			175mm			**175mm**
Size bxd	20f-E	SP		24f-E	DF	
(mm)	M_r kNm	V_r kN	$EIx10^9$ Nmm²	M_r kNm	V_r kN	$EIx10^9$ Nmm²
175x190	24.3	34.9	1,030	28.9	39.9	1,310
175x228	34.9	41.9	1,780	41.8	47.9	2,260
175x266	47.5	48.9	2,830	56.8	55.9	3,600
175x304	62.1	55.9	4,220	74.2	63.8	5,370
175x342	78.6	62.8	6,010	94.1	71.8	7,640
175x380	97.1	69.8	8,240	116	79.8	10,500
175x418	117	76.8	11,000	140	87.8	14,000
175x456	140	83.8	14,200	167	95.8	18,100
175x494	164	90.8	18,100	196	104	23,000
175x532	190	97.8	22,600	227	112	28,800
175x570	218	105	27,800	261	120	35,400
175x608	248	112	33,800	297	128	42,900
175x646	280	119	40,500	335	136	51,500
175x684	314	126	48,100	376	144	6,100
175x722	350	133	56,500	419	152	71,900
175x760	388	140	65,900	464	160	83,900
175x798	428	147	76,300	512	168	97,100
175x836	470	154	87,800	561	176	112,000
175x874	513	161	100,000	614	184	145,000
175x912	559	168	114,000	668	192	164,000
175x950	606	175	129,000	725	200	184,000
175x988	656	182	145,000	784	207	206,000

Table M22.8 Laminated beam metric properties (continued)

			215mm			215mm
Size bxd	20f-E	SP		24f-E	DF	
(mm)	M_r kNm	V_r kN	$EIx10^9$ Nmm^2	M_r kNm	V_r kN	$EIx10^9$ Nmm^2
215x266	58.4	60	3,470	69.8	68.6	4,420
215x304	76.3	68.6	5,180	91.2	78.4	6,590
215x342	96.6	77.2	7,380	115	88.2	9,390
215x380	119	85.8	10,100	143	98	12,900
215x418	144	94.4	13,500	172	108	17,100
215x456	172	103	17,500	205	118	22,300
215x494	201	112	22,200	241	127	28,300
215x532	234	120	27,800	279	137	35,300
215x570	268	129	34,200	321	147	43,500
215x608	305	137	41,500	365	157	52,800
215x646	345	146	49,700	412	167	63,300
215x684	386	154	59,100	462	176	75,100
215x722	430	163	69,500	514	186	88,300
215x760	477	172	81,000	570	196	103,000
215x798	526	180	93,800	628	206	119,000
215x836	577	189	108,000	690	216	137,000
215x874	526	197	123,000	754	225	157,000
215x912	577	206	140,000	821	235	178,000
215x950	631	214	158,000	891	245	201,000
215x988	687	223	178,000	963	255	226,000
215x1026	745	232	199,000	1,040	265	253,000

			265mm			265mm
Size bxd	20f-E	SP		24f-E	DF	
(mm)	M_r kNm	V_r kN	$EIx10^9$ Nmm^2	M_r kNm	V_r kN	$EIx10^9$ Nmm^2
265x342	119	95	9,100	142	109	11,600
265x380	147	106	12,500	176	121	15,900
265x418	178	116	16,600	213	133	21,100
265x456	212	127	21,600	253	145	27,400
265x494	248	137	27,400	297	157	34,900
265x532	288	148	34,200	344	169	43,600

Table M22.8 Laminated beam metric properties (continued)

265x570	331	159	42,100	395	181	53,600
265x608	376	169	51,100	450	193	65,000
265x646	425	180	61,300	508	205	78,000
265x684	476	190	72,800	569	218	92,600
265x722	530	201	85,600	634	230	109,000
265x760	588	211	99,800	703	242	127,000
265x798	648	222	116,000	775	254	147,000
265x836	711	233	133,000	850	266	169,000
265x874	777	243	152,000	929	278	193,000
265x912	846	254	173,000	1,010	290	219,000
265x950	918	264	195,000	1,100	302	248,000
265x988	993	275	219,000	1,190	314	279,000
265x1026	1,070	285	246,000	1,280	326	312,000
265x1064	1,150	296	274,000	1,380	338	348,000
265x1102	1,240	307	304,000	1,480	350	387,000

Example M22.1

Note: There are no equivalent metric tables, so this problem cannot be solved with this procedure.

Example M22.2

A floor timber beam with a 7.32m span, simply supported at the ends, must carry a load of 14.23kN/m. The beam is fully laterally supported. Use Western species with 24f-E stress grade.

The Glulam beam tables give the moment capacity (M_r) and shear capacity (V_r) of various beam sections.

The maximum shear in the beam is $V = (14.23kN/m)(7.32m/2) = 52.1kN$.

The maximum moment of our beam is $M = (14.23kN/m)(7.32m)^2/8 = 95.3kNm$.

A 80x532 has a moment capacity $M_r = 104kNm$ and an area $A = 42,560mm^2$.
$V_r = 51.1kN$, which is close.

The next best choice is a 130x418 with a $M_r = 104kNm$ and $V_r = 65.2kN$.

$A = 54,340mm^2$, so this section is heavier than the previous choice.

As the first is clearly the lightest of all the possible choices from the table, and the shear capacity is sufficiently close, we use an 80x532.

Example M22.3

A double span structure is made of 12.2m-long beams at 12.2m o.c. and supports 12.2m-long purlins 2.44m o.c. The beam is unbraced between purlins. 4.88m-long foam core roofing panels are glued and nailed to the purlins. The roof slope is negligible for structural calculation purposes and the beams are considered horizontal.

Purlin design

Single-span beam 12.2m at 2.44m o.c., full lateral support from roofing panels

Roof loads: snow load = 1440N/m^2

                         dead load = 576N/m^2 excluding beam

Use Western species laminated timber 24-f stress grade.

(a) Loads

   Uniformly distributed load:

   Snow = 1440N/m^2 (2.44 m) = 3514N/m

   DL = 576N/m^2 (2.44m) = 1405N/m

   Estimate of purlin weight:

    Try d = (12200mm)/25 = 488mm, b = 488/4 = 122mm.

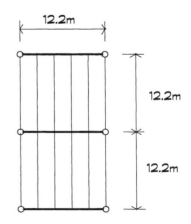

Choose on the safe side 130mm x 532mm, A = 69160x10⁻⁶m²

weight = (5498N/m³)(69160x10⁻⁶m²) = 380N/m

Total load supported by beam = 3514N/m + 1405N/m + 380N/m = 5299N/m

Critical load combination using LDF:

DL only = (1405N/m + 380N/m)/0.9 = 1983N/m

Snow LL + DL = (35299N/m)/1.15 = 4608N/m critical: use C_D = 1.15.

(b) Statics

R = (5299N/m)(12.2m)/2 = 32,324N

Maximum V = 32.324kN

Maximum M = 98.588kNm

(c) Design values

For deflection, the total load controls, since DL = 1785N/m > 50% of the snow LL (even if barely).

Allowable deflection for total load Δ = (12,200mm)/240 = 51mm

w = 5,299N/m

$$EI = \frac{5wL^3}{384\Delta} = \frac{5(5299kN/m)(12200mm)^4}{384(51mm)} = 29971 \times 10^9 Nmm^2$$

Since the roof panels provide continuous lateral support to the compressed fibers, C_L = 1.0. From the tables, a 130X608 has EI = 31,900x10⁹Nmm², M_r = 221 kNm, V_r = 94.8kN > required.

Weight = (130mm x 608mm x 10⁻⁶m²)(5500N/m³) = 435N/m

The purlin is heavier than the initial estimate, but this would cause only a slight increase of the required I calculated above, so no adjustments are necessary. Use a 130X608.

The recommended camber is 1.5 Δ_{DL}

$$\Delta_{DL} = \frac{5[(1405N/m + 435N/m)/1000mm/m](12200mm)^4}{384(31900 \times 10^9 Nmm^2)} = 16.6mm$$

Minimum camber = 1.5(16.6mm) = 25mm

Example M22.4

For the same system in Example M22.3, design the primary beam supporting the purlins. Loads directly over the supports are not considered in M, V, or deflection calculations.

Laminated timber beams are 12.2m o.c. Span of 12.2m with four purlins 2.44m o.c. supporting roof panels.

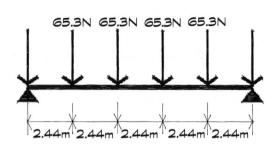

(a) Loads

Concentrated loads P from purlins, tributary area 12.2m×2.44m:

Snow load = (3,514N/m)(12.2m) =	42,870N
Dead load = (1,405N/m)(12.2m) =	1,7141N
Purlin weight = (130×608×10^{-6}m^2)(5500N/m^3)(12.2m) =	5,304N
Dead load total = 17,141N+5,304N =	22,445N
Total load P =	65,315N

Distributed load: weight of beam, Western species, 5500 N/m^3
Estimated beam size:

d = 12,200mm/10 = 1220mm

b = 1,220/4 = 305mm

Select a 315×1216 with A = 0.383m^2

Beam weight (DL) w = 5,500N/m^3(0.383 m^2) = 2,107N/m

Total beam weight W = 12.2m(2,107N/m) = 25,705N

Critical load combination:

DL/0.9 = [4(22,445N) + 25,705N]/0.9 = 128,535N   (total dead load on each beam)

(Snow + DL)/1.15 = [4(65315N) +25705N]/1.15 = 249,535 N   Critical: use C_D = 1.15.

(b) Statics

R = 4P/2 + wL/2 = 4(65315N) /2 + (25705N) /2 = 143,482N

V = 143 kN

M = aPL = 0.6(65.315kN)(12.2m) = 478kNm   (formula from AISC tables, p. 2-295)

Maximum Δ = L/240 = (12200mm)/240 = 51mm for roof without plaster ceiling.

(c) Design values

Adjustment factors:

$C_D = 1.15$

C_L to be determined after first trial section. Assume it is 1.0.

Use Western species 24f-E Stress Grade glulam.

$$EI_p = \frac{0.063(65315N)(12200mm)^3}{51mm} = 146,508 \times 10^9 Nmm^2$$

(d) Section properties

Required EI for snow load + creep point load

$P = 42870N = 0.5(22445N) = 54092N$

$$EI_p = \frac{0.063(65315N)(12200mm)^3}{51mm} = 146,508 \times 10^9 Nmm^2$$

Required EI for distributed load w:

$$EI_w = \frac{5(2.107N/mm)(12200mm)^4}{384(51mm)} = 11,917 \times 10^9 Nmm^2$$

Total required EI = $(146,508+11,917) \times 10^9 Nmm^2 = 158,425 \times 10^9 Nmm^2$

(e) Section selection

Remembering that the required $V = 143kN$, $M = 478kNm$, from glulam tables:

Size	M_r (kNm)	V_r(kN)	EI X10⁹(Nmm²)
175x912	668	192	164,000
215x874	754	225	157,000
265x836	850	266	169,000

Of the three sections, the first has the least size. Now the values must be adjusted with the duration factor C_D and then recalculated with the stability factor C_L (NDS and metric tables).

We first calculate the stability factor C_L with $l_u = 2.44m$:

$l_e = 1.68(2.44) = 4.1m$ (NDS Table 3.3.3)

Slenderness ratio $R_B = \sqrt{\dfrac{l_{ed}}{b^2}} = \sqrt{\dfrac{4100mm(912mm)}{(175mm)^2}} = 11.14 < 50$ __OK__

From the C_L table, by interpolating between 10 and 12, we obtain $C_L = 0.98$.

The new values of M_r and V_r can now be calculated for the 175x912:

$M_r = (668kNm)(1.15)(0.98) = 753kNm$

$V_r = (192kN)1.15 = 221kN$

These values are greater than required: OK. **USE 175×912 24f-E Stress Grade.**

Recommended camber = $1.5 \, \Delta_{DL}$

$$\Delta_P = \frac{0.063(22445N)(12200mm)^3}{164000Nmm^2} = 15.6mm \text{ for the point loads P;}$$

$$\Delta_w = \frac{5(0.185N/mm)(12200mm)^4}{164000Nmm^2} = 3.2mm$$

Total $\Delta_{DL} = \Delta_P + \Delta_w = 15.6mm + 3.2mm = 18.8mm$

Minimum camber = 1.5(18.8mm) = 28.2mm. Use a 30mm camber.

23 COMPRESSION AND TENSION MEMBERS

23.1 Types of compression and tension members

As we noted in steel structural systems, any member that is axially loaded in compression is a "column," a term including what is conventionally known as a column, as well as other members, such as chords and webs of trusses, studs in wall framing, and heavy-timber diagonal bracing (Fig. 23.1). Tension members are equally common and inconspicuous at the same time: diagonal bracing in walls and roof systems, truss members, and ties or "collar beams" of rafters are some examples.

The old Eastern braced frame (also called the "barn frame," because of the similarity with heavy-timber framing used in barn construction); was still in use in the 1940s, particularly in New England. This required heavy-timber corner posts continuously for two floors, with 4X6 [89X140mm] girts mortised and tenoned to the posts. This joinery work resulted in oversizing the entire column to make up for the reduction in area where the mortises were carved. As the name suggests, the posts were braced with 3X4 [63X89mm] members cutting diagonally through the studs.

A bracket supporting a canopy works like an elementary truss, the diagonal member is axially loaded in compression, whereas the horizontal member is in tension. Note that a reversal of the braces also reverses the forces in all the members. The NDS (p. 13) classifies columns according to their type of construction.

Fig. 23.1 Half-timber framing in Germany

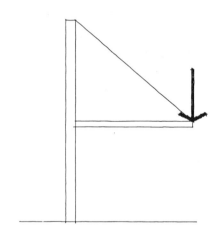

Fig. 23.2 Canopy bracing system

(a) **Solid wood columns;** consisting of a single piece of sawn lumber or laminated timber. Some examples of these columns are wall studs, truss members, and heavy-timber columns.

(b) **Spaced columns,** formed by two or more individual members with their longitudinal axis parallel, spaced by blocking at the ends and intermediate points, and connected together at the blocking. This is also common in trusses.

(c) **Built-up columns,** made of individual pieces nailed or bolted together. This technique is common for small columns, such as multiple studs to support beams in residential framing. Laminated timber is used for larger structures, especially where architectural appearance is a consideration.

Round columns are designed as solid square columns of the same cross-sectional area. These types of columns are shown in Fig. 23.3.

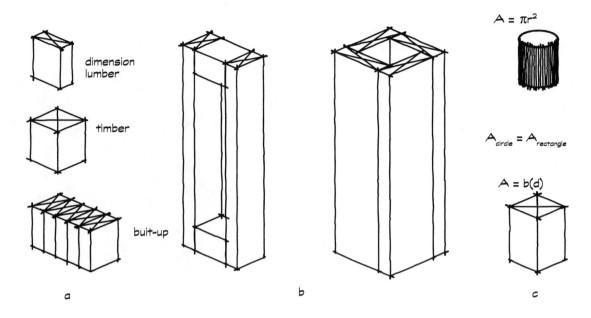

Fig. 23.3 Variations of solid columns (a), spaced columns (b), and round to square column equivalents (c)

The way columns are framed and connected at their ends affects the slenderness ratio, and therefore their load-bearing capacity, just as it does in steel members. Nailed or bolted connections, which can be used in combination with metal plates and column bases, can be considered as pinned connections. Only large laminated timber columns can, in practice, be connected to beams to form a rigid, moment-resisting joint.

23.2. Design procedure for solid columns

Conceptually, the design of columns is based on $F = P/A$. Axial load P applied to column area A creates a compression stress: $f_c = P/A$. That stress cannot be greater than the allowable stress F_c'. This allowable stress depends on the strength of the wood material and on the slenderness ratio of the column; the basic allowable compressive stress F_c is found in the NDS tables and must be multiplied by the appropriate adjustment factors, which also take slenderness into account. In practice, the design is a trial-and-error process, starting with the selection of a column size, and checking if the working stress f_c is not greater than the allowable stress F_c'. This design procedure is illustrated in the following paragraphs.

(a) **Design values.** Wood has very different strengths for compression parallel or perpendicular to the grain. Recall the discussion of steel columns; tubes make excellent column sections. Wood sections are a collection of tubes that exhibit the strength characteristics you would expect from a tube loaded parallel to its axis. At the same time, if a tube section is loaded in compression from the side, it can collapse rather easily. This is the exact parallel with a wood piece. Compressive strength values change also according to the NDS Supplement classification of dimension lumber, timbers, and laminated timbers (see Chapter 18).

(b) **Adjustment factors.** Table 23.1 indicates the factors applicable to the allowable stress for compression parallel to the grain, F_c.

Table 23.1 Adjustment factors applicable to columns (axial compression only)

Symbol	Definition and applicability	NDS Supplement reference	Adjusts
C_D	Load duration	Table 2.3.2	F_c
C_M	Wet service	Tables of design values	
C_t	Temperature	Table 2.3.4	
C_P	Column stability	Sec. 3.7.1	
C_F	Size factor	Tables of design values for dimension lumber only For timber, see Sec. 4.3.2.	

The new factor introduced here is the column stability factor, which depends on the slenderness ratio L_e/d of the column, where L_e is the effective length, and d is the least column dimension. The slenderness ratio for a solid wood column is limited to 50.

$$L_e \leq 50$$

The slenderness ratio must be taken as the larger of the ratios L_{e1}/d_1 and L_{e2}/d_2, whenever the column has different unbraced lengths with respect to the X-X and Y-Y axes. C_P is calculated with the formula (NDS, Sec. 3.7.1, Eq. 3.7-1):

$$C_P = b - \sqrt{b^2 - a/c}$$

where

a = F_{cE}/F_c^* (F_c^* = compression design value multiplied by all adjustment factors **except** C_P)

b = (1 + a)/2c

c = 0.8, for sawn lumber

c = 0.9, for laminated timber

F_{cE} is the stress calculated with a variation of Euler's formula for long columns.

An aside:

$$F_E = P_a/A = \pi^2 E/(L/r)^2$$

where

L = effective length

r = radius of gyration, divided by a safety factor

The formula becomes, with NDS symbols:

$$F_{cE} = K_{cE}E'/(L_e/d)^2$$

where

K_{cE} = 0.3, for sawn lumber

K_{cE} = 0.418, for laminated timber

This stress depends only on the modulus of elasticity of the lumber and not on its compressive strength, since the column will fail in buckling before the maximum compressive stress is reached. In other words, a very strong wood will not make a slender column strong in compression (remember the similar condition in steel).

Furniture design has an obvious analogy with wood framing, and is frequently modeled on the post-and-beam system, with legs acting as columns and the seat as a beam-diaphragm system. The Louis XV chair is an essay in elegance and stability: The legs are short and are flowing into the horizontal frame, helping to create a rigid connection. The American Shaker chair minimizes the use of material and labor, and relies on multiple horizontal members to brace the legs and reduce the slenderness. As in some buildings, the rigid joints allow the frame to resist lateral loads.

It may appear logical to use this modified Euler's stress as the allowable compressive stress, adjusted by the remaining factors, or $F_c' = (C_D C_m C_t \, C_F) F_{cE}$; in reality, this would result in a value much higher than the material can safely carry. The additional factor C_p is obtained with the help of empirical analysis.

The table of buckling length coefficients K in Table 23.2 can be used to calculate the effective length, depending on end conditions, with the formula:

$$L_e = KL$$

where L is the actual column length.

Table 23.2 Recommended K values for typical end conditions.

Case 4 is the base to which everything is related. Note that condition 6 uses a significantly different K value than we used for the same condition in Steel.

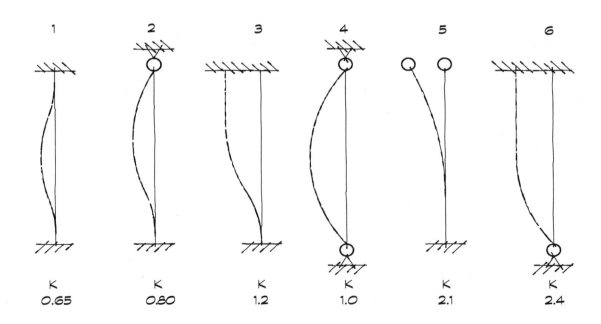

1	2	3	4	5	6
K	K	K	K	K	K
0.65	0.80	1.2	1.0	2.1	2.4

The adjusted value of the allowable compressive stress is, therefore:

$F_c' = F_c(C_D C_M C_t C_P C_F)$ for dimension lumber and timbers

$F_c' = F_c(C_D C_M C_t C_P)$ for Glulam members

Example 23.1

A two-story house is built with a platform frame system using 2X4 studs. Studs will ordinarily carry the loads from the floors and roof. We want to check, first of all, how much gravity load an 8ft-long stud, nailed to the wall plates, can safely support. We will assume the compression load is perfectly axial. Stud area $A = 5.25in^2$

Use Douglas fir-South, Stud grade: $F_c = 825psi$; $E = 1,100,000psi$.

Applicable adjustment factors: $C_D = 1.15$(snow), C_M and $C_t = 1.0$, $C_F = 1.05$ (from NDS, Table 4A), and C_P must be calculated. The slenderness ratio must be checked first. We can assume that the stud is nailed to the top and bottom (pin connections) and also to the exterior sheathing and siding, so that it will not buckle about its weak axis.

Therefore, the least dimension d that controls the buckling is the nominal 4in.

$L_e/d = (8ft)(12in/ft)/(3.5in) = 27.4 < 50$ OK!

$F_{cE} = 0.3(1,100,000psi)/27.4^2 = 438.6psi$

$F_c^* = F_c(C_D C_M C_t C_F) = (825psi)(1.15)(1.0)(1.0)(1.05) = 996psi$

$a = F_{cE}/F_c^* = (438.6psi)/(996psi) = 0.44$

$b = (1 + a)/2c = (1 + 0.44)/(2)(0.8) = 0.9$

$C_p = b - \sqrt{b^2 - a/c} = 0.9 - \sqrt{0.9^2 - 0.44/0.8} = 0.39$

$F_c' = F_c^* C_p = (996psi)(0.39) = 338psi$

$P_a = F_c'A = (338psi)(5.25in^2) = 2,040lb$

Based on our assumptions, this is the gravity load each stud can safely carry. It is unlikely that in residential construction studs spaced at 16in o.c. will have to support this load. A total floor load of 50psf with 18ft span joists would give a load on each stud $P = (50psf)(9ft)(16in)/(12in/ft) = 598lb$, and 2X4 framing could be considered safe for approximately three floors plus the roof. In reality, we will see that studs may be subject to bending combined with the axial load, and this limits their safe compression load.

Example 23.2

A laminated timber column for a sports arena in Denver supports a 28k axial compression load. The unbraced length is different in the two main directions. The conditions are dry, $C_D = 1.15$ (snow load combination critical).

Design values (NDS, Supplement, Table 5B):

    Combination 1, Douglas fir, has

    E = 1,500,000psi

    F_c = 1,550psi

Before C_p can be calculated, a trial section is selected with an area

    A = 3(P/F_c)

    A = 3(28,000lb)/(1,550psi) = 54in²

assuming arbitrarily that the C_p will reduce the F_c by around 0.33.

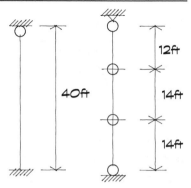

From NDS, Supplement, Table 1C:

    5-1/8in x 10.5in has A = 53.81in²

The slenderness ratio is different with respect to the X-X and Y-Y axes:

    KL_y/d_y = (1.0)(14ft)(12in/ft)/(5.125in) = 32.8 < 50

    KL_x/d_x = (0.8)(40ft)(12in/ft)/(10.5in) = 36.6

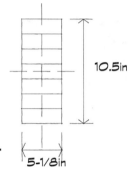

We can make the column more mathematically square by increasing d_x to 12in.

    KL_x/d_x = (0.8)(40ft)(12in/ft)/(12in) = 32, which is close to 32.8.

The slenderness ratio about the Y-Y axis is the greater.

We can now calculate C_p with the NDS equation.

    c = 0.9

    K_{cE} = 0.418

    $$F_{cE} = \frac{K_{cE}E'}{(l_e/d)^2} = \frac{0.418(1,500,000psi)}{32.8^2} = 582.8psi$$

    F_c^* = (1,550psi)1.15 = 1,782.5psi

    F_{cE}/F_c^* = (582.8psi)/(1,782.5psi) = 0.327

    $$b = \frac{1+0.327}{2(0.9)} = 0.737$$

    $$C_p = 0.737 - \sqrt{0.737^2 - 0.327/0.9} = 0.313$$

The adjusted allowable stress is

    F_c' = F_c(C_p) = (1,782.5psi)0.313 = 557.2psi

$$f_c = P/A < F_c'$$

$$f_c = (28,000\text{lb})/(61.5\text{in}^2) = 455\text{psi} < 557.2\text{psi} \qquad \text{OK!}$$

(The gross area is used for compression members even if connectors are used at joints, which may decrease the net column area.)

23.3. Built-up and spaced column design

A column created by fastening lumber pieces together with nails, bolts, or other mechanical fasteners will not carry as much load as a solid column of the same size because of the shear stresses at the fastening points between the lumber faces. This is essentially the same problem of a composite or built-up beam, where the connection between the web and flange has to resist horizontal shear. Since the column will be subjected to some bending as well as compression, it is essential to ensure that the different lumber pieces can work together. An increase of load-bearing capacity can be achieved by nailing cover plates to the edges of the layers or by boxing around a timber core. Figure 23.4 shows different techniques of building up a column using lumber pieces.

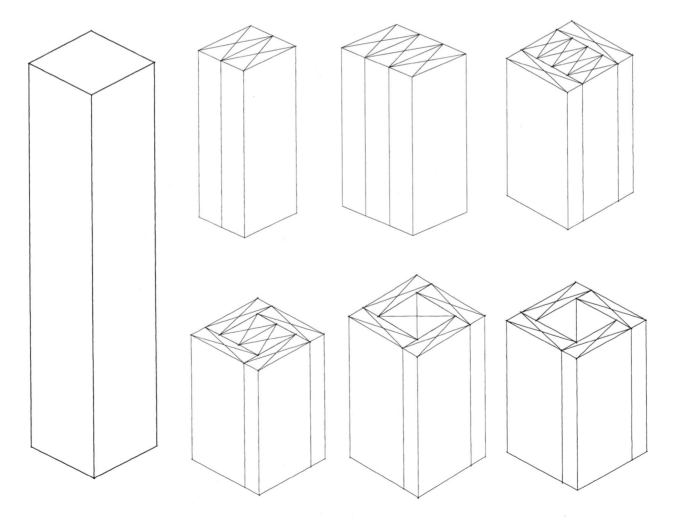

Fig. 23.4 Alternative designs to solid wood columns

Current recommendations indicate that whatever arrangement of mechanical fasteners is employed, no built-up column will have a strength equal to its equivalent solid column. The equivalent solid column refers to a one-piece solid column of the same overall dimensions, end conditions, and material as those of the built-up column. The stiffness of a built-up column is also less than that of its equivalent solid column, but considerably greater than that of an unconnected assembly in which individual members are treated as independent columns without any sharing of load between them. The strength of a built-up column can be expressed as a percentage of the strength of its equivalent solid column. These values are valid if the individual planks are not wider than 5 times the thickness, and if the spikes or nails are not spaced more than 6 times plank thickness. Although nailing is the most common technique for constructing a built-up column, laminations may also be bolted. For complete specifications see the NDS, Sec. 15.3.

Table 23.3 Strength of built-up columns as percentage of solid column strength

Slenderness ratio L_e/d	Strength compared to solid equivalent column
6	82%
10	77%
14	71%
18	65%
22	74%
26	82%

Spaced columns consist of two or more individual members with their longitudinal axes parallel and separated at the ends and at intermediate locations by blocking (Fig. 23.5). This is a common condition in residential construction at lintels or for other light column situations. The individual components are joined at the ends by connectors capable of developing the required shear between the members and blocking (end blocks). If a single spacer is provided at the center of the column, only bolts need be used. When two or more spacer blocks are used, timber connectors are required. Spaced columns are used as compression chords, compression web members of trusses, and supporting columns. When compression chords of a truss are considered as spaced columns, panel points that are braced laterally are taken as the ends of spaced columns. The truss web members between the individual components of the assembly can function as end blocks. The spacer and end blocks must be at least as thick as the members they are connecting, and must also comply with the requirement for split-ring or shear plate connectors.

The design of spaced columns is similar to that of a group of simple columns of rectangular cross-section with fixed ends. The design load is the sum of the design loads for each of the individual members (NDS, Sec. 15.2). The individual members can be considered as independent columns, each loaded with $P/2$, when P is the concentric compressive load. $P/2$ is transferred from the end block to the split ring or shear plate in the individual members (connectors are discussed in Chapter 26). This load can be considered axial. The end conditions are pins relative to the strong axis (bending in the plane of the wide face or dimension d_2) and fixed about the weak axis (bending in the plane parallel to the short dimension d_1). Figure 23.6 shows the possible buckling modes in these two planes. The capacity in the plane of the wide face is the same as that of a solid column with a dimension $d_2 \times 2d$. The allowable load P_2 depends, therefore, on the slenderness KL_2/d_2, and has a maximum slenderness ratio of 50. In the other direction, the allowable load depends on the slenderness ratio KL_1/d_1 of the individual compressed member, and has a maximum slenderness KL_1/d_1 of 80. A greater slenderness ratio is allowed for spaced columns because of the end fixity (rigidity) caused by the connectors and end blocks. This design procedure assumes a shear plate connection at the end blocks.

The allowable load is $\quad P_a = F_c'A \quad$ or $\quad A = P_a F_c'$

where F_c' is the adjusted allowable compressive stress parallel to the grain. The design value is obtained by multiplying F_c by all applicable factors, including the column stability factor C_p, which depends on

$$F_{cE} = \frac{K_{cE}K(E')}{(L_e/d)^2}$$

The difference with solid columns is in the coefficient K_x, which depends on the fixity conditions of the end restraints. The end condition can be such as to have

(a) A connector center located $L_1/20$ or less from the end of the column. In this case, $K_x = 2.5$.

(b) A connector center located between $L_1/20$ and $L_1/10$ from the end of the column. In this case, $K_x = 3.0$.

Additional requirements have to be satisfied for the connections (NDS, Sec. 15.2.2.5): refer to Chapter 26.

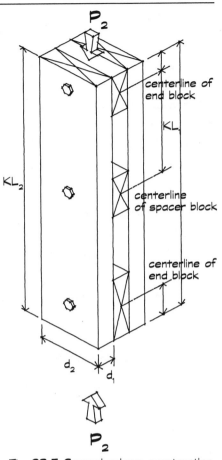

Fig. 23.5 Spaced column construction

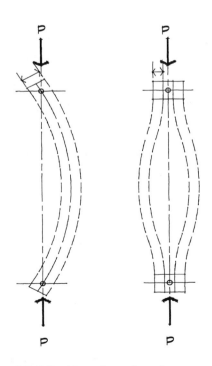

Fig. 23.6 Buckling of members in a spaced column

Example 23.3

A spaced column is constructed, as illustrated, using a pair of 2X6 1200f MSR (machine stress rated) lumber. The column has an axial load of 5,000lb applied at the top. Assume C_D, C_M, and C_t = 1.0. Determine if the column is adequately sized. We must check the entire column, as well as the single 2X6 between the spacers. C_F in this case does not apply.

The design values are $F_c^* = 1400$psi, and $E = 1,500,000$psi.

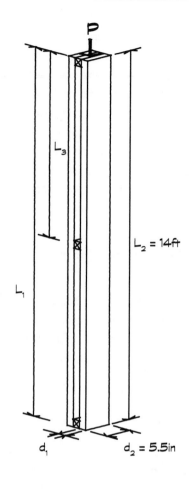

(a) The maximum slenderness ratio for the entire column assembly (potential buckling about X-X axis) is

L_2/d_2 = (14ft)(12in/ft)/(5.5in) = 30.5 < 50　OK!

(b) The maximum slenderness ratio for the entire column made of two 2X6s (potential buckling of individual 2X6 about Y-Y axis), is

$L_1/d_1 \leq 80$.

L_1/d_1 = (14ft)(12in/ft)/(1.5in) = 112　NO!

We need to increase the size of the layers to

$d_1 = L_1/80$ = (168in)/80 = 2.1in　Use 3X6 nominal lumber.

L_1/d_1 = (168in)/(2.5in) = 67.2 < 80

(c) The maximum slenderness ratio for single layers between connections is

$L_3/d_1 < 40$

Find the maximum distance between spacers:

$L_3 = 40d_1$ = 40(2.5in) = 100in = 8ft-4in

Use one spacer at midspan, 7ft from the ends.

L_3/d_1 = (7ft)(12in/ft)/(2.5in) = 33.6 < 40

Calculate the column stability factor with the maximum slenderness value:

$$C_p = b - \sqrt{b^2 - a/c}$$

$F_{cE} = K_{ce}K_x E/(L_1/d_1)^2$ = 0.3(2.5)(1,500,000psi)/67.2² = 249psi

　　$a = F_{cE}/F_c^*$ = (249psi)/(1,400psi) = 0.18

　　$b = (1 + a)/2c$ = (1 + 0.18)/(2)(0.8) = 0.74

$$C_p = 0.74 - \sqrt{0.74^2 - 0.18/0.8}　= 0.17$$

Allowable compression stress:　　F_c' = 0.17(1400psi) = 241psi

The total capacity of the column is

$P_a = F_c'A$ = (241psi)(2×13.75in²) = 6,623lb > 5,000lb　OK!

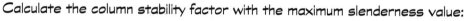

23.4 Tension members

Tension members have no buckling problems, so they are very easy to design using the direct stress formula:

$$f_t = P/A < F_t'$$

where

P = tension load

A = net area of the tension member

F_t' = allowable tension stress modified by C_D, C_M, C_t, and C_F

Tension perpendicular to the grain should be avoided for the same reason that shear, a measure of diagonal tension, is critical. The bonds between (perpendicular to) cells (grain) are much weaker than those within (parallel to) the cells. At connections, however, members can be loaded perpendicular to the grain in tension and compression; this is often the case with trusses. The design of the connection is the critical portion and it often determines the sizes of the members.

As mentioned at the beginning of this chapter, tension members are found in building structures under different disguises:

(a) suspension systems supporting floor and roof beams (a modest system of this sort is used to reduce the span of ceiling joists in roof systems)

(b) ceiling joists that act as ties for the three-hinged arches of many residential roof systems

(c) truss members

(d) diagonal bracing systems for walls and roofs

This last system is used sometimes during construction when structural panel sheathing (plywood or other panel materials) is later nailed to the rafters or trusses.

23.5 Axial compression and bending

Pure axial compression in columns and truss members is not always possible to achieve. In columns, the actual point of application of the vertical load depends on the design of the connection or support of the horizontal members. In the example of Fig. 23.7, a floor joist end is subject to bearing pressures; the resultant of these pressures is the support reaction of the wall plate. The point of application of this resultant is likely to be at a distance "e" from the axis of the stud, so that the stud experiences a vertical force P plus a bending moment M = Pe.

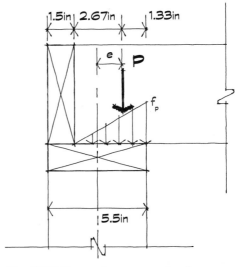

Fig. 23.7 Eccentric compression force

Studs and columns often resist wind loads as well as vertical loads. In this case, the bending moment is M = wL²/8, where w is the wind load (Fig. 23.8).

Truss members are generally subject to bending due to roof sheathing being applied continuously to the compression (upper) chord. Simple roof joists will experience both moment and compression as a result of three-hinged arch construction. In larger-scale wood structures, intermediate framing may be used to reduce decking or sheathing spans. If this framing does not always attach to panel points (joints) in the truss, it will inevitably create moments in the member it bears on. When secondary members are placed between joints, the bending moments have to be considered with the axial loads. Ceiling loads on bottom chords, including HVAC equipment, are also examples of combined axial load (in this case tension) and bending (Fig. 23.9).

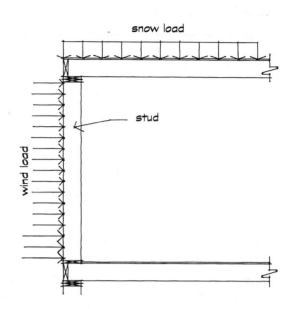

Fig. 23.8 Lateral load and compression

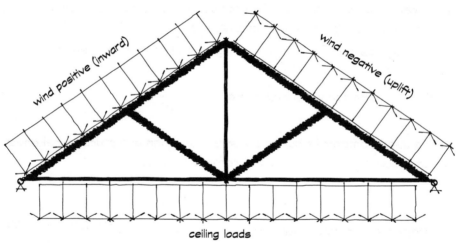

Fig. 23.9 Truss members

The general interaction formula in the NDS, Sec. 3.9.2, valid for axial compression and biaxial bending, can be simplified for bending in one plane only:

$$\frac{f_c}{(F_c')^2} + \frac{f_b}{F_b'[1.0 - (f_c/F_{cE})]} \leq 1.0$$

where

f_c = P/A, actual compression stress

F_c' = $F_c^*(C_p)$, allowable compression stress parallel to the grain based on the largest L_e/d value

$F_{cE} = K_{cE}E/(L_e/d_1)^2 > f_c$ is Euler's stress in the plane of bending, and L_e/d_1 is the slenderness ratio in the plane of bending

$f_b = M/S$, bending stress

F_b' = allowable bending stress F_b multiplied by all applicable adjustment factors, except that C_F and C_L are not applied cumulatively for sawn lumber (see Chapter 19)

For laminated timber, C_v is used to determine $f_b = M/S$.

The term $1.0/[1.0 - (f_c/F_{cE})]$ in the interaction equation is an amplification factor that considers the additional bending stresses caused by the compression load (P) when the member deflects (e) under the action of the loads perpendicular to the beam axis ($M = Pe$). You'll remember this same concept being applied to steel members under combined compression and bending.

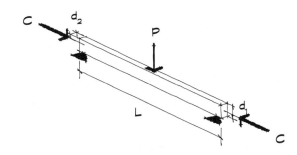

23.6 Axial tension and bending

If the member is subject to axial tension combined with bending, the amplification factor **DOES NOT APPLY** and the general interaction formula becomes:

$$\frac{f_t}{F_t^2} + \frac{f_b}{F_b'} \leq 1.0$$

where

$f_t = P/A$, actual tensile stress

F_t' = allowable tensile stress

Example 23.4

A Pratt truss is designed to carry the roof loads distributed by 19 purlins 4.5ft o.c. The purlins located at the center of each panel represent a point load P on the upper chord members:

Loads on truss at each purlin: P = 3695.7lb (3.7k).

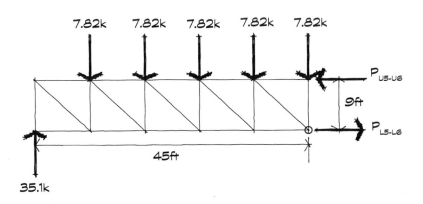

left half of truss with point loadings

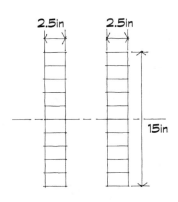

double upper chord of truss

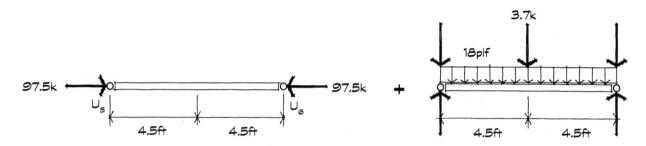

axial forces resulting from truss action   bending forces resulting from beam action

The chord member with the highest compression load is U_5U_6 with 97,525lb (97.5k). This member is also subject to bending due to the purlin load P and to the weight of the timber chord w. Assuming a chord size of 2-1/2in x 15in previously determined by the axial load and connection design, the distributed dead load due to the weight of wood is

$$w = 2.5in(15in)(35pcf)/(144in^2/ft^2) = 9plf \times 2 \text{ members} = 18plf$$

The actual compression stress is $f_c = P/A = (97,525lb)/(2)(2.5in)(15in) = 1,300psi$

The allowable compression stress is $F_c' = F_c(C_DC_p)$. This allowable stress must be calculated for the largest slenderness ratio L_e/d of the chord. The chord has two different unbraced lengths:

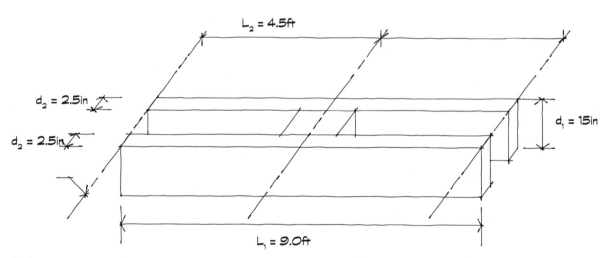

(We are assuming that the distance between the upper chord members is such that the overall width will be less than 15in.)

X-X axis (K = 1.0):

$$L_e/d_1 = (9ft)(12in/ft)/(15in) = 7.2 < 50$$

Y-Y axis (K = 1.0):

$$L_e/d_2 = (4.5ft)(12in/ft)/(2.5in) = 21.6 < 50$$

Example

413

The largest slenderness ratio is about the Y-Y axis. The adjustment factor C_p is now calculated with NDS Eq. 3.7-1. The timber must also be selected: laminated timber, Western species, combination 12. The design values are as follows:

$F_{c(perpendicular)}$ = 560psi (Group B)

F_c = 1950psi

F_b = 2200psi

E = 1,800,000psi

Adjustment factor C_D = 1.15 (snow)

F_{cE} = 0.418(1,800,000psi)/21.6² = 1,612.6psi

F_c^* = (1,950psi)1.15 = 2,242.5psi

F_{cE}/F_c^* = 1,612.6psi/2,242.5psi = 0.719

c = 0.9

$$C_p = \frac{1+0.719}{2(0.9)} - \sqrt{\left(\frac{(1+0.719)}{2(0.9)}\right)^2 - \frac{0.719}{0.9}} = 0.62$$

F_c' = (2,242.5psi)0.62 = 1,390.3psi > f_c OK!

The member works as a compression member. Now we have to check the interaction of compression with bending. The maximum bending moment at midspan is

$$M = \frac{wl^2}{8} + \frac{Pl}{4} = \frac{9plf(9ft)^2}{8} + \frac{3,695.7lb(9ft)}{4} = 91.1ftlb + 8315.3ftlb = 8406.4ftlb$$

$$S = \frac{bh^2}{6} = \frac{(2.5in + 2.5in)(15in^2)}{6} = 187.4in^3$$

$$f_b = \frac{M}{S} = \frac{8406.4ftlb(12in/ft)}{187.5in^3} = 538psi$$

$$F_b^* = 2,200psi(1.15) = 2,530psi$$

The adjustment factor C_L has to be calculated with NDS Eq. 3.3-6 (see Chapter 22). The highest bending stress in this case corresponds to the midspan joint with the purlin, which holds the compressed edge of the member in place. The unbraced length L_u = (4.5ft)(12in/ft) = 54in From NDS, Table 3.3.3:

L_e = 1.11(54in) = 60in

$$R_B = \sqrt{\frac{L_e d}{b^2}} = \sqrt{\frac{60in(15in)}{(2.5in)^2}} = 12 \leq 50$$

$$F_{bE} = 0.609(1,800,000psi)/12^2 = 7,612psi$$

$$F_{bE}/F_b^* = 3.0$$

$$C_L = 2.1 - \sqrt{2.1^2 - 3.0/0.95} = 0.98$$

$$F_b' = (2,530psi)0.98 = 2,479psi$$

The volume factor C_V does not apply simultaneously with C_L (NDS, Sec. 5.3.2). Therefore:

$F_b' = F_b^*(C_V)$ or $F_b' = F_b^*(C_L)$, whichever is the least.

In other words, C_L does not cumulate with C_V.

In this example, C_V is > 1, or see Table 23.3; therefore, C_L only applies.

In the interaction formula, the quantity F_{cE} in the amplification factor depends on the depth of the member on the plane of bending and therefore on $l_e/d_1 = 7.2$:

$$F_{cE} = 0.418(1,800,000psi)/7.2^2 = 14,514psi.$$

$$f_c/F_{cE} = (1300psi)/(14,514psi) = 0.09$$

The interaction formula becomes:

$$[(1300psi)/(1390.3psi)]^2 + (538psi)/[(2479psi)(1-0.09)] = 0.87 + 0.24 = 1.11 > 1 \quad NO!$$

It is evident that the 1300psi working stress in compression is too close to the allowable stress of 1390.3psi. We can increase the allowable stress F_c' by decreasing the slenderness ratio about the weak Y-Y axis. Instead of a 2.5in thickness, a 3.125in × 15in section is used, and the new value of F_c' is calculated:

$$L_e/d = (4.5ft)(12in/ft)/(3.125in) = 17.28$$
$$F_{cE} = 2519.7psi$$
$$F_{cE}/F_c^* = 1.12$$

$$C_p = C_p = 1.18 - \sqrt{1.18^2 - 1.12/0.9} = 0.796$$
$$F_c' = 1785 \ psi$$

It can be noticed that a 25% increase of area has produced a substantial increase (30%) in allowable stress. This also has the effect of increasing the C_L factor and consequently the F_b' value. It is not necessary to recalculate F_b', since the interaction equation is certainly satisfied:

$$(f_c/F_c')^2 = [(1300psi)/(1785psi)]^2 = 0.53$$

Inserting this in the previous interaction equation:

0.53 + 0.24 = 0.77 < 1   OK!

Use two 3-1/8in × 15in laminated timber, Western species Combination 12, for the upper chord.

An example of a compression and tension system with bending is a pavilion roof, with hip beams and perimeter tie beams. Each sloping beam is at the same time a part of an arch (the roof is made of two 3-hinged arches connected at the crown point), and thus in compression, and part of a beam system with triangular distributed loads . The perimeter needs tension elements to hold the base of the arches in place. The entire system is like a three-dimensional truss. The tension on the ties can be calculated easily using a free-body diagram.

Problem 23.1

A 4X4 column built with solid Southern pine Stud grade material has a length of 7ft-6in and is nailed to the top and bottom plates. You may consider this a pinned connection.

(a) Calculate the allowable axial load.

(b) Calculate the allowable axial load of this column that is built up using two 2X4's adequately nailed.

Problem 23.2

A wood frame load-bearing wall is loaded with a uniform load of 920plf. The studs are 2X6, Eastern hemlock stud grade, 24in o.c., 8ft long (high). Are they adequate?

Assume the studs are fully braced by sheathing about the weak axis, but are pinned about the strong axis at both top and bottom. What would be the maximum permissible spacing of the studs?

Answer:

Allowable P = 2492lb > 1840lb; maximum spacing = 32.5in

Problem 23.3

A glulam bracket supports an overhang beam, as shown in the illustration. Use a Southern pine Combination Symbol 45 (F_c = 850psi; E = 1,100,000psi) glulam, size 6-3/4inX6-7/8in, pinned at B and C, and verify the adequacy of the system. Assume C_D = 1.0 and ignore the glulam weight.

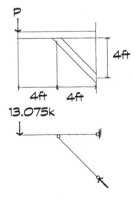

Answer:

Allowable P = 38.4k > 37k compression load

Problem 23.4

A double-leaf spaced column, 8ft long, is constructed of Southern pine No. 2 Nondense. The two leaves are 4X6's connected at the ends with spacers and shear plates. The axial load is 15.8k. Determine if the column is adequate, with C_D = 1.15.

Problem 23.5

The spaced column in the illustration is built with 1200f MSR dimension lumber. Calculate the allowable axial compression load based on the slenderness of the entire column and of the individual leaves.

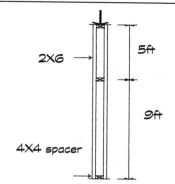

2X6

5ft

9ft

4X4 spacer

Answer:

Allowable P = 5,808lb

Problem 23.6

A 2X6 stud, 8ft long, Eastern white pine No. 3, is subject to an axial load of 720lb and a lateral load of 35plf. The sheathing is nailed to the narrow dimension of the stud and acts as bracing. Determine if the stud is adequate (ignore the effect of the sheathing on the stud for axial and bending stresses).

Problem 23.7

The glulam column in the illustration is built with a 6-3/4inX18in Western Species Combination Symbol 17. C_D = 1.15 for axial load, C_D = 1.6 for lateral load, and C_L = 1.0. Assume the ends are pinned. Determine its load capacity.

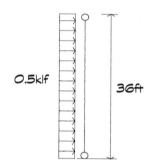

0.5klf

36ft

Problem 23.8

A braced column is built with 4X6 aspen No. 2. Assume all connections are pinned and determine the allowable axial load.

Answer:

P = 5,935lb

Problem 23.9

A tension chord of a truss is built with Douglas fir No. 1. The chord has a 17.5k axial tension and a 300plf uniform gravity load (due to ceiling and HVAC). Find the minimum required size of the member.

METRIC VERSIONS OF ALL EXAMPLES AND PROBLEMS

Example M23.1

A two-story house is built with a platform frame system using 38X89 studs. Studs will ordinarily carry the loads from the floors and roof. We want to check, first of all, how much gravity load a 2.44m-long stud, nailed to the wall plates, can safely support. We will assume the compression load is perfectly axial. Stud area A = 3382mm²

Use Douglas fir-South, Stud grade: F_c = 5.7MPa; E = 7,600MPa

Applicable adjustment factors: C_D = 1.15 (snow), C_M and C_t = 1.0, C_F = 1.05 (from NDS, Table 4A), and C_p must be calculated. The slenderness ratio must be checked first. We can assume that the stud is nailed to the top and bottom (pin connections) and also to the exterior sheathing and siding, so that it will not buckle about its weak axis.

Therefore, the least dimension d that controls the buckling is 89mm.

L_e/d = (2440mm)/(89mm) = 27.4 < 50 OK!

F_cE = 0.3(7600MPa)/27.4² = 3.0MPa

F_c^* = $F_c(C_DC_MC_tC_F)$ = 5.7MPa(1.15)(1.0)(1.0)(1.05) = 6.9MPa

a = F_cE/F_c^* = (3.0MPa)/(6.9MPa) = 0.44

b = (1 + a)/2c = (1 + 0.44)/(2)(0.8) = 0.9

$C_p = b - \sqrt{b^2 - a/c} = 0.9 - \sqrt{0.9^2 - 0.44/0.8} = 0.39$

F_c' = $F_c^*C_p$ = (6.9MPa)0.39 = 2.7MPa

P_a = $F_c'A$ = (2.7MPa)(3382mm²) = 9,100N

Based on our assumptions, this is the gravity load each stud can safely carry. It is unlikely that in residential construction studs spaced at 406mm o.c. will have to support this load. A total floor load of 2400N/m² with 5.5m span joists would give a load on each stud P = (2400N/m²)(5.5m/2)(0.406m)= 2,680N, and 38X89 framing could be considered safe for approximately three floors plus the roof. In reality, we will see that studs may be subject to bending combined with the axial load, and this limits their safe compression load.

Example M23.2

A laminated timber column for a sports arena in Denver supports a 124kN axial compression load. The unbraced length is different in the two main directions. The conditions are dry, C_D = 1.15 (snow load combination critical).

Design values

   Combination 1, Douglas fir, has

   E = 10,300MPa

   F_c = 10.7MPa

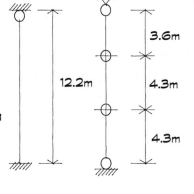

Before C_p can be calculated, a trial section is selected with an area

   A = 3(P/F_c)

   A = 3(124000N)/(10.7MPa) = 3,4767mm²

assuming arbitrarily that the C_p will reduce the F_c by around 0.33.

From tables of glulam shapes:

   130X266 has A = 34580mm²

The slenderness ratio is different with respect to the X-X and Y-Y axes:

   KL_y/d_y = (1.0)(4,300mm)/(130mm) = 33 < 50

   KL_x/d_x = (0.8)(12200mm)/(266mm) = 36.7

We can make the column more mathematically square by increasing d_x to 304mm:

   KL_x/d_x = (0.8)(12200mm)/(304mm) = 32.1, which is close to 33.

The slenderness ratio about the Y-Y axis is the greater.

We can now calculate C_p with the NDS equation.

   c = 0.9

   K_{cE} = 0.418

   $F_{cE} = \dfrac{K_{cE}E'}{(L_e/d)^2} = \dfrac{0.418(10300MPa)}{33^2} = 3.9MPa$

   F_c^* = (10.7MPa)(1.15) = 12.3MPa

   F_{cE}/F_c^* = (3.9MPa)/(12.3MPa) = 0.317

   $b = \dfrac{1 + 0.317}{2(0.9)} = 0.73$

   $C = 0.732 - \sqrt{0.732^2 - \dfrac{0.317}{0.9}} = 0.30$

The adjusted allowable stress is

   F_c' = F_c (C_D)(C_p) = 12.3MPa(0.30) = 3.7MPa

   f_c = P/A < F_c'

f_c = 124000N/34767mm² = 3.57MPa < 3.73MPa   OK!

Note that the gross area is used for compression members even if connectors are used at joints, which may decrease the net column area.

Example M23.3

A spaced column is constructed as illustrated, using a pair of 38X140, 8.3MPa MSR (mechanically graded) lumber. The column has an axial load of 22.24kN applied at the top. Assume C_D, C_M, C_t = 1.0. Determine if the column is adequately sized. We must check the entire column, as well as the single 38x140 between the spacers. C_F in this case does not apply.

The design values are F_c^* = 9.7MPa, E = 10300MPa.

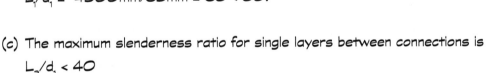

(a) The maximum slenderness ratio for the entire column assembly (potential buckling about the X-X axis)

L_2/d_2 = 4300mm/140mm = 30.7 < 50 OK !

(b) The maximum slenderness for entire column made of two 38x140s (potential buckling of individual 38x140 about the Y-Y axis) is

$L_1/d_1 \leq 80$

L_1/d_1 = 4300mm/38mm = 113 > 80   NO!

We need to increase the size of the layers to:

d_1 = $L_1/80$ = 4300mm/80 = 53.7 Use 63x140 lumber.

L_1/d_1 = 4300mm/63mm = 68 < 80.

(c) The maximum slenderness ratio for single layers between connections is

$L_3/d_1 < 40$

Find the maximum distance between spacers:

L_3 = $40d_1$ = 40(63mm) = 2520mm = 2.52m

Use one spacer at midspan = 2.1m from ends.

L_3/d_1 = 2100mm/63mm = 33.3 < 40

Calculate the column stability factor with the maximum slenderness value:

$$C_p = b - \sqrt{b^2 - a/c}$$

$$F_{cE} = K_{cE}K_xE/(L_1/d_1)^2 = 0.3(2.5)(10300MPa)/(68)^2 = 1.7MPa$$

$$a = F_{cE}/F_c^* = 1.7MPa/9.7MPa = 0.17$$

$$b = (1+a)/2c = (1 + 0.17)/2(0.8) = 0.73$$

$$C_p = 0.73 - \sqrt{0.73^2 - \frac{0.17}{0.8}} = 0.16$$

Allowable compression stress:

$F_c' = 0.16(9.7MPa) = 1.6MPa$

The total capacity of the column is:

$P_a = F_c'A = (1.6MPa)(63mm \times 140mm) = 28,224N > 22,240N$ OK!

Example M23.4

A Pratt truss is designed to carry the roof loads distributed by 19 purlins @ 1.37m o.c. The purlins located at the center of each panel represent a point load P on the upper chord members: Loads on truss at each purlin: P = 16,438N (16.5kN).

The chord member with the highest compression load is U_5U_6

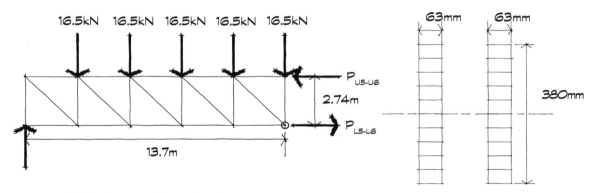

left half of truss with point loadings

double upper chord of truss

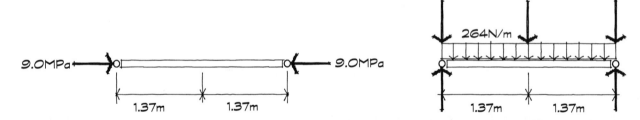

with 433.8kN This member is also subject to bending due to the purlin load P and to the weight of the timber chord w. Assuming a chord size of 63mmx380mm previously determined by the axial load and connection design, the distributed dead load due to the weight of wood is

$w = (63mm)(380mm)(10^{-6}m^2/mm^2)/(5500N/m^3) = 132N/m \times 2$ members $= 264N/m$

The actual compression stress $f_c = P/A = (433,800N)/[(2)(63mm)(380mm)] = 9.0MPa$

The allowable compression stress $F_c' = F_c(C_DC_P)$. This allowable stress must be calculated for the largest slenderness ratio L_e/d of the chord. The chord has two different unbraced lengths:

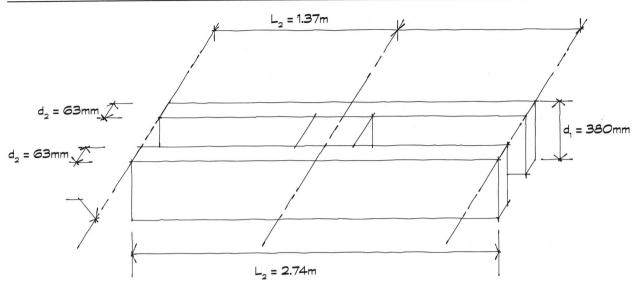

(We are assuming that the distance between the upper chord members is such that the overall width will be less than 380mm.)

X-X axis (K = 1.0)

L_e/d_1 = (2,740mm)/(380mm) = 7.2 < 50

Y-Y axis (K = 1.0)

L_e/d_2 = (1,370mm)/(63mm) = 21.7 < 50

The largest slenderness ratio is about the Y-Y axis. The adjustment factor C_p is now calculated with NDS Eq. 3.7-1 .the timber must also be selected: laminated timber, Western species, combination 12. The design values are as follows:

$F_{c\_}$ = 3.9MPa (Group B)

F_c = 13.4MPa

F_b = 15.2MPa

E = 12,400MPa

Adjustment factor C_D = 1.15 (snow).

F_{cE} = 0.418(12,400MPa)/21.7² = 11.0MPa

F_c^* = (13.4MPa)(1.15) = 15.4MPa

F_{cE}/F_c^* = (11.0MPa)/(15.4MPa) = 0.714

c = 0.9

$$C_p = \frac{1+0.719}{2(0.9)} - \sqrt{\left(\frac{(1+0.719)}{2(0.9)}\right)^2 - \frac{0.79}{0.9}} = 0.61$$

F_c' = (15.4MPa)0.61 = 9.5MPa > 9.0MPa OK

The member works as a compression member. Now we have to check the interaction of compression with bending. The maximum bending moment at midspan is

$$M = \frac{wL^2}{8} + \frac{PL}{4} = \frac{132N/m(2.74m)^2}{8} + \frac{16,438N(2.74m)}{4} = 11,384Nm$$

$$S = \frac{(63mm + 63mm)(380mm)^2}{6} = 3032.4 \times 10^3 mm^3$$

$$f_b = \frac{M}{S} = \frac{11,384Nm(1000mm/m)}{3032.4 \times 10^3} = 3.75MPa$$

$$F_b^* = (15.2MPa)(1.15) = 17.5MPa$$

The adjustment factor C_L has to be calculated with NDS Eq. 3.3-6 (see Chapter 22). The highest bending stress in this case corresponds to the midspan joint with the purlin, which holds the compressed edge of the member in place. The unbraced length $L_u = 1.37m$. From NDS, Table 3.3.3:

$$L_e = 1.11(1,370mm) = 1,521mm$$

$$R_B = \sqrt{\frac{1521mm(380mm)}{(63mm)^2}} = 12 \le 50$$

$$F_{bE} = 0.609(12,400MPa)/12^2 = 52.4MPa$$

$$F_{bE}/F_b^* = 3.0$$

$$C_L = 2.1 - \sqrt{2.1^2 - 3.0/0.95} = 0.98$$

$$F_b' = (17.5MPa)(0.98) = 17.1MPa$$

The volume factor C_V does not apply simultaneously with C_L (NDS, Sec. 5.3.2). Therefore:

     $F_b' = F_b^*(C_V)$ or $F_b' = F_b^*(C_L)$, whichever is the least.

In other words, C_L does not cumulate with C_V.

In this example, C_V is > 1.0; or see Table 23.3; therefore, C_L only applies.

In the interaction formula, the quantity F_{cE} in the amplification factor depends on the depth of the member on the plane of bending and therefore on $L_e/d_1 = 7.2$:

     $F_{cE} = 0.418(12,400MPa)/7.2^2 = 100MPa$

     $f_c/F_{cE} = (9.0MPa)/(100MPa) = 0.09$

     $C_p = 1.19 - \sqrt{1.19^2 - 1.28} = 0.82$

     $F_c' = 15.4MPa(0.82) = 12.3MPa$

The interaction formula becomes:

     $[(9.0MPa)/(9.5MPa)]^2 + (3.75MPa)/[(17.1MPa)(1 - 0.09)] = 0.897 + 0.20 = 1.1 > 1$ NO!

It is evident that the 9MPa working stress in compression is too close to the allowable stress of 9.5MPa. We can increase the allowable stress F_c' by decreasing the slenderness ratio about the weak Y-Y axis. Instead of a 63mm thickness, an 80x380 section is used, and the new value of F_c' is calculated with the same procedure used above:

$L_e/d = (1370mm)/(80mm) = 17.12$

$F_{cE} = 17.7MPa$

$F_{cE}/F_c^* = (17.7MPa)/(15.4MPa) = 1.15$

$F_c' = (15.4MPa)0.82 = 12.3MPa$

It can be noticed that a 25% increase of area has produced a substantial increase (30%) in allowable stress. This also has the effect of increasing the C_L factor, and consequently, the F_b' value. It is not necessary to recalculate F_b', since the interaction equation is certainly satisfied:

$(f_c/F_c')^2 = [(9.0MPa)/(12.32MPa)]^2 = 0.53$

Inserting this in the previous interaction equation:

$0.53 + 0.24 = 0.77 < 1$ OK.

Use two 80X380 laminated timber, Western species Combination 12, for the upper chord.

Problem M23.1

An 89mmX89mm column built with solid Southern pine Stud grade material has a length of 2238mm and is nailed to the top and bottom plates. You may consider this a pinned connection.

(a) Calculate the allowable axial load.

(b) Calculate the allowable axial load of this column that is built up using two 38X89's adequately nailed.

Problem M23.2

A wood frame load-bearing wall is loaded with a uniform load of 13,426N/m. The studs are 38X140, Eastern hemlock Stud grade, 610mm o.c., 2.44m long (high). Are they adequate?

Assume the studs are fully braced by sheathing about the weak axis, but are pinned about the strong axis at both top and bottom. What would be the maximum permissible spacing of the studs?

Answer:

Allowable P = 11,084N > 8,190N; maximum spacing = 825mm

Problem M23.3

A glulam bracket supports an overhang beam. Use a Southern pine Combination Symbol 45 (F_c = 5.9MPa; E = 7,583MPa) glulam, size 171mmX175mm, pinned at B and C, and verify the adequacy of the system. Assume C_D = 1.0 and ignore the glulam weight.

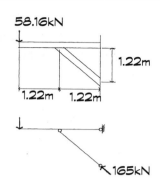

Answer:

Allowable P = 171kN > 165kN compression load

Problem M23.4

A double-leaf spaced column, 2.44m long, is constructed of Southern pine No. 2 Nondense. The two leaves are 89X140's connected at the ends with spacers and shear plates. The axial load is 70.3kN. Determine if the column is adequate, with C_D = 1.15.

Problem M23.5

The spaced column in the illustration is built with 8.3MPa MSR dimension lumber. Calculate the allowable axial compression load based on the slenderness of the entire column and of the individual leaves.

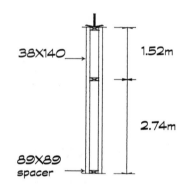

Answer:

Allowable P = 25.8kN

Problem M23.6

A 38X140 stud, 2.44m long, Eastern white pine No. 3, is subject to an axial load of 3200N and a lateral load of 511N/m. The sheathing is nailed to the narrow dimension of the stud and acts as bracing. Determine if the stud is adequate (ignore the effect of the sheathing on the stud for axial and bending stresses).

Problem M23.7

The glulam column in the illustration is built with a 171mmX457mm Western species Combination Symbol 17. C_D = 1.15 for axial load, C_D = 1.6 for lateral load, and C_L = 1.0. Assume the ends are pinned. Determine its load capacity.

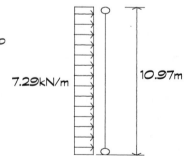

Problem M23.8

A braced column is built with 89mmX140mm aspen No. 2. Assume all connections are pinned and determine the allowable axial load.

Answer:

P = 26.4kN

24 TIMBER TRUSS DESIGN

24.1 Truss types

Timber trusses are very old systems used to cover large spans, from the porch of the Pantheon in Rome to modern bridges. A long constructional history, together with today's better understanding of structural behavior, improved materials (glulam and steel connector systems), and computerized calculation methods, have made the truss an efficient and inexpensive system even for residential roofs and floors. There are over 2,000 manufacturers that currently produce this type of component in the United States.

The geometries of manufactured light trusses most commonly built are shown in Fig. 24.1.

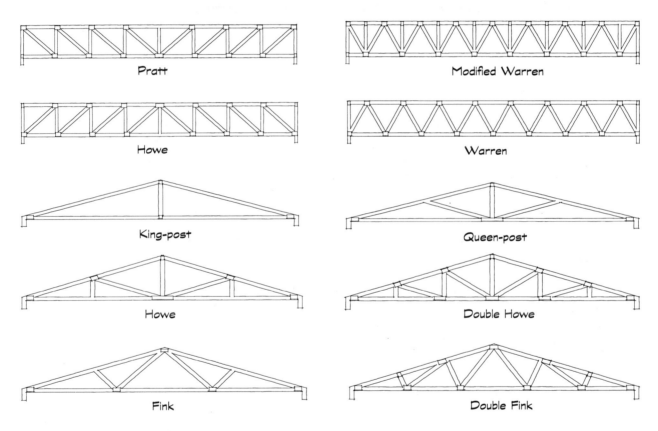

Fig. 24.1 Common premanufactured trusses

Small parallel-chord Warren trusses are used for residential floors. They are economical where clear spans of 20 to 25ft [6 to 7.6m], beyond the capacity of dimension lumber, are required. Additional considerations are the light weight and open web (similar to bar joists) that provides space for the passage of ducts. It is common to have a chase at midspan created by two vertical web members (Fig. 24.2).

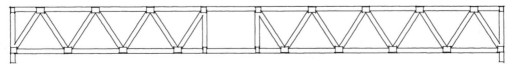

Fig. 24.2 Warren truss with mechanical chase

Span ranges for various primary truss-framing systems are given in Table 24.1. Truss members are designated in three categories: top and bottom chords and webs (Fig. 24.3).

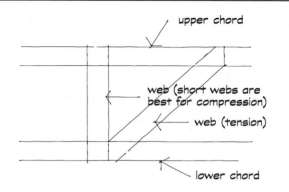

Fig. 24.3 Truss member nomenclature

Table 24.1 Recommended timber truss spans and depth-span ratios

Truss type	Material	Camber (in)[mm]; L = span(ft)[m]	
Flat Pratt, Howe, triangular or pitched	Sawn timber	$C = (1/2)(L/10)$	[4L]
	Laminated timber	$C = (3/8)(L/10)$	[3L]
Bowstring	Sawn timber	$C = (3/8)(L/10)$	[3L]
	Laminated timber	$C = (1/4)(L/10)$	[2L]

Source: American Institute of Timber Construction

Pratt trusses have the advantage that under normal loading conditions, the longer (diagonal) webs are in tension and the shorter (vertical) webs are in compression. (This is an objective of all truss design.)

Bowstring trusses usually have a radius of the top chord equal to the span. Since this particular truss is similar to an arch, the chords take the major stresses, and the web stresses are negligible; therefore, both web members and web connections are light.

Chords and webs in all truss types may be constructed as single-layer, double-layer, or multilayer members. Web locations may be dictated by selection of secondary purlin framing to minimize chord bending stresses. Remember the moments introduced by secondary framing elements bearing on members between the supports from the previous chapter. In parallel chord trusses, diagonal webs should be sloped between 45° and 60° from the horizontal for greatest economy.

Roof trusses must be braced during construction with temporary braces. In the case of light-frame construction, bracing may be continuous 2X4's [38X89] nailed to the top chords, or the simultaneous application of roof sheathing, or an end-bracing frame. Where the roof sheathing provides adequate connection between all trusses, there may be no need to brace the compression chords with diagonal lumber except during construction. Often, long web members in compression must be tied together by longitudinal bracing to reduce their unbraced length. Since it is not always possible to distinguish the difference between construction and permanent bracing, removal of bracng members within a truss system should never be done without the approval of the architect or engineer.

In a string instrument like a violin or a bass [in Fig. 24.4 (a)], the tension in the strings is maintained by the rigidity of the wood frame, which works in compression and bending. The bass behaves like a king-post truss, or single-strut trussed beam. The wood structure (the case of the bass or the horizontal timbers of the beam) bends because the tension applied at the ends by the strings has a perpendicular as well as an axial component. The strut (bridge of the bass) is in compression, and it applies a load on the beam equal to the perpendicular end components [Fig. 24.4 (b)].

Fig. 24.4 (a) A double bass as a trussed structural system

Reprinted by the permisison of the publisher frm HARVARDCONCISE DICTIONARY OF MUSIC by Don M. Randel, Cambridge, Mass.: Harvard University Press, Copyright © 1978 by the President and Fellows of Harvard College.

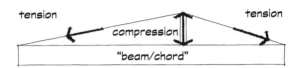

Fig. 24.4 (b) Structural diagram of forces

24.2 Deflection and camber

The formulas used for calculating deflection in bending members are not valid for a truss. In truss members, the loads are axial, corresponding to axial deformations. In simple span trusses, the bottom chord will elongate, the top chord will shorten, and sagging (deflection) will result. In addition to the elastic deformation, an inelastic deformation or slip should be considered. To avoid a sag between supports due to dead loads and slip, trusses are built with a camber, which is given, as a first approximation, in the Table 24.2.

Table 24.2 Camber in trusses

Truss type	Material	Camber (in)[mm]; L = span(ft)[m]	
Flat Pratt, Howe, triangular or pitched	Sawn timber	C = (1/2)(L/10)	[4L]
	Laminated timber	C = (3/8)(L/10)	[3L]
Bowstring	Sawn timber	C = (3/8)(L/10)	[3L]
	Laminated timber	C = (1/4)(L/10)	[2L]

Deflection can be evaluated with sufficient accuracy using beam tormulas, decreasing the modulus of elasticity by 10% to account for deformations the web members, and for movement in mechanical connections. For parallel chord trusses, the moment of inertia is easily calculated as

$$I = \sum (I_o + Ad^2)$$

Since I_o is approximately "O", when the chords have similar areas (which is often the case in lightweight trusses), the formula becomes

$$L = 2(A)(d/2)^2 = Ad^2/2$$

where A is the average area of each chord, assumed equal, and d is the distance between the two chords. For nonparallel chord trusses, the average d can be used for a conservative result. The value of the deflection found can be used to specify the camber or to ensure adequate drainage in roof systems.

24.3 General design procedure

24.3.1 Preliminary design

Most trusses have chords that are continuous across several panel points for simplicity of construction. These same chords may also have distributed loads at points other than joints due to the secondary framing system used. Secondary framing may consist of purlins attached to the chords between joints, or it may be a decking material that applies a uniform load to the chord. Member forces and architectural consideration determine the type of connections used, which could be any combiation of pinned, fixed, or semifixed. With the last two types, the force distribution in the members will not be statically determinate (i.e., the forces cannot be solved with the simple equilibrium of joints) and bending moments will be introduced. However, in order to determine the geometry of the center lines of the members, it is necessary to start with a preliminary approximate calculation, where loads are placed at panel points and all joints are assumed pinned.

Axial member forces can be calculated with a graphical or analytical method. The "stress diagram" we discussed in Part 1, Steel, is an excellent way to determine which members are critical to the design, both in magnitude and type of force. The most critical three or four members can be determined and the critical chords can be sized for the axial loads plus the approximate bending moments due to loads not located at joints. Splices should be located in chords based both on standard lengths of materials and minimal forces. **Remember that tension member sizes may have to be increased to compensate for loss of area at connections.**

24.3.2 Final design

This procedure must take into account the location of chord splices, fixed and hinged connections, and the actual loads configuration. A computer analysis or a complex manual method is required to solve for all the forces.

When both axial load and bending are present, the design values of the material must be adjusted accordingly, with the applicable adjustment factors from NDS Table 2.3.1, including buckling stiffness factor C_T. This applies to 2X4 [38X89] or smaller sawn lumber truss compression chords subjected to combined flexure and axial compression when 3/8in [9.5mm] or thicker plywood sheathing is nailed to the narrow face (NDS, Sec. 4.4.3). The factor is calculated with the formula

$$C_T = 1 + \left(\frac{K_m L_e}{K_T E} \right)$$

where the coefficients K are given in the NDS and L_e is the effective length of the member, considering buckling in the plane perpendicular to the plywood sheathing. The maximum L_e = 96in [2.4m]; therefore, the length should be taken as 96in [2.4m] when greater than that.

Example 24.1

A series of parallel-chord Pratt trusses 16ft o.c. are used over a clear span of 90ft. Purlins span between trusses at the panel points. Horizontal bracing is provided in the plane of the purlins with diagonal members at the two end bays. The purlins support 4ft-wide stressed-skin panels and a roofing membrane.

Truss geometry:

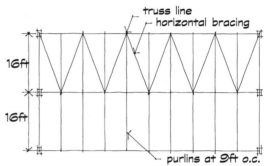

Truss depth at 10% of the span = L/10 = 9ft

Diagonal webs angle = 45°

Panel length = 9ft, giving 10 equal panels

Dead loads:

Estimated truss weight = 100plf (9ft panel)	=	900lb
Estimated purlin weight = 25plf (16ft spacing)	=	408lb
Foam core panels, 8in expanded polyst., 5/8in OSB skin	=	5.6psf
Roofing membrane with gravel	=	5.5psf
Gypsum board ceiling, 1/2in	=	2.0psf
Estimate for HVAC, electrical	=	2.0psf

Total distributed loads located at panel points:

(15.1psf)(9ft)(16ft)   = 2,174lb

Dead loads at the panel points =   3,482lb

Live loads:

Snow = (30psf)(9ft)(16ft)   = 4,320lb

Total loads at the panel points   7,802lb

Support reaction: R = (7802lb)(9loads)/2 = 35,109lb

Member forces:

The maximum tension force in bottom chord L_4-L_5 can be calculated with the method of sections.

9 loads of 7.8k each

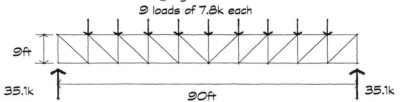

9ft

35.1k   90ft   35.1k

Equilibrium of moments about U_4:

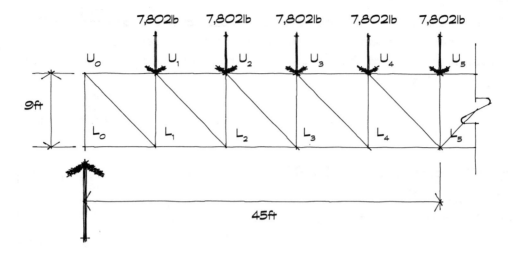

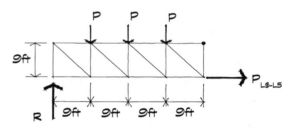

$P_{L4-L5}(9ft) = (35,109lb)(36ft) - (7,802lb)(27ft) -(7,802lb)(18ft) -(7,802lb)(9ft)$

$P_{L4-L5} = 93,624lb$ (tension)

The maximum compression force in top chord member U_4-U_5.

Equilibrium of moments about L_5:

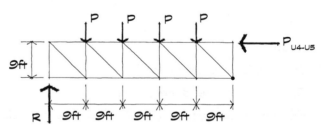

$P_{U4-U5}(9ft) = (35,109lb)(45ft)$

$\qquad - (7,802lb)(36ft) -(7,802lb)(27ft) - (7,802lb)(18ft) - (7,802lb)(9ft)$

$P_{U4-U5} = 97,525lb$ (compression)

Maximum forces in the webs:

Equilibrium of joint L_0 :

$P_{LO-UO} = 35,109lb$ (compression)

P_{UO-L1} = (35109lb)(1.414) = 49.651lb (tension)

Design of truss members:

The size of the members depends on the type of timber and on the design of the connections. The chords can be designed as double members. This works well for the compression chord, in order to economically decrease its slenderness ratio, and for the design of the connections. If a connection system is used that allows the members to overlap at joints (split rings or bolts), the double-member chords will allow single-member webs to slip between them to create a connection. If, however, plates are used to transfer the loads at connections, the members must be in the same plane (shear plates or metal connector systems) and the two pieces should probably be replaced with a single member. For this reason, if overlap joints are to be used, wide thin members provide more connection space, and if in-plane connections are used, thicker members provide penetration depth for opposing shear plates.

Laminated timber (NDS, Supplement, Table 5B):
Visually graded Western species (members stressed primarily in axial tension or compression).
Combination Symbol 1, Douglas fir:

$F_{c(perpendicular)}$ = 560psi; this corresponds to Group B in relation to connection design.
E = 1,500,000psi; F_t = 900psi; F_c = 1,550psi (parallel to grain)

The compression chord design depends on its slenderness ratio (see Chapter 23). Each layer

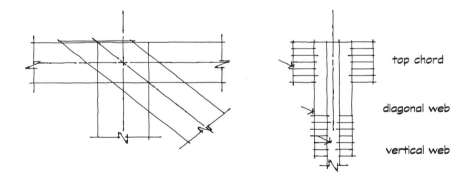

top chord

diagonal web

vertical web

has to be considered as a separate column between joints or connection points, assuming that no additional bracing is provided. The compressive load is, therefore:

P = (97,525lb)/2 = 48,626lb

Assuming the thickness to be 5-1/8in, C_D = 1.15, we can determine C_p with NDS Eq. 3.7.1:

K = 1.0

L_e/d = (9ft)(12in/ft)/(5.125in) = 21 < 50 OK!

$$F_{cE} = \frac{0.418(1,500,000psi)}{21^2} = 1421.7psi$$

$$F_c^* = (1,550psi)(1.15) = 1,782.5psi$$

$$c = 0.9$$

$$C_p = 0.999 - \sqrt{0.99^2 - 0.886} = 0.664$$

$$F_c' = (1,782.5psi)(0.664) = 1184psi$$

Required area in compression:

$$A = (48,626lb)(1,184psi) = 41in^2$$

Required width of each layer:

$$d = (41in^2)/(5.125in) = 8in$$ The closest section is 5-1/8in × 9in with six laminations > 4; therefore the correct value of F_c has been used (NDS, Table 5B).

Use a pair of 5-1/8inX9in for compression chord. $P = 2(1184psi × 46.13in^2) = 109,224lb$
Tension chord:

Required $A = (93,624lb)/(900psi ×)1.15) = 90.0in^2$

Two 5-1/8in × 9in will give $A = 2(46.12in^2) = 92.24in^2$

The sizing of this TENSION member is, however, going to be determined by the connections with the web members.

Problem 24.1

A king-post truss is made of laminated-timber sloping chords, a steel post, and a steel tension rod.

(a) What is the tension in the rods?

(b) Assuming the steel tie rod is connected to the sides of the glulam with steel plates and a number of shear plates, which formula should be used for the calculation of the design value of the timber?

(c) Timber members AC and CB are supporting a uniformly distributed gravity load. What type of stresses are they subject to?

Problem 24.2

A Warren truss has a span of 36ft and a depth of 3ft. If each upper chord panel point carries 0.6k, size the tension chord using E-rated Western species laminated timber, Combination Symbol 22F-E1.

Problem 24.3

A 34ft Howe truss has both chords built with two 2X6 No. 1 Southern pine members for architectural reasons. Find the minimum depth that is required to safely carry lower chord panel point loads of 1.2k each.

METRIC VERSIONS OF ALL EXAMPLES AND PROBLEMS

Example M24.1

A series of parallel-chord Pratt trusses 5m o.c. are used over a clear span of 27.5m. Purlins span between trusses at the panel points. Horizontal bracing is provided in the plane of the purlins with diagonal members at the two end bays. The purlins support 1.25m-wide stressed-skin panels and a roofing membrane.

Truss geometry:

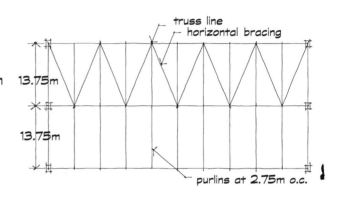

Truss depth at 10% of the span = L/10 = 2.75m

Diagonal webs angle = 45°

Panel length = 2.75m, giving 10 equal panels

Loads

Dead loads:

Estimated truss weight = 1500N/m(2.75m panel)		= 4125N
Estimated purlin weight = 375N/m(5m spacing)		= 1875N
Foam core panels, 203mm expanded polyst., 16mm OSB skin		= 268N/m²
Roofing membrane with gravel		= 263N/m²
Gypsum board ceiling, 13mm		= 96N/m²
Estimate for HVAC, electrical		= 96N/m²
Total distributed loads located at panel points:		723N/m²

$$(723N/m²)(2.75m)(5m) = 9941N$$

Dead load at panel points = 4125N + 1875N + 9941N = 15,941N

Live loads:
Snow = 1440N/m²(2.75m)(5m) =   19,800N

Total loads at panel points   35,741N

Member forces:

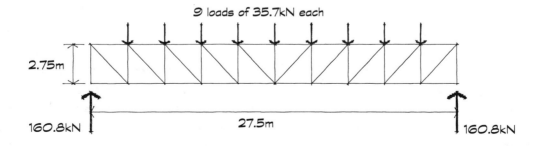

9 loads of 35.7kN each

2.75m

160.8kN   27.5m   160.8kN

The maximum tension force in bottom chord L_4L_5 can be calculated with the method of sections. Equilibrium of moments about U_4:

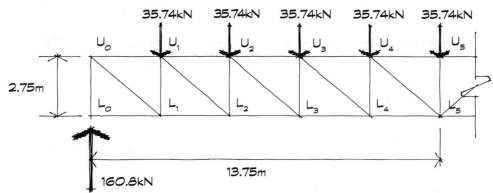

P_{L4-L5}(2.75m) = 160,834N(4X2.75m) -
35,741N(3X2.75m +2X2.75m + 2.75m)

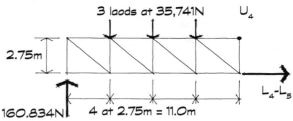

P_{L4-L5} = 1,179,447.5Nm/2.75m
= 428,890N (tension)

The maximum compression force in top chord member U_5U_6
Equilibrium of moments about L_5:

P_{U5-U6}(2.75m) = 160,834N(27.5m/2) - 35,741N(4 + 3 + 2 + 1)2.75m
P_{U5-U6} = 1,228,590Nm/2.75m =
446,760N

Maximum forces in the webs:

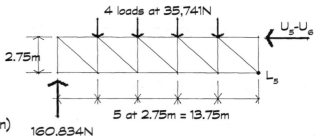

Equilibrium of joint U_0
P_{L0-U0} = R = 160,834N (compression)
P_{U0-L1} = 160,834N(1.414) = 227,451 (tension)

Design of truss members:

The size of the members depends on the type of timber and on the design of the connections.

The chords can be designed as double members. This works well for the compression chord, in order to economically decrease its slenderness ratio, and for the design of the connections. If a connection system is used that allows the members to overlap at joints (split rings or bolts), the double-member chords will allow single-member webs to slip between them to create a connection. If, however, plates are used to transfer the loads at connections, the members must be in the same plane (shear plates or metal connector systems) and the two pieces should probably be replaced with a single member. For this reason, if overlap joints are to be used, wide thin members provide more connection space, and if in-plane connections are used, thicker members

provide penetration depth for opposing shear plates.
$$F_c = 3.9 MPa, \ E = 10,300 MPa, \ F_t = 6.2 MPa, \ F_{c//} = 10.7 MPa$$

The compression chord design depends on its slenderness ratio (see Chapter 23). Each layer has to be considered as a separate column between joints or connection points, assuming that no additional bracing is provided. The compressive load is, therefore:

$$P = 446,760 N/2 = 223,380 N$$

Assuming the thickness to be 130mm, $C_D = 1.15$, we can determine C_p with NDS Eq. 3.7.1:

$$K = 1$$
$$L_e/d = 2,750mm/130mm = 21 < 50 \ OK$$
$$F_{cE} = \frac{0.418(10,300 MPa)}{21^2} = 9.8 MPa$$
$$F_c^* = 10.7 MPa(1.15) = 12.3 MPa$$
$$c = 0.9$$

$$C_p = 0.998 - \sqrt{0.998^2 - .885} = 0.664$$

$$F_c' = 12.3 MPa(0.664) = 8.2 MPa$$

Required area in compression:
$$A = 223,380 N/8.2 MPa = 27,241 mm^2$$

Required width of each layer:
$$d = 27,241 mm^2/130mm = 209mm$$

The closest section is 130mmX228mm with six laminations > 4, therefore the correct value of F_c has been used (NDS, Table 5B).

Tension chord:
$$Req. \ A = 428,890 N/(6.2 MPaX1.15) = 60,153 mm^2$$

Two 130X228mm will give $A = 2(29,640 mm^2) = 59,280 mm^2$ OK within 5%

The sizing of this TENSION member is, however, going to be determined by the connections with the web members.

Problem M24.1

A king-post truss is made of laminated-timber sloping chords, a steel post, and a steel tension rod.

(a) What is the tension in the rods?

(b) Assuming the steel tie rod is connected to the sides of the glulam with steel plates and a number of shear plates, which formula should be used for the calculation of the design value of the timber?

(c) Timber members AC and CB are supporting a uniformly distributed gravity load. What type

Problem M24.2

A Warren truss has a span of 11m and a depth of 0.9m. If each upper chord panel point carries 2.67kN, size the tension chord using E-rated Western species laminated timber, Combination Symbol 22F-E1.

Problem M24.3

A 10.4m Howe truss has both chords built with two 38X140 No. 1 Southern pine members for architectural reasons. Find the minimum depth that is required to safely carry lower chord panel point loads of 5.34kN each.

25 ARCHES, DOMES, AND VAULTS

25.1 Two- and three-hinged arches

The category of arches includes portal frames, rectilinear and curvilinear three-hinged arches, and trussed arches. Domes and other roof structures designed as three-hinged arches, with two sloping members and a tie at the base, are essentially trusses, but the bending stresses on the sloping compression members are a significant aspect of the design. These systems are common for one-story large-span structures encountered in bridges, industrial buildings, and sports halls, but are also used in more sophisticated solutions, where the structure plays an architectural role (Fig. 25.1).

Where the arches rest on vertical supports, a large horizontal thrust is generated. This is resisted with tension members connecting one side of the base of the arch or dome to the opposite side. Where the arch is supported directly on the foundations, the horizontal reactions may be resisted by the ground using buttresses or angled foundations, as in the example of Fig. 25.1.

Fig. 25.1 Wooden dome sports arena, Tacoma, Washington

Portal frames are frames where the columns are rigidly connected to or continuous with the beams. If the beam is continuous, the portal is normally hinged at the foot of the columns. If there is a hinge at midspan, the portal is effectively a three-hinged arch with knees between the wall arm and the roof. Table 25.1 lists typical spans of laminated-timber frames and arches to be used for outline design. One of the primary advantages of glulam is that the laminations can be bent into a wide range of curves. Although the designer has a great degree of freedom in deciding the profile of the members, limits are imposed by the minimum radius of curvature based on the thickness of the laminations used and the species of wood. The very tight

Fig. 25.2 Arch highway bridge in Colorado

curves used in laminated-wood furniture design are achieved by using very thin laminations (put glue between all the pages of this book, bend it to whatever curve you like and hold it there until it sets; after all, it is your book).

A very common portal frame is the three-hinged arch (Fig. 25.2) or the three-hinged Tudor arch (Fig. 25.3). Most arches are symmetrical about the center line, but the designer has considerable freedom in the placement of the interior hinge, which can result in nonsymmetrical spans or unequal leg heights and pitches.

The constraints of the shape and size of the arch are primarily those of manufacturing and transportation. It is possible to utilize moment splices at interior joints for this purpose, but this is expensive and can increase the deflection of the structure.

Fig. 25.3 Tudor arch (half) geometry

Three-hinged curved arches are economical for large spans since the moment-resisting splice is not required. They can be designed with a constant radius, or can be parabolic, cycloidal, or pointed. Supports for arches must resist the horizontal thrust caused by gravity and lateral loads, and possibly uplift of the foundation. An example of a pinned base hinge is shown in Fig. 25.4.

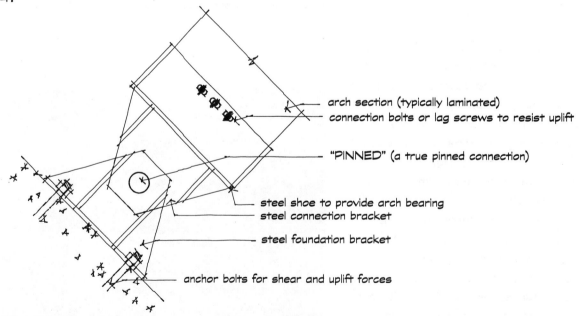

arch section (typically laminated)
connection bolts or lag screws to resist uplift

"PINNED" (a true pinned connection)

steel shoe to provide arch bearing
steel connection bracket

steel foundation bracket

anchor bolts for shear and uplift forces

Fig. 25.4 Pinned arch base

25.2 Preliminary design of laminated-timber arches

The statics of an arch are conceptually simple, but in practice, the procedure is lengthy due to the curved axis of the structure and to the different load combinations that have to be considered (unbalanced live or snow load, dead loads plus wind, etc.). It is easier today to solve the statics with a computer program and find the value of the maximum shear at the base of the arch, the maximum moment (not necessarily at midspan), and the deflection for alternative loads.

For preliminary design purposes, just to have a ballpark figure of the size that may be required, some tables can be handy. Table 25.1 compares different arches and beam types and systems built with laminated timber.

Table 25.1 Common design dimensions of laminated timber system

Description of system	Usual span L = feet	Spacing e = feet	Beam depth d = feet
Simply supported beam parallel edges	30-130	16-32	0.06xL
Simply supported beam top edge pitched	30-130	16-32	0.07xL
Simply supported beam hunched maximum slope 1:4	30-115	16-32	0.07xL
Tied three-hinged arch parallel edge members	60-200	16-32	0.03xL
Tied three-hinged arch curved members	60-300	16-32	0.025xL
Three-hinged arch curved members	75-450	16-32	0.02xL
Three-hinged portal spliced rigid knees	30-150	16-32	0.06xL
Three-hinged Tudor arch	30-250	16-32	0.05xL
Two-hinged portal spliced at knees	30-100	16-32	0.05xL
Continuous beam any slope	30-100	16-32	0.05xL
Continuous beam with variable section, any slope	30-100	16-32	0.04xL
Overhang beam with variable section l/k = 1:3	k = 30-100	16-32	.01xK

The NDS and the AITC require, in the full procedure, taking into account the bending stresses that are induced and retained in the member when the laminations of the timber arch are glued in a curved form. In order to avoid overstresses, the NDS imposes limitations on the radius of curvature of the arch in the following way:

  R = 100t for Southern pine and hardwood species

  R = 125t for other softwoods

where R is the radius of curvature, and t is the lamination thickness.

Since the standard lamination thickness for Southern pine is 1-3/8in [35mm] and for Western species 1-1/2in [38mm], we obtain the respective minimum curvature radii:

  R = 100(1.375in) = 137.5in = 11.45ft (Southern pine)

    R = 100[35mm] = 3,500mm

  R = 125(1.5in) = 187.5in = 15.62ft   (Western species)

    R = 125[38mm] = 4,745mm

A three-hinged arch with a pin connection at the crown is statically determinate. We will simplify the design procedure by projecting the total load along the span of the arch, although we will calculate the weight of the structure along the length of the arch and assume that the lateral loads do not control the design. This means that we will design for gravity loads only in this case.

With these simplifications, the statics of a circular arch can be easily solved with equation of moments about the pins. If W is the total load and r is the design radius:

  $R_v = W/2$ (vertical reaction)

  $R_h = W/4$ (horizontal reaction)

$$M = -\left(\frac{W}{4}\right)\sqrt{2rx - x^2} + \left(\frac{W}{2}\right)x - \frac{\left(\frac{W}{4}\right)x^2}{r}$$

Where M = maximum moment at a distance from the support

  x = 0.135r, corresponding to an angle $\Theta = 30°$ and $h_x = 0.5r$

The term (W/2)x represents the moment of the vertical support reaction, which is shown positive (opposite to M).

The allowable bending stress F_b' takes into account the curvature overstress on the laminations with a curvature factor C_c:

  $C_c = 1 - 2000(t/R)^2$

with t and R as previously defined.

Thus, we can calculate the approximate value of the allowable bending stress (considering the timber section as solid lumber and therefore not applying the volume factor for simplicity):

  $F_b' = F_b(C_f C_c)$

We can now calculate the required section modulus S, assuming the section rectangular and constant along the arch:

$$S = M/F_b'$$

By using the equation of equilibrium for the portion of the arch to the left or right of the maximum moment, the compressive force (perpendicular to the radius) also can be determined. The arch should then be designed as a combined axial load and bending member.

Example 25.1

A hall is designed to be roofed with circular arches with a span L = 90ft and a rise h = 30ft. The arches are spaced 24ft o.c. and for ease of construction are built with a crown hinge (forming a three-hinged arch). The snow load is 30psf and the dead load of the roofing structure is 20psf (excluding the arch).

From Table 25.1, we can assume a depth of section

   d = 0.02L = 0.02(90ft)(12in/ft) = 21.6in

If we use Western species, the closest size is 8-3/4in × 24in, weighing 51plf.

The minimum R is about 15ft, which is much smaller than the design radius r = L/2 = 45ft.

Total dead load: Length of arch = 3.14(45ft) = 141.3ft

Weight of arch: W_a = (141.3ft)(51plf)	=	7,206lb	
Weight of roofing: W_r = (141.3ft)(24ft)(20psf)	=	67,824lb	
Weight of snow: W_s = (90ft)(24ft)(30psf)	=	64,800lb	
	Total =	139,830lb = 140k	

Support reactions components are

   R_v = (140k)/2 = 70k

   R_h = W/4 = (140k)/4 = 35k

The moment is maximum at x = 0.135(45ft) = 6.1ft.

$$M = -\frac{140k}{4}\sqrt{2(45ft)6.1ft - (6.1ft)^2} + \left(\frac{140k}{2}\right)6.1ft - \frac{\frac{140k}{4}(6.1ft)^2}{45ft} = -393.74ftk$$

   M = -393.74ftk

The minus sign indicates that M is clockwise on the left half of the arch, i.e., the extreme outer fibers are in tension.

The required S is

   $$S = M/F_b'$$

$C_f = (12/h)^{1/9} = (12/24\text{in})^{1/9} = 0.93$

$C_c = 1 - 2{,}000(t/r)^2 = 1 - 2{,}000[(1.5\text{in})/(540\text{in})]^2 = 0.98$

$F_b' = (2{,}400\text{psi})(0.93)(0.98) = 2{,}187\text{psi}$

$S = (393{,}740\text{ftlb})(12\text{in/ft})/(2{,}187\text{psi}) = 2{,}160\text{in}^3$

A possible choice is a 10-3/4in x 36in with $S_x = 2{,}322\text{in}^3$. The proportion of the section is between 1:4 and 1:3; therefore, it is not too slender. This preliminary size has not yet taken into consideration the axial load that will also exist. We could conservatively use the horizontal reaction at the base of the arch to determine the area required: $A = (35{,}000\text{lb})/(2400\text{psi}) = 14.6\text{in}^2$. If we use this component in the interaction formula, it would look something like this:

$$\frac{F_b}{F_{b'}} + \frac{A_{(reqd)}}{A_{(provided)}} = \frac{S_{(reqd)}}{S_{(provided)}} + \frac{A_{(reqd)}}{A_{(provided)}}$$

$$\frac{2160\text{in}^3}{2322\text{in}^3} + \frac{14.6\text{in}^2}{387\text{in}^2} = .93 + .03 = .97 \leq 1.0 \text{ it WORKS!}$$

The C_p factor is not used because of the marginal influence of compression in the arch.

Table 25.2 provides a preliminary design for Tudor arches made of Southern pine (F_b = 2400psi [16.5MPa]) under wind and gravity loads, with the assumptions indicated.

Table 25.2 Design table for laminated arches

Vertical and wind loadings   For preliminary design only

vertical dead load = 240plf and horizontal wind load = 240plf

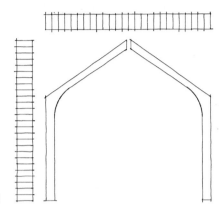

Loading diagram

Roof pitch	Wall height	Span 40ft					Span 50ft				
		Width in.	Base in.	Tangent pts. (in) lower	upper	Crown in.	Width in.	Base in.	Tangent pts. (in) lower	upper	Crown in.
10:12	8ft	5	7-1/2	10-1/4	11-1/4	11-1/4	5	7-1/2	12-1/4	13-1/2	13-1/2
	10ft	5	7-1/2	12	10-3/4	10-3/4	5	7-1/2	13-1/2	15	12-3/4
	12ft	5	7-1/2	14-3/4	9-3/4	9-3/4	5	7-1/2	15	16-1/2	11-1/4
13:12	8ft	5	7-1/2	11-1/2	12-3/4	12-3/4	5	7-1/2	14-1/4	15-3/4	15-3/4
	10ft	5	7-1/2	13-1/2	15	9-3/4	5	7-1/2	15-3/4	17-1/2	12-1/4
	12ft	5	7-1/2	15-3/4	17-1/2	8-1/2	5	7-1/2	19	21	10-1/4
14:12	8ft	5	7-1/2	13-1/4	14-3/4	11-3/4	5	8	16-1/4	18	14-1/2
	10ft	5	7-1/2	15-3/4	17-1/2	10	5	7-1/2	18-3/4	20-3/4	12-1/2
	12ft	5	7-1/2	18-3/4	20-3/4	8-3/4	5	7-1/2	22	24-1/4	11

Source: American Institute of Timber Construction, Laminated Timber Design Guide.

Note: These tables use a simplified loading condition that may not adequately represent actual conditions. For this reason, the dimensions indicated should be used **ONLY FOR PRELIMINARY DESIGN**. They will provide a reasonable size approximation for architectural design purposes. Final designs must be evaluated using the critical-load conditions specified by the applicable building code.

Table 25.3 Design table for laminated arches
Vertical and wind loadings　　　　　For preliminary design only

vertical dead load = 320plf and horizontal wind load = 320plf

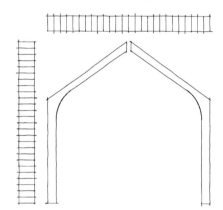

Loading diagram

Roof pitch	Wall height	Span 40ft						Span 50ft					
		Width in.	Base in.	Tangent pts. (in)		Crown in.		Width in.	Base in.	Tangent pts. (in)		Crown in.	
				lower	upper					lower	upper		
10:12	8ft	5	7-1/2	11-1/2	12-3/4	12-3/4		5	9-1/4	14-1/4	15-3/4	15-3/4	
	10ft	5	7-1/2	13-1/2	15	10-3/4		5	7-1/2	16-1/2	18-1/4	12-1/2	
	12ft	5	7-1/2	16-1/4	18	9		5	8	19-1/2	21-1/4	11	
13:12	8ft	5	7-1/2	14-3/4	16-1/4	11		5	9	17-1/4	19	15	
	10ft	5	7-1/2	17-3/4	19-3/4	8-3/4		5	9-1/4	20-1/2	22-3/4	12-1/4	
	12ft	5	7-1/2	20-1/2	22-3/4	8-3/4		5	10-1/4	18-1/2	20-1/2	10-1/2	
14:12	8ft	5	7-1/2	17-1/4	19	11-1/4		5	12-1/2	20-3/4	23	14-1/4	
	10ft	5	7-1/2	20-1/2	22-3/4	9-3/4		5	10-1/4	18-1/4	20-1/4	12-1/2	
	12ft	5	7-1/2	18-1/4	20-1/4	10-1/4		5	10-1/4	21-1/2	23-3/4	11	

Source: American Institute of Timber Construction, Laminated Timber Design Guide.

Note: These tables use a simplified loading condition which may not adequately represent actual conditions. For this reason, the dimensions indicated should be used **ONLY FOR PRELIMINARY DESIGN**. They will provide a reasonable size approximation for architectural design purposes. Final designs must be evaluated using the critical-load conditions specified by the applicable building code.

Table 25.4 Design table for laminated arches

Vertical and wind loadings   For preliminary design only

vertical dead load = 400plf and horizontal wind load = 400plf

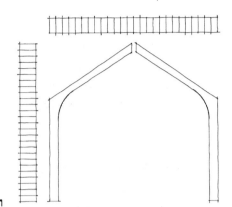

Loading diagram

Roof pitch	Wall height	Span 40ft					Span 50ft				
		Width in.	Base in.	Tangent pts. (in) lower	upper	Crown in.	Width in.	Base in.	Tangent pts. (in) lower	upper	Crown in.
10:12	8ft	5	7-1/2	13-3/4	15-1/4	12-1/4	5	10-1/2	17	18-3/4	14-1/2
	10ft	5	7-1/2	16-3/4	18-1/2	10-1/4	5	10-1/4	20-1/4	22-1/2	12-1/4
	12ft	5	8	20-1/4	22-1/2	8-3/4	5	10-1/4	18	20	11
13:12	8ft	5	9-3/4	18	20	10-3/4	5	13-1/4	21-1/2	23-3/4	13-1/2
	10ft	5	9-1/2	21-1/4	23-1/2	10-1/4	5	10-1/4	19-1/2	21-1/2	11-3/4
	12ft	6-3/4	10-1/4	19	21	10-1/4	5	10-1/4	22-1/2	25	10-1/2
14:12	8ft	5	12-1/2	20-3/4	23	11-1/4	5	11-1/4	19-1/2	21-1/2	14-1/2
	10ft	6-3/4	10-1/4	19	21	10-1/4	5	10-3/4	22-1/2	24-3/4	12-1/2
	12ft	6-3/4	10-1/4	22-1/2	24-3/4	10-1/4	5	11-1/4	26-1/4	29	11

Source: American Institute of Timber Construction, Laminated Timber Design Guide.

These tables use a simplified loading condition which may not adequately represent actual conditions. For this reason, the dimensions indicated should be used **ONLY FOR PRELIMINARY DESIGN**. They will provide a reasonable size approximation for architectural design purposes. Final designs must be evaluated using the critical load conditions specified by the applicable building code.

25.3 Construction of timber arches

Two-hinged arches are limited in span by transport. They can be spliced in two or three segments, but this adds to the cost, although the member depth will be smaller than for an equivalent three-hinged arch.

The arches themselves can be built with different materials and by different technologies: lumber, laminated timber, or curved trusses. Wood arches for large spans have been used since the ninteenth century, when wood laminations were connected by mechanical fasteners instead of glue. These bolts through the laminations take the horizontal shear stresses associated with bending in the arch.

Laminated timber, as we have seen, is the descendant of these early composite arches, and it is very common and cost-effective for long spans. However, trussed arches can be viable alternatives due to their light weight and the simplicity of calculation relative to solid arches. Exceptionally large-span arches have been built with the truss technique, such as in the University of Idaho stadium (Fig. 25.5), where the arch components consist of microlaminated wood ribs (tension and compression chords) and steel webs.

Fig 25.5 Arch section under costruction at University of Idaho stadium

It must be clear by now that wood is in a lot of composite structural members, such as engineered trusses with steel webs and structural insulated panels made of oriented strand board skins and polystyrene core. We give here two examples of arch systems using composite members. The first is a curved stressed-skin panel made of plywood, with curved wood ribs that can be partially replaced by a lightweight web of insulation material like paper honeycomb or polyurethane foam.

The second example is a very sophisticated system for traveling exhibitions designed by Renzo Piano for IBM (Fig. 25.6). Each arch is made of three chords of laminated wood, with demountable connections of cast aluminum. The wood chords are connected together by Plexiglas prisms, also demountable, that constitute the web of the trussed arch.

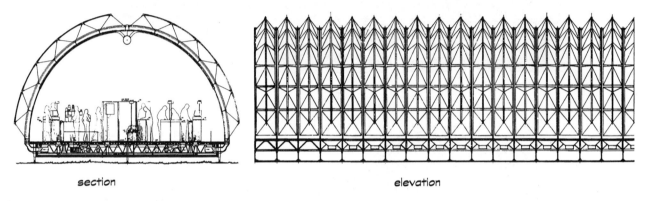

section   elevation

Fig. 25.6 IBM traveling exhibition pavilion

25.4 Domes and vaulted roofs

Domes can be built with three types of systems:

Arches connected at the crown and tied at the base with tension members (or a reinforced concrete foundation), as shown in Fig. 25.4. Each half-arch forms a rib of the dome and supports the roofing systems.

Reticular shells made of a mesh of thin wood members in different geometric configurations. Most systems have a triangular grid: Geodesic domes are made of triangles based on pentagons and hexagons; lamella domes can be arranged in other patterns.

Lamella roof systems are constructed of straight, relatively short pieces of wood that may vary in size from 2X8 to 3X16 [38X184 to 63X394], to 8 to 16ft [25 to 50m] long. These pieces, called lamellas, are connected in a triangular plan pattern with metal connectors.

Complex roof forms are possible with reticular shells, such at the Multihalle in Mannheim, Germany (Fig. 25.7). This large and complex structure is a double-layer grid shell designed by the Institute for Lightweight Structures directed by Frei Otto. The idea is to achieve pure compression stresses in the wood members forming the mesh when dead loads only are applied. The initial form was in fact arrived at by building a model of a suspended wire mesh (where the wires

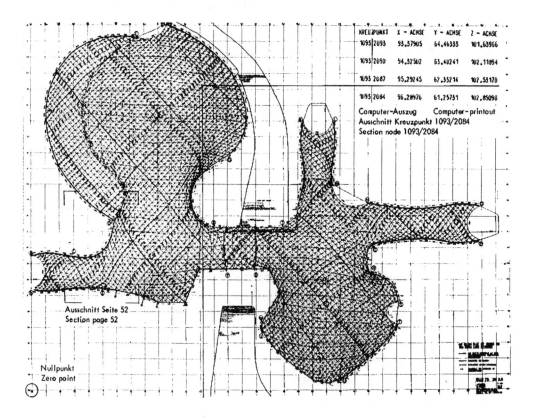

Fig 25.7 Multihalle (multipurpose exhibition hall), Mannheim, Germany, concourse and framing plan

were in tension under their own weight) and turning it upside down to make it work in compression. The hemlock lattice members of the double shell are only 2inX2in [38mmX38mm] in cross-section, and develop a maximum length of 300ft [100m] for the arches forming the main dome.

METRIC VERSION OF EXAMPLE AND TABLES

Table M25.1 Common design dimensions of laminated timber system

Description of System	Usual span L = m	Spacing e = m	Beam depth d = m
Simply supported beam parallel edges	10-40	5-10	0.06xL
Simply supported beam top edge pitched	10-40	5-10	0.07xL
Simply supported beam hunched maximum slope 1:4	10-35	5-10	0.07xL
Tied three-hinged arch parallel edge members	20-60	5-10	0.03xL
Tied three-hinged arch curved members	20-100	5-10	.025xL
Three-hinged arch curved members	25-150	5-10	0.02xL
Three-hinged portal spliced rigid knees	10-50	5-10	0.06xL
Three-hinged Tudor arch	10-80	5-10	0.05xL
Two-hinged portal spliced at knees	10-30	5-10	0.05xL
Continuous beam any slope	10-30	5-10	0.05xL
Continuous beam with variable section, any slope	10-30	5-10	0.04xL
Overhang beam with variable section l/k = 1:3	k = 10-30	5-10	.01xk

Example M25.1

A hall is designed to be roofed with circular arches with a span L = 27.45m and a rise h = 9.15m. The arches are spaced 7.2m o.c. and for ease of construction are built with a crown hinge (forming a three-hinged arch). The snow load is 1440N/m² and the dead load of the roofing structure is 960N/m² (excluding the arch).

From Table 25.1, we can assume a depth of section:

$$d = 0.02L = 0.02(27,450mm) = 550mm$$

If we use glulam, the closest size in the tables is 215x570, weighing

$$w = (215mm \times 570mm \times 10^6 m^2) = 675N/m$$

The minimum R is about 4.5m, which is much smaller than the design radius r = L/2 = 13.725m.

Total dead load: Length of arch = 3.14(13.725m) = 43m

Weight of arch: W_a = (43m)(675N/m)	=	29,025N
Weight of roofing: W_r = (43m)(7.2m)(960N/m²)	=	297,216N
Weight of snow: W_s = (43m)(7.2m)(1440N/m²)	=	284,602N
Total	=	610,843N = 611kN

Support reactions components are

$$R_v = (611kN)/2 = 305kN$$
$$R_h = W/4 = (611kN)/4 = 153kN$$

The moment is maximum at x = 0.135(13.725m) = 1.85m

$$M = -\frac{611kN}{4}\sqrt{2(13.725m)(1.85m)-(1.85m)^2} + (\frac{611kN}{2})(1.85m) - \frac{(611kN/4)(1.85m)^2}{13.725m} = -524kNr$$

The minus sign indicates that M is clockwise on the left half of the arch, i.e., the extreme outer fibers are in tension.

From the tables of the moment capacity for 24f-DF (Douglas fir), the following sections are adequate:

Shape (mm)	M_r
215X760	570kNm
265X684	569kNm
175X836	561kNm

175X836 has the smallest area of the three (about 146,000mm²). The proportion of the section is between 1:4 and 1:5; therefore, it is not too slender for lateral stability. This preliminary size has not yet taken into consideration the axial load that accompanies bending. We could conservatively use the reaction at the base of the arch to determine the area required in compression:

$R_v = 305kN$

$A = (305,000N)/(16.5MPa) = 18.5 \times 10^3 mm^2.$

If we use this component in the interaction formula, it would look something like this:

$$\frac{M}{M_r} + \frac{A_{required}}{A} = \frac{524kNm}{61} + \frac{18.5 \times 10^3 mm^2}{146 \times 10^3 mm^2} = 0.93 + 0.13 = 1.06 \quad \text{higher than 1. 0.}$$

A larger 215x760 can be selected, with $A = 163 \times 10^3 mm^2$. The previous interaction formula becomes sufficiently close to 1.0 (+3%). The C_p factor is not used because of the marginal influence of compression in the arch.

Table M25.2 Design table for laminated arches

Vertical and wind loadings   For preliminary design only

vertical dead load = 3,500N/m and horizontal wind load = 3,500N/m

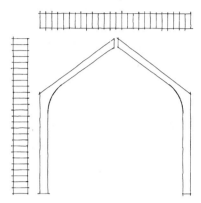

Loading diagram

Roof pitch	Wall height		Span 12m						Span 15m			
	m	Width mm	Base mm	Tangent pts. (mm) lower	upper	Crown mm	Width mm	Base mm	Tangent pts. (mm) lower	upper	Crown mm
10:12	2.4	127	190	267	286	286	127	190	311	343	343
	3	127	190	305	273	273	127	190	343	381	324
	3.6	127	190	375	248	248	127	190	381	419	286
13:12	2.4	127	190	292	324	324	127	190	362	400	400
	3	127	190	349	381	248	127	190	400	444	311
	3.6	127	190	400	444	215	127	190	483	533	260
14:12	2.4	127	190	336	375	298	127	190	413	457	368
	3	127	190	400	444	254	127	190	476	527	317
	3.6	127	190	476	527	222	127	190	559	616	279

These tables use a simplified loading condition which may not adequately represent actual conditions. For this reason, the dimensions indicated should be used **ONLY FOR PRELIMINARY DESIGN**. They will provide a reasonable size approximation for architectural design purposes. Final designs must be evaluated using the critical load conditions specified by the applicable building code.

Table M25.3 Design table for laminated arches
Vertical and wind loadings   For preliminary design only

vertical dead load = 4,670N/m and horizontal wind load = 4,670N/m

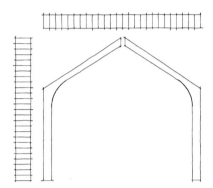

Loading diagram

Roof pitch	Wall height	Span 12m					Span 15m				
	m	Width mm	Base mm	Tangent pts. (mm) lower	upper	Crown mm	Width mm	Base mm	Tangent pts. (mm) lower	upper	Crown mm
10:12	2.4	127	190	292	324	324	127	235	362	400	400
	3	127	190	343	381	273	127	190	419	463	317
	3.6	127	190	413	457	229	127	203	495	540	279
13:12	2.4	127	190	375	413	279	127	229	438	483	381
	3	127	190	451	502	222	127	235	521	578	311
	3.6	127	190	521	578	222	127	260	470	521	267
14:12	2.4	127	190	438	483	286	127	317	527	284	362
	3	127	190	521	578	248	127	260	463	514	317
	3.6	127	190	463	514	260	127	260	546	603	279

These tables use a simplified loading condition which may not adequately represent actual conditions. For this reason, the dimensions indicated should be used **ONLY FOR PRELIMINARY DESIGN**. They will provide a reasonable size approximation for architectural design purposes. Final designs must be evaluated using the critical load conditions specified by the applicable building code.

Table M25.4 Design table for laminated arches
Vertical and wind loadings   For preliminary design only

vertical dead load = 5,840N/m and horizontal wind load = 45,840N/m

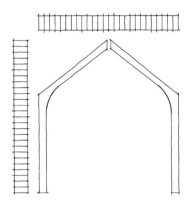

Loading diagram

Roof pitch	Wall height	Width	Base	Tangent pts. (mm)		Crown	Width	Base	Tangent pts. (mm)		Crown
				Span 12m					Span 15m		
	m	mm	mm	lower	upper	mm	mm	mm	lower	upper	mm
10:12	2.4	127	190	349	387	311	127	267	432	476	368
	3	127	190	425	470	260	127	260	514	571	311
	3.6	127	203	514	571	222	127	260	457	508	279
13:12	2.4	127	248	457	508	273	127	387	546	603	343
	3	127	241	540	597	260	127	260	495	546	298
	3.6	171	260	483	533	260	127	260	571	635	267
14:12	2.4	127	317	527	584	286	127	286	495	546	368
	3	171	260	483	533	260	127	273	571	629	317
	3.6	171	260	571	629	260	127	286	667	737	279

These tables use a simplified loading condition which may not adequately represent actual conditions. For this reason, the dimensions indicated should be used **ONLY FOR PRELIMINARY DESIGN.** They will provide a reasonable size approximation for architectural design purposes. Final designs must be evaluated using the critical load conditions specified by the applicable building code.

26 CONNECTIONS

26.1 Connecting wood members

Connections and joints for wood members can be built conceptually in three ways as in Figure 26.1 (a), (b), and (c):

1. With notches, lap joints, and mortise-and-tenon joints that transfer the stress directly from member to member, with the possibility of wood dowels or pegs to provide stability; a Tinker toy is a good example of this system.

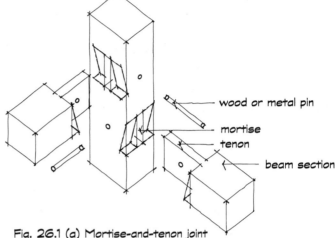

Fig. 26.1 (a) Mortise-and-tenon joint

wood or metal pin
mortise
tenon
beam section

2. With steel or iron mechanical fasteners, such as nails, screws, and bolts that transfer the stress usually by shear through the fastener from one member to the other; an Erector set illustrates this system.

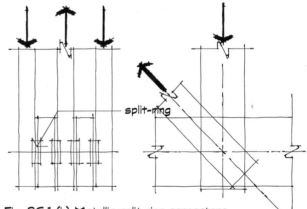

split-ring

Fig. 26.1 (b) Metallic split-ring connectors

3. With steel connectors, such as plates and specially shaped steel supports (joist hangers, column bases, framing anchors, etc.), that transfer the load from one wood member to the other through the metal and the fasteners (bolts, nails, etc.) that secure the metal piece to the wood member.

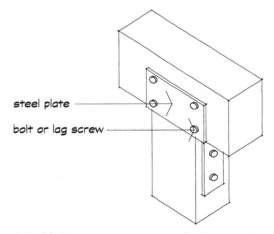

steel plate
bolt or lag screw

Fig. 26.1 (c) Steel plate connectors for heavy timber

Today, wood members are connected almost exclusively with the last two systems, and a large number of proprietary plates and metal supports exist, the best known probably being the toothed plate used in engineered trusses. Various forms of tension connectors (anchors, hold-downs, ties, and straps) are necessary as well as the metal gusset plate and toothed-plate connectors.

Connectors used for wind and earthquake loads are shown in Fig. 26.2. Galvanized steel joist hangers are also a very common type of metal plate connector that can be nailed to other wood members, or anchored to a masonry or concrete wall. Glues are used to produce reconstituted wood like plywood and glulam, but they cannot replace mechanical fasteners entirely in primary structural connections. The use of adhesives is introduced in Sec. 26.3.

We will deal with the most common fasteners: nails, screws, bolts, shear plates, and split rings. There are other widely used fasteners, such as staples and truss metal plates, that are subject to more specialized recommendations and tests, but these are not found in the NDS.

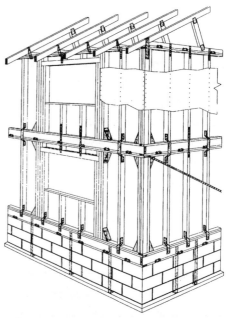

Fig 26.2 Metal connectors commonly used in light-frame wood construction

26.2 Nails and spikes

Nails and spikes, (Fig. 26.3), are part of the class of mechanical fasteners or connectors for timber joints. When the structural design depends on specific design values of nails (engineered construction), specifications must clearly describe the desired fastener.

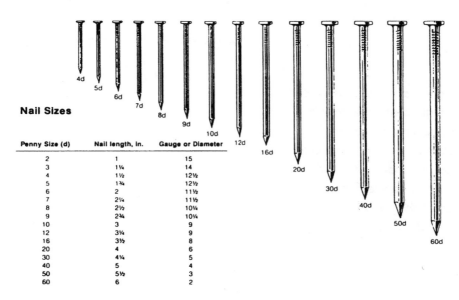

Nail Sizes

Penny Size (d)	Nail length, in.	Gauge or Diameter
2	1	15
3	1¼	14
4	1½	12½
5	1¾	12½
6	2	11½
7	2¼	11½
8	2½	10¼
9	2¾	10¼
10	3	9
12	3¼	9
16	3½	8
20	4	6
30	4¼	5
40	5	4
50	5½	3
60	6	2

Fig. 26.3 Standard nail and spike dimensions

The length is indicated in pennyweight (e.g., a 3in [76mm] long nail or spike is called 10 pennyweight, written 10d). The diameter, on the other hand, distinguishes the various types of nails, and nails from spikes. For instance, a 20d box nail has a diameter of 0.148in [3.8mm], and a 20d spike has a diameter of 0.225in [5.7mm], and both are 4in [102mm] long.

Let's answer the big question: what is a pennyweight? A pennyweight is a unit of troy measurement that equals 1/20 of a troy ounce, with a pound containing 12 troy ounces. This suggests that early nails were sized by weight. It's a piece of history trivia. Its meaning as a measurement of nails has long since lost relevance since a 16d nail and a 16d spike certainly do not weigh the same. They're both the same length, but of significantly different diameters.

Nails and spikes are made of common steel (wire nails, box nails) or hardened high-carbon steel wire (heat-treated and tempered, annularly or helically threaded). As for other mechanically fastened joints, the strength of the connection depends on the nail type, as well as on the lumber species or its specific gravity G. Values of G for different species are given in the NDS (Table 12A).

26.2.1 Adjustment factors and design values

Nailed connections work in withdrawal (nail or spike pull-out from the wood) and in shear (single or double). Joints using nails or spikes in withdrawal should be avoided since the structure is essentially constantly attempting to take itself apart. The only real exception to this recommendation is if the nails penetrate through the material and are "clinched" (bent over) on the other side. This significantly reduces the possibility of their being pulled back out through the hole.

The respective design values W and Z are given in the NDS (Tables 12.2 and 12.3). The design values must be multiplied by the applicable adjustment factors, as indicated in the NDS (Table 7.3.1). The adjusted shear design value is

$$Z' = Z(C_D C_M C_t C_d C_{di} C_{tn})$$

C_D, C_M, and C_t are already known, as we used them earlier; the three new factors are C_d, C_{di}, and C_{tn} (see Table 26.1).

Table 26.1 Adjustment factors applicable to nails and spikes

C_D	Load duration	
C_M	Wet service	
C_t	Temperature	
C_d	Penetration depth	NDS 12.3.4
C_{di}	Diaphragm	NDS 12.3.6
C_{tn}	Toe nail	NDS 12.3.7

Although nailing into the end grain is permissible, it is more common and safer to connect wood on the lateral grain. For end-grain connections, the end-grain factor C_{eg} applies (NDS). Think again of the wood as a bundle of straws. A nail driven parallel to the fibers (into the end grain) will tend to displace a few cells and consequently has little holding power. Grip a pencil parallel to your fingers and pull; now grip the pencil perpendicular to your fingers and pull. This illustrates why end-grain connections are so weak.

The design values given in the NDS tables are based on nail or spike penetration into the main member of 12 times the nail or spike diameter,

$$p = 12D$$

where D is the diameter of the nail or spike.

The nail or spike can be driven less than that into the wood but not less than the minimum

$$p_{min.} = 6D$$

For all intermediate penetration depths between 12D and 6D, the bearing capacity of the connection must be reduced by multiplying the tabulated design values by the penetration depth factor, calculated as

$$C_d = p/12D \leq 1$$

which is simply a proportion of the actual penetration p to the required one, reducing the allowable load of the connection.

C_{di} applies in diaphragm construction, which uses plywood (or equivalent) sheathing to resist and transmit lateral loads. The design values for these connections have to be multiplied by $C_{di} = 1.1$.

Example 26.1

A 3/4in plywood subfloor is nailed to Sitka-spruce joists, G = 0.43. Dry conditions, ordinary temperature. 8d threaded hardened-steel nails are used (0.120in diameter and 2.5in long), and $C_D = 1.0$. The required penetration for full-shear strength is (NDS, Table 12.3C):

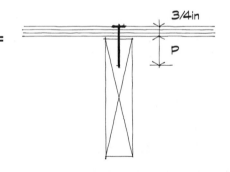

$$12D = 12(0.12in) = 1.44in$$

Determine the design shear value of each nail.

In the NDS (Table 12.3C), G = 0.43 corresponds to the value of hem-fir. In this column, for side-member thickness = 3/4in and pennyweight 8d, we find Z = 68lb.

The actual penetration is

$$p = 2.5in - 0.75in = 1.75in > 12D \quad OK!$$

Therefore, the reduction factor $C_d = 1.0$.

The shear design value of each nail is

$$Z' = Z(C_D C_d C_{di}) = (68lb)(1.0)(1.0)(1.1) = 74.8lb$$

Example 26.2

The toe-nail connection in the illustration, designed according to the NDS (Sec. 12.1.3.2), connects a stud to a sole plate. Determine the load capacity of this connection. Use Western cedar (North) and 16d common wire nails. C_D = 1.6.

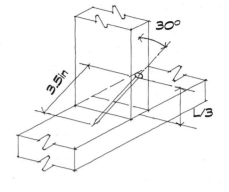

Since the length of the nail is 3.5in, the "side-member" thickness or portion of the spike in the stud is (3.5in)/3 = 1.17in.

Required penetration depth:

12D = 12(0.162in) = 1.9in

Actual p = 2(3.5in)/3 = 2.33in > 12D

Therefore: C_d = 1.0
From the NDS (Table 12.3B):
Side-member thickness 1in: Z = 85lb
Side-member thickness 1.25in: Z = 95lb

Interpolating: Z = (1.17in)(95lb)/(1.25in) = 88.9lb

Z' = Z $(C_D C_d C_m)$ = (88.9lb)(1.6)(1.0)(0.83) = 118lb

26.3 Adhesives

Adhesives are used to manufacture engineered lumber, but also for connections on-site. It is becoming common practice to nail and glue the sheathing to the joists to ensure a rigid floor, but in these cases the glue cannot replace the mechanical fasteners [Fig. 26.3(b)]. Special resin-based adhesives are used for repairs of structurally damaged beams.

Adhesives are potentially very efficient connectors, since they allow a full utilization of the lumber section without the need for holes (as in bolted connections), and they eliminate the problem of splitting of the grain caused by nails and screws. The limitation is that they cannot be used for a major structural connection, but only to make a connection built with mechanical fasteners more rigid for a period of time (e.g., during transportation) or to produce some form of engineered wood material.

Types of adhesives for structural use today are all synthetic and can be classified in two groups.

26.3.1 Synthetic thermoset

Resorcinol-formaldehyde resin

Properties: Normally sold as a two-component system, with a liquid resin and a hardener. Members must be kept under pressure for several hours at room temperature. The bond is resistant to water and chemical attack, and suitable for treated wood.

Use: Production of glulam, plywood, and engineered components also for exterior use.

Melamine-formaldehyde resin

Properties: Usually requires mixing a powder with water and a catalyst. Requires heat for setting. Water resistant.

Use: Scarf and finger joints for engineered wood products.

Epoxy

Properties: Prepared by mixing a liquid resin with a hardener. Can be mixed with various additives or aggregates.

Use: Bonding wood with other materials; structural repairs.

26.3.2 Synthetic thermoplastic

Resin emulsions

Properties: These are sold ready to use in the form of a white liquid. They harden quickly and are colorless. Creep and water-resistance problems.

Use: Connections requiring temporary strength, such as frames in mobile or modular homes for transportation.

Elastomeric mastic

Properties: The paste is easy to apply also to coarse surface. Gives bond of moderate strength, subject to creep and not resistant to water and intense heat. Does not require much pressure during hardening, but emits flammable fumes.

Use: Connections of sheathing to joists and studs in combination with nails, and other similar semistructural uses.

The most common adhesives used to connect members on-site are elastomeric mastics, because of their ease of application. There are no codes or standards that provide allowable design values for glued connections, and the strength of these is determined empirically by each manufacturer. Tests also have to take into account the change in properties in the long term, load duration, and environmental effects.

26.4 Glue-line stresses: Rolling shear

Structural glue is always stronger than the timber it connects, and it generally is designed to transfer load in shear. The glued joint is therefore calculated as the allowable shear stress on the faces of the timber and plywood being glued. Applicable adjustment factors for wood design values must be used.

The rolling shear has this name because shear stresses perpendicular to the grain tend to separate the fibers and make them roll (the same way that a bundle of pencils will roll between your hands when the hands are pushed in opposite directions). The rolling shear has to be checked at the contact surfaces between lumber and plywood or between two lumber members joined with the grain perpendicular to each other. Since plies are bonded with the grain perpendicular to each other, plywood has to be checked for rolling shear irrespective of the direction of the face grain. Plywood used with the grain at 45° is therefore stronger. The rolling shear LOAD is calculated using the shear flow formula:

$$P_v = \frac{VQ}{I} = \frac{VA\tilde{y}}{I}$$

where Q (the moment of the area) = $A\,\tilde{y}$, V is the shear force at the point of investigation, and I is the **total** moment of inertia of the section. P_v is expressed in lb/in or kip/inch [*N/mm or kN/mm*] of length, and it is the force required to hold the flange to the web. This formula could be used to design the mechanical fasteners that would assume the responsibility of transferring this load between two components. The following formula, which determines the horizontal shear STRESS, could be utilized to determine if the remaining wood at a partial horizontal joint was sufficient to transfer the load. It could be used to check the glue stress at a joint. Finally, it could be used to determine the rolling shear stress developed between adjacent plies at a joint where you are attempting to transfer the load from plywood to adjacent wood through a glue joint. The value f_v is the horizontal shear stress at the location within a component at a location experiencing a specific shear value (V) at that point. The unit shearing stress is force P_v divided by the contact width of material (b).

$$f_v = \frac{P_v}{b} = \frac{VA\tilde{y}}{Ib}$$

As an analogy, in a laminated-timber beam, P_v is the shear FORCE between laminations. On a T-beam combining plywood sheathing and lumber joist, $f_v = (VA y)/(bI)$ is the rolling shear stress experienced by the glued plywood (see Figure 26.4).

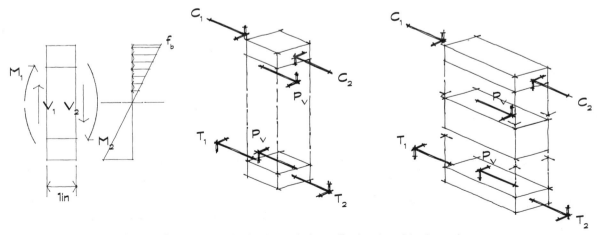

Fig. 26.4 Equilibrium of beam flanges and the horizontal shear P_v developed in the web.

Example 26.3

As an example of the design of a glued-and-nailed connection, we will use the box beam designed in Example 26.2.

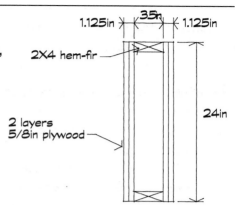

Allowable rolling shear stress for S-2 plywood: $F_s = 53$psi.

Maximum shear value: $V = 4,800$lb

$$b = 1.5\text{in} + 1.5\text{in} = 3\text{in}$$

$$I_x = \frac{3.5\text{in}\left[(24\text{in})^3 - (21\text{in})^3\right]}{12} + 2880\text{in}^4 = 4210.9\text{in}^4$$

$$f_v = \frac{VA\tilde{y}}{Ib} = \frac{(4800\text{lb})(59.1\text{in}^3)}{(4210.9\text{in}^4)(3\text{in})} = 22.5\text{psi} \leq 53\text{psi} \qquad \text{OK!}$$

The nailed connection must carry the entire shear load at the joint in case of failure of the glue. The shear flow is

$$P_v = \frac{VA\tilde{y}}{I} = \frac{4800\text{lb} \times 59.1\text{in}^3}{4210.9\text{in}^4} = 67.4\text{lb/in}$$

Hem-fir is used for the flanges. The NDS (Table 12A) gives $G = 0.43$ for this species. If 12d box nails are used, Table 12.3A should be consulted for data. The required penetration is

$$12D = 12(0.128\text{in}) = 1.536\text{in}$$

The actual penetration depth, using a double 5/8in plywood web on each side, is

$$p = 3.25\text{in} - 1.25\text{in} = 2\text{in} > 12D$$

Therefore, the full tabulated strength of the nail can be used.

For $G = 0.43$ (hem-fir), Table 12.3A gives $Z = 80$lb.

Adjusted design value:

$$Z' = Z(C_D C_d C_{di}) = (80\text{lb})(1.0)(1.0)(1.1) = 88\text{lb}$$

This is the shear capacity of each nail.

The force to be supported by the nails per inch on each side of the flange is

$$P_v/2 = (67.4\text{lb/in})/2 = 33.7\text{lb/in on each side}$$

The spacing of the nails is

$$\text{Spacing} = Z'/(P_v/2) = (88\text{lb})/(33.7\text{lb/in}) = 2.61\text{in}$$

Use 12d box nails spaced 2-1/2in.

In the central portion of the beam, where the shear value decreases to half the maximum (see design of stiffeners, Sec. 21.5), space the nails 5in o.c. This is a considerable amount of nails; larger (and longer) nails would result in longer spacing, but since we are using a low-grade 2X4 for the flanges, the grain could split. A nailing gun would be convenient to execute the connection on-site.

Example 26.4

In order to increase the moment capacity of floor joists, 23/32in (0.72in) Sturd-I-Floor ply-wood sheathing is glued and nailed to 2X12 Eastern pine joists. The joists must be designed for a maximum shear of V = 260lb. The joists are 16in o.c.

Section properties:

Flange A = (16in)(23/32in) = 11.5in²;   y = 11.25in + 0.5(23/32in) = 11.61in

Web A = (1.5in)(11.25in) = 16.87in²;   y = (11.25in)/2 = 5.625in

Total A = 28.37in²

The distance of the neutral axis from the bottom is found from the moment of the areas equation:

$$A\widetilde{Y} = \Sigma a\widetilde{y}$$

[(Total area) × (distance from the bottom to the centroid of the total area) is equal to the (sum of each individual area) × (distance from the bottom to the centroid of that area).]

$$28.37\text{in}^2\widetilde{Y} = 11.5\text{in}^2(11.61\text{in}) + 16.87\text{in}^2(5.625\text{in})$$

which is the distance from the bottom to the neutral axis of the total section.

$$\widetilde{Y} = 8.05\text{in}$$

Rolling shear stress:

$$f_v = \frac{VQ}{Ib} = \frac{V(A\widetilde{y})}{Ib}$$

In this case, the shear stress is a function of the area × (the distance from the neutral axis of the total section to the neutral axis of this area). We are explaining this carefully since there can easily be confusion about which area and which distance to use.

$$I_{flange} = \Sigma(I_o + Ad^2) = 1\times\left[\frac{16\text{in}(0.72\text{in})^3}{12} + 11.5\text{in}^2(11.61\text{in}^2 - 8.05\text{in})^2\right]+$$

$$I_{web} = \Sigma(I_o + Ad^2) = 1\times\left[\frac{1.5\text{in}(11.25\text{in})^3}{12} + 16.87\text{in}^2\left(8.05\text{in} - \frac{11.25\text{in}}{2}\right)^2\right]$$

$$I_{total} = 146.24\text{in}^4 + 277.19\text{in}^4 = 423.43\text{in}^4$$

Rolling shear stress:

$$f_s = \frac{VA\widetilde{y}}{Ib} = \frac{(260\text{lb})(40.94\text{in}^3)}{(423.23\text{in}^4)(1.5\text{in})} = 16.8\text{psi} \leq 53\text{psi} \qquad \text{OK for rolling shear.}$$

To design alternative nailing, we can use this previous calculation and simply not divide by 1.5in or take the answer and multiply it by 1.5 to determine P_v (horizontal shear LOAD):

$$P_v = f_s b = (16.8\text{psi})(1.5\text{in}) = 25.2\text{lb/in}$$

Nails: 12d common-wire spikes, 12D = 2.3in

Actual penetration depth: p = 3.25in - 0.72in = 2.53in > 12D

Wood: G = 0.36

For side-member thickness:

> 5/8in   Z = 91lb
>
> 3/4in   Z = 93lb
>
> 23/32in   Z = 92lb (approximate interpolation)

Nail spacing = (92lb)/(25.2lb/in) = 3.6in

Use 12d common-wire spikes spaced 3-1/2in for the entire length of the joist. Obviously, since the shear value (V) changes constantly, we could also vary the spacing of the nails, but this becomes more calculation hassle than it's worth. We could recognize, however, that the shear value at the 1/4 point of the span is only 1/2 the value at the end (for a uniform load) and realize that if the shear value is 1/2, the nail load will also be 1/2. At this location, we could double the spacing of the nails. This technique could be used as many times across the span as you have the patience for, but too many changes in nail spacing causes more coordination problems than it saves in nails.

26.5 Bolts

Bolts by themselves are not particularly efficient structural connectors in wood. They are so much stronger than the wood they connect that they cannot effectively transfer their full capacity to the wood through a standard bearing connection. However, they are very useful in a temporary structure or any other situation where a system must be repetitively disassembled and reassembled. They do not rely on threads in the wood as screw systems do, and they are not damaged by removal of the connector as nailed or spiked systems might be. In more permanent structures, bolts are more effectively used in conjunction with shear plates or split rings that have large bearing areas relative to the amount of material that they use (see Fig.26.5).

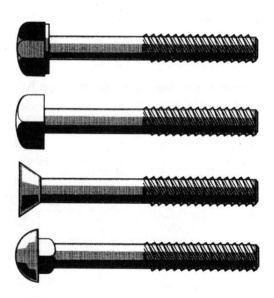

Fig. 26.5 Common bolts

As in steel, bolted connections in wood load the bolts in single or multiple shear. Bolt holes are drilled larger than the bolt shank for assembly purposes, a minimum of 1/32in [1mm] to a maximum of 1/16in [2mm]. The bolts should not be forced into tight holes, since this may split the wood grain. Washers are always used in wood bolted connections on the head and nut sides to avoid cutting into the wood grain when tightening; washers are not necessary when metal plates are used. Carriage bolts are used without washers, and the heads tend to be drawn into the wood. They are therefore used only in minor structural applications.

A bolted connection is likely to loosen due to wood shrinkage; this is not dangerous, since the bolts will still work in shear as before, but tightening the connection again will ensure that the members are kept in line.

The design values given in the NDS allow for the inclusion of the thread in the shear plane, but in practice, it is recommended to keep the threaded length as short as is necessary to tighten the nut snugly on the washer or plate, allowing also for some shrinkage. Threaded bars, or bolts where the shank is fully threaded (such as a stove bolt), do not have a good bearing on the hole and should not be used in shear connections.

Bolts are often used to connect wood members to other structures: concrete foundations, masonry walls, or steel members. In some cases, proprietary anchor bolts are used in concrete, and the manufacturers' tables will provide the load-bearing capacities.

Anchor bolts should be embedded in the concrete or masonry grout to a minimum of 4in [102mm]. The UBC (Table 24-M) gives allowable shear values for bolts embedded in masonry, based on bolt diameter and embedment length. Similarly, the UBC (Table 26-E) gives allowable loads for bolts embedded in concrete. These values are valid for masonry, but have to be checked for the wood connection with the NDS tables.

We will consider only the bolted connections that work in shear, since, as in steel construction, these represent the majority of cases. The bolt shear design values have to take into account the bearing stress of the bolt on the wood (Fig. 26.6). These values are given in the NDS (Tables 8.2.A and 8.2.B) and have to be adjusted with a number of factors, as shown in Table 26.2.

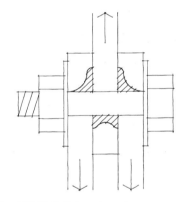

Fig. 26.6 Bolt bearing stresses

Table 26.2 Adjustment factors applicable to bolts

C_D	Load duration factor	
C_M	Wet service factor	NDS Table 7.3.3
C_t	Temperature factor	NDS Table 7.3.4
C_g	Group action factor, for lag screws in a row	NDS Table 7.3.6A
C_{delta}	Geometry factor	
	(same as for bolts with equal shank diameter)	NDS Art 9.4

The group-action factor takes into account the fact that in multiple-bolt connections, the distribution of stress is not equal for each bolt. This is true for various types of mechanical fasteners. In general, the fasteners at the end of a row parallel to the direction of applied force assume more load than the ones in the center. One consequence of this fact is that by adding more bolts in the same row, the strength of the connection does not increase appreciably, since the stress in the end connectors controls the total strength of the connection.

From the NDS (Table 7.3.6A), it can be seen that when using a small number of bolts in a row (two or three) connecting members of equal cross-sectional area, the group-action factor is in many cases equal to 1.0, so this factor is not crucial for the outline design of small connections. In our examples, we will assume that this factor is equal to 1.0, but it must be remembered that for a correct design solution of a connection with two or more bolts, the group action must be considered.

Example 26.5

A 4×4 Douglas-fir knee brace is connected to a column with 1/4in steel plates and bolts. The structure is exposed to the exterior. The portion of wind load to be resisted by this frame is W = 1200lb.

We can assume that both corner braces will work together, one in tension and the other in compression. The horizontal reactions in A and B are therefore equal to (1,200lb)/2 = 600lb. The equation of moments about C for the column gives us the horizontal component:

$$V = (P/2)(h/b) = (1,200lb)(10ft)/2(3ft) = 2,000lb$$

The tension in the brace is

$$T = \frac{V}{\cos\theta} = \frac{2000lb}{\cos 45°} = 2828lb$$

Using the NDS (Table 8.3B) for bolts in double shear, we try a 5/8in bolt.

$C_M = 0.75$ (Table 7.3.3)

$C_D = 1.6$ (wind); therefore:

$Z' = 1.6(0.75)(2,250lb) = 2,700lb$

This is within 5% of the required capacity, so we would say . . . OK!

Knee braces are often connected without steel plates with a through bolt. In this case, the NDS requires us to consider the equivalent shear area, L_s, being the equivalent thickness of the side member for the purpose of determining the shear design value from the NDS (Table 8.2A). The component of the tension load parallel to the bolt axis generates a pressure under the washer that has to be considered in the design of the connection. An alternative solution is to use a wood wedge in place of the notch to accommodate the bolt head.

26.6 Lag screws

Lag screws are effectively very large wood screws that have hex or square heads so they may be inserted with a wrench (a screwdriver of this size would be unusable). They are used where it is not necessary, undesirable, or impossible for the bolts to penetrate completely through the member. For example, they may be used for large timbers connected with metal side plates or to attach a large laminated wood beam into a steel stirrup-shaped connection. Typical dimensions are found in the NDS (Appendix L); the nominal length (exclusive of the head) varies from 1 to 12in [25 to 305mm], with diameters of the unthreaded shank from 1/4 to 1-1/4in [6.3 to 31.7mm].

The NDS (Tables 9.3A and 9.3B) gives the design value for single shear in the case of, respectively, two wood members of identical species and connections with steel side plates. These design values, as usual, have to be adjusted with the following applicable factors (from NDS Table 7.3.1):

Table 26.3 Adjustment factors applicable to lag screws

C_D = Load-duration factor
C_M = Wet-service factor (Table 7.3.3)
C_t = Temperature factor (Table 7.3.4)
C_g = Group-action factor, for lag screws in a row (Table 7.3.6A)
C_Δ = Geometry factor (same as for bolts with equal shank diameter; Art. 9.4)
C_d = Penetration factor (full and minimum penetration defined in Art. 9.3.3)
C_{eg} = End-grain factor (Art. 9.3.4)

Lag screws have a similar load-bearing capacity to bolts of the same diameter in single shear.

Example 26.6

The connection in Example 26.5 could be made with 5/8in lag screws, one on each side of the connection. From the NDS (Table 9.3B),

$Z = 1,140\text{lb}$

$Z' = 2(1.6)(0.75)(1,140\text{lb}) = 2,736\text{lb}$ OK!

26.7 Wood screws

Screws work in a similar way to nails, with the advantage of a superior withdrawal resistance. Screws differ in the length, diameter, and head shape. The NDS (Table 11.2A) gives withdrawal design values for wood of different specific gravity. The applicable adjustment factors are the already known C_D, C_M, C_t, C_d, and C_{eg}. There is no group-action factor.

C_d is the penetration factor, defined as the ratio between the actual penetration and the full penetration of the screw, or

$$C_d = p/7D$$

where

p = actual penetration of the threaded part of the shank in the main member

D = diameter (unthreaded) of the shank, called also gage

The range of gages listed in the NDS is from 6 (minimum length of 1/2in [38mm]) to 24(lengths of 3-1/2 or 4in [89 or 102mm]).

Example 26.7

A tension member made of a 2X12 spruce-pine-fir, accessible only on one side, has to be spliced using a 10-gage steel plate and screws. The tension load is 750lb due to wind; therefore, C_D = 1.6.

Try to use four screws each side of the joint. Since it must be (4 screws)(1.6)Z = 750lb, the required shear capacity of a single screw is

Z = (750lb)/(4)(1.6) = 117lb

By using 8g screws, Z = 122lb

Ten-gage steel has a thickness t = 0.134in, the 8g screw has a gage of 0.164in, so the full penetration is 7D = 7(0.164in) = 1.15in. With a 1.5in screw, the actual penetration corresponds to the length of the threaded part, or approximately (2/3)(1.5in) = 1.0in. Therefore:

C_d = 1.0/1.15 = 0.87

Z' = 1.6(0.87)(122lb) = 170lb

Splice capacity P = 4(170lb) = 679lb, which is insufficient!

We can use five screws:

Splice capacity P = 5(170lb) = 850lb OK!

26.8 Split-ring and shear plate connectors

Split rings and shear plates are shown in Fig. 26.7. These connectors are the most efficient type for load transfer between wood members, and work essentially by bearing on the wood grain with a surface much larger than any bolt or spike. In split rings, the bolt is used mainly to keep the parts together, and in shear plates, the bolt works in single shear.

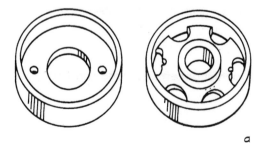
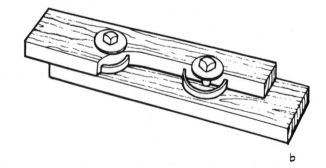

a

b

Fig. 26.7 Shear plates (a) and split rings (b)

A split-ring unit consists of one split ring with its bolt. Spit rings are driven snug into a groove, and the "split" makes them somewhat flexible and able to adjust to distortions or shrinkage of the wood.

A shear plate unit consists of either

(a)   one shear plate connector with its bolt in shear, used with a steel side plate, or

(b)   two shear plates used back to back with a bolt

A special tool is required to cut out the seat for the plate. An assembly tolerance is left, so that a certain amount of slip may occur after construction. Dimensions of split rings and shear plates are given in the NDS (Appendix K). They are made in two diameters and are used with standard bolt sizes.

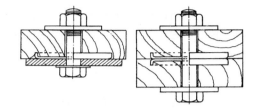

Fig. 26.8 Shear plate connection detail

Although a shear plate unit has a load capacity only slightly higher than a split ring, it can be used with steel side plates, which allows the members themselves to be smaller and easier to assemble on-site.

Table 26.4 Standard sizes of shear plates and split rings

Connector	Diameter (in) [mm]	Bolt diameter (in) [mm]	Material
Split ring	2-1/2 [63.5]	1/2 [12.7]	Steel
	4 [102]	3/4 [19]	Steel
Shear plate	2-5/8 [66.7]	3/4 [19]	Pressed steel
	4 [102]	3/4 or 7/8 [19 or 22.2]	Malleable iron

Source: Adapted from the NDS

The transfer of loads of split-ring and shear plate connections differ considerably. The depth of the split ring is shared equally between the two members connected face to face. This requires that the members overlap, and the location where connectors may be placed is in the overlap zone. For this reason, split rings are most effectively used in relatively thin and wide members. The necessity to provide overlap areas will frequently be the most critical criterion in determining member size. A 4in [89mm] member that will adequately carry the load may have to be increased to a 6 or 8in [140 or 191mm] member simply to provide enough area to place the split rings. On the other hand, shear plates do not transfer their load directly from wood to wood. A shear plate is a combination connector that first transfers its load to a steel plate and then from the steel plate back into the wood. For this reason, the members may be designed based on their specific load (and not on connector space requirements), and the steel plate will be increased in size to provide space for connectors. Although in both cases some component is enlarged, to enlarge the end of a wood member generally requires that the entire member be larger; but to enlarge a steel connector plate requires only that component be enlarged. The loads on either connection cause compression stresses between the ring and the wood both inside and outside the groove. These compression forces must be resisted by shear forces in the grain close to the connector; because of timbers' low resistance to horizontal shear, the distance of the connector from the end is critical.

The action of a shear plate is similar to the way steel bears on the wood, but the shear plate transfers the load from its "web" to the bolt in bearing; the bolt then transfers the load to the connection plate in shear.

The load capacity of a split-ring or shear plate unit depends primarily on the species group of timber, design configuration, conditions of use (adjustment factors), and the angle of the load with the grain.

1. Species groups. Wood is classified in four species groups, in decreasing order of strength: A, B, C, and D (NDS, Table 10A). Laminated timber is classified according to compression design values perpendicular to the grain, F_c, according to the NDS (Supplement, Table 5A, footnote 12).

2. Adjustment factors. NDS Table 7.3.1 gives the applicable factors for shear plates and split rings.

Table 26.5 Adjustment factors applicable to shear plates and split rings

C_D Load-duration
C_M Wet service
C_t Temperature
C_g Group Action   NDS 7.3.6 and Table 7.3.6.
         When using only two or three fasteners in a row, in most cases C_g can be taken equal to 1.0. When using more fasteners and when connecting members with different cross-sectional areas, it should be checked.

C_Δ Geometry   NDS Table 10.3
         Applies if edge distance, end distance, or spacing of connectors has to be reduced to the corresponding value in the table.

C_d Penetration Depth   NDS 10.2.3 and Table 10.2.3
         Applies to lag screws used in place of bolts.

C_{st} Metal Side Plate   NDS 10.2.4 and Table 10.2.4
         Modifies design value parallel to the grain in connections between wood and steel plates.

Source: Adapted from the NDS, Table 7.3.1.

3. Angle of load to grain. Two sets of design values are given:

   P = load capacity parallel to the grain

   Q = load capacity perpendicular to the grain

   (Always sketch your situation; errors here are very common and can be avoided with a simple sketch.)

Fig. 26.9 Loads at angles to the grain

When the load is at an angle Θ with the grain, the design value is in between the two (P, which is $0°$ to the grain, and Q, which is $90°$ to the grain) and can be calculated with Hankinson's formula (NDS, Sec. 10.2.5 and Appendix J), which gives the design value N' as a function of adjusted values P' and Q'.

The design values can be found in NDS Table 10.2A and Table 10.2B. In addition to the factors already mentioned, these values depend also on the thickness of the member and the number of faces (1 or 2) with connectors on the same bolt (Fig. 26.10).

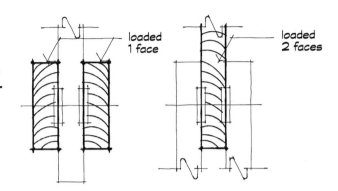

Fig. 26.10 Number of loaded faces of members

26.9 Design of shear plate and split-ring connectors

Once the design value has been defined, the required number of connectors is given by the ratio of the load W to be transferred by the connector to the load capacity or to the adjusted design value of each connector:

Required connectors = W/P' for load parallel to the grain, or

Required connectors = W/Q' for load perpendicular to the grain

Required connectors = W/N' for load at an angle to the grain

The total number of connectors used must be an even number for the joint design to be symmetrical. The connectors are used in pairs on opposite sides of the members. A symmetrical connection will transfer the loads along the geometrical axis of the members, thereby not inducing torsion stresses. More important, the geometric center of the group of connectors must lie on the crossing of the geometric axes of the members, in order to avoid eccentric loads with the consequently induced bending moments.

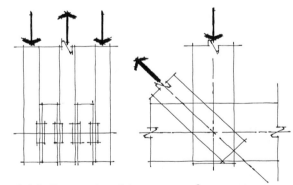

Fig. 26.11 Concentric axial geometry of connections

The geometry of the connection must comply with the minimum edge distance, end distance, and spacing mentioned earlier in relation to C_Δ. The definition of these dimensions is illustrated in Fig. 26.12. The following conditions should be considered.

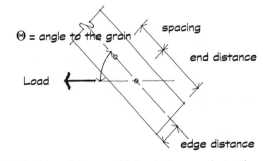

Fig 26.12 Edge distance with load at an angle to the grain

When members are loaded at an angle to the grain of $\Theta > 45°$, the edge distance is the same as for the load perpendicular to the grain. When $\Theta < 45°$, the minimum loaded edge distance must be interpolated between the two minimum values given in the NDS.

Table 26.6 Connector edge distances

Connector diameter	Edge distance
2-5/8in Shear plate 2-1/2in Split ring	$C = \Theta/45° + 1.75\text{in}$
4in S. P. or S. R.	$C = \Theta/45° + 2.75\text{in}$

Source: Adapted from the NDS

Finally, the connection of bending members must take into account a reduced section in shear according to the NDS (Sec. 3.4.5). This is essentially the same procedure used to design notched bending members. The design procedure for a given thickness b, with C as the unloaded edge distance, and d as the depth of member, is as follows:

(a) $L_e < 5d$:

Required $d = (1.5)(V)/(b)(F_v')$

Round up the required value of d.

Check if $5d > L_e$. Find: $d_e = d - C$.

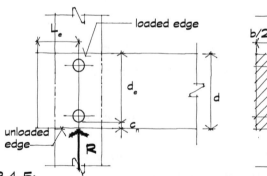

Fig. 26.13 Connection of a bending member

Check the working stress with NDS Eq. 3.4-5:

$$f_v = (1.5)(V)(d)/(b)(d_e^2) < F_v'$$

(b) $L_e > 5d$:

Required $d = (1.5)(V)/(b)(F_v')$

Check if $5d > L_e$. Find: $d_e = d - C$.

Check the working stress with NDS Eq. 3.4-6:

$$f_v = V/(b)(d_e) < F_v$$

Example 26.8

A double, laminated-timber beam system, span 28ft, spaced 20ft o.c., grade 16F-V3 DF/DF, supports an engineered wood joist floor. The beam is connected to a laminated-timber column Combination Symbol 6.

1. Loads:

   Load on connection

   Total floor load = 50psf + 14psf = 64psf

   Load on beam

   w = (64psf)(20ft) = 1280plf

   Estimated beam w = (24in)(6in)(35pcf)/(144in²/ft²) = 35plf

   C_D = 1.0 (occupancy)

   Total w = 1,280plf + 35plf = 1,315plf

   Support reaction at column A

   R_A = (1,315plf)(28ft)(14ft)/(24ft) = 21,479lb

2. Design criteria for connection on the beam:

   Timber species group: F_c = 560psi, group B.

   Use a shear plate diameter of 2-5/8in.

   Number of faces of a member with connectors on the same bolt: 1.

   Beam thickness is over 1-1/2in (use 3-1/8in or thicker, depending on V and M requirements).

   Load is perpendicular to the grain.

   Design value (NDS, Table 10-2B)

   Q = 1,860lb

   Number of connectors required = R_A/Q = (21,479lb)/(1,860lb) = 11.5 connectors

   This number is too large and could affect the column dimension.

   Try 4in shear plates: Q = 3,040lb

   Number of connectors required = R_A/Q = (21,479lb)/(3,040lb) = 7.1 connectors

   Use 8 connectors, 4 on each side.

   | Spacing for loading: | perpendicular to the grain 6in |
   | | parallel to the grain 5in |

   | Edge distance: | unloaded edge 2-3/4in |
   | | loaded edge 3-3/4in |

   | End distance: | The connection is "far" from the ends. |

Width of beam required: d = 3.75in + 6in + 2.75in = 12.5in

3. Design criteria for connection on the column:

   Timber species group: F_c = 470 psi, group C.

   Use 4in shear plates.

   Number of faces of a member with connectors on the saem bolt: 2.

   Thickness > 3-1/2in.

   Load parallel to the grain.

   Design value P = 3,600lb

   Number of connectors required = (21,479lb)/(3,600lb) = 6.0 connectors

   Use 6 connectors, 3 on each side minimum.

   Since 4 connectors each side are used for the beam, the beam connection will be adequate for the column.

   Spacing for loading: parallel to the grain 9in

                                 perpendicular to the grain 5in

   Edge distance: unloaded edge 2-3/4in

   End distance: The connection is, again, "far" from the ends.

Since the spacing parallel to grain is larger than the spacing on the beam, we can try to reduce it.

The load capacity of the 8 connectors on the column is

   P = (3,600lb)(8) = 28,800lb

This design capacity is used only partially and precisely:

Capacity usage = (21,479lb)/(28,800lb) = 0.75, or 75%.

This can be expressed also by saying that the geometry factor C_Δ = 0.75. By interpolating the C_Δ values (NDS, Sec. 10.3.5 and Table 10.3):

Spacing = [(4in)(0.75-0.5)/0.5] + 5in = 7in

Make the spacing 7in. This changes the minimum required beam dimension for the connection to 13.5in.

Minimum column width = 2.75in + 5in + 2.75in = 10.5in

The design of the column will have to be consistent with this requirement.

4. Design of the beam for shear at the support:

Shear:

Left of support A:

$V = (1,315lb)4 = 5,260lb$

Right of support A:

$V = 21,479lb - 5,260lb = 16,219lb$

$F_v = F_v' = 165psi$

First trial section:

Required $A = 1.5(16,219lb)/(165psi) = 147.4in^2$

Two 5-1/8in × 15in have

$A = 2(76.88in^2) = 153.76in^2 >$ required A

Check shear stress.

$d_e = 15in - (2.75in - 2in) = 14.25in$

$f_v = 1.5(V)(d)/(b)(d_e)(d_e)$

$= 1.5(16,219lb)(15in)/[(5.125in+5.125in)(14.25in)(14.25in)] = 175.3psi > 165psi$

A larger section is required for shear. Try 5-1/8in × 16-1/2in:

$d_e = 15.75in$; $f_v = 143.5$ psi < 165 OK!

Example 26.9

The mezzanine floor of a Frank L. Wright Usonian house is suspended from the roof structure with two-layer sawn lumber members. The floor loads on the beam, including all L and D loads, are a uniform load of 490plf and a concentrated 230lb parapet load at the end of the overhang balcony. The wood is Eastern hemlock; the beam is a nominal 6X12. Check if a connection with two bolts and 2-1/2in split rings works.

The load on the connection is the support reaction of the suspension members.

$R = [(490plf)(15ft)(7.5ft) + (230lb)(15ft)]/12ft = 4,881lb$

Tension members:

1-1/2in thick members, connected on one face:

$P = 1960lb$

Number of connectors required $= (4881lb)/(1960lb) = 2.5$ connectors

Use four split rings, although this is more than required. We always want to create connections without eccentricity which means we must use them in opposing pairs.

Spacing: load parallel to the grain (see the figure).

Beam:

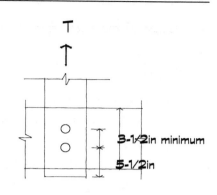

    5.5in-thick member, connected on two faces:

    Q = 1390lb

Number of connectors required =

(4881lb)/(1390lb) = 3.5 connectors

Use 4 split rings.

Spacing: load perpendicular to the grain (see the figure).

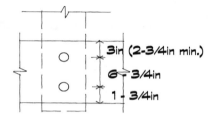

The distances for full load on the beam are

    D = 2.75in + 6.75in + 1.75in = 11.25in

Since the actual beam dimensions are 5.5in × 11.5in, the connection is adequate. The connection design is shown in the illustration.

Shear at the support.

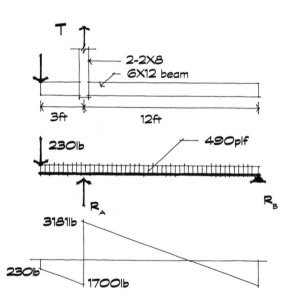

From shear diagram: V = 3181lb

$d_e = 11.5in - [1.75in - (2.5in)/2] = 11in$

$f_v = 1.5(V)(d)/(b)(d_e)^2 =$

    $= 1.5(3181lb)(11.5in)/(5.5in)(11in)^2$

    $= 82.4psi = 80psi + 3\%$ OK!

METRIC VERSIONS OF ALL EXAMPLES

The following tables are produced only for the purposes of these example problems, please consult the product manufacturers for the most current metric versions.

Table M26.4 Standard metric sizes of shear plates and split rings

Connector	Diameter (mm)	Bolt diameter (mm)	Material
Split rings	63.5	12.7	Steel
	102	19	Steel
Shear plates	66.7	19	Pressed steel
	102	19 or 22.2	Malleable iron

Table M26.6 Connector edge distances

Connector diameter	Edge distance
66.7mm shear plate 63.5mm split ring	$C = \Theta/45° + 44.5mm$

Example M26.1

A 19mm plywood subfloor is nailed to Sitka-spruce joists, G = 0.43. Dry conditions, ordinary temperature, 8d threaded hardened-steel nails are used (3.0mm diameter and 63.5mm long), and C_D = 1.0. The required penetration for full shear strength is (NDS, Table 12.3C).

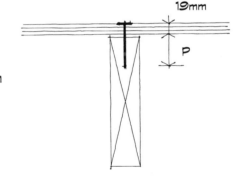

$$12D = 12(3.0mm) = 36mm$$

Determine the design shear value of each nail:

In the NDS (Table 12.3C), G = 0.43 corresponds to the value of hem-fir. in this column, for side-member thickness = 19mm and pennyweight 8d, we find Z = 302N.

The actual penetration is:

$$p = 63.5mm - 19mm = 44.5mm > 12D \quad OK!$$

Therefore, the reduction factor C_d = 1.0.

The shear design value of each nail is:

$$Z' = Z(C_D C_d C_{di}) = 302N(1.0)(1.0)(1.1) = 332N$$

Example M26.2

A toe-nail connection, designed according to the NDS (Sec. 12.1.3.2), connects a stud to a sole plate. Determine the load capacity of this connection. Use Western cedar (North) and 16d common wire nails. $C_D = 1.6$.

Since the length of the nail is 89mm, the "side-member" thickness or portion of the spike in the stud is (89mm)/3 = 29.6mm

Required penetration depth:

$$12D = 12(4.1mm) = 49mm$$

Actual p = 2(89mm)/3 = 59.3mm > 12D

Therefore: $C_d = 1.0$.

From the NDS (Table 12.3B):

Side-member thickness: 25.4mm　　　　Z = 378N

Side-member thickness: 31.7mm　　　　Z = 422N

Interpolating Z = (29.6mm)(422N)/(31/7mm) = 394N

$$Z' = Z(C_D C_d C_{tm}) = 394N(1.6)(1.0)(0.83) = 523N$$

Example M26.3

As an example of the design of a glued-and-nailed connection, we will use the box beam designed in Example M26.2.

Allowable rolling shear stress for S-2 plywood: $F_s = 0.366MPa$.

Maximum shear value: V = 21,350N.

$$b = 38mm + 38mm = 76mm$$

$$I = I_{flanges} + I_{web} = \frac{89mm[(610mm)^3 - (534mm)^3]}{12} + 1078 \times 10^6 mm^4 = 1632 \times 10^6 mm^4$$

$$Q = A\tilde{y} = 89mm (38mm)(305mm - 19mm) = 967,252mm^3$$

$$f_v = \frac{VQ}{bI} = \frac{21350N(967,252mm^3)}{76mm(1632 \times 10^6 mm^4)} = 12.65N/mm^2 \qquad OK!$$

The nailed connection must carry the entire shear load at the joint in case of failure of the glue. The shear flow is

$$P_v = \frac{VQ}{I} = f_v b = 12.65N/mm^2(76mm) = 12.65N/mm$$

Hem-fir is used for the flanges. The NDS (Table 12A) gives G = 0.43 for this species. If 12d box nails are used, Table 12.3A should be consulted for data. The required penetration is

$$12D = 12(3.25mm) = 39mm$$

The actual penetration depth, using a double 16mm plywood web on each side, is

$$p = 82.5mm - 32mm = 50.5mm > 12D$$

Therefore, the full tabulated strength of the nail can be used.

For G = 0.43 (hem-fir), Table 12.3A gives Z = 356N.

Adjusted design value:

$$Z' = Z(C_D C_d C_{di}) = (356N)(1.0)(1.0)(1.1) = 391N$$

This is the shear capacity of each nail.

The force to be supported by the nails per inch on each side of the flange is:

$$P_v/2 = (12.65N/mm)/2 = 6.32N/mm \text{ on each side}$$

The spacing of the nails is

$$\text{Spacing} = Z'/(P_v/2) = (391N)/(6.32N/mm) = 61.9mm$$

Use 12d box nails spaced 60mm.

In the central portion of the beam, where the shear value decreases to half the maximum (see the design of stiffeners, Sec. 21.5), space the nails 120mm. This is a considerable amount of nails; larger (and longer) nails would result in longer spacing, but since we are using a low-grade 38X89 for the flanges, the grain could split. A nailing gun would be convenient to execute the connection on-site.

Example M26.4

In order to increase the moment capacity of floor joists, 18mm Sturd-I-Floor plywood sheathing is glued and nailed to 38x286 Eastern pine joists. The joist must be designed for a maximum shear of V = 1156N. The joists are 406mm o.c.

Section properties:

Flange A = (406mm)(18mm) = 7315mm²	y = 286mm + 0.5(18mm) = 295mm
Web A = (38mm)(286mm) = 10,868mm²	y = (286mm)/2 = 143mm
Total A = 18,183mm²	

The distance of the neutral axis from the bottom is found from the moment of the areas equation:

$$A\widetilde{Y} = \Sigma a\widetilde{y}$$

[(Total area) × (distance from the bottom to the centroid of the total area) is equal to the (sum of each individual area) × (distance from the bottom to the centroid of that area).]

$$18,183\text{mm}^2\widetilde{Y} = 7315\text{mm}^2(295\text{mm}) + 10,868\text{mm}^2(143\text{mm})$$

$$\widetilde{Y} = 204\text{mm} \text{ The distance from the bottom to the neutral axis of the total section}$$

Rolling shear stress:

$$f_v = \frac{VQ}{Ib} = \frac{V(A\widetilde{y})}{Ib}$$

In this case, the shear stress is a function of the area × (the distance from the neutral axis of the total section to the neutral axis of this area). We are explaining this carefully since there can easily be confusion about which area and which distance to use.

$$V = 1,156\text{N}$$

$$Q = A\widetilde{y} = 7315\text{mm}^2(295\text{mm} - 204\text{mm}) = 665,665\text{mm}^3$$

$$I_{flange} = \Sigma(I_o + Ad^2) = Ad^2 = \frac{406\text{mm}(18\text{mm})}{12} + 7315\text{mm}^2(295\text{mm} - 204\text{mm})^2 = 59.28\text{x}10^6\text{mm}^4$$

$$I_{web} = \Sigma(I_o + Ad^2) = \frac{38\text{mm}(286\text{mm})^3}{12} + 10,868\text{mm}(204\text{mm} - \frac{286\text{mm}}{2})^2 = 114.52\text{x}10^6\text{mm}^4$$

$$I_{total} = 59.28\text{x}10^6\text{mm}^4 + 114.52\text{x}10^6\text{mm}^4 = 173.8\text{x}10^6\text{mm}^4$$

Rolling shear stress:

$$f_s = \frac{VA\widetilde{y}}{Ib} = \frac{1156\text{N}(665,665\text{mm}^3)}{38\text{mm}(173.8\text{x}10^6\text{mm}^4)} = 0.116\text{MPa} < 0.366\text{MPa}   \text{OK}$$

To design alternative nailing, we can use this previous calculation and simply not divide by 38mm or take the answer and multiply it by 1.5 to determine P_v (horizontal shear LOAD):

$$P_v = f_s b = (0.116\text{MPa})(38\text{mm}) = 4.41\text{N/mm}$$

Nails: 12d common-wire spikes, 12D = 58.4mm

Actual penetration depth: p = 82.5mm - 18mm = 64.5mm > 12D

Wood: G = 0.36

For side-member thickness:

16mm $Z = 405N$

19mm $Z = 414N$

18mm $Z = 409N$ (approximate interpolation)

Nail spacing $\overset{!}{=}$ (409N)/(4.41N/mm) = 93mm

Use 12d common-wire spikes spaced 90mm for the entire length of the joist. Obviously, since the shear value (V) changes constantly, we could also vary the spacing of the nails, but this becomes more calculation hassle than it's worth. We could recognize, however, that the shear value at the 1/4 point of the span is only 1/2 the value at the end (for a uniform load) and realize that if the shear value is 1/2, the nail load will also be 1/2. At this location, we could double the spacing of the nails. This technique could be used as many times across the span as you have patience for, but too many changes in nail spacing causes more coordination problems than it saves in nails.

Example M26.5

An 89X89 Douglas-fir knee brace is connected to a column with 6mm steel plates and bolts. The structure is exposed to the exterior. The portion of the wind load to be resisted by this frame is $W = 5340N$.

We can assume that both corner braces will work together, one in tension and the other in compression. The horizontal reactions in A and B are therefore equal to (5,340N)/2 = 2,670N. The equation of moments about C for the column gives us the horizontal component:

$V = (P/2)(h/b) = (5,340N)(3m)/(2)(0.9m) = 8,900N$

The tension in the brace is

$$T = \frac{V}{\cos\theta} = \frac{8900N}{\cos 45°} = 12,586N$$

Using the NDS (Table 8.3B) for bolts in double shear, we try a 16M bolt.

$C_M = 0.75$ (Table 7.3.3)

$C_D = 1.6$ (wind); therefore:

$Z' = 1.6(0.75)(10,008N) = 12,010N$

This is within 5% of the required capacity so we would say . . . OK!

Example M26.6

The connection in Example M26.5 could be made with 16mm lag screws, one on each side of the connection. From the NDS (Table 9.3B),

$Z = 5,073N$

$Z' = 2(1.6)(0.75)(5,073N) = 12,175N$ OK

Example M26.7

A tension member made of a 38X286 spruce-pine-fir, accessible only on one side, has to be spliced using a 10-gage steel plate and screws. The tension load is 3336N due to wind; therefore, $C_D = 1.6$.

Try to use four screws each side of the joint. Since it must be (4 screws)(1.6)Z = 3,336N, the required shear capacity of a single screw is

$$Z = (3,336N)/(4)(1.6) = 521N$$

By using 8g screws, $Z = 543N$

Ten-gage steel has a thickness t = 3.4mm, the 8g screw has a gage of 4.2mm, so the full penetration is 7D = 7(4.2mm) = 29.4mm. With a 38mm screw, the actual penetration corresponds to the length of the threaded part, or approximately (2/3)(38mm) = 25.4mm. Therefore

$$C_d = 1.0/1.15 = 0.87$$

$$Z' = 1.6(0.87)(543N) = 756N$$

Splice capacity P = 4(756N) = 3024N, which is insufficient!

We can use five screws:

Splice capacity P = 5(756N) = 3,780N > 3,336N OK!

Example M26.8

A double, laminated-timber beam system, 8.54m span, spaced 6.1m o.c., grade 24f-E (DF), supports an engineered wood joist floor. The beam is connected to a laminated-timber column Combination Symbol 6.

1. Loads

   Load on connection

   Total floor load = 2,400N/m² + 670N/m² = 3,070N/m²

   Load on beam:

   $$w = (3,070N/m^2)(6.1m) = 18,727N/m$$

   Estimated beam w = (175mm)(608mm$\times10^{-6}$m²)/(5,500N/m³) = 585N/m

   $C_D = 1.0$ (occupancy)

   Total w = 18,727N/m + 585N/m = 19,312N/m

   Support reaction at column A

   $$R_A = (19,312N/m)(8.54m)(4.27m)/(7.32m) = 96,206N$$

2. Design criteria for connection on the beam:

   Timber species group: F_c = 3.9MPa, group B.

   Use a shear plate diameter of 66.7mm.

   Number of faces of a member with connectors on the same bolt: 1.

Beam thickness is over 38mm (use 79mm or thicker, depending on V and M requirements). Load is perpendicular to the grain.

Design value (NDS, Table 10-2B)

$$Q = 8273N$$

Number of connectors required = R_A/Q = (96,206N)/(8273N) = 11.63 or 12 connectors
This number is too large and could affect the column dimension.
Try 102mm shear plates:

$$Q = 13,522N$$

Number of connectors required = R_A/Q = (96,206N)/(13522N) = 7.1 connectors
Use 8 connectors, 4 on each side.

Spacing for loading: perpendicular to the grain 152mm
                               parallel to the grain 127mm

Edge distance: unloaded edge 70mm
                               loaded edge 95mm

End distance: The connection is "far" from the ends.

Width of beam required d = 95mm + 152mm + 70mm = 317mm

3. Design criteria for connection on the column:

Timber species group: F_c = 3.2MPa, group C.

Use 102mm shear plates.

Number of faces of a member with connectors on the same bolt: 2.

Thickness > 89mm.

Load parallel to the grain.

Design value P = 16,013N
Number of connectors required = (96,206N)/(16,013N) = 6.0 connectors
Use 6 connectors, 3 on each side minimum.
Since 4 connectors each side are used for the beam, the beam connection will be adequate for the column.

Spacing for loading: parallel to grain 229mm
                               perpendicular to grain 127mm

Edge distance: unloaded edge 70mm

End distance: The connection is, again, "far" from the ends.

Since the spacing parallel to grain is larger than the spacing on the beam, we can try to reduce it.

The load capacity of the 8 connectors on the column is

    P = (16,013N)(8) = 128,104N.

This design capacity is used only partially and precisely:

Capacity usage = (96,206N)/(128,104N) = 0.75, or 75%.

This can be expressed also by saying that the geometry factor C_Δ = 0.75. Interpolating the C_Δ values (NDS, Sec. 10.3.5 and Table 10.3):

Spacing = (102mm)(0.75-0.5)/(0.5) + 127mm = 178mm

Make the spacing 170mm. This changes the minimum required beam dimension for the connection to 343mm.

Minimum column width = 70mm + 127mm + 70mm = 267mm

The design of the column will have to be consistent with this requirement.

4. Design of the beam for shear at the support:

    Shear:

    Left of support A:

        V = 23,561N

    Right of support A:

        V = 72,645N

F_v = F_v' = 1.1MPa

First trial section:

Required A = 1.5(72,645N)/(1.1MPa) = 99,061mm²

Two 130X380 have A = 2(49,400mm²) = 98,800mm² $\geq$ Required A 99,000mm²
        (within our 5% margin)

Check shear stress.

        d_e = 380mm - (70mm - 51mm) = 361mm

        f_v = (1.5)(V)(d)/(b)(d_e)(d_e)

         = (1.5)(72,645N)(380mm)/[(130mm + 130mm)(361mm)²] = 1.2MPa > 1.1MPa

    A larger section is required for shear. Try 130X418:

        d_e = 418mm -19mm = 399mm; f_v = 1.1MPa OK.

Example M26.9

The mezzanine floor of a Frank L. Wright Usonian house is suspended from the roof structure with two-layer sawn lumber members. The floor loads on the beam, including all L and D loads, are a uniform load of 7,100N/m and a concentrated 1,000N parapet load at the end of the overhang balcony. The wood is Eastern hemlock; the beam is a nominal 140X292. Check if a connection with two bolts and 63.5mm split rings works.

The load on the connection is the support reaction of the suspension members.

$$R = [(7100N/m)(4.5m)(2.25m) + (1000N)(4.5m)]/(3.6m) = 21,219N$$

Tension members:

        38mm-thick members, connected on 1 face:

        P = 8718N

Number of connectors required = (21,219N)/(8,718N) = 2.4 connectors

Use four split rings, although this is more than required. We always want to create connections without eccentricity which means we must use them in opposing pairs.

Spacing: load parallel to the grain (see the figure).
Beam:

        140mm-thick member, connected on two faces:

        Q = 6183N

Number of connectors required

        = (21,219N)/(6183N) = 3.4 connectors

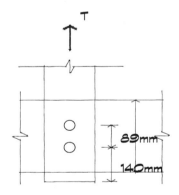

Use 4 split rings.

Spacing: load perpendicular to the grain (see figure).

The distances for full load on the beam are:

        D = 70mm + 171mm + 44mm = 285mm < 292mm

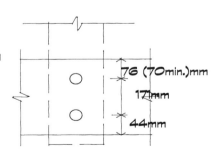

Shear at the support.

From shear diagram: V = 13,829N

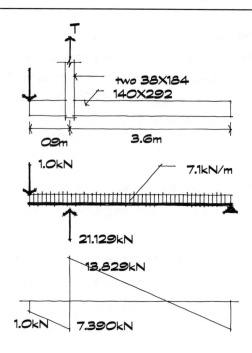

two 38X184
140X292

0.9m   3.6m

1.0kN   7.1kN/m

21.129kN

13.829kN

1.0kN   7.390kN

d_e = 292mm - [44mm - (63.5mm)/2] = 280mm

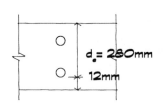

d_e = 280mm

12mm

$f_v = (1.5)(V)(d)/(b)(d_e)(d_e)$

$\quad$ = (1.5)(13,829N)(292mm)/(140mm)(280mm)(280mm)

$\quad$ = 0.55MPa = 0.55MPa

27 PERMANENT WOOD FOUNDATIONS

27.1 Types of wood foundations

Wood foundations are as old as timber buildings, but modern treatment systems have provided them with greater decay resistance under moist use conditions. As discussed in Chapter 18, wood decay is prevented or at least slowed by environmental conditions. Consequently, timber that is totally submerged below the water table will last for centuries. As an example, buildings constructed in Venice in the fourteenth century on timber piles are still standing. Settlement problems are, in fact, due more to the recent lowering of the underground aquifers than to failure of the wood piles. Unfortunately, untreated or unprotected wood that experiences alternating periods of wetting and drying is highly vulnerable to decay. This is the real problem to be addressed.

The types of wood foundations are similar to the systems built with other materials today. Although these are more commonly constructed of concrete, the types of foundations built with timber are as follows:

(a) Raft or mat foundations, using layers of logs laid flat on wet, soft ground, such as encountered in marshlands. The timbers are kept under water by masonry walls or by stone ballast, used as the base for the building.

(b) Pile or pier foundations, built by driving logs vertically into the ground. This technique can be used to consolidate soft soils or to bear on a stronger soil stratum below. The timbers can be capped with a masonry foundation, or directly support a wood superstructure, if the partially submerged wood is pressure-treated.

(c) Footings, consisting of treated sawn lumber laid flat on the bottom of a prepared excavation, and wood framing erected up to first floor level. These footings require a level support and good drainage.

In the case of all foundations, they must provide an area of support capable of distributing the loads being transferred from the superstructure to the soil. Pile foundations transfer load two ways. An "end-bearing" pile penetrates through a soft soil layer to rest on a stronger layer below, much like a simple column. A "friction" pile works like a nail to transfer load; it relies on soil displacement to develop frictional forces along its length. The "friction" pile can remain entirely in soft material, but must be driven into the stratum far enough to develop adequate surface area to transfer its load through friction.

In this chapter, we'll examine how simple wood foundations can be built to "float" or "spread" their loads on the surface of the soil. We will also look at techniques for designing basement walls to be built with wood products.

27.2 Permanent wood foundation

The permanent wood foundation (PWF) system is a load-bearing, lumber-framed foundation wall sheathed with plywood; it is essentially the same construction as a platform-frame stud wall, using pressure-treated material capable of resisting moisture and termite attack. This system has been employed since the 1940s and has become common practice in the coldest regions of the world, where masonry or concrete systems are difficult to construct during winter months. The PWF does not need frost protection, and it can be used with a perimeter-insulated raised floor, where insulation is applied only to the inside of the perimeter foundation wall, eliminating the need for underfloor insulation. Crawl-space vents are eliminated, and a vapor barrier is used under the floor to avoid summer condensation on an air-conditioned floor structure. Figures 27.1 and 27.2 illustrate alternative wood foundation systems.

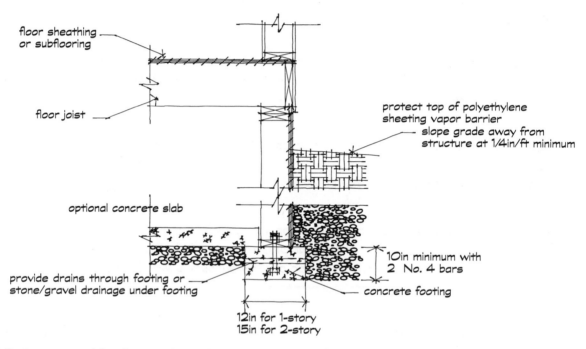

Fig 27.1 Permanent wood foundation with crawl space and concrete footing

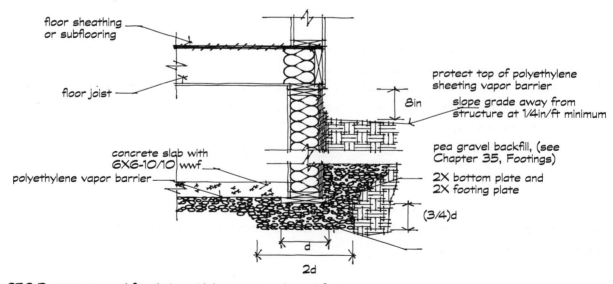

Fig 27.2 Permanent wood foundation with basement and wood footing

The basement version is designed to resist lateral soil pressures. Waterproofing and drainage are very important in this situation.

27.3 Design of footing foundations

In foundation design, the first factors to consider are the soil type and the general grading conditions. Soils are classified by their chemical composition and by particle size. Both of these characteristics affect the strength, absorption of water, and drainage characteristics. Essentially, soils range from cohesive to granular. Clay soils are cohesive, plastic (easily moldable), and dimensionally unstable. They absorb water and change structurally, drain poorly, and are unsatisfactory for foundations in situations where the subsoil moisture content will vary significantly. On the other hand, clay is a very good foundation material if the moisture content can be held relatively stable (and low). Sand and gravel display the opposite qualities and are generally the best for foundations as long as they do not have water "flowing" through them. Unfortunately, you will seldom find any soils in pure form; consequently, you should engage a soils testing professional to take borings and evaluate the soil's structural properties prior to any architectural design work. Table 27.1 gives some common soil characteristics and related allowable bearing pressures.

Table 27.1 Types of soils and their design properties

Soil type	Allowable bearing		Drainage	Frost heave	Expansion
Gravels, gravel/sand little or no fines	8.0ksf	383N/m²	Good	Low	Low
Well-graded sands, gravely sands little or no fines	6.0ksf	287N/m²	Good	Low	Low
Poorly graded sands, gravely sands little or no fines	5.0ksf	239N/m²	Good	Low	Low
Silty gravels, gravel/sand/silt	4.0ksf	191N/m²	Good	Medium	Low
Silty sand, sand/silt	4.0ksf	191N/m²	Good	Medium	Low
Clayey gravels, gravel/sand/clay	4.0ksf	191N/m²	Medium	Medium	Low
Clayey sands, sand/clay	4.0ksf	191N/m²	Medium	Medium	Low
Inorganic silts/very fine sands, silty/clayey fine sands, clayey silts with some plasticity	2.0ksf	96N/m²	Medium	High	Low
Inoganic clays (low/medium plasticity) gravely/sandy/silty clays	2.0ksf	96N/m²	Medium	Medium	Medium
Inorganic clays (high plasticity)	2.0ksf	96N/m²	Poor	Medium	High
Inorganic silts	2.0ksf	96N/m²	Poor	High	High
Organic silts, organic silty clays (low plasticity) [a]	0.4ksf	19N/m²	Poor	Medium	Medium

[a] Organic materials in general are NOT good substrata. Great care should be taken to remove all organic materials from beneath any foundation or slab. Over a period of time, this material will decompose and significant settlement may occur. At the very least, it has a tendency to change volume under different moisture conditions.

The design of a wood footing consists, first, of determining the appropriate width of the lumber to be placed with the wide face down. This will typically be a 2X [38X]dimension lumber, its width depending both on soil-bearing capacity and on the width of the studs supported on it. For instance, a basement wall may require 2X8 [38X184] studs, and the lumber footing should therefore be at least a 2X10 [38X235] to accommodate a 2X8 [38X184] plate.

The design procedure is the following.

1.   Determine the vertical load from the structure above at the foundation level per foot of length of the foundation.

2.   The required width B of the footing is determined with the direct stress formula:
     $$A = P/F_p$$

     where   A = required area of the footing (in/ft)[mm/m] = B (width) times 1.0ft [1m]

             P = vertical load on the footing (lb/ft)[N/m]

             F_p = allowable bearing pressure (psi)[MPa or N/mm²]

     Therefore, solving for the required width (ft)[m]:

     $$B(1ft)[1m] = P/F_p$$

Table 27.2. gives some indicative footing sizes for one- and two-story buildings. The soil conditions are assumed average, and granular fill such as pea gravel (not pea fill) is required under the footing to ensure good pressure distribution in poor soils. The gravel or crushed stone bed must be at least three-fourths the width of the footing selected in Table 27.2 (e.g., 2X10 [38X235] footing, gravel bed 7.5in [190mm] deep). Figure 27.3 shows recommended dimensions for wood footings.

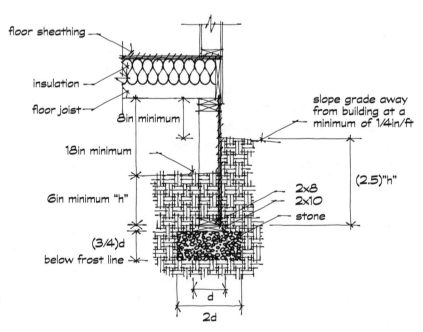

Fig. 27.3 Recommended dimensions of a wood foundation

Table 27.2 Minimum recommended footing plate size

House width (ft)		Loadings: Roof LL+DL = 50psf Second floor LL+DL = 50psf First floor LL+DL = 50psf				Loadings: Roof LL+DL = 40psf Second floor LL+DL = 50psf First floor LL+DL = 50psf			
ft	[m]	2 story		1 story		2 story		1 story	
36	[10.98m]	2X10	[38X235]	2X10	[38X235]	2X10	[38X235]	2X8	[38X184]
32	[9.76m]	2X10	[38X235]	2X8	[38X184]	2X10	[38X235]	2X8	[38X184]
28	[8.54m]	2X10	[38X235]	2X8	[38X184]	2X8	[38X184]	2X6	[38X153]
24	[7.32m]	2X8	[38X184]	2X6	[38X153]	2X8	[38X184]	2X6	[38X153]

27.4 Basement walls

Basement walls are subject to lateral soil pressure, which is conventionally transformed into an equivalent soil fluid-density pressure, acting like a hydrostatic load. The wall works like a vertical beam, simply supported by the foundation and basement floor at the bottom, and by the first-floor structure at the top. Tables 27.3 and 27.4, to be used for preliminary design, give, respectively, stud specifications for two-story construction and plywood specifications. All wood and plywood materials are assumed to be pressure-treated for PWF.

Table 27.3 Minimum structural requirements for basement foundations (two-story construction)

Maximum backfill combination		Stud & Plate size		Stud spacing		Species grade				
in	[mm]	in	[mm]	inches o.c.		B-1	C-S	B-2	C-1	C-2
86in	[2184mm]	2X6¹	[38X153]	12	[305mm]	a*				
86in	[2184mm]	2X8¹	[38X184]	16	[406mm]	*	b	b*		
86in	[2184mm]	2X8	[38X184]	12	[305mm]					
72in	[1829mm]	2X6	[38X153]	16	[406mm]	c				
72in	[1829mm]	2X6	[38X153]	12	[305mm]				d	d
72in	[1829mm]	2X8	[38X184]	16	[406mm]					
60in	[1524mm]	2X6	[38X153]	16	[406mm]				d	d
60in	1524mm]	2X6	[38X153]	12	[305mm]					
48in or less	[1219mm]	2X6²	[38X153]	16	[406mm]					

¹Where indicated by*, length of end splits of checks at lower end of studs shall not exceed the width of the stud.

²For backfill heights of 48in [1220mm] or less, certain species-grade combinations of 2X4 [38X89] studs spaced 12in or 16in [305mm or 406mm] o.c. may be used for PWF basement walls in two-story construction.

a Maximum building width 24ft [7.32m], maximum roof live load 20psf [957N/m²]

b Maximum building width 28ft [8.54m], maximum roof live load 40psf [1915N/m²]
   Maximum building width 32ft [9.76m], maximum roof live load 30psf [1436N/m²]
   Maximum building width 36ft [10.98m], maximum roof live load 20psf [957N/m²]

c Maximum building width 24ft [7.32m], maximum roof live load 30psf [1436N/m²]
   Maximum building width 28ft [8.54m], maximum roof live load 20psf [957N/m²]

d Maximum building width 32ft [9.76m], maximum roof live load 40psf [1915N/m²]

Lumber species-grade combination:

B-1   Douglas fir No. 1 or Southern pine No. 1 dry

B-2   Douglas fir No. 2 or Southern pine No. 2 dry

C-S   Hem-fir select structural

C-1   Hem-fir No. 1

C-2   Hem-fir No. 2

Tables 27.1, 27.2, 27.3.27.4 provided courtesy of Southern Pine Marketing Council, Kenner, Louisiana, and American Forest &Paper Association, Washington, D.C.

Example 493

Table 27.4 Plywood grade and thickness for basement foundations

Height of fill	Stud spacing on center	Face grain across studs			Face grain parallel to studs		
		Grade	Min. thick	Span rating	Grade	Min. thick	Span rating
86in	12in	B	19/32	40/20	A	19/32	40/20
					A	23/32	48/24
86in	16in	B	23/32	48/24	...	...	...
72in	12in	B	15/32	32/16	A	19/32	40/20
					B	23/32	48/24
72in	16in	B	23/32	48/24	...	...	...
60in	12in	B	15/32	32/16	A	19/32	40/20
					B	19/32(5ply)	40/20
					B	23/32	48/24
60in	16in	B	19/32	40/20	A	23/32	48/24
48in	12in	B	15/32	32/16	A	15/32	32/16
					B	19/32(4ply)	40/20
48in	16in	B	19/32	40/20	A	19/32	40/20
					A	23/32	48/24
36in	12in	B	15/32	32/16	A	15/32	32/16
					B	15/32(4ply)	32/16
					B	19/32(4ply)	40/20
36in	16in	B	15/32	32/16	A	19/32	40/20
					B	23/32	48/24
24in	12in	B	15/32	32/16	A	15/32	32/16
					B	15/32	32/16
24in	16in`	B	15/32	32/16	A	15/32	32/16
					B	19/32(4ply)	40/20

Example 27.1

A basement wall for a two-story house is backfilled for a height of 84in with gravel. The loads from the load-bearing walls above grade on the foundations are LL = 1,010plf (snow + occupancy) and a DL = 466plf excluding the foundations.

Table 27.3 recommends 2X8 studs at 16in o.c. with any of the species listed (or equivalent from NDS tables), with the exclusion of hem-fir No. 2.

The basement must be nonload-bearing, that is, the floor joists have to bear on the outer walls; in this case, the effect of compression loads (due to the exterior walls only) is not very significant and is included in the tabulated solution. The treated plywood can be selected from Table 27.4 for 86in backfill height, 16in stud spacing, Grade B, which is APA-rated sheathing, 23/32in thick.

We can check the stud size for the load-bearing basement wall with a design procedure based on the considerations of the NDS (Sec. 23.2), and on what we know about axial compression and bending (Sec. 23.5).

Assuming an equivalent fluid pressure of the soil of 30pcf, the maximum pressure at the base of the wall is f_p = (30pcf)(7ft) = 210psf. For ease of calculation, we can assume this triangular load distributed on the entire length of 8ft.

This corresponds to a load on each stud:

$$W_{max} = (210psf)(16in/12in/ft) = 280plf$$

From the beam tables,

$$M = 0.1283WL = 0.1283(1,120lb)(8ft) = 1,150ftlb$$

The axial load is

$$P = (1,010plf + 466plf)(16in/12in/ft) = 1,968lb$$

Using hem-fir No. 1, F_b = 950psi; F_c = 1,300psi; E = 1,500,000psi

C_D = 1.0 for bending, but the value is 1.15 for compression (snow). C_r = 1.15 for bending.

(a) Design for bending:

$$F_b' = (950psi)1.15 = 1,092psi$$

$$\text{Required } S_x = M/F_b' = (1,150ftlb)(12in/ft)/(1,092psi) = 12.63in^3$$

A 2X8 has S_x = 13.14in³ and A = 10.88in²

Actual bending stress:

$$f_b = M/S_x = (1,150ftlb)(12in/ft)/(13.14in^3) = 1,050psi$$

(b)   Design for compression:

$$f_c = P/A = (1,968lb)/(10.88in^2) = 181psi << 1,300psi$$

which is allowable for hem-fir No. 1. The factors to be entered in the bending-compression interaction formula are:

$$F_{cE} = 0.3(1,500,000)/(96in/8in)^2 = 3,125psi$$

$$F_{cE}/F_c^* = (3,125psi)/(1,300psi) = 2.4$$

$$C_p = a - \sqrt{a^2 - b/c}$$

$$a = \frac{1 + \frac{F_{cE}}{F_c^*}}{2c} = 2.13$$

$$b = \frac{F_{cE}}{F_c^*} = 2.4$$

$$c = .8$$

$$C_p = 2.13 - \sqrt{2.13^2 - 3.0} = 0.67$$

Example 495

$F_c' = (1,300psi)(0.67)(1.15) = 999psi$

$f_c/F_c' = (181psi)/(999psi) = 0.18$

(c) Solving the interaction formula:

$0.18^2 + [(1050psi)/(1092psi)][1 - 0.18] = 0.18 + 1.02 = 1.20$ OHHH, too small!

The studs are too small and a 2X10 must be used. A 2X10 has $S_x = 21.39in^3$ and $A = 13.88in^2$.
Actual bending stress:

$f_b = M/S_x = (1,150ftlb)(12in/ft)/(21.39in^3) = 645psi$

$f_b/F_b' = (645psi)/(1,092psi) = 0.59$

So, if you refer to the previous interaction formula, this is obviously going to work. An alternative may be to space the 2X8 studs at 12in o.c.

The studs are in reality going to be more resistant to bending due to the glued-and-nailed plywood sheathing, so our design is actual conservative.

Use 2X10 studs, 16in o.c., hem-fir No. 1.

METRIC VERSION OF EXAMPLE

Table M27.4 Plywood grade and thickness for basement foundations

Height of fill	Stud spacing on center	Face grain across studs			Face grain parallel to studs		
		Grade	Min. thick	Span rating	Grade	Min. thick	Span rating
2184mm	305mm	B	15.1mm	1016/508	A	15.1mm	1016/508
					A	18.3mm	1220/610
2184mm	406mm	B	18.3mm	1220/610	...	...	...
1829mm	305mm	B	11.9mm	813/406	A	15.1mm	1016/508
					B	18.3mm	1220/610
1829mm	406mm	B	18.3mm	1220/610	...	...	...
1524mm	305mm	B	11.9mm	813/406	A	15.1mm	1016/508
					B	15.1mm(5ply)	1016/508
					B	18.3mm	1220/610
1524mm	406mm	B	15.1mm	1016/508	A	18.3mm	1220/610
1219mm	305mm	B	11.9mm	813/406	A	11.9mm	813/406
					B	15.1mm(4ply)	1016/508
1219mm	406mm	B	15.1mm	1016/508	A	15.1mm	1016/508
					A	18.3mm	1220/610
914mm	305mm	B	11.9mm	813/406	A	11.9mm	813/406
					B	11.9mm(4ply)	813/406
					B	15.1mm(4ply)	1016/508
914mm	406mm	B	11.9mm	813/406	A	15.1mm	1016/508
					B	18.3mm	1220/610
610mm	305mm	B	11.9mm	813/406	A	11.9mm	813/406
					B	11.9mm	813/406
610mm	406mm`	B	11.9mm	813/406	A	11.9mm	813/406
					B	15.1mm(4ply)	1016/508

Example M27.1

A basement wall for a two-story house is backfilled for a height of 2.1m with gravel. The loads from the load-bearing walls above grade on the foundations are LL = 14,736N/m (snow + occupancy) and a DL = 46,800N/m excluding the foundations.

Table 27.3 recommends 38x184 studs at 406mm o.c. with any of the species listed (or equivalent from NDS tables), with the exclusion of hem-fir No. 2.

The basement must be nonload-bearing, that is, the floor joists have to bear on the outer walls; in this case, the effect of compression loads (due to the exterior walls only) is not very significant and is included in the tabulated solution. The treated plywood can be selected from Table M27.4 for 2,184mm backfill height, 406mm stud spacing, Grade B, which is APA-rated sheathing, 18mm thick.

We can check the stud size for the load-bearing basement wall with a design procedure based on the considerations of the NDS (Sec. 23.2), and on what we know about axial compression and bending (Sec. 23.5).

Assuming an equivalent fluid pressure of the soil of 4710N/m^3, the maximum pressure at the base of the wall is f_p = (4710N/m^3)(2.1m) = 9,891N/m^2. For ease of calculation, we can assume this triangular load distributed on the entire length of 2.4m.

This corresponds to a load on each stud:

w_{max} = (9,891N/m^2)(406mm)/(1,000mm/m) = 4,016N/m

W = (4,016N/m)(2.4m)/2 = 4,819N

From the beam tables,

M = 0.1283Wl = 0.1283(4,819N)(2.4m) = 1,484Nm

The axial load is P = (14,736N/m + 6,800N/m)(406mm/1,000mm/m) = 8,744N

Using hem-fir No. 1, F_b = 6.5MPa; F_c = 9.0MPa; E = 10,300MPa

C_D = 1.0 for bending, but the value is 1.15 for compression (snow). C_r = 1.15 for bending.

(a) Design for bending:

F_b' = (6.5MPa)1.15 = 7.47MPa

Required S_x = M/F_b' = (1,484Nm)(1,000mm/m)/(7.47MPa) = 198,528mm^3

A 38X184 has S_x = 214,420^3 and A = 6,992mm^2

Actual bending stress:

f_b = M/S_x = (1,484Nm)(1000mm/m)/(214,420mm^3) = 6.9MPa < F_b' OK!

(b) Design for compression:

f_c = P/A = (8,744N)/(6,992mm^2) = 1.25MPa << 9.0MPa

which is allowable for hem-fir No. 1.

The factors to be entered in the bending-compression interaction formula are

F_{cE} = 0.3(10,300MPa)/(2400mm/184mm)^2 = 18.16MPa

F_{cE}/F_c^* = (18.16MPa)/(9.0MPa) = 2.0

$$C_p = a - \sqrt{a^2 - b/c}$$

$c = 0.8$

$$a = \frac{1 + 2.02}{2(0.8)} = 1.9$$

$$b = \frac{F_{cE}}{F_c^*} = 2.02$$

$$C_p = 1.9 - \sqrt{1.9^2 - 2.5/0.8} = 0.86$$

F_c' = (9.0MPa)(0.86)(1.15) = 8.9MPa

f_c/F_c' = (1.25MPa)/(8.9MPa) = 0.14

(c) Solving the interaction formula:

$f_c/F_c' + f_b/F_b' [1 - f_c/F_{cE}]$ = 0.14 + [(6.9MPa)/(7.47MPa)][1 - (1.25MPa)/18.16]

$\qquad\qquad$ = 0.14 + 0.99 = 1.13 > 1.0

The studs are too small, and a 38X235 should be used. A 38X235 has S_x = 349,760mm^3 and A = 8930mm^2.

Actual bending stress:

f_b = M/S_x = (1,484Nm)(1000mm/m)/(349,760mm^3) = 4.24MPa

f_b/F_b' = (4.24MPa)/(7.47MPa) = 0.58

So, if you refer to the previous interaction formula, this is obviously going to work. An alternative maybe to space the studs at 305mm o.c.

The studs are in reality going to be more resistant to bending due to the glued-and-nailed plywood sheathing, so our design is very conservative.

Use 38X235 suds, 406mm o.c., hem-fir No.1.

PART THREE

CONCRETE

Kenzo Tange, Olympic Gymnasium, Tokyo (photo by Michele Chiuini)

28 MATERIALS AND PROPERTIES

28.1 Structural concrete materials

Concrete is a mix of three basic components: cement, water, and aggregate. The aggregate should be a mixture of sand and gravel, both coarse and fine crushed stone, or other similar combinations of coarse and fine material. Several admixtures can be added to improve concrete quality.

28.1.1 Cement

The most common type is Portland, obtained from grinding calcareous stone (like limestone and chalk) with substances containing silica, alumina, and iron oxide (such as clay or shale), heating in a kiln, and grinding to a fine powder. Cement content in the mix is measured in bags or cubic feet per cubic yard of concrete (e.g., 5 ft^3/yd^3 [0.18m^3/m^3] cement-to-concrete ratio). Portland cement must conform to ASTM C 150. Blended hydraulic cements is the other group used for structural concrete.

There are five types of Portland cement classified with a Roman numeral from I to V, plus some letters indicating special admixtures (e.g., Type I P contains pozzolan; Type IA is air-entrained; see "Admixtures"). Type I is the standard cement for structural use, Type II is a cement with moderate sulfate resistance, Type IIA has moderate sulfate resistance with air entraining, Type III achieves 28-day strength in 7 days; consequently it is referred to as high early strength cement, Type IIIA is high early strength, air-entrained, Type IV is a cement with a low heat of hydration that produces less heat during the curing process and consequently produces less thermal stresses, making it particularly well suited for large masses of concrete, Type V is a cement with a high sulfate resistance. Some soils, particularly in the Southwest, have high levels of sulfates that will leach out of the soil and will have a destructive effect on concrete.

28.1.2 Water

The water must be free of organic contaminants. Potable water is generally acceptable. Water-cement ratios affect the strength of concrete: Low ratios generally make concrete stronger. A water-cement content by weight between 0.40 and 0.50, or about 35 gallons of water per cubic yard [200L/m^3], is typical of medium-strength concrete. Table 28.1 shows allowable water-cement ratios for ordinary concrete and air-entrained concrete.

Table 28.1 Maximum permissible water:cement ratios

Compressive strength, f_c		Water:cement ratio by weight	
		Non-air-entrained concrete	Air-entrained concrete
2,500 psi	[17MPa]	0.67	0.54
3,000 psi	[20MPa]	0.58	0.46
3,500 psi	[24MPa]	0.51	0.40
4,000 psi	[28MPa]	0.44	0.35
4,500 psi	[32MPa]	0.38	....

28.1.3 Aggregate

Fine: Natural sand or crushed stone with a minimum size of 0.003in [0.076mm]. Particles that are too fine fill all spaces, and the cement cannot bond to the larger particles. Unfortunately, at the same time, too small a proportion of fine aggregate will leave too much space between the large aggregate particles for the cement paste to effectively bridge.

Coarse: Gravel or crushed stone with a shape and maximum size depending on construction requirements (spacing of steel reinforcement, workability, etc.). Particular care must be taken when using gravel since some gravel deposits may be unsuitable, for example, if quantities of chert are present. Chert is a highly expansive stone that when used in concrete exposed to weather will cause **pop outs** in areas where the aggregate actually expands and cracks out of the mixture, leaving a void.

The maximum size of aggregate is specified in the Building Code as

(a) 1/5 of the narrowest dimension between sides of forms

(b) 1/3 the depth of slabs

(c) 3/4 the minimum space between reinforcing bars

The size and shape of the aggregate affects the cement content and the workability of the mix (Fig. 28.1). The percentage of particles of different sizes that will give a required concrete strength with minimum cement content is called grading. For concrete to achieve full strength, a full range of aggregate sizes must be present. This allows solid packing and does not leave large voids in the mass that must be filled with cement paste. Gap-graded mixes (mixes that have certain sizes missing from the range) are sometimes used for special purposes, but care should be taken in their use. Specifications for aggregate are in ASTM C 33.

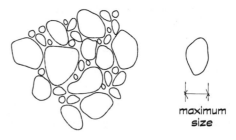

Fig. 28.1 Well-graded gravel mix
(full range of sizes present)

The type of aggregate determines the density and weight of the concrete. Lightweight concrete can be obtained by adding to the grading materials like expanded polystyrene beads, pumice, or expanded clay. Specifications for lightweight aggregate are in ASTM C 330.

28.1.4 Admixtures

Substances that are added to improve workability accelerate hardening and increase the waterproof quality. Typical admixtures are as follows.

Air-entraining: Substances that improve the workability, permit lower water-cement ratios, and improve the durability of concrete by creating smaller pores (voids), therefore avoiding frost damage. Between 0.35 and 0.75 of concrete volume can be entrained air, depending on exposure conditions.

Hydrated lime, calcium cloride: These substances increase workability and produce a finer pore structure and therefore a lower permeability. Proportions are limited to 0.25 of cementitious materials for concrete exposed to deicing chemicals. Calcium chloride accelerates the hydration rate (producing more heat of hydration) which makes it attractive for use during cold weather, but it also increases the probability of future corrosion of the reinforcing bars.

These materials are combined to form:

Cementitious material = cement + admixtures

Paste = cementitious materials + water

Concrete = cementitious material + water + aggregates

28.2 Structural concrete properties

The proportions of the mix, the type of materials, and the external conditions affect the principal properties of concrete:

28.2.1 Workability

The ease with which the concrete can be placed in the formwork and finished. The addition of water makes concrete more workable, but decreases its strength. Workability is measured with the **slump test** (Fig. 28.2) by filling a conical mold with fresh concrete, compacting it in a specific manner, and removing the cone. The amount that the concrete drops after removal of the form is referred to as slump. An average slump is around 2in [50mm], but can be less if the aggregate maximum size is 0.5in [13mm] or less.

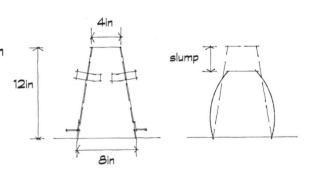

Fig. 28.2 Slump test

28.2.2 Weight

Weight depends largely on the density of the coarse aggregate: Lightweight aggregates (such as expanded clay) will obviously reduce the overall density of the concrete, but can also adversely affect the strength. The addition of lightweight nonstructural aggregate (e.g., polystyrene beads) or air cells has the same effect. Typical weights are given in Table 28.2.

Table 28.2 Weights of common types of concrete (lb/ft³) [Kg/m³]

Plain concrete, stone aggregate	145lb/ft³	2320Kg/m³
Reinforced concrete, stone aggregate	150lb/ft³	2400Kg/m³
Plain lightweight concrete: aerated	50-80lb/ft³	800-1200Kg/m³
expanded clay aggregate	85-100lb/ft³	1360-1600Kg/m³
pumice aggregate	60-90lb/ft³	960-1400Kg/m³

28.2.3 Strength

The ultimate compressive strength, f_c', is the primary index of designation of a concrete mixture. Typical strengths range from 2,500 to 5,000psi [17 to 35MPa], but 10,000psi [69MPa] concrete or stronger is used for special structures and 20,000psi [140MPa] can be achieved. The strength of concrete increases almost indefinitely with time if adequate moisture is present. After the initial setting period, the concrete will harden at a rate depending on the cement type, on the water-cement ratio, and on the curing environment. Curing is the process of hydration of the cement, and may be accelerated. Steam curing is the usual method of achieving early maturity, but requires factory facilities and is most commonly used for precast sections. Keeping concrete moist and warm with plastic covers and using the hydration heat with insulating formwork is sufficient to speed hardening and to avoid excessive evaporation of water from the surface (Fig. 28.3). It is normally considered that the specified strength will be reached after 28 days for ordinary concrete (Fig. 28.4). It is interesting to note at this point that the strength of the concrete does not always reflect the relative importance of the component. For example, sidewalks are frequently 3,500psi [24MPa] air-entrained concrete and building foundations may be as low as 2,500 or 3,000psi [17 or20MPa] standard concrete. The explanation centers on the fact that concrete durability increases as its strength increases. A foundation may be more cost-effective using a lower grade of concrete (slightly more, naturally) in an area not subjected to severe weathering, whereas the lowly sidewalk needs the weather resistance provided both by higher strength and by air entraining.

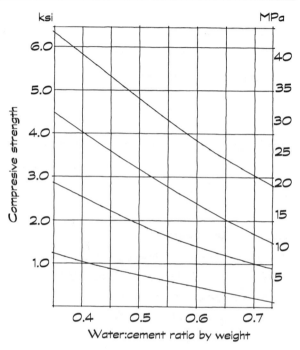

Fig. 28.3 Age-strength relationships for different water:cement ratios

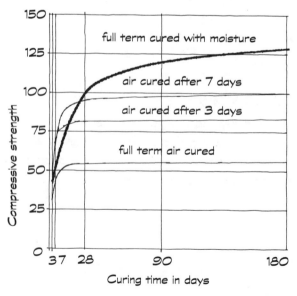

Fig. 28.4 Effect of curing time on strength of concrete as % of 28-day moist cured

The value of the modulus of elasticity E_c depends on the density w and on the compressive strength f_c' according to the formula:

$$E_c = 33 \, w^{1.5} \sqrt{f_c'} \text{ psi and for concrete with } w = 145 \text{ lb/ft}^3$$

$$E_c = 57000 \sqrt{f_c'} \text{ psi}$$

$$E_c = 0.043 \, w^{1.5} \sqrt{f_c'} \text{ MPa and for concrete with } w = 2320 \text{ Kg/m}^3$$

$$E_c = 4700 \sqrt{f_c'} \text{ MPa}$$

28.2.4 Creep

Long-duration stress produces an increase in strain over time and consequently causes permanent deformations. This phenomenon can be reduced through the use of reinforcment in the compression zone of the member.

28.2.5 Fire resistance

Concrete is incombustible and somewhat insulative, but long exposure to fire will damage it. Density of the concrete and the type of aggregate affect fire resistance.

28.2.6 Shrinkage

Ordinary concrete shrinks a significant amount during the curing process. This may account for as much potential movement as it would experience over a 100°F [55°C] temperature differential. For thin casts, expansive admixtures can be used, but these shrinkage stresses will cause cracks, and therefore this movement **must be controlled with joints and steel reinforcement.**

28.2.7 Hardness

This is a property of the surface that is related to durability as well as to appearance. A slab can be troweled or floated to draw more paste to the surface, but the harder superficial layer has a tendency to spall due to the different coefficients of expansion from the core material. Curing with carbon monoxide vapors (such as produced by burning fossil fuels) will form a calcium carbonate (limestone) surface, decreasing hardness, and also increasing the **dusting** character of the material.

28.2.8 Porosity

Size of pores left during the hydration process or created by excessive evaporation and shrinkage cracks affect the water tightness of concrete. Admixtures and low water content can have a beneficial effect. Also *vibration* can be used to obtain higher density and lower porosity, but excessive vibration can cause segregation of the materials and loss of strength of the concrete.

28.3 Reinforcing steel

Reinforced concrete (R/C) members are made of concrete and steel bars. When concrete hardens, it bonds well to steel. The coefficients of thermal expansion of concrete and steel are virtually the same, which allows the two materials to work together. The function of reinforcement in structural concrete is to provide the following:

28.3.1 Tensile strength for bending members

A 3000psi [20MPa] concrete has a tensile strength of only 150psi [1.0MPa]; consequently, tensile strength is not considered in R/C calculations, and steel reinforcement must be designed to take all the tension in bending members.

28.3.2 Resistance to shrinkage stresses

As mentioned earlier, shrinkage induces tension in the concrete mass as it attempts to pull adjacent material to itself, causing tensile cracks. A minimum shrinkage reinforcement is always required.

28.3.3 Ductility (mode of failure)

Concrete is a brittle material that will resist loads with comparatively small deformations. Once the ultimate strength is reached, sudden collapse may follow any load increase. Steel bars will hold concrete together should it fail in tension, compression, or shear.

Reinforcing bars are made in many grades, classified according to the yield stress in tension of steel, F_y. The most common are

Grade 40	40ksi	276MPa
Grade 50	50ksi	345MPa
Grade 60	60ksi	414MPa

R/C structures almost exclusively use deformed bars, with projections on the round surface to increase the bond between steel and concrete, much like the threads on a bolt or screw. Deformed steel bars must conform to ASTM A 615, 616, and 617. Bars in chemically aggressive or high-moisture environments can be coated with zinc or epoxy resin.

Bars are produced in many standard diameters, listed in Tables 28.3 and 28.4 . Bar diameters are given in multiples of 1/8in [3.17mm]. The smallest bar is No. 2, which stands for a 2/8in diameter or 0.25in [6.35mm].

Table 28.3 Properties of reinforcing bars

Bar No.	2	3	4	5	6	7	8	9	10	11	14S	18S
Area (in^2)	0.05	0.11	0.2	0.31	0.44	0.6	0.79	100	1.27	1.56	2.25	4
Diameter (in)	0.25	0.38	0.5	0.63	0.75	0.88	1	1.13	1.27	1.41	1.69	2.26
Wt/ft (lb/ft)	0.17	0.38	0.67	1.04	1.5	2.04	2.67	3.4	4.3	5.31	7.65	13.6

REMEMBER OR MARK THE LOCATION OF THIS TABLE, YOU'LL USE IT A LOT!

Table 28.4 Areas (in²) of bars in 1ft sections (slabs, walls, etc.)

Bar Spacing	No. 2	No. 3	No. 4	No. 5	No. 6	No. 7	No. 8	No. 9	No. 10	No.11
2" o.c.	0.3	0.66	1.2	1.86	2.64					
2-1/2" o.c.	0.24	0.53	0.96	1.49	2.11	2.88	3.79			
3" o.c.	0.2	0.44	0.8	1.24	1.76	2.4	3.16	4		
3-1/2" o.c.	0.17	0.38	0.69	1.06	1.51	2.06	2.71	3.43	4.36	
4" o.c.	0.15	0.33	60	0.93	1.32	1.8	2.37	3	3.81	4.68
4-1/2" o.c.	0.13	0.29	0.53	0.83	1.17	1.6	2.11	2.67	3.39	4.16
5" o.c.	0.12	0.26	0.48	0.74	1.06	1.44	1.9	2.4	3.05	3.74
5-1/2" o.c.	0.11	0.24	0.44	0.68	0.96	1.31	1.72	2.18	2.77	3.4
6" o.c.	0.1	0.22	0.4	0.62	0.88	1.2	1.58	2	2.54	3.12
6-1/2" o.c.	0.09	0.2	0.37	0.57	0.81	1.11	1.46	1.85	2.35	2.88
7" o.c.	0.09	0.19	0.34	0.53	0.75	1.03	1.35	1.71	2.18	2.67
7-1/2" o.c.	0.08	0.18	0.32	0.5	0.7	0.96	1.26	1.6	2.03	2.5
8" o.c.	0.08	0.17	0.3	0.47	0.66	0.9	1.19	1.5	1.91	2.34
8-1/2" o.c.	0.07	0.16	0.28	0.44	0.62	0.85	1.12	1.41	1.79	2.2
9" o.c.	0.07	0.15	0.27	0.41	0.59	0.8	1.05	1.33	1.69	2.08
9-1/2" o.c.	0.06	0.14	0.25	0.39	0.56	0.76	1	1.26	1.6	1.97
10" o.c.	0.06	0.13	0.24	0.37	0.53	0.72	0.95	1.2	1.52	1.87
10-1/2" o.c.	0.06	0.13	0.23	0.35	0.5	0.69	0.9	1.14	1.45	1.78
11" o.c.	0.05	0.12	0.22	0.34	0.48	0.65	0.86	1.09	1.39	1.7
11-1/2" o.c.	0.05	0.11	0.21	0.32	0.46	0.63	0.82	1.04	1.33	1.63
12" o.c.	0.05	0.11	0.2	0.31	0.44	0.6	0.79	1	1.27	1.56
13" o.c.			0.18	0.29	0.41	0.55	0.73	0.92	1.17	1.44
14" o.c.			0.17	0.27	0.38	0.51	0.68	0.86	1.09	1.34
15" o.c.			0.16	0.25	0.35	0.48	0.63	0.8	1.02	1.25
16" o.c.			0.15	0.23	0.33	0.45	0.59	0.75	0.95	1.17
17" o.c.			0.14	0.22	0.31	0.42	0.56	0.71	0.9	1.1
18" o.c.			0.13	0.21	0.29	0.4	0.53	0.67	0.85	1.04

Welded wire fabric or mesh is frequently used in slab reinforcement. Fabric is made of wires running in orthogonal directions forming a grid and welded together at the intersections. The most common w.w.f. or w.w.m. is 6x6-10/10. This is the standard reinforcing for a 4in [100mm] slab on grade (shrinkage reinforcement).

Table 28.5 Common stock sizes of welded wire fabric (welded wire mesh)

Steel designation (rolls)		Area of steel provided in²/ft		Weight
New designation	Old designation	Longitudinal	Transverse	pounds/100sqft
6X6-W1.4XW1.4	6X6-10/10	0.029	0.029	21
6X6-W2.0XW2.0	6X6-8/8	0.041	0.041	30
6X6-W2.9XW2.9	6X6-6/6	0.058	0.058	42
6X6-W4.0XW4.0	6X6-4/4	0.080	0.080	58
4X4-W1.4XW1.4	4X4-10/10	0.043	0.043	31
4X4-W2.0XW2.0	4X4-8/8	0.062	0.062	44
4X4-W2.9XW2.9	4X4-6/6	0.087	0.087	62
4X4-W4.0XW4.0	4X4-4/4	0.120	0.120	85
Steel designation (sheets)		Area of steel provided in²/ft		Weight
New designation	Old designation	Longitudinal	Transverse	pounds/100sqft
6X6-W2.9XW2.9	6X6-6/6	0.058	0.058	42
6X6-W4.0XW4.0	6X6-4/4	0.080	0.080	58
6X6-W5.5XW5.5	6X6-2/2	0.110	0.110	80
4X4-W4.0XW4.0	4X4-4/4	0.120	0.120	85

The main requirements for the placement of the reinforcing bars depend on the bar diameter d_b and are shown in Table 28.6:

Table 28.6 Spacing and cover requirements for reinforcing bars

Minimum spacing between:			
	parallel bars in a layer	d_b or 1in	25mm
	layers of parallel bars	1 in.	
	longitudinal bars of columns	$1.5d_b$ or 1.5in	38mm
Minimum cover (concrete protection for reinforcing):			
	Concrete cast against earth	3in	76mm
Concrete exposed to weather or a hostile environment:			
	No. 6 through No. 18 bars	2in	51mm
	No. 5 bar or smaller	1.5in	38mm
Concrete not exposed to the weather:			
	Slabs, walls, and joists		
	No. 14 and No. 18 bars	1.5in	38mm
	No. 11 bar and smaller	0.75in	19mm
	Beams and columns	1.5in	38mm

Ties are also subject to spacing requirements, as indicated in Chapters 31 and 36 dealing with columns and beams.

28.4 Placement of concrete

After the initial placement or as part of the initial placement process, the following steps will occur (steps vary depending on the type of member being cast). When placing concrete, fill forms from the outside to the interior. This will cause any water that has been driven out of the mix to be reabsorbed and not trapped against the forms, causing honeycombs (like Swiss cheese voids in the corners or at the edges).

28.4.1 Vibrating

Deep or thick pours of concrete should be vibrated to cause the concrete to flow into all of the voids in the form and to encourage any trapped air bubbles to float to the surface. Even thin slabs are frequently vibrated. This will occur immediately as the concrete is placed. Excessive vibration will cause segregation of the aggregate, much as you do when you shake your popcorn to cause Mr. Redenbocker's unpopped kernels to sink to the bottom of the lighter popped portions. Water weighs $62.4 lb/ft^3$ [$1000 kg/m^3$], cement weighs $90 lb/ft^3$ [$1440 kg/m^3$], aggregate weighs 90 to $160 lb/ft^3$ [1440 to $2569 kg/m^3$], and these different density materials will seek their respective levels in the concrete if overvibrated. Of course, if the concrete is not vibrated adequately, the form will not be completely filled and **honeycombs** will be apparent when the forms are removed.

In slabs, a series of additional steps are generally required.

28.4.2 Screeding

In simplest terms, screeding is rough grading (in slabs or other flat surfaces). Typically, a 2X4 [38X89] or 2x6 [38X140]is used to span between **screeds** or members previously set at the level of the finished surface. The board is pulled back and forth with someone either pulling away excess concrete or adding concrete in low areas in front of the board.

28.4.3 Floating

Floating immediately follows the screeding process, and its purpose is to work exposed aggregate down into the concrete mass and to "float" a small amount of the cement-water paste to the surface. Excessive floating will bring too much cement and water to the surface, which will then expand and contract at a slightly different rate than the base mass, and the surface will form a maze of tiny cracks called **crazing**. These tiny cracks will allow water to penetrate into the surface and may ultimately cause portions of the surface to spall off. Floating is accomplished by tipping, pushing, and pulling a 3 to 4ft [0.9 to 1.2m] magnesium plate over the surface of the newly placed concrete. The attachment of a long handle, 8-16ft [2.4-4.8m] long, allows the worker to stand outside the form and work the concrete.

28.4.4 Darbying

Darbying is floating with a hand tool (3 to 4ft [0.9 to 1.2m] long, with two handles) along edges or around columns or other obstructions that make floating with a large tool impractical.

28.4.5 Trowelling

After the concrete has reached a stage of hydration that will allow it to support weight without significant material displacement, the surface is trowelled. This is usually done with a combination of a powered device that looks a bit like a lawn mower with four adjustable blades. The blades can be easily sloped at different angles to cause a greater or lesser amount of lateral force to be applied to the surface of the concrete. This does two things; it further smoothes the surface of the material and it consolidates it (packs it), which makes it less permeable to water and a bit more dense. Some finishing around the edges or adjacent to obstructions must be typically done by hand.

At this point, the concrete may be "finished" and the only remaining consideration is to keep it moist enough and warm or cool enough to allow hydration to continue to the desired strength level. This can take as much as 28 days for normal concrete.

28.4.6 Curing

The concrete must be maintained at 70°F [21°C] for 3 days or 50°F [10°C] for a period of 7 days following placement for standard concrete or for 3 days for high early strength concrete. Varying time and temperature combinations are possible, but this is a crucial time in the life of the concrete and extreme care must be taken. Maintaining adequate moisture is also crucial for development of concrete strength. This can be a problem during the hottest periods of the year. If the surface of the concrete is not maintained in a moist condition or sealed with a curing agent to retain the moisture in the concrete, it will evaporate and at least the surface will dry. If this occurs, the concrete will not be able to hydrate (the chemical reaction between water and cement will not take place completely) and, consequently, the concrete surface will not develop its strength potential.

28.4.7 Finishes

A textured finish to increase traction on sidewalks is often created by broom finishing the surface. This consists merely of pulling a broom across the surface to roughen it after it has begun to set up. This does create a more abrasive surface; however, it also pulls the surface apart and makes it more susceptible to water penetration and ultimately to surface spalling. A more desirable method might be to trowel an abrasive material (carbide particles) into the surface, creating the slip-resistant quality without the detrimental effects of broom finishing. Architectural textures are created by the surface of the formwork. Virtually any desired texture can be created by lining the formwork (usually steel or plywood) with embossed sheets of rubber, fiberglass, wood, and so on. Care must be taken with heavy textures so that the cover of the steel reinforcing remains adequate.

Special finishes such as exposed aggregate finishes can be obtained by using a retarder on the concrete to slow or stop hydration of a thin surface layer of cement. Later, after the base mass of concrete has adequately cured, this surface layer can be scrubbed to remove a small amount of the cement-sand mixture and expose the underlying aggregate. This obviously affects the water-resisting characteristics of the concrete.

NEVER:

Apply raw **cement** to the surface of concrete that has been excessively floated in an attempt to absorb the excess water. This will guarantee crazing and spalling of the surface. To solve this problem, fans will probably need to be brought in to evaporate the excess water, but it is likely that the surface will still have an excess of cement and will still craze.

Add **water** to the concrete when it is delivered to the site, unless it is water that was initially designed into the mix. This is not likely to ever be the case. Addition of water to the concrete changes the water-cement ratio and may dramatically affect the strength of the final product.

Accept **concrete** that has been mixed (cement in contact with water) for over an hour. By this time, the hydration process has begun and the strength of the mix will be adversely affected by additional mixing or agitation. If delivery distances are excessive, it may be necessary to wait to add water until the truck has reached the site. This may make the water-cement ratio harder to control unless the mixer has precise water quantities. Hydration retarders can be used if excessive transportation time is expected.

METRIC VERSIONS OF REINFORCING TABLES

Table M28.3 Properties of reinforcing bars

U.S. Bar No.	2	3	4	5	6	7	8	9	10	11	14S	18S
Metric No.	6	10	13	16	19	22	25	29	32	36	43	57
Area (in^2)	0.05	0.11	0.2	0.31	0.44	0.6	0.79	100	1.27	1.56	2.25	4
Diameter (in)	0.25	0.38	0.5	0.63	0.75	0.88	1	1.13	1.27	1.41	1.69	2.26
Wt/ft (lb/ft)	0.17	0.38	0.67	1.04	1.5	2.04	2.67	3.4	4.3	5.31	7.65	13.6

Table M28.4 Areas (mm^2) of bars in 1m sections (slabs, walls, etc.)

Bar designation	M10	M15	M20	M25	M30	M35	M45
Diameter (mm)	10	15	20	25	30	35	45
Area (mm^2)	78.5	176.6	314.2	490.6	706.5	962.2	1,590.6
Bar spacing							
50mm o.c.	1,571	3,534	6,283	9,817	14,137	19,242	31,809
60mm o.c.	1,309	2,945	5,236	8,181	11,781	16,035	26,507
70mm o.c.	1,122	2,524	4,488	7,012	10,098	13,744	22,720
80mm o.c.	982	2,209	3,927	6,136	8,836	12,026	19,880
90mm o.c.	873	1,963	3,491	5,454	7,854	10,690	17,671
100mm o.c.	785	1,767	3,142	4,909	7,069	9,621	15,904
110mm o.c.	714	1,606	2,856	4,462	6,426	8,746	14,458
120mm o.c.	654	1,473	2,618	4,091	5,890	8,018	13,254
130mm o.c.	604	1,359	2,417	3,776	5,437	7,401	12,234
140mm o.c.	561	1,262	2,244	3,506	5,049	6,872	11,360
150mm o.c.	524	1,178	2,094	3,272	4,712	6,414	10,603
160mm o.c.	491	1,104	1,963	3,068	4,418	6,013	9,940
170mm o.c.	462	1,039	1,848	2,887	4,158	5,659	9,355
180mm o.c.	436	982	1,745	2,727	3,927	5,345	8,836
190mm o.c.	413	930	1,653	2,583	3,720	5,064	8,371
200mm o.c.	393	883	1,571	2,454	3,534	4,810	7,952
210mm o.c.	374	841	1,496	2,337	3,366	4,581	7,573
220mm o.c.	357	803	1,428	2,231	3,213	4,373	7,229
230mm o.c.	341	768	1,366	2,134	3,073	4,183	6,915
250mm o.c.	314	707	1,257	1,963	2,827	3,848	6,362
300mm o.c.	262	589	1,047	1,636	2,356	3,207	5,301
350mm o.c.	224	505	898	1,402	2,020	2,749	4,544
400mm o.c.	196	441	785	1,227	1,767	2,405	3,976
450mm o.c.	174	393	698	1,091	1,571	2,138	3,534

29 REINFORCED CONCRETE IN ARCHITECTURE

29.1 Structural forms

Concrete is a relatively inexpensive material that can be poured in place in virtually any form limited only by the characteristics of the form material. For this reason, reinforced concrete has been employed during the twentieth century for a large variety of buildings and engineering structures, ranging from long-span bridges to small residential structures. In smaller buildings, reinforced or prestressed concrete floor joist and slab systems have become common, particularly where fireproofing is a consideration. Earth-sheltered or underground facilities are prime candidates for concrete, which can create structural support and enclosure all in one process.

Because of the labor-intensive formwork required for site casting, prefabrication techniques are frequently employed to produce components in a plant, where experienced labor and ideal conditions for curing can be achieved. This can allow production to continue year-round regardless of the local weather conditions. In a plant, the same permanent formwork can be used repeatedly. This approach is typically used for site-cast concrete as well, and under those conditions, construction sequencing becomes a major concern. Finally, prestressing can be achieved more simply and efficiently in the comfort and convenience of a production plant than attempting to utilize post tensioning (prestressing after the fact) on-site.

In addition to basic commercial products such as joists and floor slabs, many complex forms have been created in reinforced and prestressed concrete, either with on-site formwork or as prefabricated components. Double-curvature shells, such as in the TWA Terminal by Eero Saarinen (Fig. 29.1), can be built with a small thickness to span large spaces.

Fig. 29.1 TWA Terminal, New York, NY

Double-curvature shells require expensive formwork, and for this reason, the hyperbolic parabo-loid has a geometry preferable to other shells. This surface is generated by straight lines, and the formwork consequently can be built with straight wood components (Fig. 29.2). Felix Can-dela, a Mexican engineer, is one the many designers who have employed hyperbolic paraboloid shells (curves built with straight form members) for different building types.

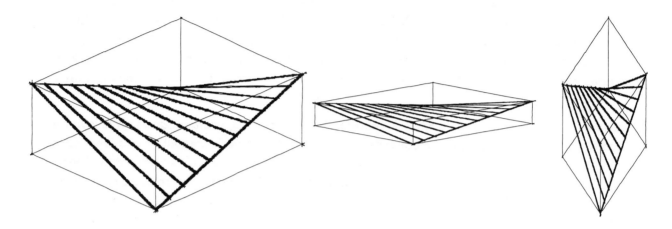

Fig. 29.2 Hyperbolic paraboloid geometry

Geodesic vaults and domes are other structural systems that can be used for long-span build-ings. Pier Luigi Nervi was the first to use this system with reinforced concrete for airplane hangars and, later, in the form of domes for sports buildings. Other types of thin concrete members can be used for vaults, as in Nervi's Turin Exhibition Hall (Fig. 29.3), or in the form of prestressed beams, as in Louis Kahn's Kimbell Museum (Fig. 29.4).

Fig. 29.3 Turin Exhibition Hall

Fig. 29.4 Kimbell Museum vaults, Fort Worth, and Palasport dome, Rome

A unique characteristic of reinforced concrete is the continuity that can be achieved between separate castings, for instance, between a column and a beam cast subsequently on it, after the column has already partially cured. This allows the relatively easy creation of moment-resisting connections, a particularly important consideration for multistory frames. The combination of concrete (with good compressive strength, but brittle) with tension-resisting steel bars produces frames of great ductility, able to absorb large deformations before collapsing under unusual loading conditions, such as those encountered during earthquakes. The mass of concrete also has a dampening effect on seismic oscillation; consequently, reinforced concrete shear walls are sometimes employed even in steel-framed buildings.

At the same time, this characteristic can cause moment transfer to occur in cases where it is NOT desirable. Consequently, joint design and detailing in reinforced-concrete systems is a crucial consideration.

Architecturally, the reinforced-concrete frame has become part of the language of modern architecture as much as the multistory steel frame. Developed primarily in France at the end of the ninteenth century, the reinforced-concrete (R/C) frame was used by Perret for the first time as part of architectural expression. Le Corbusier turned it into an icon of his architectural manifesto with the Domino house, where the frame, set back from the exterior envelope, frees both the elevation and the plan from the constraints of the load-bearing wall.

Interestingly enough, the plastic qualities of the material have lent themselves to express the organic philosophy of Frank Lloyd Wright. Wright probably avoided frames for the reason that he hated the rigidity of the Victorian box and the classical language, but he saw in the continuity of reinforced concrete an analogy with the continuity of organic forms of plants and trees. This idea can be seen in the cantilevering structures of the Johnson Wax offices and labs, roofs of the Robie House, and in the classic "Falling Water."

29.2 Structural design issues

As the great Mexican engineer Eduardo Torroja said, the process of visualizing or conceiving a structure is an art that requires intuition and cannot result from a predetermined logical reasoning. However, this intuition is developed by experience on the basis of a clear understanding of all properties of reinforced concrete (such as shrinkage and thermal performance) and of all factors that may affect these properties during the process of production and the conditions of use. Concrete is appropriate for structures that are required to be permanent, not subject to frequent alterations, fire-resistant, and maintenance-free. Road bridges typically use concrete, but museums, offices, and laboratories are also building types where prestressed or ordinary reinforced concrete has been successfully employed. These building types require complex environmental systems, which are often concealed in a suspended ceiling. This technique, however, hides the aesthetic potential of concrete systems (as Kahn suggests in the opening quote of this section), and many innovative ideas have been used to achieve integration of systems and structure.

In some cases, the design of the environmental systems prevails over structural considerations. This is the case in the Lloyd's headquarters in London designed by the Richard Rogers Partnership, without question an important and complex structure in precast concrete. The floor assembly (Fig. 29.5) consists of a precast grid, supported on inverted U-beams. These have the same depth as the floor system, allowing flexibility in the division of space. *"In pure structural engineering terms it is not an especially elegant or economical solution. Such considerations are secondary. It is best thought of not as a simple load bearing structure, but as a framework for services, a distinction which may well be applied to the whole building"* (Colin Davies, "Lloyd's," Architectural Review, October 1986). In Fig. 29.5, the circular concrete columns support, on brackets, prestressed concrete beams. The floor grid is open for the passage of systems. Raised stub columns at each intersection support the floor, built by casting a concrete screed over metal trays used for permanent shuttering.

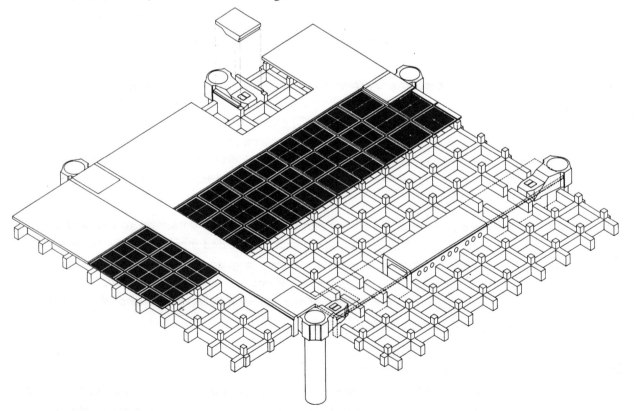

Fig. 29.5 Lloyds of London integrated structural and mechanical systems

One of the architecturally most creative solutions of this question is the Richards Medical Laboratories, Philadelphia, by Louis Kahn (Fig. 29.6, where the architect realized the idea of "servant" and "served" spaces. The horizontal structure, in precast reinforced concrete, has hollows to run the ductwork, which is exposed at ceiling level.

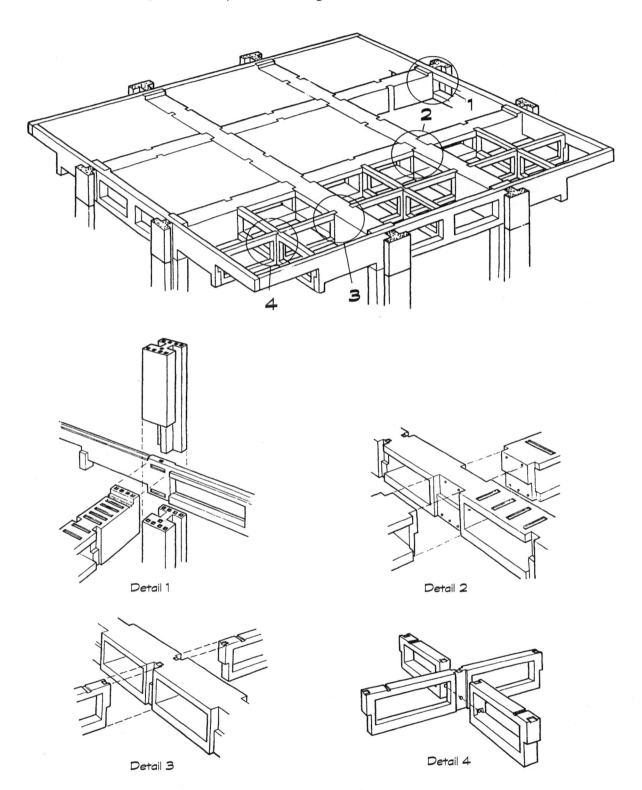

Detail 1

Detail 2

Detail 3

Detail 4

Fig. 29.6 Richards Medical Laboratories, precast modular system with integrated mechanical systems

This text analyzes only the simpliest types of reinforced-concrete structures: compression members (columns) and bending members (beams and slabs, retaining walls and foundations). Most structures, frames in particular, make use of the monolithic nature of concrete connections to obtain statically indeterminate systems. Although this is definitely an advantage for structural safety and economy, statics calculations can be complex, even for a simple low-rise structure, and are extremely laborious. For multistory structures subject to lateral loads, the indeterminacy creates the necessity to consider a wide range of alternative loading conditions, making the problems even more lengthy and complicated. Only rarely can systems be represented as simply supported beams or axially loaded columns. Even a common floor slab on masonry walls can become a statically indeterminate structure when both ends are fixed (Fig. 29.7). Having said this, we WILL design simple beams and columns as elementary examples of reinforced-concrete design.

Floor and roof slabs, made with precast panels or cast-in-place concrete, are normally connected at their ends to the beams, creating a slab continuous over more supports. These cases still can be calculated easily with the help of tables of shear and moments for common loading conditions. The case of an axially loaded column is even more rare than a simply supported beam, since in single-story or multistory buildings, beams and columns are rigidly connected. Gravity loads and lateral loads generate bending moments on the columns (Fig. 29.8), that are therefore subjected to compression and bending. The ability to design simple reinforced-concrete structures is, however, important to develop a structural intuition for this material, and also to outline the dimension and specifications of some types of structure, such as small retaining walls and footings, that occur frequently in the construction of any building. Once the principles are well understood, it will be easier to conceive the structure as an integral part of the building design, in relation to finishes and mechanical systems, as well as to the architectural conception.

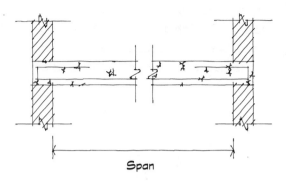

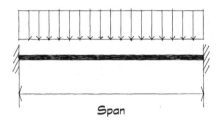

Fig. 29.7 R/C construction and statics diagram for floor slab on bearing walls

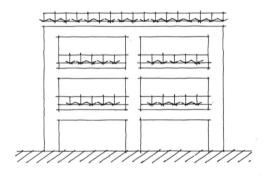

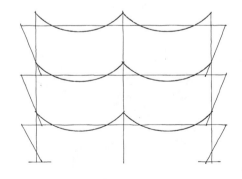

Fig. 29.8 Gravity loads acting on a multistory frame and corresponding moment diagram

The many advantages of R/C are partially offset by the weight of the structure. A steel structural system (beams, columns, decking, etc.) might have a weight of 7 to 10psf [*335 to 479N/m²*], but a concrete system will typically weigh 100 to 150psf [*4790 to 7185N/m²*]. This suggests that the supporting capabilities of the soil may become a major determinant in selecting concrete as a structural material. On the other hand, concrete structures can have excellent fire-resistance characteristics without the addition of other systems that might be necessary with other structural materials.

29.3 System selection

You were warned; here it is again: The selection of a concrete system involves a complex set of interrelated issues, so there is no single answer. You need to at least consider the following:

Spatial requirements

What are the volumetric requirements of the function that the space will house?

Is there a planning module or a leasing module that should be column-free or could be easily isolated? Parking garages, office structures, schools, hospitals, and retail establishments all have identifiable modules that are more desirable. Unfortunately, these are not absolute, perhaps with the exception of parking facilities, and must be identified during programming and schematic design stages.

Expansion

Is the spatial system likely to require frequent remodeling?

What is the likelihood of expansion of the system?

Integration of systems

Will the environmental/lighting systems require frequent change?

Does the spatial system require isolated volumes of space: for temperature, sound, security, utilities management, and so on?

Ease of erection/construction

What is the availability of skilled labor?

Are material or component production facilities easily accessible?

Fire resistance

What are the fire-rating requirements of the building as a whole?

What are the isolation characteristics of zones of the building?

What are the patterns of access required for emergency egress?

Soil conditions

What is the nature of the soil on the site?

      Chemical reactions with concrete products?

      Bearing capacities? Concrete structures are inherently heavy structures: This may be an advantage for thermal balancing, and a disadvantage for foundation design. A concrete structure may weigh several times its equivalent in steel. This could be a major consideration if the building is supported by weak soils which would require an excessive investment in foundations.

Spans and loadings

Some common combinations of span/loading/mechanical integration characteristics for simple systems would include the following.

Concrete flat-slab systems

Flat slabs are generally used for spans of 10 to 30ft [3 to 9m] and relatively light loads; beyond 30ft [9m], the dead load of a solid flat slab becomes a controlling factor. They have little flexibility for mechanical systems if integrated into the slab and are generally used with a suspended system to accommodate systems. This system is the simplest of the cast-in-place systems, since the formwork consists of easily built flat planes and the introduction of uncomplicated reinforcing. Flat-slab construction will have reinforcing placed in both directions regardless of its "spanning" capabilities. In a one-way slab, the reinforcing carries moment-developed tensile stress in one direction and temperature/shrinkage-induced stresses in the other direction. In a two-way slab, the reinforcing carries moment-induced stresses in both directions. So, what's the difference?

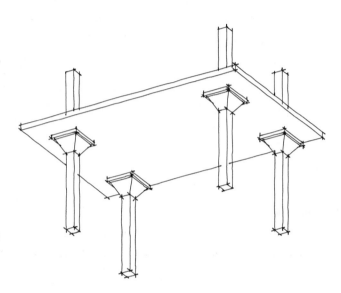

Fig. 29.9 One-way flat slab with shear heads at columns; note that the bay proportions exceed 1:1.5

> One-way slabs: One-way slabs are designed to span in one direction only and can be used in almost any bay proportion (Fig. 29.9).

> Two-way slabs: Two-way slabs are designed to span (surprise) two directions simultaneously, and to accomplish this, the bay proportions are most desirably 1:1 up to 1.5:1.

Concrete joist systems

Joist systems are generally used for spans of 15 to 40ft [4.5 to 12.2m], and can be used for "moderate" loads. The system is relatively lightweight since most of the concrete in the tensile zone has been eliminated, and has some possibility for mechanical integration in one direction only—parallel to the span. The formwork is a bit more complicated, but standard systems have been developed that reduce cost and time of construction. HOWEVER, they are constructed on a strict module in one direction. This means that very early in the design stage this system must be taken into account and walls, columns, vertical openings, and so on, should be considered relative to this module (Fig. 29.10).

Slab thicknesses of 4 to 6in
Stem spacings of 24 to 36in o.c.
Stem dimensions of 4 to 8 inches wide, 8 to 24in deep

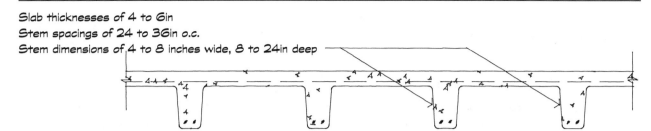

Fig. 29.10 Concrete joist system

Concrete waffle systems

Waffle systems are really two-way joist systems, which means that the proportions identified for two-way flat slabs, 1:1 to 1.5:1, are critical and the strict modular planning identified with joist systems must be acknowledged. This system is relatively light and spans 25 to 55ft [7.6 to 16.8m] with moderate to heavy loading conditions. Formwork is more complex as well, as reinforcement placement and mechanical integration can be quite difficult. Holes can be cast in the stems of the waffles to allow mechanical integration, which complicates form-work. Even this option, however, provides minimal space for manipulation of the systems. Flexible ductwork is frequently used to allow it to be "threaded" into the stem openings.

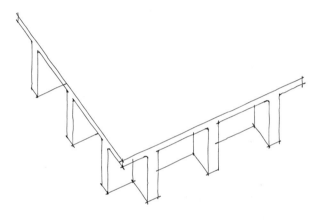

Fig. 29.11 Waffle system (2-way joist system)

Concrete "T" systems (single or double)

These "precast," frequently "prestressed," systems are used for spans of 30 to 120ft [9.1 to 36.6m] and because they are premanufactured, provide extremely rapid erection times. They are generally 4 to 8ft [1.2 to 2.4m] wide, although other flange widths can be accommodated during manufacturing, and they provide the potential for mechanical integration in one direction; parallel to the span. Prestressed "T" sections will have an upward curvature, "camber" induced by the prestressing strands, which can result in coordination difficulties when short spans are used adjacent to long spans. The same problem exists with other systems, but is usually associated with deflection, which occurs after construction.

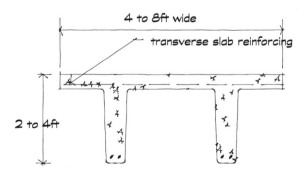

Fig. 29.12 Precast/prestressed single or double "T"s

Short spans immediately adjacent to long spans will show differential deflection between the spans, which must be accounted for by expansion details in finish materials.

Concrete cellular deck systems

Precast and/or prestressed cellular deck systems are utilized for spans of 20 to 50ft [6.1 to 15.3m] with moderate loads. They are also premanufactured, offering rapid erection, and have some potential for mechanical integration in one direction within the open interior cells. These are basically lightweight, strong, and rapidly constructed flat-slab systems.

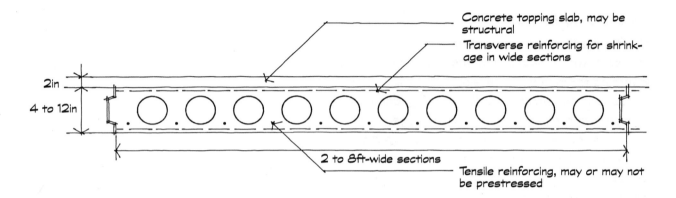

Fig. 29.13 Precast concrete cellular deck system

Flat-slab and cellular deck systems generally require a dropped ceiling for integration of environmental control systems, whereas the joist, "T," and waffle systems have some minimal integration possibilities. An additional architectural consideration, which is frequently the governing one, is the final finished quality of the system. If a waffle system is used, it is commonly exposed. The intricate repetitive modules of the structure can be architecturally integrated into systems of lighting, windows, partitions, and even furniture layouts. Structural slab systems have none of the organizational constraints of joist or waffle systems; however, the simple flat planes tend to show irregularities and flaws in workmanship more than the highly articulated forms of waffles.

30 BEAM STRENGTH THEORY

30.1 Stress and strain in flexure members

When a beam is subject to bending (Fig. 30.1), the internal stresses resisting the external actions are bending and shear stresses. In a beam made of material with significant tensile strength, as we've just seen in steel, the resultant of these stresses can be visualized on a cut section of the beam: Vertical equilibrium of the free body requires V, and horizontal equilibrium requires C = T. For the equilibrium of moments, the moment couple generated by C and T is equal and opposite to the bending moment. In concrete, unfortunately, tensile resistance is very limited, and Fig. 30.1 is applicable only for tension stresses not greater than a few hundred psi [MPa]. Beyond these stress levels, the concrete tensile zone cracks (fails) and the beam collapses instantaneously. In concrete-beam design, any tensile resistance must be disregarded and all tensile stresses must be designed to be taken by the steel reinforcement (Fig. 30.2). The stress diagram shows compression stresses increasing proportional to the distance from the neutral axis, and zero tension on the concrete. In the **working-stress method**, the maximum compression stress f_c must remain within the elastic limit, well below the ultimate strength, and the tension stress in the steel must not be greater than F_y reduced by a safety factor.

As the stress approaches the ultimate value, its variation is not linear. If we use the more common **strength design method**, we assume the form of stress distribution shown in Fig. 30.3(a); for practical purposes, the diagram of compressive stress is approximated to a rectangle, with the limit for the concrete stress set at $0.85f_c'$ as shown in Fig. 30.3 (b). The material below this rectangular

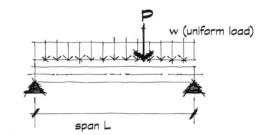

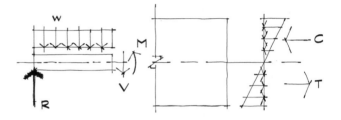

Fig. 30.1 Beam with loading, free body diagram and generalized stress distribution diagram

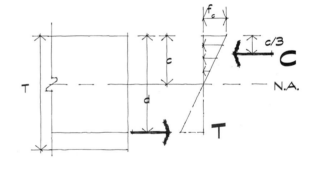

Fig. 30.2 Stress distribution in R/C "working stress" design method (similar to that used for steel and wood)

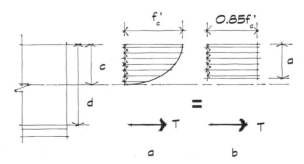

Fig. 30.3 Stress distribution in R/C "strength" design method (a new approach compared to steel and wood)

assumption must remain in place since in reality it provides some compressive resistance. The rectangular approximation provides a much simpler mathematical solution for both the magnitude of the resultant force and its location.

The distance a (to the bottom of the rectangular approximation from the extreme fiber) can be determined by the expression:

$$a = \beta_1 c$$

where β_1 is a factor that varies with the compressive strength of concrete according to Table 30.1, and c is the distance from the extreme fiber to the neutral axis.

Table 30.1 Values of β_1 (beta) for reinforced-concrete beams

Concrete strength f_c (psi)	[MPa]	b_1 beta
≤ 4000	28	0.85
5000	35	0.80
6000	42	0.75
7000	49	0.70

Returning once more to the basic formula of F = P/A, we can determine the total forces produced by both the compression zone of the beam and the tensile reinforcement.

If F = P/A, then P = FxA.

Looking at the nomenclature of the R/C beam, we see that the maximum stress level of the compression block is $0.85f_c'$ and the area is b (the width of the beam) x a (the depth of the rectangular assumption).

Therefore, the resultants of the compressive and tensile stresses can be written

$$C = 0.85f_c'(ba) \qquad and \qquad T = F_y A_s$$

Since equilibrium requires that $\Sigma F_H = 0$, consequently C = T, and we can equate the two expressions for force

$$0.85f_c'(ba) = F_y A_s$$

which can be solved for a:

$$a = \frac{F_y A_s}{0.85f_c' b}$$

The percentage of steel in an R/C beam is based on A_s/A_g (the area of steel divided by the gross area of the cross-section). So, if $p_g = A_s/A_g$, then $A_s = p_g A_g = p_g(bd)$ (again using the beam nomenclature identifed in Figure 30.3.

Replacing this expression in the "a" equation yields

$$a = \frac{F_y p_g bd}{0.85f'_c b} = \frac{F_y p_g d}{0.85f'}$$

The percentage of steel corresponding to the condition when C = T (balanced section) is called p_b. Since a and d are unknown, it is not possible to use the preceding formula as a design formula. In order to eliminate one of the unknowns, we need to replace a with the independent equation previously introduced:

$$a = \beta_1 c$$

By equating this with the previous expression:

$$\beta_1 c = \frac{F_y p_g d}{0.85f'_c}$$

The distance c is also unknown, but can be easily derived from the strain diagram. Unlike the stress diagram, strain is proportional to the distance from the neutral axis (Fig. 30.4). The maximum strain the concrete can reach at the compression fibers is given by the Uniform Building Code (Sec. 2610, Art. c) as $\varepsilon_c = 0.003$ in/in [mm/mm]. (This is a good number to remember since it is a common question on the Architectural Registration Examination.)

Note: The strain is the ratio of deformation to initial length, in this case, of the extreme compressive fiber. For instance, if the concrete in a 20ft [6100mm] beam subject to bending reaches the maximum strain, the deformation of the extreme compressive fiber is $\delta = \varepsilon_c L$.

$$\varepsilon_c = 0.003(20ft)(12in/ft) = 0.72in \quad [0.003(6100mm) = 18mm]$$

From the strain diagram, we can identify similar triangles and write the following proportion:

$$\frac{c}{0.003} = \frac{d}{\delta_c + 0.003}$$

where ε_c is the strain in the steel.

The modulus of elasticity of the steel is by definition (Fig. 30.5):

$$E_s = F_s/\delta_c \quad \text{therefore,} \quad \delta_c = F_s/E_s$$

Replacing ε_c in the previous proportion:

$$\frac{c}{0.003} = \frac{d}{\frac{F_s}{E_s} + 0.003}$$

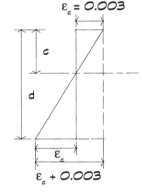

Fig. 30.4 Similar triangles

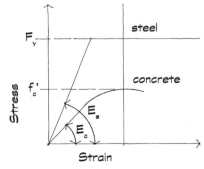

Fig. 30.5 Stress-strain diagram

Since $E_s = 29{,}000{,}000 \text{psi } [200{,}000 \text{MPa}]$ and $F_s = F_y$, we can solve the proportion for c:

$$c = \frac{0.003(29{,}000{,}000\text{psi})d}{F_y + 0.003(29{,}000{,}000\text{psi})} \quad \text{or} \quad c = \frac{0.003(200{,}000\text{MPa})d}{F_y + 0.003(200{,}000\text{MPa})}$$

$$c = \frac{87{,}000(d)}{87{,}000 + F_y} \quad\quad \text{or} \quad c = \frac{600d}{600 + F_y}$$

We have now found a formula for c that can be substituted in the previous equation:

$$\frac{F_y p_g d}{0.85 f'_c} = \beta_1 c$$

$$\frac{F_y p_g d}{0.85 f'_c} = \beta_1 \frac{87{,}000}{87{,}000 + F_y} d$$

(For metric, replace 87,000 with 600, the stresses are MPa, and dimensions are measured in mm both here and in equation 30.1).

Solving for p_g and canceling the d's on each side gives the percentage p_b for a balanced section:

$$p_b = \beta_1 \frac{87{,}000}{87{,}000 + F_y} \frac{0.85 f'_c}{F_y} \quad\quad\quad\quad \textbf{(Eq. 30.1)}$$

These values, which are tabulated for different values of F_y and f'_c, are the basis for calculating the steel area. The ACI requires:

Maximum $p_g = 0.75 p_b$
Recommended $p_g = 0.5 p_g$ maximum
Minimum $p_g = 200/F_y$

If you follow this recommendation, you will **not** even approach a condition that might lead to compression failure. The maximum p_g you'll have at 1/0.75 is an additional factor of safety of 133% for compression failure. If you use the recommended value at $0.5p_g$ you'll have a 1/(0.5)(0.75) or 166% factor of safety. So even though we are theoretically utilizing the full "strength" of the concrete, recommended steel percentages will provide a substantial "factor of safety" against failure in a catastrophic (compression) mode.

30.2 Beam design formula

Finally, in order to derive a design formula, let us consider the moment capacity M of the beam, which must be equal to the moment couple. By taking the moments about the point of application of C, the moment arm is d - a/2. The design moment based on tensile forces is

$$M = T(d - a/2)$$

and since $T = A_s F_y$,

$$M = A_s F_y(d - a/2)$$

where the strength reduction factor $\Phi = 0.9$.

As previously found, $a = F_y p_g d/0.85 f_c'$, and replacing this in the M formula:

$$M = \Phi A_s F_y\left(d - \frac{F_y p_g d}{2(0.85 f_c')}\right)$$

Factoring d:

$$M = \Phi A_s F_y d\left(1 - \frac{0.59 F_y p_g}{f_c'}\right)$$

Replacing A_s with $p_g bd$:

$$M = \Phi p_g F_y bd\left(1 - \frac{0.59 F_y p_g}{f_c'}\right)$$

By dividing by Φbd^2, the right member becomes a constant, independent from the section dimensions:

$$\frac{M}{\Phi bd^2} = p_g F_y\left(1 - \frac{0.59 F_y p_g}{f_c'}\right)$$

$$\text{where } R = p_g F_y\left(1 - \frac{0.59 F_y p_g}{f_c'}\right) \qquad \text{(Eq. 30.2)}$$

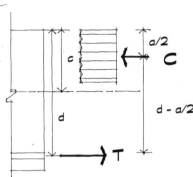

Fig. 30.6 Force equilibrium

R is a constant tabulated as a function of the materials and the steel percentage.
We therefore can write

$$M = \frac{R}{\Phi bd^2} \qquad \text{(Eq. 30.3)}$$

If a value of b is assumed, this equation can be solved for d and used as a design formula:

$$d = \sqrt{\frac{M}{\Phi Rb}} \qquad \text{(Eq. 30.4)}$$

Values of **R** (in psi) are tabulated for different values of F_y and f_c' and for values of p_g corresponding to p_g max., $0.5 p_g$ max., and p_g min. in the R value table. The R value corresponding to $0.5 p_g$ max. is usually selected to calculate the required d. Once the actual dimensions are assigned to the section, R has to be recalculated with Eq. (30.3). This new value of R corresponds to a value of p_g according to Eq. (30.2). It is not necessary to calculate p_g, since values are tabulated in Tables 30.1 to 30.6. Once this is determined, the steel reinforcement can be designed.

sponds to a value of p_g according to Eq. (30.2). It is not necessary to calculate p_g, since values are tabulated in Tables 30.1 to 30.6. Once this is determined, the steel reinforcement can be designed.

Alternatively, we could rewrite the formula in terms of an assumed width b and determine the required structural depth d for that situation. We could also solve for the maximum moment capacity of a particular geometry of a beam (**the width, b, and the structural depth, d**). These are not common situations in practice, but are equally valid possibilities.

Now that we've looked at both columns and beams, you might take note that high-strength concrete is more cost-effective in columns, which use 100% of the cross-section in compression, than in beams, which use 30 to 40% of the section in compression. This is not true, however, for "T" beams, which have a geometry that places a majority of the total concrete in the compression zone.

Table 30.2 Balanced ratio of reinforcement

p_b For rectangular sections with tension reinforcement only

f_c'		2,500 psi [17Mpa]	3,000 psi [21MPa]	4,000 psi [28MPa]	5,000 psi [35MPa]	6,000 psi [42MPa]
F_y		$\beta = 0.85$	$\beta = 0.85$	$\beta = 0.85$	$\beta = 0.80$	$\beta = 0.75$
GRADE 40	p_b (balanced)	0.0309	0.0371	0.0495	0.0582	0.0655
F_y=40,000psi	$p_{max} = 0.75p_b$	0.0232	0.0278	0.0371	0.0437	0.0492
[276MPa]	$p_{rec.} = 0.5\ p_{max}$	0.0116	0.0139	0.0185	0.0218	0.0246
	$p_{min} = 200$	0.0050	0.0050	0.0050	0.0050	0.0050
GRADE 50	p_b (balanced)	0.0229	0.0275	0.0367	0.0432	0.0486
F_y=50,000psi	$p_{max} = 0.75p_b$	0.0172	0.0206	0.0275	0.0324	0.0365
[345MPa]	$p_{rec.} = 0.5\ p_{max}$	0.0086	0.0103	0.0137	0.0162	0.0182
	$p_{min} = 200$	0.0040	0.0040	0.0040	0.0040	0.0040
GRADE 60	p_b (balanced)	0.0178	0.0214	0.0285	0.0335	0.0377
F_y=60,000psi	$p_{max} = 0.75p_b$	0.0134	0.0161	0.0214	0.0252	0.0283
[414MPa]	$p_{rec.} = 0.5\ p_{max}$	0.0067	0.0080	0.0107	0.0126	0.0141
	$p_{min} = 200$	0.0033	0.0033	0.0033	0.0033	0.0033
GRADE 75	p_b (balanced)	0.0129	0.0155	0.0207	0.0243	0.0274
F_y=75,000psi	$p_{max} = 0.75p_b$	0.0097	0.0116	0.0155	0.0182	0.0205
[518MPa]	$p_{rec.} = 0.5\ p_{max}$	0.0048	0.0058	0.0077	0.0091	0.0102

Table 30.3 Beam-design constants: R value

$$R = M/(\Phi bd^2) = p_g F_y [1 - (0.59 p_g F_y / f_c')]$$

f_c'		3,000 psi [21MPa]	4,000 psi [28MPa]	5,000 psi [35MPa]
F_y				
GRADE 40	@ p_{max}	868.8 psi	1159.2 psi	1387.4 psi
F_y=40,000psi	@0.5 p_{max}	495.2 ps	660.7 psi	782.3 psi
[276MPa]	@ p_{min}	192.1 psi	194.1 psi	195.3 psi
GRADE 50	@ p_{max}	821.3 psi	1096.1 psi	1310.3 psi
F_y=50,000psi	@0.5 p_{max}	462.8 psi	617.8 psi	732.3 psi
[345MPa]	@ p_{min}	192.1 psi	194.1 psi	195.3 psi
GRADE 60	@ p_{max}	782.5 psi	1040.8 psi	1242.2 psi
F_y=60,000psi	@0.5 p_{max}	437.0 psi	581.2 psi	688.6 psi
[414MPa]	@ p_{min}	190.3 psi	192.2 psi	193.4 psi

Table 30.4 Beam-design constants F_y = 50ksi [345MPa] f_c' = 3ksi [21MPa]

P_g	R psi	P_g	R psi	P_g	R psi	P_g	R psi
0.0040	192.1	0.0082	376.9	0.0124	544.4	0.0166	694.5
0.0041	196.7	0.0083	381.1	0.0125	548.2	0.0167	697.9
0.0042	201.3	0.0084	385.3	0.0126	551.9	0.0168	701.2
0.0043	205.9	0.0085	389.5	0.0127	555.7	0.0169	704.6
0.0044	210.5	0.0086	393.6	0.0128	559.4	0.0170	707.9
0.0045	215.0	0.0087	397.8	0.0129	563.2	0.0171	711.2
0.0046	219.6	0.0088	401.9	0.0130	566.9	0.0172	714.5
0.0047	224.1	0.0089	406.1	0.0131	570.6	0.0173	717.8
0.0048	228.7	0.0090	410.2	0.0132	574.3	0.0174	721.1
0.0049	233.2	0.0091	414.3	0.0133	578.0	0.0175	724.4
0.0050	237.7	0.0092	418.4	0.0134	581.7	0.0176	727.7
0.0051	242.2	0.0093	422.5	0.0135	585.4	0.0177	731.0
0.0052	246.7	0.0094	426.6	0.0136	589.1	0.0178	734.2
0.0053	251.2	0.0095	430.6	0.0137	592.7	0.0179	737.5
0.0054	255.7	0.0096	434.7	0.0138	596.4	0.0180	740.7
0.0055	260.1	0.0097	438.7	0.0139	600.0	0.0181	743.9
0.0056	264.6	0.0098	442.8	0.0140	603.6	0.0182	747.1
0.0057	269.0	0.0099	446.8	0.0141	607.2	0.0183	750.3
0.0058	273.5	0.0100	450.8	0.0142	610.9	0.0184	753.5
0.0059	277.9	0.0101	454.8	0.0143	614.5	0.0185	756.7
0.0060	282.3	0.0102	458.8	0.0144	618.0	0.0186	759.9
0.0061	286.7	0.0103	462.8	0.0145	621.6	0.0187	763.1
0.0062	291.1	0.0104	466.8	0.0146	625.2	0.0188	766.2
0.0063	295.5	0.0105	470.8	0.0147	628.7	0.0188	769.4
0.0064	299.9	0.0106	474.8	0.0148	632.3	0.0190	772.5
0.0065	304.2	0.0107	478.7	0.0149	635.8	0.0191	775.6
0.0066	308.6	0.0108	482.6	0.0150	639.4	0.0192	778.7
0.0067	312.9	0.0109	486.6	0.0151	642.9	0.0193	781.8
0.0068	317.3	0.0110	490.5	0.0152	646.4	0.0194	784.9
0.0069	321.6	0.0111	494.4	0.0153	649.9	0.0195	788.0
0.0070	325.9	0.0112	498.3	0.0154	653.4	0.0196	791.1
0.0071	330.2	0.0113	502.2	0.0155	656.9	0.0197	794.2
0.0072	334.5	0.0114	506.1	0.0156	660.3	0.0198	797.2
0.0073	338.8	0.0115	510.0	0.0157	663.8	0.0199	800.3
0.0074	343.1	0.0116	513.8	0.0158	667.3	0.0200	803.3
0.0075	347.3	0.0117	517.7	0.0159	670.7	0.0210	906.4
0.0076	351.6	0.0118	521.5	0.0160	674.1	0.0202	909.4
0.0077	355.8	0.0119	525.4	0.0161	677.5	0.0203	812.4
0.0078	360.1	0.0120	529.2	0.0162	681.0	0.0204	815.4
0.0079	364.3	0.0121	533.0	0.0163	684.4	0.0205	818.4
0.0080	368.5	0.0122	536.8	0.0164	687.8	0.0206	821.3
0.0081	372.7	0.0123	540.6	0.0165	691.1		

Conversion to metric would be [X0.0069MPa]. The simplest way is to get your metric answer and convert it to inlb to use the existing tables.

Table 30.5 Beam-design constants $F_y = 50$ksi [345MPa] $f_c' = 4$ksi [28MPa]

P_g	R psi	P_g	R psi	P_g	R psi	P_g	R psi	P_g	R psi	P_g	R psi
0.0040	194.1	0.0089	415.8	0.0137	615.8	0.0185	798.8	0.0234	968.1		
0.0041	198.8	0.0090	420.1	0.0138	619.8	0.0186	802.4	0.0235	971.3		
0.0042	203.5	0.0091	424.5	0.0139	623.7	0.0187	806.0	0.0236	974.6		
0.0043	208.2	0.0092	428.8	0.0140	627.7	0.0188	809.7	0.0237	977.9		
0.0044	212.9	0.0093	433.1	0.0141	631.7	0.0189	813.3	0.0238	981.1		
0.0045	217.5	0.0094	437.4	0.0142	635.6	0.0190	816.9	0.0239	984.4		
0.0046	222.2	0.0095	441.7	0.0143	639.6	0.0191	820.5	0.0240	987.6		
0.0048	231.5	0.0096	446.0	0.0145	647.5	0.0193	827.6	0.0242	994.0		
0.0049	236.1	0.0097	450.3	0.0146	651.4	0.0194	831.2	0.0243	997.2		
0.0050	240.8	0.0098	454.6	0.0147	655.3	0.0195	834.8	0.0244	1000.4		
0.0051	245.4	0.0099	458.9	0.0148	659.2	0.0196	838.3	0.0245	1003.6		
0.0052	250.0	0.0100	463.1	0.0149	663.1	0.0197	841.9	0.0246	1006.8		
0.0053	254.6	0.0101	467.4	0.0150	667.0	0.0198	845.4	0.0247	1010.0		
0.0054	259.2	0.0102	471.6	0.0151	670.9	0.0199	849.0	0.0248	1013.2		
0.0055	263.8	0.0103	475.9	0.0152	674.8	0.0200	852.5	0.0249	1016.4		
0.0056	268.4	0.0104	480.1	0.0153	678.7	0.0201	856.0	0.0250	1019.5		
0.0057	273.0	0.0105	484.3	0.0154	682.5	0.0202	859.5	0.0251	1022.7		
0.0058	277.6	0.0106	488.6	0.0155	686.4	0.0203	863.0	0.0252	1025.8		
0.0059	282.2	0.0107	492.8	0.0156	690.3	0.0204	866.5	0.0253	1029.0		
0.0060	286.7	0.0108	497.0	0.0157	694.1	0.0205	870.0	0.0254	1032.1		
0.0061	291.3	0.0109	501.2	0.0158	697.9	0.0206	873.5	0.0255	1035.2		
0.0062	295.8	0.0110	505.4	0.0159	701.8	0.0207	877.0	0.0256	1038.3		
0.0063	300.4	0.0111	509.6	0.0160	705.6	0.0208	880.5	0.0257	1041.4		
0.0064	304.9	0.0112	513.7	0.0161	709.4	0.0209	883.9	0.0258	1044.5		
0.0066	313.9	0.0114	522.1	0.0162	713.2	0.0211	890.8	0.0260	1050.7		
0.0067	318.4	0.0115	526.2	0.0163	717.0	0.0212	894.3	0.0261	1053.8		
0.0068	322.9	0.0116	530.4	0.0164	720.8	0.0213	897.7	0.0262	1056.9		
0.0069	327.4	0.0117	534.5	0.0165	724.6	0.0214	901.1	0.0263	1059.9		
0.0070	331.9	0.0118	538.6	0.0166	728.4	0.0215	904.5	0.0264	1063.0		
0.0071	336.4	0.0119	542.8	0.0167	732.1	0.0216	907.9	0.0265	066.0		
0.0072	340.9	0.0120	546.9	0.0168	735.9	0.0217	911.3	0.0266	1069.1		
0.0073	345.3	0.0121	551.0	0.0169	739.7	0.0218	914.7	0.0267	1072.1		
0.0074	349.8	0.0122	555.1	0.0170	743.4	0.0219	918.1	0.0268	1075.0		
0.0075	354.3	0.0123	559.2	0.0171	747.2	0.0220	921.5	0.0269	1078.2		
0.0076	358.7	0.0124	563.3	0.0172	750.9	0.0221	924.9	0.0270	1081.2		
0.0077	363.1	0.0125	567.4	0.0173	754.6	0.0222	928.3	0.0271	1084.2		
0.0078	367.6	0.0126	571.4	0.0174	758.3	0.0223	931.6	0.0272	1087.2		
0.0079	372.0	0.0127	575.5	0.0175	762.1	0.0224	935.0	0.0723	1090.2		
0.0080	376.4	0.0128	579.6	0.0176	765.8	0.0225	938.3	0.0274	1093.1		
0.0081	380.8	0.0129	583.6	0.0177	769.5	0.0226	941.6	0.0275	1096.1		
0.0082	385.2	0.0130	587.7	0.0178	773.2	0.0227	945.0				
0.0083	389.6	0.0131	591.7	0.0179	776.8	0.0228	948.3				
0.0084	394.0	0.0132	595.7	0.0180	780.5	0.0229	951.6				
0.0085	398.4	0.0133	599.8	0.0181	784.2	0.0230	954.9				
0.0086	402.7	0.0134	603.8	0.0182	787.8	0.0231	958.2				
0.0087	407.1	0.0135	607.8	0.0183	791.5	0.0232	961.5				
0.0088	411.4	0.0136	611.8	0.0184	795.1	0.0233	964.8				

Conversion to metric would be [X0.0069MPa]. The simplest way is to get your metric answer and convert it to inlb to use the existing tables.

Table 30.6 Beam-design constants F_y = 60ksi [414MPa] F_c' = 3ksi [21MPa]

P_g	R psi	P_g	R psi	P_g	R psi
0.0033	190.3	0.0076	415.1	0.0119	613.7
0.0034	195.8	0.0077	420.0	0.0120	618.0
0.0035	201.3	0.0078	424.9	0.0121	622.3
0.0036	206.8	0.0079	429.8	0.0122	626.6
0.0037	212.3	0.0080	434.7	0.0123	630.9
0.0038	217.8	0.0081	439.5	0.0124	635.1
0.0039	223.2	0.0082	444.4	0.0125	639.4
0.0040	228.7	0.0083	449.2	0.0126	643.6
0.0041	234.1	0.0084	454.0	0.0127	647.8
0.0042	239.5	0.0085	458.8	0.0128	652.0
0.0043	244.9	0.0086	463.6	0.0129	656.2
0.0044	250.3	0.0087	468.4	0.0130	660.3
0.0045	255.7	0.0088	473.2	0.0131	664.5
0.0046	261.0	0.0089	477.9	0.0132	668.6
0.0047	266.4	0.0090	482.6	0.0133	672.8
0.0048	271.7	0.0091	487.4	0.0134	676.9
0.0049	277.0	0.0092	492.1	0.0135	681.0
0.0050	282.3	0.0093	496.8	0.0136	685.0
0.0051	287.6	0.0094	501.4	0.0137	689.1
0.0052	292.9	0.0095	506.1	0.0138	693.2
0.0053	298.1	0.0096	510.7	0.0139	697.2
0.0054	303.4	0.0097	515.4	0.0140	701.2
0.0055	308.6	0.0098	520.0	0.0141	705.2
0.0056	313.8	0.0099	524.6	0.0142	709.2
0.0057	319.0	0.0100	529.2	0.0143	713.2
0.0058	324.2	0.0101	533.8	0.0144	717.2
0.0059	329.4	0.0102	538.3	0.0145	721.1
0.0060	334.5	0.0103	542.9	0.0146	725.1
0.0061	339.7	0.0104	547.4	0.0147	729.0
0.0062	344.8	0.0105	551.9	0.0148	732.9
0.0063	349.9	0.0106	556.4	0.0149	736.8
0.0064	355.0	0.0107	560.9	0.0150	740.7
0.0065	360.1	0.0108	565.4	0.0151	744.6
0.0066	365.2	0.0109	569.9	0.0152	748.4
0.0067	370.2	0.0110	574.3	0.0153	752.3
0.0068	375.3	0.0111	578.8	0.0154	756.1
0.0069	380.3	0.0112	583.2	0.0155	759.9
0.0070	385.3	0.0113	587.6	0.0156	763.7
0.0071	390.3	0.0114	592.0	0.0157	767.5
0.0072	395.3	0.0115	596.4	0.0158	771.2
0.0073	400.3	0.0116	600.7	0.0159	775.0
0.0074	405.2	0.0117	605.1	0.0160	778.7
0.0075	410.2	0.0118	609.4	0.0161	782.5

Conversion to metric would be [X0.0069MPa]. The simplest way is to get your metric answer and convert it to inlb to use the existing tables.

Table 30.7 Beam-design constants F_y = 60ksi [414MPa] f_c' = 4.0ksi [28MPa]

P_g	R psi	P_g	R psi	P_g	R psi	P_g	R psi
0.0033	192.2	0.0079	440.9	0.0124	662.3	0.0169	862.3
0.0034	197.9	0.0080	446.0	0.0125	667.0	0.0170	866.5
0.0035	203.5	0.0081	451.2	0.0126	671.7	0.0171	870.7
0.0036	209.1	0.0082	456.3	0.0127	676.3	0.0172	874.9
0.0037	214.7	0.0083	461.4	0.0128	681.0	0.0173	879.1
0.0038	220.3	0.0084	466.5	0.0129	685.6	0.0174	883.2
0.0039	225.9	0.0085	471.6	0.0130	690.3	0.0175	887.4
0.0040	231.5	0.0086	476.7	0.0131	694.9	0.0176	891.5
0.0041	237.1	0.0087	481.8	0.0132	699.5	0.0177	895.6
0.0042	242.6	0.0088	486.9	0.0133	704.1	0.0178	899.7
0.0043	248.2	0.0089	491.9	0.0134	708.6	0.0179	903.9
0.0044	253.7	0.0090	497.0	0.0135	713.2	0.0180	907.9
0.0045	259.2	0.0091	502.0	0.0136	717.8	0.0181	912.0
0.0046	264.8	0.0092	507.1	0.0137	722.3	0.0182	916.1
0.0047	270.3	0.0093	512.1	0.0138	726.9	0.0183	920.2
0.0048	275.8	0.0094	517.1	0.0139	731.4	0.0184	924.2
0.0049	281.2	0.0095	522.1	0.0140	735.9	0.0185	928.3
0.0050	286.7	0.0096	527.1	0.0141	740.4	0.0186	932.3
0.0051	292.2	0.0097	532.0	0.0142	744.9	0.0187	936.3
0.0052	297.6	0.0098	537.0	0.0143	749.4	0.0188	940.3
0.0053	303.1	0.0099	542.0	0.0144	753.9	0.0189	944.3
0.0054	308.5	0.0100	546.9	0.0145	758.3	0.0190	948.3
0.0055	313.9	0.0101	551.8	0.0146	762.8	0.0191	952.3
0.0056	319.3	0.0102	556.7	0.0147	767.2	0.0192	956.2
0.0057	324.7	0.0103	561.7	0.0148	771.7	0.0193	960.2
0.0058	330.1	0.0104	566.6	0.0149	776.1	0.0194	964.1
0.0059	335.5	0.0105	571.5	0.0150	780.5	0.0195	968.1
0.0060	340.9	0.0106	576.3	0.0151	784.9	0.0196	972.0
0.0061	346.2	0.0107	581.2	0.0152	789.3	0.0197	975.9
0.0062	351.6	0.0108	586.1	0.0153	793.7	0.0198	979.8
0.0063	356.9	0.0109	590.9	0.0154	798.1	0.0199	983.7
0.0064	362.2	0.0110	595.7	0.0155	802.4	0.0200	987.6
0.0065	367.6	0.0111	600.6	0.0156	806.8	0.0201	991.5
0.0066	372.9	0.0112	605.4	0.0157	811.1	0.0202	995.3
0.0067	378.2	0.0113	610.2	0.0158	815.4	0.0203	999.2
0.0068	383.4	0.0114	615.0	0.0159	819.7	0.0204	1003.0
0.0069	388.7	0.0115	619.8	0.0160	824.1	0.0205	1006.8
0.0070	394.0	0.0116	624.5	0.0161	828.3	0.0206	1010.7
0.0071	399.2	0.0117	629.3	0.0162	832.6	0.0207	1014.5
0.0072	404.5	0.0118	634.1	0.0163	836.9	0.0208	1018.3
0.0073	409.7	0.0119	638.8	0.0164	841.2	0.0209	1022.0
0.0074	414.9	0.0120	643.5	0.0165	845.4	0.0210	1025.8
0.0075	420.1	0.0121	648.2	0.0166	849.7	0.0211	1029.6
0.0076	425.3	0.0122	653.0	0.0167	853.9	0.0212	1033.3
0.0077	430.5	0.0123	657.7	0.0168	858.1	0.0213	1037.1
0.0078	435.7					0.0214	1040.8

Conversion to metric would be [X0.0069MPa]. The simplest way is to get your metric answer and convert it to inlb to use the existing tables.

31 BEAM DESIGN

31.1 Design for bending moment

The working-stress method (where the actual stress in members is limited by the allowable stress) has been replaced by the widely used **strength method**. The basic procedure in strength design is to design the members based on maximum material properties. The ultimate strength of the member at failure (called its **design strength**) is the only type of resistance considered. Safety is not provided by limiting stresses, but by using a factored design load, called the **required strength**, that is greater than the expected service load.

The Uniform Building Code establishes the value of the required strength U as

$$U = 1.4D + 1.7L$$

where

D = dead load

L = live load

In reality this makes a lot of sense; applying a safety factor to the strength of a material regardless of the predictability of the loads is an overly conservative method. You'll notice when developing the loads for ultimate-strength design, that a higher "factor of safety" is applied to the live load than to the more predictable dead load. It is reasonable to assume that in the future, all materials will be designed using this approach. The required strength U must not be greater than the design strength P_{max}, or $P_{max} \geq U$. This means that a beam designed using this criterion will fail at loads much greater than the actual service live and dead loads.

In the Building Code formulas, adopted from ACI 318 standards, the **beam design strength** is reduced by various factors listed in Table 31.1. These factors are used to reflect several issues:

(a) the fact that concrete is subject to a wide range of potential flaws; ranging from mix variations (water-cement ratios), improper placement (drop height, distribution, and vibration), and curing (freezing or high evaporation conditions).

(b) potentional inaccuracies in the design formulas; Concrete design still relies on an acknowledgement of observed responses as well as mathematical theory.

(c) the desirable reliability and importance of a particular member under actual use.

Table 31.1 Design strength reduction factors Φ for beams

$\Phi = 0.9$	Strength reduction factor for members under pure flexural (bending) stresses
$\Phi = 0.85$	Strength reduction factor for shear and torsion

Fig. 31.1 Tapered cantilevered concrete beams reflecting the moment variation along the span. Note the parallel with the structural proportions of your outstretched arm.

The reinforced-concrete beam-design procedure is as follows.

1. Calculate the required strength by factoring the loads. If the required "strength" and "factoring" are a bit confusing, rest assured, all we're doing is determining the critical loads and their respective maximum moment and shear values just as we have done in STEEL and WOOD and "factoring" the loads is merely a technique for introducing factors of safety into the design.

   The estimated beam weight must be added to the dead loads (DL). THIS IS CRITICAL! CONCRETE BEAMS ARE EXTREMELY HEAVY. A good estimate for the beam dimensions can be obtained by using the depth as d = 1in/ft of span (L) (or L/12 if you prefer). The proportion of breadth to depth is 1:2. (e.g., for a 30ft span, try d = 30in and b = 15in [9m span, d = (9000mm)/12 = 750mm, b = (750mm)/2 = 375mm]. The cross-sectional area is A_g = bd. The weight of the beam per foot of span is

   $$w = bd \times \frac{150\text{lb/ft}^3}{144\text{in}^2/\text{ft}^2} \quad \text{or} \quad w = bd \times \frac{24,000\text{N/m}^3}{1,000,000\text{mm}^2/\text{m}^2}$$

   We won't typically design or specify a beam with an odd inch width, but this proportion will give us a good initial approximation of weight. In the design calculations, we would make initial checks for a 14 or 16in [350 or 400mm] beam. Don't confuse weight approximations with the actual design calculations.

2. Calculate the maximum bending moment M and shear V. Note that the moment must be calculated for every portion of the beam where the moment changes sign (positive or negative), such as in a cantilever beam, in order to place the reinforcement at all locations where tensile stresses exist in the concrete.

Remember also that you may need to investigate several different loading conditions to determine the maximum value for these two different maximums, especially in the case of cantilevers.

3. In order to select a trial value of the design constant R, assume a recommended $p_g = 0.5p_{max}$. Find R in Tables 30.3 to 30.6 for the appropriate values of F_y and f_c'. This will give us a relationship between b and d for this recommended R value. Don't forget that R has to be recalculated once the b and d dimensions are selected.

Calculate the depth with the R/C beam formula (see Chapter 30)

$$d = \sqrt{\frac{M}{\Phi Rb}}$$

with the strength-reduction factor $\Phi \cong 0.9$.

Specify the design depth in inches rounded to the closest inch and try to keep $b \geq d/3$.

4. Calculate the area of steel:
   (a) Find the actual beam design constant $R = M/(\Phi bd^2)$
   (b) Find the value of p_g corresponding to R in Tables 30.3 to 30.6 for the appropriate F_y and f_c' values.
   (c) Calculate $A_s = p_g A_g = p_g bd$.
   (d) Decide a trial bar size and find number of bars = A_s/A_{bar}.

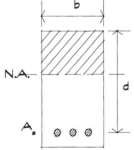

5. Check the cover and spacing according to Table 28.6. If necessary, change the bar size. Note that the cover requirements apply to stirrups, usually No. 3 bars, as well.

Example 31.1

A floor system is built with 28ft span R/C beams, 9ft-4in o.c., supporting 6in R/C precast planks with a 2in nonstructural topping.

   f_c' = 4,000psi
   F_y = 60,000psi
   Code-specified live load = 100psf

---

1. Required strength:

    Distributed live load on the beam:

    $$w_L = (100psf)(9.33ft) = 933plf$$

    In order to find the dead load, the beam dimensions are estimated:

    d = 28in, b = d/2 = 14in. The distributed load is

Beam weight: $(14in)(28in)(150lb/ft^3)/(144in^2/ft^2)$ =		408plf
Slab weight: $(6in+2in)(12.5psf/in)(9.33ft)$	=	933plf
Total w_D	=	1,341plf

    Required strength:

    $$U = 1.4(1341plf) + 1.7(933plf)$$
    $$U = 3,463.5plf$$

2. Shear and moment:

    Load diagram: w = 3.463klf

    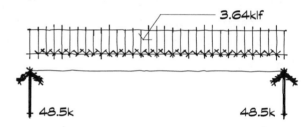

    Shear diagram: maximum V = 48.5k

    Moment diagram:

    Maximum M = (14ft)(48.5k)/2 = 339.5ftk

    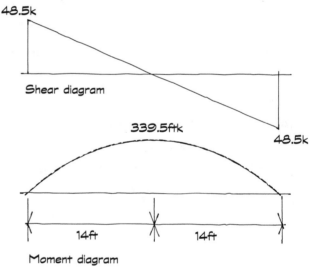

    Shear diagram

3. Beam dimensions:

    From the tables, for the recommended
    $p_g = 0.5(p_{g\,max.})$, R = 581.2psi
    Assume b = 12in

    $$d = \sqrt{\frac{339,500ftlb(12in/ft)}{0.9(581.2psi)12in}} = 25.5in$$

    Moment diagram

    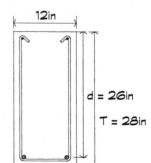

    Beam total depth:

Bars	=	0.5in
Cover	=	1.5in
d	=	25.5in
T	=	27.5in

    Make T= 28in; therefore, d = 26in

    (The beam diagram indicates "stirrups" which are not being designed at this time.)

4. Steel area:

From the beam formula:

$$R = M/(\Phi bd^2) = (339,500 ftlb)(12in/ft)/[(0.9)(12in)(26in)^2] = 559 psi$$

In Table 30.6, we find the closest value is $R = 561.7$, corresponding to

$p_g = 0.0103$

$A_s = p_g bd = 0.0103(12in)(26in) = 3.21in^2$

Two No. 11 bars $= 2(1.56in^2) = 3.1in^2 \pm 5\%$ of required area, or

Four No. 8 bars $= 4(0.79in^2) = 3.16in^2$

5. Spacing:

Minimum spacing of 1in, or the bar diameter $d_b = 1.0in$ with No. 8 bars, the same in this case.

Required width of beam:

Bars 4(1.0in)   = 4.0in

Spacing 3(1.0in)   = 3.0in

Cover 2(1.5in)   = <u>3.0in</u>

     Minimum b   = 10.0in

The final bar placement is shown in the sketch.

As previously indicated, the illustrated stirrups are not calculated for this example. We will design these components in Chapter 32.

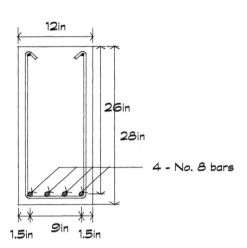

31.2 Development of reinforcement

Tension forces in reinforcement bars would pull the bars out of the concrete if the bond between the two materials was not sufficient. Bond stresses act on the surface of the steel bars, encased in concrete, for their entire length. The ACI code defines development length L_d as the length of the embedment required to develop the design strength of the concrete at a critical section. For a uniformly loaded simple-span beam, one critical section is at midspan, and the bars must extend on both sides beyond the critical section for a length $L \geq L_d$. Bar anchorage is a function of bar size, reinforcement surface finishes, and concrete strength and weight, so we may need to check more than one anchorage criterion. The ACI Building Code Requirements for structural concrete specifies a minimum length for development for uncoated reinforcing in standard-weight concrete as:

For No.6 [*No. 19M*] or smaller bars: $L_d = \dfrac{0.04(d_b)F_y}{\sqrt{f'_c}}{}^*$ or $L_d = 12\text{in minimum}$

$$\left[L_d = \frac{0.75(d_b)F_y}{\sqrt{f'_c}} \right]$$ or $L_d = 305\text{mm minimum}$

For No. 7 [*No. 22M*] or larger bars: $L_d = \dfrac{0.05(d_b)F_y}{\sqrt{f'_c}}{}^*$ or $L_d = 12\text{in minimum}$

$$\left[L_d = \frac{0.937(d_b)F_y}{\sqrt{f'_c}} \right]$$ or $L_d = 305\text{mm minimum}$

*In no case can the value for $\sqrt{f'_c}$ exceed 100psi [*0.7MPa*].

(The preceding equations are simplified versions of the ACI base formulas.)

For the preceding equations, use whichever value of L_d is the greatest, where d_b is the bar diameter. Since the bars are extended to the ends of the beam, where the moment is zero, the length will ordinarily be more than sufficient (Fig. 31.2). In the case of a cantilever, the reinforcement bars for negative moment can be theoretically cut off at the point of zero moment (Fig. 31.3). The development length of top bars (bars with at least 12in [*305mm*] of concrete cast under them) is calculated with the previous formulas multiplied by a factor of 1.3. If necessary, the bars will be extended beyond the theoretical cutoff point. Top bars tend to be a collection point for bubbles from the vibration process. This area of bubbles has little bond strength; consequently, we need 30% more anchorage to compensate.

Between midspan and the supports of a uniformly loaded beam, some bars can be cut off because the bending moment decreases. Such terminations create peak stress in the remaining bars that extend the full length of the beam. For terminated bars (bars cut off at points of

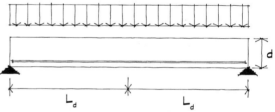
Fig. 31.2 Simple beam potential development length

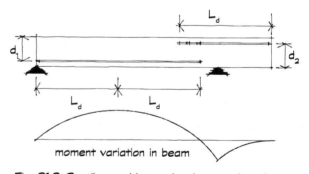
Fig. 31.3 Cantilevered beam development lengths

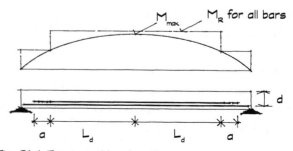
Fig. 31.4 Terminated bar lengths

nonzero moments), the code requires that they extend beyond the point at which they are no longer needed in bending for a distance of

$a = d$ (structural depth)   or $a = 12d_b$

whichever is greater, and where d is the beam effective depth. and d_b is the bar diameter.

Example 31.2

A uniformly loaded rectangular beam on a simple span of 16ft has b = 10in, d = 14in, F_y = 60ksi, and f_c' = 4ksi.

The moment diagram is illustrated on the right. The tension reinforcement consists of four No. 8 bars, two extending the full length of the beam and two cut off where they are no longer required. The line on the M diagram indicates the moment capacity with two No. 8 bars. Where the line intersects the M diagram, two continuing bars provide sufficient moment capacity. This distance is approximately L/7 = 28in from the support. This point is a critical section for the continuing bars, which must extend for L_d from this section.

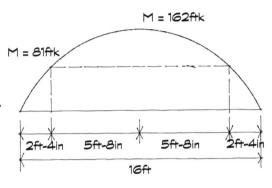

$$L_d = \frac{0.05d_bF_y}{\sqrt{f_c'}}$$

$$L_d = \frac{0.05(1.0in^2)(60,000psi)}{63.24psi} = 47.43in$$

L_d = 12in minimum

By using the controlling maximum anchorage length: L_d = 47.5in

The bar, therefore, must be bent to form a standard hook to provide the additional length of

47in - 28in = 19in

Additional length of the terminated bars:

$a = d = 14in$

$a = 12d_b = 12in$

The two terminated bars must extend 14in on both sides of the theoretical cutoff points; therefore, their total length is 164in.

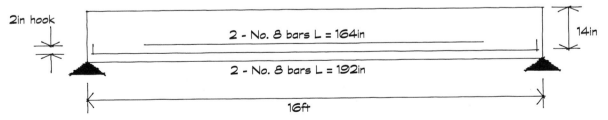

Note that these bars would be in the same horizontal layer but are shown one above the other for clarity.

Problem 31.1

A simply supported R/C beam, 26ft span, has the loads indicated in the illustration.

Design the beam for maximum moment using f_c' = 3.0ksi and F_y = 60ksi.

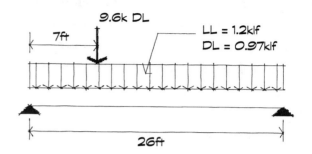

Answer:

With b = 14in and d = 30in, use 4 No. 8 bars.

Problem 31.2

A beam with a 20ft span and 48in overhang is reinforced with No. 6 top bars and No. 8 bottom bars. Using f_c' = 4ksi and F_y = 50ksi, determine if the development length is adequate with bars cut at the point of zero moment, as illustrated.

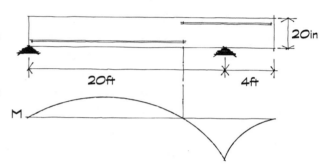

Answer:

By investigating the M diagram, the bottom bars must extend 115in from the section of maximum moment. The top bars must extend 9.6in left of the support. By assuming d = 20in, the required development length is as follows:

No. 8 bars: L_d = 39.53in < 115in OK

No. 6 bars: L_d = 23.72in, min. L_d = 12in   23.72in minimum required left of the support.

The 23.72in < 48in on the right side.   OK.!

Total length of bottom bars = 115in + 115in = 230in from support A.

Total length of top bars = 48in + 23.72in = 71.72in from the end of the overhang.

METRIC VERSIONS OF ALL EXAMPLES AND PROBLEMS

Example M31.1

A floor system is built with 8.5m span R/C beams, 2.84m o.c., supporting 150mm R/C precast planks with a 50mm nonstructural topping.

$f_c' = 28MPa$

$F_y = 414MPa$

Code-specified live load = $4,800N/m^2$

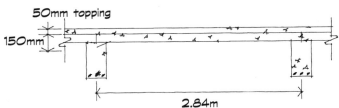

1. Required strength:

   Distributed live load on the beam:

   $w_L = (4800N/m^2)(2.84m) = 13,632N/m$

   In order to find the dead load, the beam dimensions are estimated:

   d = 0.8(8500mm) = 680mm, b = d/2 = 340mm. The distributed load is:

   Beam weight: $(340 \times 680mm)(24,000N/m^3)/(10^6mm^2/m^2)$ = 5,549N/m

   Slab weight: $(150mm+50mm)(24,000N/m^3)/(1000mm/m)(2.84m)$ = <u>13,632N/m</u>

   Total w_D = 19,181N/m

   Required strength:

   U = 1.4 (19,181N/m) + 1.7 (13,632N/m) = 50,027N

2. Shear and moment:

   Load diagram: w = 50,027N

   Shear diagram: maximum V = 212,616N

   Moment diagram: maximum M = 451,809Nm

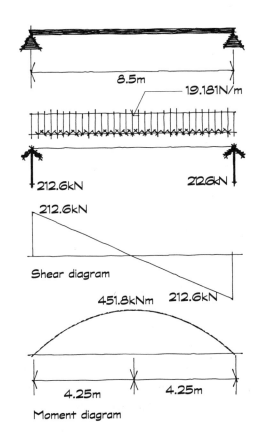

8.5m

19.181N/m

212.6kN

212.6kN

212.6kN

Shear diagram

451.8kNm

212.6kN

4.25m   4.25m

Moment diagram

3. Beam dimensions:

   From the tables, for the recommended

   $p_g = 0.5(p_{g\,max.})$,

   R = 4.0MPa

   Assume b = 300mm

   $$d = \sqrt{\frac{415,809Nm(1000mm/m)}{0.9(4.0MPa)(300mm)}} = 647mm$$

Beam total depth:

Bars   = 12mm

Cover   = 38mm

d   = <u>647mm</u>

   T   = 697mm

Make T = 700; therefore,

d = 700mm - (12mm + 38mm) = 650mm

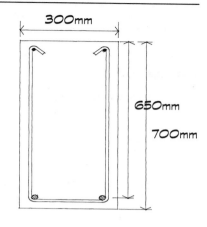

4. Steel area:

From the beam formula:

$R = M/(bd^2) = (451,809Nm)(1,000mm/m)/[0.9(300mm)(650mm)^2] = 3.96MPa$

   This converts to a psi value of (3.96MPa)/0.006894 = 574.4psi

In Table 30.6 we find the closest value is R = (574.4psi) 3.93MPa, corresponding to

   $p_g = 0.0106$ $A_s = p_g bd = 0.0106(300mm)(650mm) = 2,067mm^2$

2 No. 36M bars = 2 (1006mm²) = 2012mm² , or

4 No. 25M bars = 4 (510mm²) = 2040mm² OK!, within 5%

5. Spacing:

Minimum spacing of 25.4mm, or bar diameter d_b = 25.4mm with No. 36M bars.

Required width of beam:

Bars 4(25.4mm)   = 101.6mm

Spacing 3(25.4mm)   = 76.2mm

Cover 2(38mm)   = <u>76.0mm</u>

Minimum b   = 253.8mm < 300mm OK.

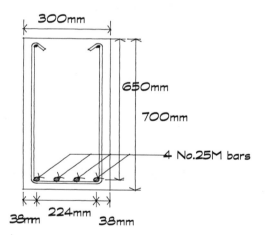

The final bar placement is shown in the sketch.

As previously indicated, the illustrated stirrups are not calculated for this example. We will design these components in Chapter 32.

Example M31.2

A uniformly loaded rectangular beam on a simple span of 4.88m has b = 250mm, d = 356mm, F_y = 414MPa, and f_c'= 28MPa.

The moment diagram is illustrated on the following page. The tension reinforcement consists of four No. 25M bars, two extending the full length of the beam and two cut off where they are no longer required. The line on the M diagram indicates the moment capacity with two No. 25M bars. Where the line intersects the M diagram, two bars provide sufficient moment capacity. This distance is approximately L/7 = 700mm from the support. This point is a critical section for the continuing bars, which must extend for L_d from this section. The bars are larger than No. 22M, thus

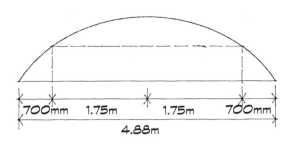

$$L_d = \frac{0.937(414 \text{MPa})(25.4\text{mm})}{\sqrt{28 \text{MPa}}} = 1863 \text{mm}$$

Both sets of bars must extend this length from the critical section.

Since the distance from midspan to the theoretical cut-off points is (3,450mm)/2 = 1,725mm, the bars should be extended farther; in practice it is convenient to make all four as long as the beam. The minimum length of the terminated bars is L = 2(L_d + a) = 2(1863mm + 356mm) = 4438mm.

Problem M31.1

A simply supported R/C beam, 7.9m span, has the loads indicated in the illustration.

(LL = 17,500N/m, DL = 14,000N/m,

P = 42.7kN).

Design the beam for maximum moment using f_c' = 21MPa and F_y = 414MPa.

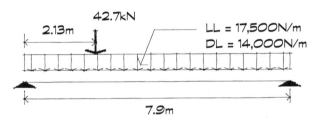

Answer:

With b = 350mm and d = 760mm, use four No. 25M.

Problem M31.2

A beam with a 6.1m span and 1.2m overhang is reinforced with No. 19M top bars and No. 25M bottom bars. Using f_c' = 28MPa and

F_y = 345MPa, determine if the development length is adequate with bars cut at the point of zero moment, as illustrated.

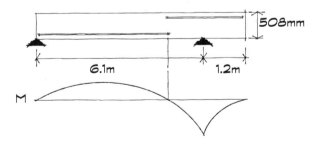

Solution:

By investigating the M diagram, the bottom bars must extend 2920mm from the section of max. M. Top bars must extend 244mm left of support. Assuming d = 508mm, the required development length is:

25M bars L_d = 1527mm < 2920mm OK

20M bars L_d = 978mm, min L_d = 305mm, so 1221mm required left of support.

978mm < 1200mm right side OK!

Total length of bottom bars = 2920mm + 2920mm = 5840mm from support A.

Total length of top bars = 1200mm + 978mm = 2178mm from end of overhang.

32 SHEAR IN BEAMS

32.1 Shear strength of concrete

Bending moment is always accompanied by shear stresses with both horizontal and vertical components at the point of maximum shear; the stress in concrete approaches a 45° direction. In the strength method, the shear strength is given by

$$f_v' = \Phi 2\sqrt{f_c'} \qquad \qquad \left[f_v' = \Phi\left(\frac{1}{6}\right)\sqrt{f_c'}\ \text{MPa} \right] \text{(metric version)}$$

where $\Phi = 0.85$ is the strength-reduction factor. Concrete is allowed to take this amount of shear stress. For example, with

$$f_c' = 4000 \text{psi} \qquad [28\text{MPa}]$$

$$f_v' = 0.85(2)\sqrt{4000} = 107.5\text{psi}$$

$$\left[f_v' = 0.85\left(\frac{1}{6}\right)\sqrt{28\text{MPa}} = \right] 0.75\text{MPa}$$

Shear reinforcement is required when the actual shear stress $f_v > f_v'$, or when the concrete shear capacity $V_R = f_v'(bd)$ is lower than the maximum V. The difference $\tilde{v} = f_v - f_v'$, called *overstress*, is the portion of shear stress taken by the shear reinforcement (Fig. 32.1). Since the shear forces are creating tensile forces in the concrete at 45° angles, the most efficient way to use stirrups would be to place them at 90° (perpendicular) to these forces or 45° in the opposite direction. Whereas this may be more efficient, it is both mathematically and constructionally more difficult; consequently, the most common types of shear reinforcement are stirrups in the form of U-shaped bent bars placed vertically and spaced along the beam span (Fig. 32.2).

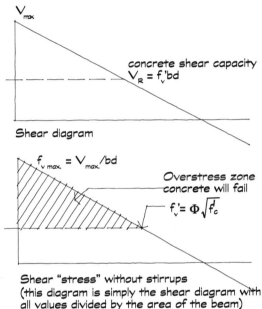

Shear diagram

Shear "stress" without stirrups
(this diagram is simply the shear diagram with all values divided by the area of the beam)

theoretical length where stirrups are required

Fig. 32.1 Shear resistance in a R/C beam

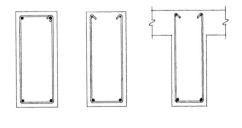

Fig. 32.2 Forms and placement of vertical stirrups

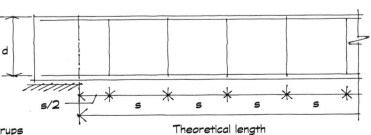

Theoretical length

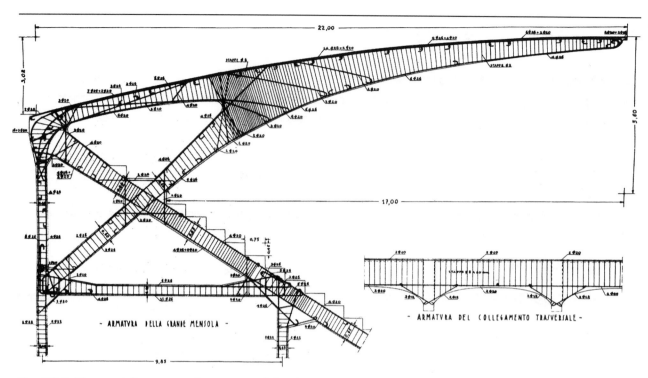

Fig. 32.3 Shear reinforcement of structure of the Florence Stadium, Italy

32.2 Design of shear reinforcement

The actual shear stress is calculated as

$$f_v = V/bd$$

This equation allows us to calculate the minimum depth of the beam able to resist V without stirrups:

$$d = \frac{V}{b\Phi 2\sqrt{f'_c}} \qquad \left[d = V / \left[b\Phi \left(\frac{1}{6}\right)\sqrt{f'_c} \right] \right]$$

Although the maximum shear value occurs at the end of the beam, the code permits the use of the shear stress at a distance d (effective beam depth) from the beam end as the critical maximum for stirrup design (Fig. 32.4). Once the actual depth d has been decided on the basis of the moment or shear, a shear "stress" diagram can be constructed with values given by $f_v = V/bd$ (Fig. 32.5). The shear strength of the concrete with stirrups is given by

$$f'_v = \Phi 2\sqrt{f'_c} \qquad \left[f'_v = \Phi \left(\frac{1}{6}\right)\sqrt{f'_c} \right]$$

Fig. 32.4 Critical shear load location

Therefore, the critical f_v at d inches from the support is

$$f_{vd} = (V @ d)/(bd)$$

(V @ d inches from the support can be found from the shear diagram.)

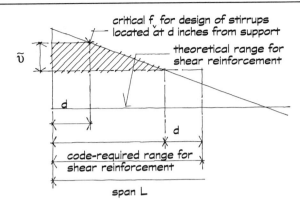

Fig. 32.5 Shear "stress" diagram

The portion of the diagram above the f_v' value indicates the overstress , where shear exceeds the capacity of the concrete and where stirrups are needed along the beam.

The maximum spacing is given in Table 32.1.

Table 32.1 Maximum spacing of stirrups

Shear overstress		Spacing
$\tilde{v} \leq 4\sqrt{f'_c}$ psi or $\left[\frac{1}{3}\sqrt{f'_c}\text{ MPa}\right]$		d/2
$\tilde{v} > 4\sqrt{f'_c}$ psi or $\left[\frac{1}{3}\sqrt{f'_c}\text{ MPa}\right]$		d/4

Stirrups must be spaced and sized according to the amount of shear they have to absorb. The resistance of a single stirrup with two legs is equal to

$$V = A_v f_y$$

where A_v = area of the steel, and f_y = yield stress of stirrup steel. If s is the spacing of the stirrups, the shear force distributed (as tension) to each stirrup is (Fig. 32.6)

$$V = \tilde{v}(bs)$$

From the two preceeding equations, we can derive an expression for the required (maximum) spacing if the steel area is known:

$$A_v F_y = \tilde{v}(bs) \quad \text{or:}$$

$$s = \frac{A_v F_y}{\tilde{v}b}$$

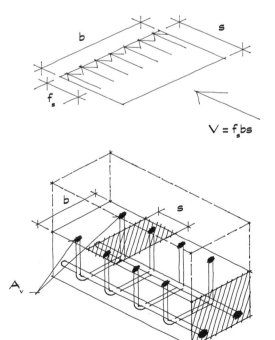

Fig. 32.6 Shear resisted by each stirrup

When s is known, the required cross-sectional area of stirrup is:

$$A_v = \frac{\tilde{v}(bs)}{F_y}$$

Thus, the required steel bar has an area $A_b = A_v/2$ for a two-leg stirrup.

A minimum area required by the code is

$$A_v = 50(bs)/F_y$$

where $\tilde{v} = 50$psi [0.35MPa] represents the minimum overstress to be used in the calculations. This is an unnecessary calculation if you look ahead and use either 50psi [0.35MPa] or the overstress if it exceeds 50psi [0.35MPa]. One calculation solves the problem.

Example 32.1

A 34ft beam has a uniform load and a point load, as shown in the illustration. Design the beam for bending moment and shear using concrete $f_c' = 4.0$ksi and reinforcement steel $F_y = 60.0$ksi.

Solve for M and V.
Beam weight estimate at 1.0in/ft of span and a width of approximately (1/2)d

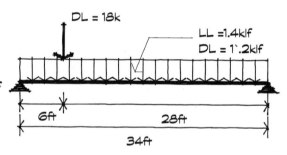

    d = 34in, b = 16in
    w = (150lb/ft³)(17in)(34in)/(144in²/ft²) = 602plf

Required strength:
    U_w = 1.7L + 1.4D =
        1.7(1.4klf) + 1.4(1.2klf + 0.602klf)
    U_w = 4.9klf
    U_p = 1.4(18k) = 25.2k

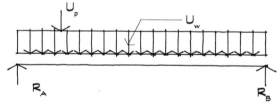

Loading :
    Shear $V_A = R_A$ = (4.9klf)(34ft)(34ft)/2
        +(25.2k)(28ft/34ft) = 104.05k

    R_A = 104.05k
    $R_B = V_B$ = 87.8k
    x = (87.8k)/(4.9k/ft) = 17.9ft

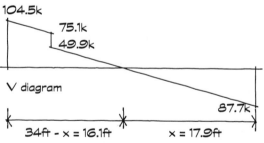

    Bending moment
    M = (17.9ft)(87.8k)/2 = 786.8ftk

Beam depth based on moment d_m:
    Critical at the face of support.
    R = 581.2psi

$$d_m = \sqrt{\frac{M}{\Phi Rb}} = \sqrt{\frac{(786,800\text{ftlb})(12\text{in/ft})}{0.9(581.2\text{psi})(16\text{in})}} = 33.58\text{in}$$

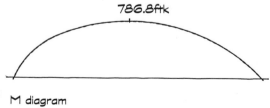

M diagram

Example **551**

Shear stress diagram d_{vb}:

Critical at d inches from the face of the support

The shear stress is calculated as $f_v = V/bd$.

$f_{v\,max.} = (104,050lb)/(16in)(34in) = 191.3psi$

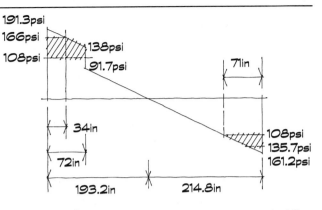

The concrete will carry

$f_v' = 2\Phi\sqrt{f_c'} = 2\,(53.7psi) = 108psi$

The section is critical at 34in from the supports:

Support A:

$f_v = 191.3psi - (53.3psi)(34in/72in) = 166psi$

$\tilde{v}_s = 166psi - 108psi = 58psi$

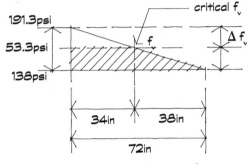

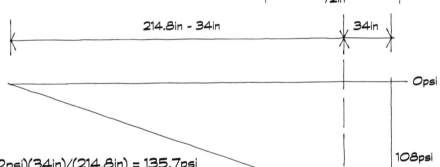

Support B:

$f_v = 161.2psi - (161.2psi)(34in)/(214.8in) = 135.7psi$

$\tilde{v}_s = 135.7psi - 108psi = 27.7psi < 50psi$

Use a 50psi minimum overstress.

Design of stirrups:

$F_y = 60ksi$

Maximum spacing: Overstress at A = $4\sqrt{f_c'} = 253psi > 58psi$; therefore, from Table 32.1:

$s = d/2 = (34in)/2 = 17in$

Required steel area, left end:

$A_v = \dfrac{\tilde{v}(bs)}{F_y} = \dfrac{(58psi)(16in)(17in)}{60,000psi} = 0.26in^2$

Required bar area = $(0.26in^2)/2 = 0.13in^2$.

This is slightly larger than a No. 3 bar which has A = 0.11in². Instead of using a No. 4 bar, the spacing can be adjusted. Required spacing with No. 3 bars:

$$s = \frac{A_v F_y}{\tilde{v}b} = \frac{2(0.11in^2)(60,000psi)}{(58psi)(16in)} = 14.2in$$

Left end: Use No. 3 stirrups 14in o.c. for a length of 72in.
Actual length with stirrups = 7in + 5 (14in) = 77in

Since the actual overstress of 58psi is only slightly greater than the minimum overstress of 50psi, which exists at the right end, it's not worth recalculating.

Use No. 3 stirrups 14in o.c. for an overall length of 71in from right support or 7in + 5(14in) = 77in, the actual dimension from both ends.

The placement of stirrups is shown in the illustration.

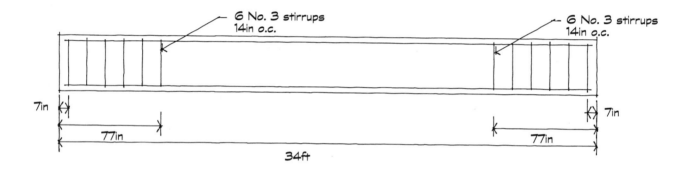

Tensile reinforcement:
M = 786.8ftk.
R = (768,800ftlb)(12in/ft)/[(0.9)(16in)(34in)²] = 567.19psi

From beam constants Table 30.6, with F_y = 60,000psi and f_c' = 4,000psi, R corresponds to a steel percentage of p_g = 0.0104.

A_s = (0.0104)(16in)(34in) = 5.66in²
Six No. 9 = 6(1.0in²) = 6.0in². Use six No. 9 bars, with 1.125in diameter.

Spacing: The minimum required b:
Cover: 2(1.5in)   = 3.0in
Bars: 6(1.125in)   = 6.75in
Spaces: 5(1.125in)   = 5.62in
        Total   = 15.37in < 16in   OK!

Although the tensile bars will be held slightly higher (one full stirrup diameter) at the ends, this is in an area of less than maximum moment, so it is of little consequence. Each case must be evaluated based on its particular moment distribution.

Problem 32.1

Design the illustrated beam using F_y = 50ksi steel and f_c' = 4.0ksi concrete.

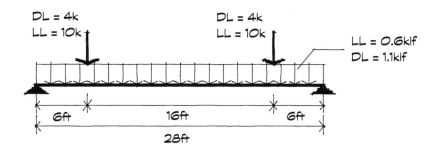

Answer:

d = 28in and b = 14in.

Three No. 11 bottom bars for the entire length.

Six No. 3 stirrups 14in o.c. for a length of 77in from both ends.

Problem 32.2

The illustrated cantilever beam is to be designed with F_y = 50ksi steel and f_c' = 4ksi concrete. Size the steel bars for maximum moment and calculate the required bar length including development length.

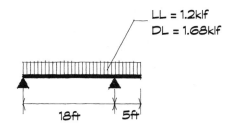

Answer:

b = 8in, d = 22in;

Use three No. 8 bars between the supports, L = 16ft-6in.

Use two No. 10 bars at overhang, L = 9ft-2in.

Problem 32.3

Design this R/C beam that has a maximum depth T (total) of 30in. Determine if you need shear reinforcement. F = 50ksi steel, f_c' = 4ksi

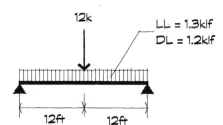

Answer:

b = 14in, d = 28in

Use four No. 10 bars for moment.

Use No. 3 stirrups at 14in o.c. for shear.

METRIC VERSIONS OF ALL EXAMPLES AND PROBLEMS

Example M32.1
A 10.36m beam has a uniform load and a point load at 1.8m from the left, as shown in the illustration (DL = 17.5kN/m, LL = 20.4kN/m, P = 80kN). Design the beam for bending moment and shear using concrete $f_c' = 20$MPa and reinforcement steel $F_y = 414$MPa.

Solve for M and V.

Beam weight estimate:

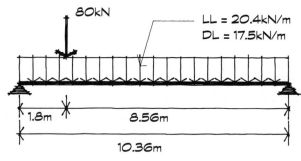

$d = 0.08L = 0.08(10,360\text{mm}) = 830\text{mm}$,

$b = d/2 = (830\text{mm})/2 = 415\text{mm}$

$w = (24,000\text{N/m}^3)(415\text{mm})(830\text{mm})/$
$(10^6\text{mm}^3/\text{m}^3) = 8,267\text{N/m} = 8.3\text{kN/m}$

Required strength:

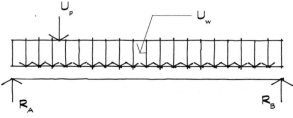

$U_w = 1.7L + 1.4D =$
$\quad 1.7(20.4\text{kN/m}) + 1.4(17.5\text{kN/m} +$
$\quad 8.3\text{kN/m}) = 70.8\text{kN/m}$

$U_p = 1.4(80\text{kN}) = 112\text{kN}$

Loading :

Shear $V_A = R_A =$
$\quad [(70.8\text{kN/m})(10.36\text{m})(10.36\text{m}/2) +$
$\quad (80\text{kN})(8.53\text{m})]/(10.36\text{m}) = 432.6\text{kN}$

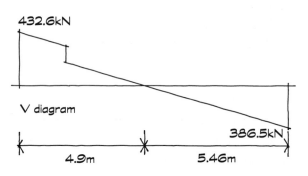

$V_B = 386.5\text{kN}$

$x = (386.5\text{kN})/(70.8\text{kN/m}) = 5.46\text{m}$

Bending moment

$M = (5.46\text{m})(386.5\text{kN})/2 = 1,055\text{kNm}$

Beam depth based on moment d_m:
Critical at d inches from the face of the support.

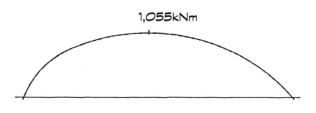

M diagram

$R = 4.01$MPa

$$d_m = \sqrt{\frac{M}{\Phi Rb}} = \sqrt{\frac{(1055000\text{Nm})(1000\text{mm/m})}{0.9(4.01\text{MPa})(415\text{mm})}}$$

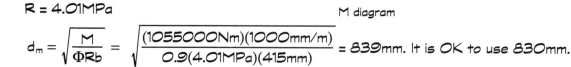

= 839mm. It is OK to use 830mm.

Shear stress diagram d$_{vb}$:

Critical at d mm from face of support: the shear stress is calculated as f$_v$ = V/bd.

f$_{v \, max.}$ = (432.6kN)/(415mm)(830mm)
          = 1.25MPa

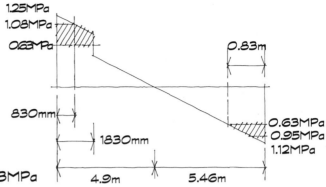

The concrete will carry

$$f_v' = (1/6)\Phi\sqrt{f_c'} = \frac{0.85\sqrt{20MPa}}{6} = 0.63MPa$$

The section is critical at 830mm from the supports:

Support A:

f$_v$ = 1.25MPa - 0.37MPa(830mm/1830mm)
      = 1.08MPa

$\tilde{v}_s$ = 1.08MPa - 0.63MPa = 0.45MPa

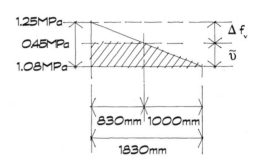

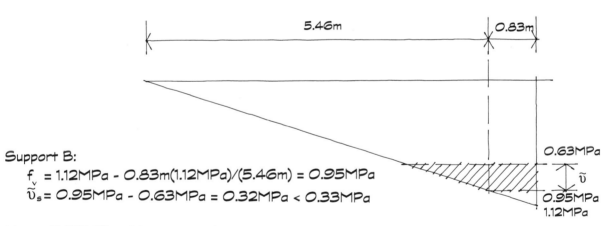

Support B:

f$_v$ = 1.12MPa - 0.83m(1.12MPa)/(5.46m) = 0.95MPa

$\tilde{v}_s$ = 0.95MPa - 0.63MPa = 0.32MPa < 0.33MPa

Use a 0.33MPa minimum overstress.

Design of stirrups:

F$_y$ = 414MPa

Maximum spacing: overstress at A:

Therefore, from Table M32.1:

$$v = (1/3)\sqrt{f_c'} = \frac{\sqrt{20MPa}}{3} = 1.49MPa$$

therefore, from Table M32.1:

s = d/2 = 830mm/2 = 415mm

Required steel area, left end:

$$A_v = \frac{\tilde{v}(bs)}{F_y} = \frac{(0.45MPa)(415mm)(415mm)}{414MPa} = 187mm^2$$

Required bar area = $(187mm^2)/2 = 93.5mm^2$.

This is slightly larger than a No. 10M bar, which has $A = 71mm^2$. Instead of using a larger bar, the spacing can be adjusted. Required spacing with No. 10M bars:

$$s = \frac{A_v F_y}{\tilde{v}b} = \frac{2(71mm^2)(414MPa)}{(0.45MPa)(415mm)} = 315mm$$

Left end: Use No. 10M stirrups 300mm o.c. for a length of 1.8m minimum.
Actual length with stirrups = $(300mm)/2 + 6(300mm) = 1950mm$

Since the actual overstress of 0.45 is only slightly greater than the minimum overstress of 0.33MPa, which exists at the right end, it's not worth recalculating.

Use No. 10M stirrups 300mm o.c. for an overall length of 2.39m from right support or actual length with stirrups = $(300mm)/2 + 8(300mm) = 2550mm$.

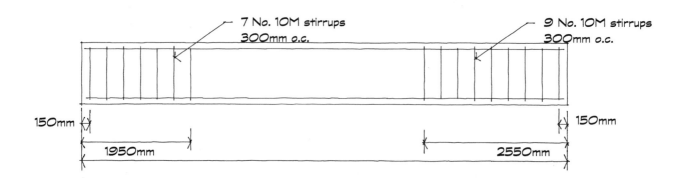

The placement of stirrups is shown in the illustration.

Tensile reinforcement:
    M = 1,055kNm
    R = (1,055,000Nm)(1000mm/m)/[(0.9)(415mm)(830mm)²] = 4.1MPa

From beam constants Table 30.6 for F_y = 414MPa and f_c' = 20MPa, R corresponds to a steel percentage of $p_g = 0.0155$.

A_s = (0.0155)(415mm)(830mm) = 5339mm²
Four No. 32M + two No. 36M = 4(819mm²) + 2(1006mm²) = 5288mm².

Spacing: Minimum required b:
Bars: 4(32mm)   = 128mm
          2(36mm)   = 72mm
Cover: 2(38mm)   = 76mm
Spaces: 5(36mm) = <u>180mm</u>
          Total   = 456mm > 415mm   NO good!

The beam width b must be increased to 460mm. This will not require redesign.

Although the tensile bars will be held slightly higher (one full stirrup diameter) at the ends, this is in an area of less than maximum moment, so it is of little consequence. Each case must be evaluated based on its particular moment distribution.

Problem M32.1

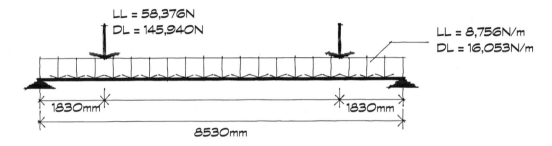

Design the beam shown in the illustration using F_y = 345MPa and f_c' = 20MPa.

Answer:
d = 710mm and b = 360mm
Use three No. 35M bottom bars for the entire length, six No.10M stirrups 360mm o.c. for a length of 1980mm from both ends.

Problem M32.2
The illustrated cantilever beam is to be designed with F_y = 345MPa steel and f_c' = 27.6MPa concrete. Size the steel bars for maximum moment and calculate the required bar length including development length.

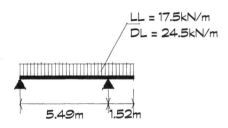

Answer:
b = 210mm, d = 560mm
Use three No. 25M bars between the supports, L = 5030mm.
Use two No. 20M bars at overhang, L = 2800mm.

Problem M32.3

Design this R/C beam that has a maximum depth T (total) of 720mm. Determine if you need shear reinforcement. F = 345MPa steel, f_c' = 27.6MPa

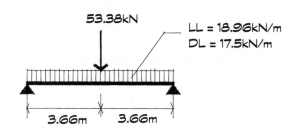

53.38kN

LL = 18.96kN/m
DL = 17.5kN/m

3.66m   3.66m

Answer:

b = 360mm, d = 720mm

Use four No. M30 bars for moment.

Use No. 10 stirrups at 360mm o.c. for shear.

33 SLABS

33.1 Flat spanning systems

Floor and roof systems in reinforced concrete can be broadly classified as one-way and two-way spanning. In other systems, the slab is supported directly on columns without beams (although a portion of the slab may be reinforced to emulate a beam); these include treelike columns with cantilevered slabs.

33.1.1 One-way spanning: Support on two ends

Solid slabs with one-way reinforcing (Fig. 33.1). This system is essentially a series of beams placed side by side; slabs cast in place are typically solid. Precast, prestressed planks are flat and made lighter with longitudinal hollows (Fig. 33.2). These planks, ranging from 2 to 8ft [1.2 to 2.4m] wide have a tongue-and-groove edge to facilitate alignment. Even though they are generally produced under high-

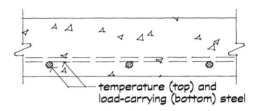

temperature (top) and
load-carrying (bottom) steel

Fig. 33.1 Cast-in-place reinforced concrete slab

quality control standards, the finish surfaces require a minimum 2 inch [50mm] topping slab which may or may not contribute to the load-carrying capacity of the system. Systems have been developed that use the open core for mechanical chases. This is an effective system in that it reduces weight by removing material from an area that doesn't contribute a great deal to the moment capacity of the system.

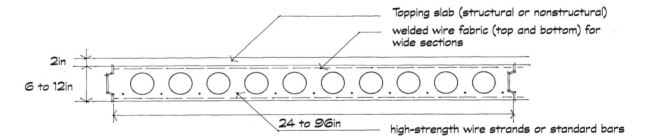

Topping slab (structural or nonstructural)

welded wire fabric (top and bottom) for wide sections

2in

6 to 12in

24 to 96in — high-strength wire strands or standard bars

Fig. 33.2 Precast/prestressed hollow slab section

Joist construction, ribbed slabs. Some systems make use of hollow clay or concrete blocks as permanent formwork for the ribs. The ribs also can be cast in place with reusable metal formwork. Ribbed planks or tees for long spans are precast, prestressed components (Fig. 33.3).

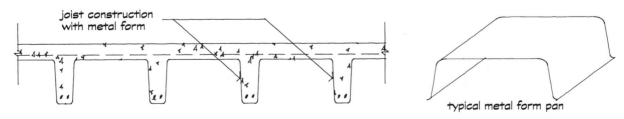

joist construction
with metal form

typical metal form pan

Fig. 33.3 Joist construction using removable metal formwork

33.1.2 Two-way spanning: Support on four edges.

The following systems must be utilized in a maximum structural bay configuration of roughly 1.5:1 for them to effectively transfer load in both directions. You may use the system in a more rectangular bay configuration (2:1 or 3:1), but you have largely negated its potential to transfer load and it simply turns into a one-way system.

Two-way solid slab. The reinforcement is placed in the two orthogonal directions. The supports on the edges can be dropped beams or soffit beams if a continuous ceiling surface is required (Fig. 33.4). In the case of a one-way slab, the long beams would be eliminated and the slab would span between the short beams.

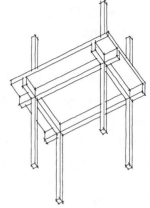

Fig. 33.4 Flat slab with beams

Two-way ribbed slab or waffle slab. This type of construction is made of a grid of "joists" cast in place with intersecting stems and supported on column-line beams (Fig. 33.5). Due to the heavy shear loads around the columns, the waffle must either be reinforced with shear reinforcement, or in some cases, the voids are filled and the shear forces are resisted by concrete mass. Either system works structurally; the decision may be an architectrual one.

Fig. 33.5 Waffle slab with and w/o shear heads at column points

33.1.3 Slabs without beams

Continuous solid slabs. These are sometimes built with drop panels and column capitals to resist shear in the slab around the column (Fig. 33.6). This problem may also be solved with additional shear reinforcement at the column locations, but if space is available, an additional mass of concrete may be more economical. The idea is a lot like the base of an old automobile bumper jack. The expanded end on the jack distributes the load over a larger area and keeps it from "punching" into the ground. In a similar manner the shear head keeps the column from "punching" up through the slab.

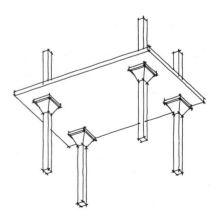

Fig. 33.6 Flat slab with dropped shear heads at columns

Ribbed slabs without beams. These can be waffle slabs (Fig. 33.5) or other types such as the statically determinate ribbed slab designed by Pier Luigi Nervi in 1951 (Fig. 33.7).

Fig. 33.7 Slab with curvilinear ribs

33.2 Flat slab design

Solid flat slabs are a cheaper construction than other types up to a span of 20 to 30ft [6 to 9m] depending on the loads. For preliminary calculations, the thickness can be assumed to be 1/2 inch per foot of span. The depth also can be determined by deflection criteria. The code considers deflection negligible in slabs when the thickness is not smaller than 1/20 of the span, or $T \geq L/20$. Therefore, the thickness can be assumed to be $T = L/20$ in the initial steps of slab design. The design procedure is essentially the same followed in beam design. A 12in-wide strip of slab is considered.

Slabs are often designed without stirrups, with $f'_v = \Phi 2\sqrt{f'_c} \quad \left[\Phi\frac{1}{6}\sqrt{f'_c} \right]$

Code requirements that apply to reinforcing are as follows:

      Maximum bar spacing = 18in [457mm] or 3 × slab thickness

      Minimum reinforcement area for shrinkage:

      $A_s = 0.002bT$ for $F_y \leq 50$ksi [345MPa] steel

      $A_s = 0.0018bT$ for $F_y \geq 60$ksi [414MPa] steel

Example 33.1

A one-way slab spans 18ft between beams. $f_c' = 4ksi$, $F_y = 50ksi$.

Assume thickness $T = 9in$ and the design section is 9inx12in.

R = 617.8psi (recommended p_g)

L = 50psf

D = (9in)(12.5psf/in) = 112.5psf

U = 1.4(112.5psf) + 1.7(50psf) = 242.5psf

Design a typical 1ft-wide strip:

The design load is (242.5psf)(1ft) = 242.5plf

$$V_{max.} = (242.5plf)(18ft)/2 = 2,182lb$$

$$M_{max.} = (2,182lb)(9ft)/2 = 9,821ftlb$$

Required depth for moment: d_m

Critical at the face of the support

$$d_M = \sqrt{\frac{(9821ftlb)(12in/ft)}{0.9(617.8psi)12in}} = 4.2in$$

Required depth for shear: d_{vb}

Critical at d inches from the face of the support.

Allowable $f_v' = \Phi2\sqrt{f_c'} = 0.85(2)\sqrt{4000} = 108psi$

Find the thickness required for $f_v \leq 108psi$.

The critical section is at d inches from the support.

$f_v' =$ (V@d from support)/A

54psi = [2,182lb - d(242.5plf)/(12in/ft)]/(12in)(d)

(648pli)(d) = 2,182lb - 20.2pli(d)

d = 3.26in required depth without stirrups for shear

The d of 3.26in < 4.2in previously found for the maximum M controls.

Since it is common to make the total thickness a full inch increment, that is, a minimum of one-half inch per foot of span, if T = 6in and the cover is 3/4in + 1/2 bar diameter (estimate 1/4in), d = 6in - 1in = 5in. This would be more than the 4.2in required for moment, so it could work.

Unfortunately, this slab will have deflection problems since it exceeds the minimum thickness required to allow us to ignore deflection calculations which would require that $T \geq L/20 = (18ft)(12in/ft)/20 = 10.8in$.

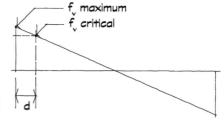

It is possible to use compression steel or simply make the slab meet this deflection criteria:

With T = 11in, d = 11in - 1in = 10in

R = (982ftlb)(1)(12in/ft)/[(0.9)(12in)(10in)^2] = 109.12psi

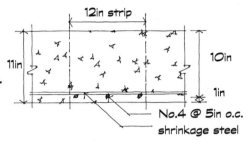

From Table 30.5, p_g = 0.004

A_s = (0.004)(12in)(10in) = 0.48in²/ft of slab width

Table 28.4 gives the area of steel provided by bars at various spacings.

Use No. 4 bars at 5in o.c. = 0.48in²/ft.

Note that the maximum spacing listed in the Table 28.4 is 18in. This is not arbitrary; the maximum spacing of bars in a slab is three x the slab thickness or 18in, whichever is less.

Shrinkage requires at least minimum reinforcement in the two directions.

Perpendicular to the span:

A_s = 0.002bT = 0.002(12in)(11in) = 0.264in²/ft

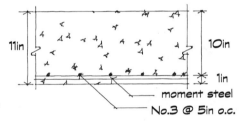

From Table 28.4: No. 3 bars at 5in o.c. = 0.26in²/ft

Use No. 3 bars at 5in o.c. as longitudinal shrinkage reinforcement placed on top of the layer of tensile reinforcement. This reinforcing would ideally be placed in the center of the slab, but the tensile steel provides a convenient surface and the difference in its ability to carry thermal and shrinkage stresses is not significantly altered.

33.3 Slabs on grade

Slabs on grade are poured on a gravel or pea gravel base after all topsoil is removed. The organic matter in the topsoil will deteriorate and cause long-term settlement. Pea gravel is the preferred base since it has no **fines** (fine material) in it and is as dense as it will ever be when placed in the excavation. Consequently, no compaction is required. This is significantly different from pea fill, which does have fines and must be compacted. Compaction is generally achieved with a small unit that will vibrate thin layers of the granular material into place. If the substrata are not adequately compacted, you can expect settlement and slab cracking over the life of the structure. Cohesive materials such as clay are generally not used as fill under slabs because they are more difficult to compact and require a "sheepsfoot" roller or its equivalent. (A sheepsfoot roller has projections or points that will penetrate into the material and "knead" out the voids, much as a potter prepares clays for making pots.) A vapor barrier is recommended in all cases, and it is placed directly below the slab. The vapor barrier will keep ground moisture vapor from moving up through the slab and causing dampness under flooring or carpeting. This barrier will also keep concrete moisture from being drawn away when the concrete is initially placed. Remember, this moisture is essential for the proper curing of the concrete.

Residential slab thickness is 4in [100mm], reinforced with steel fabric or welded wire mesh. Types of fabric are given in Table 33.2. Slabs can support residential loads and concentrated loads such as nonbearing partitions, although a 4in [100mm] thick area is required under even nonbearing locations with a No. 3 [10M] bar minimum. Under unusual conditions, a "structural" slab may be used to support bearing walls or columns, but they should more commonly penetrate through the slab and rest on independent foundations.

Table 33.2 Common stock types of welded wire fabric

| Steel Designation (rolls) | | Area of steel provided mm²/m | | Weight |
New designation	Old designation	Longitudinal	Transverse	N/m²
6X6-W1.4XW1.4	6X6-10/10	18.71	18.71	10.06
6X6-W2.0XW2.0	6X6-8/8	26.45	26.45	14.37
6X6-W2.9XW2.9	6X6-6/6	37.42	37.42	20.12
6X6-W4.0XW4.0	6X6-4/4	51.62	51.62	27.78
4X4-W1.4XW1.4	4X4-10/10	27.74	27.74	14.85
4X4-W2.0XW2.0	4X4-8/8	40.0	40.0	21.08
4X4-W2.9XW2.9	4X4-6/6	56.13	56.13	29.70
4X4-W4.0XW4.0	4X4-4/4	77.42	77.42	40.72
Steel Designation (sheets)		Area of steel provided mm²/m		Weight
New designation	Old designation	Longitudinal	Transverse	N/m²
6X6-W2.9XW2.9	6X6-6/6	37.42	37.42	20.12
6X6-W4.0XW4.0	6X6-4/4	51.62	51.62	27.78
6X6-W5.5XW5.5	6X6-2/2	70.97	70.97	38.32
4X4-W4.0XW4.0	4X4-4/4	77.42	77.42	40.72

33.4 Composite sections

Cast-in-place slabs and beams are usually built to act as composite members. A portion of the slab can be considered a compression flange of the beam, therefore forming a T-beam. This is effective, however, only in areas for positive moments (where the slab would be in compression). The effective flange width to be used in the design of symmetrical T-beams is limited to 1/4 of the span, or $B = L_n/4$, and $a = 8T$ or $L_n/2$, where a is the overhang width on either side of the web (Fig. 33.8).

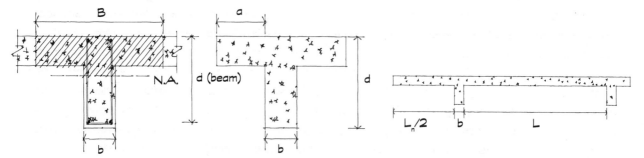

Fig. 33.8 Cast-in-place T-beam flange limitations

The same concept is used in composite construction built with a steel beam and a concrete slab (Fig. 33.9). The section can be designed by transforming the area of steel into an equivalent area of concrete by means of the "modular ratio" between the moduli of elasticity: $n = E_s/E_c$.

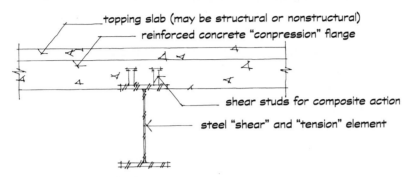

Fig. 33.9 Composite section using concrete and steel in their most effective manner.

Many steel decks are manufactured to act in composite action with the concrete topping, providing a formwork for casting that becomes tensile reinforcement once the concrete has cured. In this case, bonding is ensured by deformations on the deck or by welded steel shear connectors.

METRIC VERSION OF EXAMPLE

Table 33.2 Common stock types of welded wire fabric

| Steel Designation (rolls) | | Area of steel provided mm²/m | | Weight |
New designation	Old designation	Longitudinal	Transverse	N/m²
6X6-W1.4XW1.4	6X6-10/10	18.71	18.71	10.06
6X6-W2.0XW2.0	6X6-8/8	26.45	26.45	14.37
6X6-W2.9XW2.9	6X6-6/6	37.42	37.42	20.12
6X6-W4.0XW4.0	6X6-4/4	51.62	51.62	27.78
4X4-W1.4XW1.4	4X4-10/10	27.74	27.74	14.85
4X4-W2.0XW2.0	4X4-8/8	40.0	40.0	21.08
4X4-W2.9XW2.9	4X4-6/6	56.13	56.13	29.70
4X4-W4.0XW4.0	4X4-4/4	77.42	77.42	40.72
Steel Designation (sheets)		**Area of steel provided mm²/m**		**Weight**
New designation	Old designation	Longitudinal	Transverse	N/m²
6X6-W2.9XW2.9	6X6-6/6	37.42	37.42	20.12
6X6-W4.0XW4.0	6X6-4/4	51.62	51.62	27.78
6X6-W5.5XW5.5	6X6-2/2	70.97	70.97	38.32
4X4-W4.0XW4.0	4X4-4/4	77.42	77.42	40.72

Example M33.1

A one-way slab spans 5.5m between beams. $f_c' = 28MPa$, $F_y = 345MPa$.

Assume thickness = 23mm, and the design section is 230mm × 1,000mm.

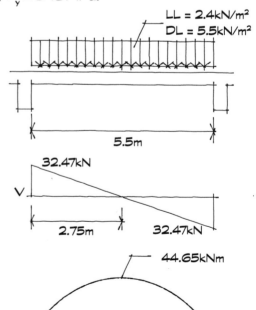

$R = 4.26MPa$ (recommended p_g)

$LL = 2,400N/m^2$

$DL = (0.23m)(24,000N/m^3) = 5,520N/m^2$

$U = 1.4(5520N/m^2) + 1.7(2400N/m^2) = 11,808N/m^2$

Design a typical 1m-wide strip:

The design load is $(11,808N/m^2)(1m) = 11,808N/m$

$V_{max.} = (11,808N/m)(5.5m/2) = 32,472N$

$M_{max.} = 0.5(32,472N)(5.5m/2) = 44,649Nm$

Required depth for moment: d_M

Critical at the face of the support

$$d_M = \sqrt{\frac{(44,649\text{Nm})(1000\text{mm/m})}{0.9(4.26\text{MPa})(1000\text{mm})}} = 108\text{mm}$$

Required depth for shear: d_{vb}

Critical at d mm from the face of the support.

allowable $f_v' = \Phi(1/6)\sqrt{f_c} = 0.85\,(1/6)\sqrt{28\text{MPa}}$

$= 0.75\text{MPa}$

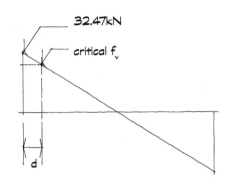

32.47kN

critical f_v

d

Find the thickness required for $f_v \leq 0.75$MPa. Critical section at d inches from the support.

$f_v' = (V@d \text{ from support})/A$

$0.75\text{MPa} = 32,472\text{N} - (d)(11,808\text{N/m})(1000\text{mm/m})/(1000\text{mm})(d)$

$(750\text{N/mm})(d) = 32,472\text{N} - 11.808\text{N/mm}(d)$

$d = 42.64\text{mm}$

Required depth without stirrups for shear < 108mm previously found, therefore, maximum M controls with a depth of 108mm.

Since it is convenient to make the total thickness T = 150mm and the cover is 19mm + 1/2 bar dia. (estimate 6mm) , d = 150mm - 25mm = 125mm. Unfortunately, this slab will have deflection problems since it exceeds the minimum thickness required to allow us to ignore deflection calculations which would require that T $\geq$ L/20 = (5,500mm)/20 = 275mm.

This slab will have deflection problems. It is possible to use compression steel or make the slab:

T = L/20 = (5,500mm)/20 = 275mm

With T = 275mm, d = 275mm - 25mm = 250mm

R = $(44,649\text{Nm})(1000\text{mm/m})/[(0.9)(1000\text{mm})(250\text{mm})^2]$ = 0.79MPa

From Table 30.5, p_g = 0.004

A_s = 0.004(1000mm)(250mm) = 1000mm²/m of width.

Use No. 13M bars at 130mm o.c.

= (129mm²)/(130mm)(1000mm)

= 992mm²/m of width.

Note that the maximum spacing = 457mm.

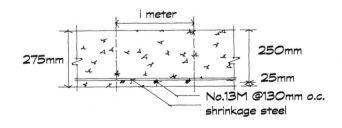

i meter

275mm

250mm

25mm

No.13M @130mm o.c.
shrinkage steel

Shrinkage requires at least minimum reinforcement in both directions.

Perpendicular to the span:

A_s = 0.002bT = 0.002(1000mm)(250mm) = 560mm²/m

No. 10M at 120mm o.c. = 71mm²(1000mm/120mm) = 592mm²

Use No. 10M bars at 120mm o.c. as longitudinal shrinkage reinforcement placed on top of the layer of tensile reinforcement. This reinforcing would ideally be placed in the center of the slab, but the tensile steel provides a convenient surface and the difference in its ability to carry thermal and shrinkage stresses is not significantly altered.

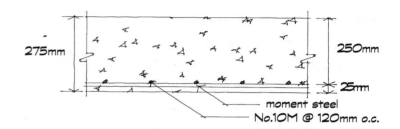

275mm

250mm

25mm

moment steel
No.10M @ 120mm o.c.

34 DEFLECTION

34.1 Creep and deflection

Deflection in bending of R/C beams increases over time with sustained loads. This phenomenon, in which concrete strain increases over time under constant stress, is called **creep**. The dead loads have a significant effect on creep, as well as loads placed before the concrete has developed its full strength at 28 days. This problem, similar to that discussed in regard to wood, is much more prevalent in concrete since a larger percentage of its total load is dead load, a sustained load. Creep is reduced if the structure is not loaded during the first 3 months of life, but greatly increased by full loading prior to 3 months. You remember that creep was more significant in green wood as well, a similar material condition.

Depth-to-span ratios can be used to take creep of concrete into account. The UBC stipulates that one-way construction not supporting or attached to partitions or other construction likely to be damaged by large deflections, unless the bending member is designed for maximum deflection, should have a minimum thickness, indicated in Table 34.1.

Table 34.1 Minimum thickness of simply supported bending members unless deflections are computed

Minimum thickness T (total thickness including cover)				
	L = span	F_y = 40ksi[276MPa]	F_y = 50ksi[345MPa]	F_y = 60ksia[414MPa]
Beams	Simple support	L/20	L/18	L/16
	Cantilever	L/10	L/9	L/8
Slabs	Simple support	L/25	L/22	L/20
	Cantilever	L/12	L/11	L/10

Source: Adapted from ACI 318-95.

Creep deflection can be reduced by **compressive reinforcing consisting of continuous top bars.** Since steel has a much higher E value than concrete, a small amount of steel (which does not creep) in the compression zone of the beam has a large effect and will tend to stabilize the concrete. In addition to the instantaneous deflection due to live loads, long-duration-loads deflection must be controlled. **Since we are calculating allowable deflections based on actual loads, the loads used in the following calculations are actual loads (without load factors).** The total deflection due to instantaneous loading (LL) and creep can be calculated by adding a proportion of the deflection due to the sustained loads multiplied by a factor λ:

$$\Delta_T = \Delta_L + \lambda(\Delta_D + q\Delta_L)$$

where q is a percentage of the live load that would be "sustained" depending on the type of occupancy loads:

q = 0.3, for residential loads

q = 0.8, for warehouses

$$\lambda = T/(1 + 50p')$$

where

p' = A_s'/A_g, the percentage of compressive reinforcement at the point of maximum moment

T = time factor for sustained loads taken as:

5 years or more	2.0
12 months	1.4
6 months	1.2
3 months	1.0

34.2 Deflection computations

Deflections that occur immediately on application of loads must be computed with equations for elastic deflection, considering the effect of cracking and reinforcement on the member stiffness. These deflections are then increased according to creep criteria, as discussed in Sec. 34.1. Cracks occur in beams in the portions of maximum stress, effectively reducing the cross-sectional area of concrete in tension. The moment of inertia I of the beam therefore changes from uncracked portions to cracked ones (Fig. 34.1).

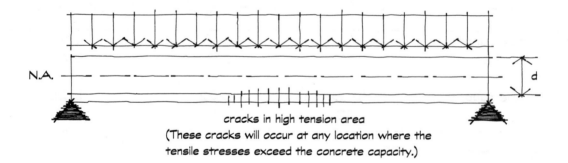

cracks in high tension area
(These cracks will occur at any location where the
tensile stresses exceed the concrete capacity.)

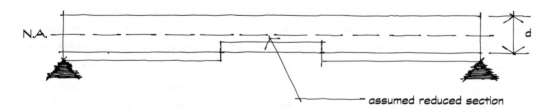

assumed reduced section

Fig. 34.1 Effective beam cross-section for deflection calculations

The immediate deflection is calculated with the usual deflection formulas where the modulus of elasticity of concrete E_c and the effective moment of inertia I_e are used. I_e is the moment of inertia for the cracked section computed with the formula:

$$I_e = (M_{cr}/M)^3[I_g + (1 - (M_{cr}/M)^3)]I_{cr}$$

where

> M_{cr} = cracking moment
>
> M = maximum moment
>
> I_g = moment of inertia of gross concrete section about the centroidal axis, neglecting reinforcement
>
> I_{cr} = moment of inertia of cracked section transformed to concrete

The cracking moment is calculated with the beam formula, $M = FI/c$, which is

$$M_{cr} = f_r I_g / y_t$$

where

> $f_r = 7.5 \sqrt{f_c'}$ is the *modulus of rupture of concrete (psi)*   [$f_r = 0.7 \sqrt{f_c'}$ MPa]
>
> y_t = distance from centroidal axis of the gross section, neglecting reinforcement, to extreme fiber in tension

In order to find I_{cr}, we must transform the moment of inertia of the cracked section into concrete using the modulus of elasticity of concrete. $E_c = 33w^{1.5}$ psi [$E_c = 0.043w^{1.5}$] for values of concrete specific weight w between 90 and 155 pcf [1440 and 1480 Kg/m³].

For normal-weight concrete $E_c = 57000 \sqrt{f_c'}$ [$E_c = 4700 \sqrt{f_c'}$]

Since for steel $E_s = 29,000,000$ psi, the modular ratio is:

$$n = E_s / E_c = 29,000,000 \text{psi} / 57,000 \sqrt{f_c'} \qquad [n = 200,000 \text{MPa} / 4700 \sqrt{f_c'}]$$

The neutral axis passes through the centroid of the section and can be located by equating the moment of the areas of the section with the steel transformed into an equivalent concrete area.

$$A_t y = \Sigma A y$$

which, for the beam in Fig. 34.2 (disregarding the top bars), becomes

$$by(y/2) = nA_s(d-y)$$
or
$$(b/2)y^2 + nA_s y = nA_s d$$

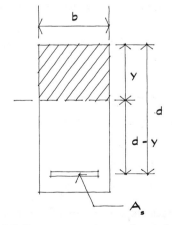

Fig. 34.2 Beam nomenclature for deflection

Note that the equation for c derived in Chapter 30 cannot be applied, since it corresponds to a balanced percentage of steel p_b, whereas a different percentage value has been used for the design. Once y is found, the moment of inertia of the section can be calculated with the transfer formula (used for composite sections):

$$I_{cr} = \sum (I_o + Ay^2) = \frac{by^3}{12} + by\left(\frac{y}{2}\right)^2 + nA_s(y-a)^2 + nA_s(d-y)^2$$

Example 34.1

For the beam shown in the illustration, the allowable deflection is:

$\Delta = L/360 = (28\text{ft})(12\text{in/ft})/360$

$= 0.93\text{in for live loads.}$

Assume residential use of the structure.

$f_c' = 4,000\text{psi}, \quad F_y = 60,000\text{psi}$

$w_L = 933\text{plf} \quad w_D = 1341\text{plf}$

The deflection equations for the ultimate design load are, for live and dead loads, respectively:

$$\Delta_L = \frac{5w_L L^4}{384E_c I_e}$$

$$\Delta_D = \frac{5w_D L^4}{384E_c I_e}$$

and the total deflection, according to the creep equation, is

$$\Delta_T = \Delta_L + \lambda(\Delta_D + q\Delta_L)$$

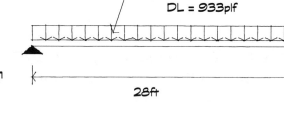

Since the live and dead loads are both uniform loads on a simple span, we can combine them into one calculation; if one were uniform and the other a point load, separate calculations would be required.

They must be the same loading condition to be combined.

$$\Delta_T = \frac{5[w_L + \lambda(w_D + qw_L)]L^4}{384E_c I_e}$$

$\lambda = T/(1 + 50p')$

Disregarding the top bars:

$p' = 0$

$\lambda = 2.0$

$E_c = 57,000\sqrt{4,000\text{psi}} = 3,604,996\text{psi}$

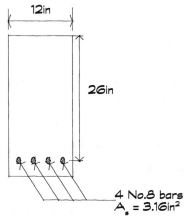

The required moment of inertia is, therefore (remember to use nonfactored loads):

Example   573

$$I_x = \frac{5[w_L + \lambda(w_D + qw_L)]L^4}{384E_c\Delta}$$

$$I_x = \frac{5\left[\frac{933lb/ft}{12in/ft} + 2.0\left(1341plf + 0.3\left[\frac{933lb/ft}{12in/ft}\right]\right)\right](28ft \times 12in/ft)^4}{384(3,604,996psi)(0.93in)}$$

$$I_x = 14,829in^4$$

The effective moment of inertia of the beam must not be smaller than this value, or $I_e \geq I$.

METRIC VERSION OF EXAMPLE

Example M34.1

For the beam shown in the illustration, the allowable deflection is

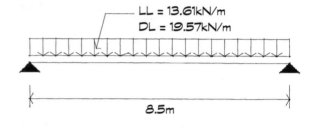

LL = 13.61kN/m
DL = 19.57kN/m

8.5m

$\Delta = L/360$

$= (8,500mm)/360$

$= 23.6mm$ for live loads.

Assume residential use of the structure.

$f'_c = 28MPa, \quad F_y = 414MPa$

$w_L = 13,612N\backslash m \quad w_D = 19,570N/m$

The deflection equations for the ultimate design load are, for live and dead loads, respectively:

$$\Delta_L = \frac{5w_L L^4}{384E_c I_e}$$

$$\Delta_D = \frac{5w_D L^4}{384E_c I_e}$$

and the total deflection, according to the creep equation, is

$$\Delta_T = \Delta_L + \lambda(\Delta_D + q\Delta_L)$$

Since the live and dead loads are both uniform loads on a simple span, we can combine them into one calculation; if one were uniform and the other a point load, separate calculations would be required. **They must be the same loading condition to be combined.**

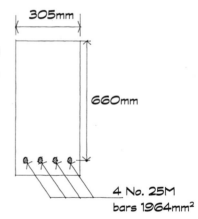

305mm

660mm

4 No. 25M
bars 1964mm²

$$\Delta_T = \frac{5[w_L + \lambda(w_D + qw_L)]L^4}{384E_c I_e}$$

$\lambda = T/(1 + 50p')$

Disregarding the top bars:

$p' = 0$

$\lambda = 2.0$

$E_c = 4700\sqrt{28MPa} = 24,870MPa$

The required moment of inertia is, therefore (remember to use nonfactored loads):

$$I_x = \frac{5[w_L + \lambda(w_D + qw_L)]L^4}{384E_c\Delta}$$

$$I_x = \frac{5[13.612 \text{N/mm} + 2.0(19.570 \text{N/mm} + 0.3(13.62 \text{N/mm})](8500 \text{mm})^4}{384(24,870 \text{MPa})(23.6 \text{mm})}$$

$$I_x = 7,054,737,965 \text{mm}^4 = 7,055 \times 10^6 \text{mm}^4$$

The effective moment of inertia of the beam must not be smaller than this value, or $I_e \geq I$.

35 FOOTINGS

35.1 Foundation design criteria

A soil survey is necessary in most cases to determine the appropriate type of foundations for a building. This is unfortunately one of the items on smaller-scale buildings that clients will frequently try to avoid; they mistakenly believe that "dirt is dirt" and are not aware that subtle changes in soil type or moisture content may dramatically change structural capabilities. In addition, they fail to recognize that the surface conditions are not necessarily representative of the conditions of the soils that will be supporting their building. Glaciated areas have some of the most critical conditions: perched water tables (thin layers of impermeable material that will hold water above the natural water table), buried organic matter (both within and below the building zone), areas of exceptionally expansive clays, and even flowing water. Any of these conditions not identified prior to building design can be disastrous. For example:

(a) In areas of exceptionally high water tables or highly variable water tables, you may wish to avoid basement construction.

(b) In areas of highly expansive clays, unusual methodologies may be necessary when backfilling around levels below grade.

(c) In areas of weak subsoil, your selection of building materials may focus on building weight. Concrete buildings may be excessively heavy for this situation.

(d) In areas where the subsoil strata have flowing water, you might expect quicksand conditions to develop if excavation occurs within the water table.

(e) In areas of highly organic soils, this material will decay over time and settle; consequently, foundations might have to be placed at a level below these strata. Even conventional "slabs on grade" may have to be replaced with structural (spanning) slabs.

A good soil survey will identify conditions that may adversely affect building design. This information is most effectively obtained at the beginning of project design.

The survey can determine the soil characteristics such as bearing capacity, cohesiveness, saturation, and compaction, as well as the location of the water table and its fluctuations during the year. Sand and gravel, or a mixture of the two, are not cohesive soils; when compacted and not saturated with water, these soils have a very good bearing capacity. Clay when dry is very hard, but moisture decreases its bearing capacity and makes it expand. This is due to the very fine particles of which pure clay is composed: When surrounded by molecules of water, the friction between particles is lost.

Soils are often a mixture of sand, clay, and silt. Moisture effect, compressibility, and expansiveness affect foundation design on these soils in particular, as opposed to more stable bedrock or compacted gravel strata. When strata of different soil types are present, as we might expect to find in glacial deposits, the soil survey must investigate the depth of the bearing stratum to make sure it will support the pressure of the foundation (Fig. 35.1). The bearing pressure must be controlled within recommended limits to minimize settlement due to the compression of the soil.

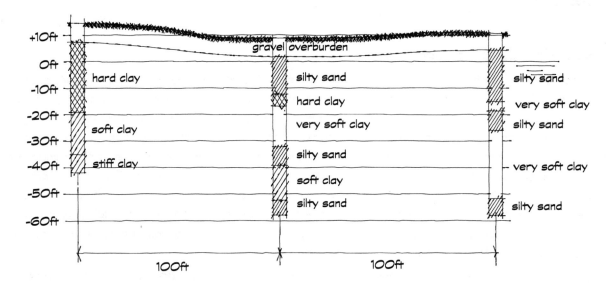

Fig. 35.1 Soil survey showing erratic glaciated deposits

Differential settlement, or settlement of different magnitude in parts of the same building, is the type of foundation failure that causes serious damage, and it is important to ensure that the pressures on the soils are equal under all support points. This is an area where a supposedly conservative approach of overdesigning foundations can be as harmful as underdesigning. **Specifically, the foundations of the structure should have equal pressure;** this may mean that all the foundations are of different sizes. **Standardization, which would cause these pressures to be widely different, is not a desirable condition.** In addition, the resultant of the loads from the superstructure should correspond to the centroid of the foundation to avoid nonuniform pressure distribution or tilting of the foundation (Fig. 35.2).

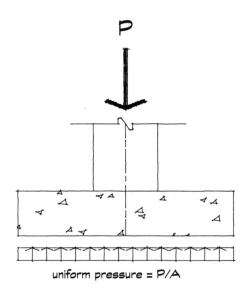

Fig. 35.2 Desired pressure distribution under a footing

A soil survey requires boring exploratory holes spaced at about 25 to 50ft [7.5 to 15m] or more in both principal directions, depending on the size of the building and on the geology of the site (Fig. 35.3). The depth of boring is equally based on geological considerations and the building type. If it is known that the subsoil does not contain any stratum of clay or soft silt, it is sufficient to bore to a depth of 20 to 30ft [6 to 9m] below the foundation bearing level. As a general rule, the depth of boring can be taken as 10ft [3m] per story of building height . On the other hand, if the subsoil contains soft strata, the seat of settlement may be located at a great depth, because even at a depth of 200ft [60m], a moderate increase of the pressure on a thick stratum of soft clay may produce a settlement of more than 1ft [0.3m]. The borings in this case must be made to a depth of about 10ft [3m] below the base of the lowest clay stratum that can influence settlement. Hard to know what to do? This is why you employ a good soils engineering company to conduct your tests.

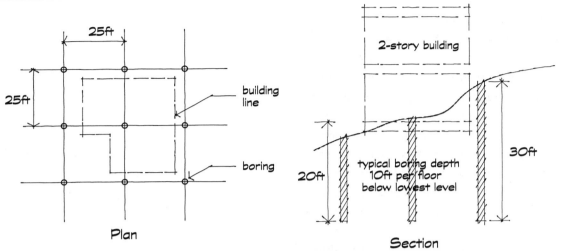

Fig. 35.3 Suggested drill pattern for soil borings

The bearing capacity of an average soil can be 2,500 or 3,000psf [120,000 or 144,000Pa]. The Uniform Building Code gives minimum standards in the absence of soil investigations. Other indicative values for soils can be found in Table 35.1

Table 35.1 General properties of soils

Material	Bearing capacity [x 47.88kN/m²]	Angle of repose[a]	Coefficient of friction
Sound rock	80ksf		
Poor rock	30ksf		
Gravel and coarse sand	10-12ksf	37deg	0.6-0.7
Sand (dry)	6-8ksf	33deg	0.4-0.6
Sand (confined wet)	4 ksf	25deg	0.3-0.4
Clay and sand (mix)	4-5ksf	36deg	0.4-0.5
Hard clay	5-6ksf	36deg	0.4-0.6

[a]The angle at which the soil will stand unsupported for long periods of time. It is sometimes referred to as the internal friction of the soil. As you might expect, the angle of repose decreases with the introduction of moisture.

35.2 Footings

You have already dealt with the principle of a footing when you designed steel bearing plates and base plates. A footing is merely another interpretation of that principle. A strong steel column rests on a weaker concrete foundation, requiring a base plate to distribute the load out on the weaker material. A footing is a "structure" that spreads the relatively high load that the concrete will support over a much larger area to be distributed into the soil. (The steel will support roughly 27,000psi [190MPa] in crushing, the concrete about 1,150psi [8MPa], and the soil only about 21psi [0.145MPa].)

In older masonry construction, foundations
consisted of widening the walls below grade.
This same principle is used today, except in
most cases foundations of masonry or wood-
framed wall-bearing buildings are made of
reinforced-concrete **continuous footings** (Fig.
35.4). R/C or steel columns rest on single
footings or, if spaced closely together, on a
combined footing. Deep foundations, which
must penetrate a greater distance into the
ground to find support, may be built with **piles**
capped with footings or continuous beams.
Piles may be either end-bearing (piles that
penetrate through a very soft layer and bear
on more solid material beneath) or friction piles
(piles which are driven into relatively soft
material until the frictional resistance along
their surface is sufficient to support the weight
of the structure). Friction piles transfer their
load much the same way nails transfer their
load in wood.

The depth of the foundation **must be below
the frost** line, which is provided by state
codes. Figure 35.6 shows the map of Indiana
divided into north, central, and south, containing
frost-line depths. The foundation depth in any
case must be a minimum of 12in [0.3m] below
the finish grade.

Low-strength concrete is usually employed for
footings, with $f_c' = 2,500$ or 3,000psi [17 or
21 MPa]. Although higher-strength concrete
may slightly reduce concrete quantity, the cost
will be higher. This is an area where life-cycle
costing doesn't come into play and a few extra
passes with the excavator bucket is not a
major consideration. Admixtures (see Chapter
27) can improve the frost resistance or the
resistance to sulfates present in the soil. It is
good practice to drain the ground at foundation
level and to protect the structure below grade
with insulating material (Fig. 35.7).

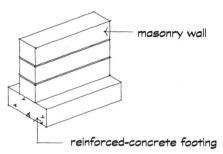

Fig. 35.4 Typical reinforced-concrete wall footing

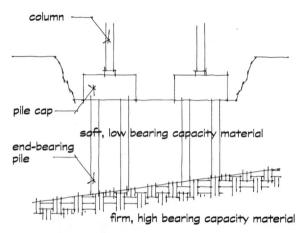

Fig. 35.5 End-bearing file foundation

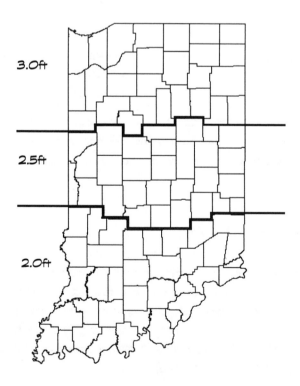

Fig. 35.6 Recommended minimum frost penetration
design depth for Indiana

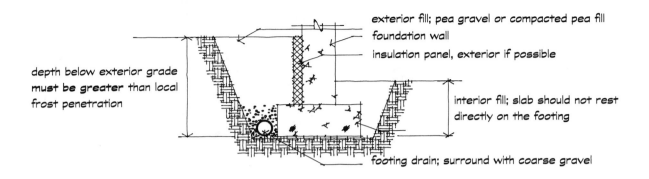

depth below exterior grade **must be greater** than local frost penetration

exterior fill; pea gravel or compacted pea fill

foundation wall

insulation panel, exterior if possible

interior fill; slab should not rest directly on the footing

footing drain; surround with coarse gravel

Fig. 35.7 Typical foundation design

Foundation drains should be provided for perimeter foundations, both to drain away potential water surcharges on basement or retaining walls and to maintain a relatively constant moisture content in the soil.

(Remember, water dramatically affects the bearing capabilities of cohesive soils.)

In designing the steel reinforcement, the applicable code requirements are found in Table 35.2.

Table 35.2 Cover, spacing, and minimum steel requirements

Minimum cover:
    Concrete cast against earth   3in   [76mm]
    Concrete exposed to weather   2in   [51mm]
Minimum spacing: See Table 28.4
Maximum spacing: 18in   [457mm]
Minimum reinforcement for shrinkage is required for continuous footings as for slabs:
        $A_s = 0.002bT$,   for $F_y \leq 50$ksi steel   [344.7MPa]
        $A_s = 0.0018bT$,   for $F_y > 60$ksi steel   [413.7MPa]

35.3 Design procedure for footings

Individual footings are designed in a manner similar to steel bearing plates or column base plates, essentially as cantilever beams distributing the relatively high structural loads out on the relatively weak supporting soil. The size of the footing is determined by the loads and by the bearing capacity of the soil. Once the required footing area is found, in order to calculate the steel reinforcing area, the length of the cantilevers must be determined. These lengths depend on the **support dimension** or theoretical bearing area of the superstructure on the footing, which differs for concrete, masonry, and steel, as shown in Fig. 35.8. The most efficient use of material suggests that the cantilevers should be equal in both directions. This will generally result in a square footing. If dimensions vary by an inch or two, it is desirable to use the larger of the two and consider the footing to be square in terms of the structural design. Foundations are specified in increments of 3in [76mm] in plan and 1in [25mm] in thickness. The process of constructing the foundations makes finer distinctions unnecessary.

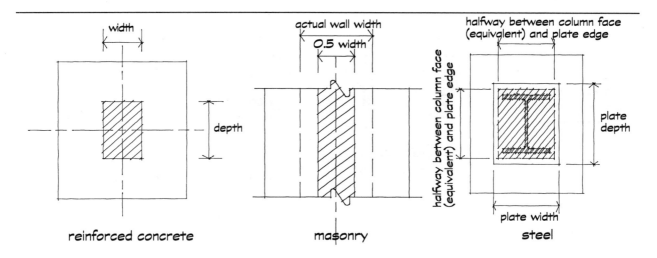

Fig. 35.8 Definition of "support" for typical conditions

For reinforced concrete columns, the support dimensions are taken as the dimensions of the column. You should use the actual column dimensions for all calculations. For masonry piers or walls, the support dimensions are taken as half the dimension of the wall or pier; i.e., an 8in masonry wall would have a 4in support dimension and a 16in × 24in masonry pier would have a support dimension of (16/2)in × (24/2)in or 8in × 12in. These dimensions would be used for all calculations related to a masonry wall or pier. For a steel column/base plate condition, the support dimensions are taken as halfway between the edge of the base plate and a rectangle which is 0.8 × the flange width and 0.95 × the depth. This assumes that the edges of the flanges do not adequately stiffen the bearing plate. (In most cases, you can use the actual column dimensions as the interior rectangle reference for this calculation). After you determine the steel "support," use those dimensions for all subsequent calculations.

The steel selected for moment reinforcement in the footing could be different since it occurs in two layers that have different structural depths. However, the differences will be relatively small, and during construction, the prefabricated mats can be easily placed in the excavation upside down. To avoid this hard-to-detect problem, the lower layer is designed and simply reproduced on the upper layer, making errors impossible.

In rectangular footings, these layers generally will be significantly different, and to use reinforcing most effectively, the longest cantilevers should have steel in the lowest layer (greatest structural depth). This is an easy inspection issue: If the longest bars are placed on top, the mat is probably upside down; have the contractor check its placement.

Design procedure for reinforced-concrete footings

1. Determine the loads:
   (a) Determine the required bearing area $A = P/f_p$, and design area A_f.

      In this instance, you **use actual loads for P** since the soil capacities are generally stated in terms of their allowable stresses and have factors of safety already considered.

(b) Concrete design load U (codes call this the "required strength").

(c) "Design" soil pressure (based on factored loads and F = P/A), $p_s = U/A_f$

In this case, you should **use the U ("required strength") loads** since this is a reflection of the load that the concrete must resist and the "factors of safety" are acknowledged in the development of the load itself.

2. Find the effective depth:

Determine support dimensions:

d_{vb} for **beam shear**, with $f'_v = 2\Phi\sqrt{f'_c}$ psi $\qquad \left[f'_v = \frac{1}{6}\Phi\sqrt{f'_c}\ \text{MPa}\right]$ metric version

d_m for **moment**

d_{vp} for **peripheral shear**, with $f'_{vp} = \Phi4\sqrt{f'_c}\text{psi}$ $\qquad \left[f'_{vp} = \frac{1}{3}\Phi\sqrt{f'_c}\ \text{MPa}\right]$ metric version

3. Reinforcing:

Area of steel in each direction $A_s = p_g A_g$, where p_g is found in the R Tables 30.4 through 30.7 for $R = M/\Phi bd^2$.

4. Design the connections between the footing and superstructure.

35.4 Peripheral shear

In addition to beam shear, the footing must be checked for shear failure in the proximity of the column it supports. The column can punch through the footing, pushing out a truncated pyramidal block with base b = 2d + t (Fig. 35.9). Instead of a 45° fracture surface, we can consider an

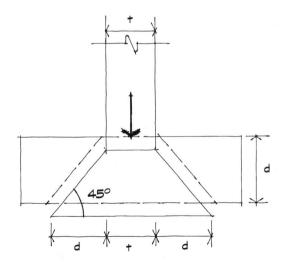
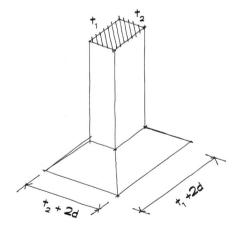

Fig. 35.9 Peripheral shear section; failure "shape" for "punch-out"

equivalent simple rectangular surface with base dimension b = 2(d/2) + t, as illustrated in Fig. 35.10. This simplification reduces the area of soil pressure acting upward, but the error is conservative. The upward reaction resultant of the soil pressure is obviously smaller than P, since the "punch-out" area is only a fraction of the entire footing. In order to satisfy the vertical equilibrium of this part, the difference must be provided by shear stresses on the vertical faces of the prism.

$$F = \frac{P@d}{A} \quad or \quad f'_{vp} = \frac{U - p_s A_b}{A_p} \qquad \textbf{(Eq. 35.1)}$$

In simplest terms, this equation represents the force acting down on the "punch-out" area minus the soil pressure directly under this area acting up. This would be the net force to be resisted by the concrete. This force is then divided by the **total area of surface** resisting the **net force**. (F = P/A) Is that clearer?

where

$f'_{vp} = \Phi 4 \sqrt{f'_c}$ peripheral shear strength;

A_p = peripheral area

U = required design strength

p_s = design soil pressure

A_b = area of base, calculated at a distance d/2 from the "support" in all directions

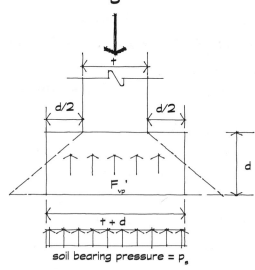

Fig. 35.10 Model of peripheral shear failure

For a square column:

A_p = total perimeter of "punch-out" area × depth
    = $A_p = 4(t + d)d$

$A_b = (t + d)(t + d) = (t + d)^2$

This particular equation is true ONLY for a square column (support configuration)! The approach is the same, however, for all support configurations.

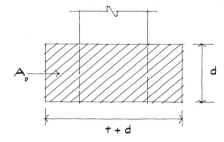

Fig. 35.11 Simplified peripheral shear surface

Equation 35.1 can be solved for d, which is the minimum effective depth for peripheral shear. This value then can be compared to the required depth for beam shear and moment, and the largest value must be taken. Once d is known, the steel area can be calculated. Do NOT use a previously calculated d value to check peripheral shear. Although it may seem easier, if it doesn't work, you'll have to do the full calculation anyway, so as Nike says, "Just do it."

Note that in all calculations, the weight of the footing or the weight of the soil above is not considered, which is a conservative procedure acceptable for shallow footings.

Example 35.1

Design the footing for the 10inx10in R/C column shown in the illustration.

Axial load: P = 100k LL + 86k DL

Allowable soil pressure: p_{all} = 3,000psf

Concrete f_c' = 3,000psi

Steel F_y = 50,000psi

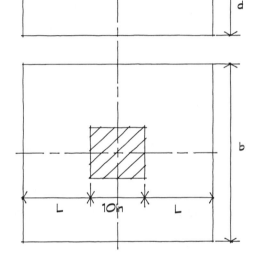

1. **Loading:**

   Bearing area:

   $$A = P/f_p = (100k + 86k)/(3.0ksf) = 62ft^2$$

   Square footing base:

   $$b = \sqrt{62sqft} = 7.9ft = 8ft = 96in$$

   Required design strength:

   $$U = 1.7(100k) + 1.4(86k) = 290.4k$$

   Design soil pressure:

   $$p_s = (290,400lb)/(96in)^2 = 31.5psi \text{ or}$$

   31.5pli on a typical 1in strip of footing

2. **Footing effective depth:**

   (a) Length of cantilever:

   $$L = (96in-10in)/2 = 43in$$

   Maximum moment:

   $$M = (1,354.5lb/in)(43in)^2/2 = 29,121.75inlb \text{ on a 1in strip of footing width}$$

   Maximum shear:

   $$V = (31.5lb/in)(43in) = 1,354.5lb \text{ on a 1in strip of footing width}$$

   (b) Depth due to beam shear (d_{vb}):

   d_{vb} is critical at d from the face of the support (column face).

   "Beam shear" strength:

   $$f_v' = \Phi 2\sqrt{f_c'} = 0.85(2)\sqrt{3,000psi} = 93.1psi$$

   The shear stress at the critical distance d from the column is

   $$f_v = P@d/A \text{ or}$$

   f_v = (load @ support - change over a distance d)/(area of section resisting load)

   $$93.1psi = [(31.5lb/in)(43in) - (31.5lb/in)(d)]/[(1in)(d)]$$

   $$(93.1 \times d) + (31.5 \times d) = 1354.5lb$$

   $$d_{vb} = (1354.5lb)/(124.6lb/in) = 10.87in \qquad \text{**required depth due to "beam" shear**}$$

(c) Depth due to moment (d_m):

 d_m is critical at the face of the support (column face).

$$d_m = \sqrt{\frac{M}{\Phi Rb}} = \sqrt{\frac{29,121.75 \text{inlb}}{0.9(462.8\text{psi})1\text{in}}} = 8.36\text{in}$$ required depth based on moment

(d) Depth due to peripheral shear (d_{vp}):

 d_{vp} is critical at d/2 from the face of the support (column face).

 "Peripheral shear" strength $f_{vp}' = \Phi 4\sqrt{f_c} = 0.85\ (4)\sqrt{3000\text{psi}} = 186.2\text{psi}$

 Lateral area of peripheral shear prism:

 $A_p = 4[d(d + 10\text{in})] = 4d^2 + 40d$

 Area of base of peripheral shear prism:

 $A_b = (d+10\text{in})^2$

From equation 35.1:

 $f_{vp}'A_p = U - pA_b$

 $(186.2\text{psi})(4d^2 + 40d) = 290,400\text{lb} - (31.5\text{psi})(d+10\text{in})^2$

 $(186.2\text{psi})(4d^2 + 40d) = 290,400\text{lb} - (31.5\text{psi})(d^2 + 20d + 100)$

 $744.8d^2 + 7,448d = 290,400 - 31.5d^2 - 630d - 3150$

 $776.3d^2 + 8,078d = 287,250$

 $d^2 + 10.4d = 370.4$

By using "completing the square" as the technique for solving the quadratic,

$(10.4/2)^2$ is added to both sides of the equality, which maintains the equality:

 $d^2 + 10.4d + (10.4/2)^2 = (10.4/2)^2 + 370.4$

 $(d + 5.2)^2 = 397.44$

 $d + 5.2 = 19.94$

 $d_{vp} = 14.74$ in   required depth due to "peripheral" shear

Peripheral shear controls: (You can also use the quadratic equation for the solution; it is just more time-consuming.)

Total depth: structural d = 14.7in
        + bars, upper layer = 1.0in
        + bars, lower layer = 0.5in
        + cover   = 3.0in
            Total T   = 19.2in

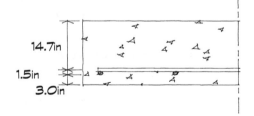

Make T = 20in; d = 20in - 4.5in = 15.5in effective depth.

3. **Area of steel:**

$$R = \frac{M}{\Phi bd^2} = \frac{29,121.75 \text{inlb}}{0.9(1\text{in})(15.5\text{in})^2} = 134.7\text{psi}$$

This is below the lowest value in Table 30.4; use the minimum p_g value:

$p_g = 0.004$

$A_{s''} = 0.004(1\text{in})(15.5\text{in}) = 0.062\text{in}^2$ per inch of width

$A_s = 0.062(96\text{in}) = 5.95\text{in}^2$ total steel in both directions

Number of bars: Maximum spacing of 18in

Space for bars = width - cover = 96in - 6in = 90in

Number of spaces = 90in/18o.c. = 5 spaces, or 6 bars

Area of bar = 5.95in²/6 = 0.99 in² /bar

Use six No. 9 bars = 6(1.0in²) = 6in² in each direction.

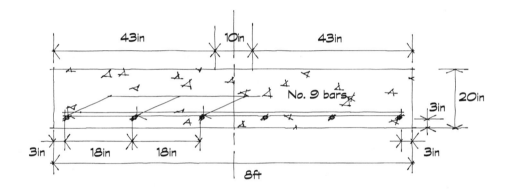

35.5 Rectangular footings

The length of the footing cantilever could be different in the two principal directions because the column is not square (dramatically) or because there are restrictions on the footing area (adjacent buildings, sewer or utility lines, property lines, etc.). The column should always be centered on the rectangular footing; in general, for footings of ir-regular shape, the column load should be applied at the centroid of the foundation in order to ensure uniform distribution of the soil pressure. Note that with rectangular columns, the width of the two column faces has to be considered in calculating the area of peripheral shear. (It is possible to place columns in eccentric locations; however, this will induce overturning moments into the system that must be addressed.)

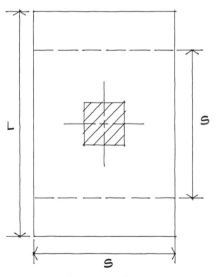

Fig. 35.12 Rectangular footing with "band" dimension referenced.

In rectangular footings, the reinforcing steel is not placed uniformly in the short direction, but is concentrated below the column in a band as wide as the short direction (Fig. 35.12).

The proportion of steel in the bandwidth S is:

$$p_b = \ 2/(B + 1)$$

where

$$B = L/S$$

If A_s is the total area of steel designed in the short (cantilever) direction, the steel quantity in the band S is $p_b A_s$.

Example 35.2

In a rectangular footing 126in x 60in the steel required in the short direction is $A_s = 7.31\text{in}^2$. Find the distribution of short bars in the long direction.

$$B = L/S = 126\text{in}/60\text{in} = 2.1$$
$$p_r = \ 2/(B + 1) = 2/(2.1 + 1) = 0.65$$

Area of steel in the band = $0.65(7.31\text{in}^2) = 4.72\text{in}^2$ Use six No. 8 = 4.74in^2.
Area of steel at each side of the bandwidth = $(0.35/2)(7.31\text{in}^2) = 1.28\text{in}^2$.
Use two No. 8 = 1.58in

Verify the bar spacing at each side of the band: (33in)/(18in o.c.) = 1.8, or 2 spaces. In this case, the two spaces are provided by starting at a "bandwidth bar" and providing two additional spaces, which means the two bars we designed meet spacing requirements.

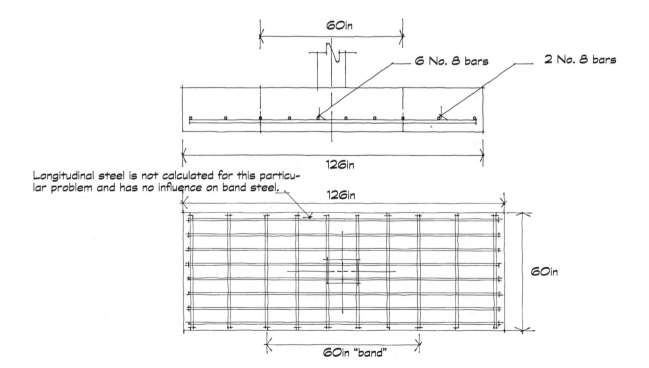

Longitudinal steel is not calculated for this particular problem and has no influence on band steel.

35.6 Simple wall footings

Contrary to our normal technique, we're ending with the most common footing situation, a continuous wall footing. Although many moderately and even large-scaled buildings have some wall footings that support structural loads, the typical use is in residential construction. Where to start?

First: Soil reports typically specify that wall footings not be constructed less than 12in [0.3m] wide. The concrete code specifies that concrete poured against the soil must provide a minimum of 3in [76mm] of concrete cover (protection) for the steel and there must be a minimum of 6in [152mm] of concrete poured above the steel (in footings). When you add these together with some small allowance for the steel, you find that a wall footing will be at least 10in [250mm] thick.

Second: When you recognize that in most situations, allowable soil pressures will range from about 2,500 to 4,000psi [120,000 to 190,000 Pa], and typical wall loads for small to moderately scaled buildings are in the 2,000 to 5,000plf [29 to 73KN/m] range, then the footing needs to be somewhere between 1 to 2ft [0.3 to 0.6m] wide.

Third: If your foundation wall is constructed of a minimum of 8in [203mm] concrete wall, frequently thicker, then the maximum amount of cantilever for moment and shear is around (24in - 8in)/2 = 8in [(610mm - 203mm)/2 = 203mm], support dimension for a masonry wall is one half the width of the wall = 4in [102mm]. This makes the cantilever on each side 10in [254mm]. If you remember our previous calculations, a 10in [254mm] cantilever and a 6in [152mm] minimum structural depth (see "first" preceding), you immediately realize that the footing is carrying a maximum of 4in [103mm] of load in moment and even less in shear (since we evaluate shear at a distance d from the face of the support).

The bottom line is that unless your design requires a footing width of 30 to 36in [762 to 914mm], there is typically no reason to do any calculations; the footing will simply be 10in [254mm] thick with minimum temperature/shrinkage reinforcement. In addition, regardless of the load, wall footings are typically 3 to 4in [76 to 100mm] wider than the wall on each side to simplify construction. This would make the minimum footing for an 8in [203mm] foundation wall 14 to 16in [356 to 406mm] wide. This is further simplified in many cases to a standard width of a backhoe bucket. One simple cut by an experienced operator will create the "final" unformed footing.

Some of the metric conversions noted in the text defeat the purpose of simplifying the dimensions of foundation construction. When metric units become the standard in the United States, some of these "simplifications" will undoubtedly become more reasonable; but until then, we'll have to live with "nonstandard simplifications."

METRIC VERSIONS OF EXAMPLES

Example M35.1

Design the footing for the 250x250mm R/C column shown in the illustration.

Axial load:

P = 445kN LL + 382kN DL = 827kN

Allowable soil pressure:

p_{all} = 143.6kN/m²

Concrete f_c' = 21MPa (N/mm²)

Steel F_y = 344.7MPa

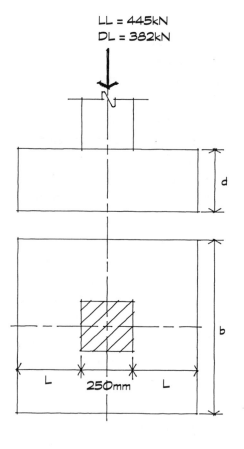

LL = 445kN
DL = 382kN

1. Loading:

    Bearing area:

    A = P/f_p = (827kN)/(143.6kN/m²) = 5.76m²

    Square footing base:

    b = $\sqrt{5.76m^2}$ = 2.4m

    Required design strength:

    U = 1.7(445kN) + 1.4(382kN) = 1,291.3kN

    Design soil pressure:

    p_s = (1,291,300N)/(2400mm)² = 0.22418MPa
    or

    0.22418N/mm on a typical 1mm strip of footing

2. Footing effective depth:

    (a) Length of cantilever L = (2,400mm - 250mm)/2 = 1,075mm

    Maximum shear:

    V = (0.22418N/mm)(1075mm) = 241N on a 1mm strip of footing width

    Maximum moment:

    M = (241N)(1075mm)/2 = 129500N/mm on a 1mm strip of footing width

    (b) Depth due to beam shear (d_{vb}):

    d_{vb} is critical at d from the face of the support (column face)

    "Beam shear" strength $f_v' = \Phi(1/6)\sqrt{f_c'} = 0.85(1/6)\sqrt{21MPa}$ = 0.65MPa (N/mm²)

    The shear stress at the critical distance d from the column is $f_v = \dfrac{P@d}{A}$ or

    f_v = (load @ support - change out at distance d)/(area of section resisting load)

$$0.65\text{MPa} = [241\text{N} - (0.22418\text{N/mm})d]/(1\text{mm})(d)$$

$$(0.65\text{N/mm} + 0.22418\text{N/mm})d = 241\text{N}$$

$$d_{vb} = 241\text{N} / 0.87418\text{N/mm} = 275.7\text{mm} \qquad \text{required depth for "beam" shear}$$

(c) Depth due to moment (d_m):

   d_m is critical at the face of the support (column face)

$$d_m = \sqrt{\frac{M}{\Phi Rb}} = \sqrt{\frac{129500\text{Nmm}}{0.9(3.19\text{N/mm}^2)(1\text{mm})}} = 212.4\text{mm} \quad \text{required depth based on moment}$$

(d) Depth due to peripheral shear (d_{vp}):

   d_{vp} is critical at d/2 from the face of the support (column face).
   "Peripheral shear" strength $f_{vp}' = \Phi(1/3)\sqrt{f_c'} = 0.85(1/3)\sqrt{21\text{MPa}} = 1.3\text{MPa}$

   Lateral area of peripheral shear prism:
   $$A_p = 4[d(d + 250\text{mm})] = 4d^2 + 1000d$$
   Area of base of peripheral shear prism:
   $$A_b = (d+250\text{mm})^2$$

From equation 38.1:

$$f_{vp}'A_p = U - pA_b$$

$$(1.3\text{N/mm}^2)(4d^2 + 1000d) = 1{,}291{,}300\text{N} - (0.224\text{N/mm}^2)(d+250\text{mm})^2$$

$$(1.3\text{N/mm}^2)(4d^2 + 1000d) = 1{,}291{,}300\text{N} - (0.224\text{N/mm}^2)(d^2 + 500d+ 250\text{mm})$$

$$5.424d^2 + 1412d = 1{,}291{,}244$$

$$d^2 + 260.3d = 238{,}061$$

Using " completing the square" as the technique for solving the quadratic equation, $(260.3/2)^2$ is added to both sides of the equality, which maintains the equality:

$$d^2 + 260.3d + (260.3/2)^2 = (260.3/2)^2 + 238{,}061$$

$$(d + 260.3/2)^2 = 255{,}000$$

$$d + 130.2 = 505$$

$$d_{vp} = 374.8\text{mm} \qquad \text{required depth due to "peripheral" shear}$$

Peripheral shear controls

Total depth: structural d   = 375mm
                + bars, upper layer   = 25.4mm
                + bars, lower layer   = 12.7mm
                + cover   = <u>76mm</u>
                    Total T   = 489mm

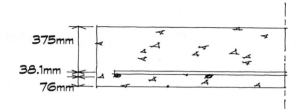

Make T = 500mm; d = 500mm - 114mm = 386mm effective depth.

3. Area of steel:

$$R = M/\Phi bd^2 = (129500Nmm)/[(0.9(1mm)(386mm)^2]$$

$$= 0.966MPa = 0.966MPa/0.006894MPa/psi = 140psi$$

This is below the lowest value in Table 30.4 (psi); use minimum p_g value:

$p_g = 0.004$

$A_s = 0.004(1mm)(386mm) = 1.54mm^2$ per mm of width

$A_s = (1.54mm^2)(2,400mm) = 3,706mm^2$ total steel in both directions

Number of bars: maximum spacing of 457mm

Space for bars = width - cover = 2,400mm - 2(76mm) = 2,248mm

Number of spaces = (2248mm)/(457mm) = 5 spaces, or 6 bars

Area of bar = (3706mm²)/6 = 618mm² /bar

Use six No. 30M = 6(645mm²) = 3,870mm² in each direction.

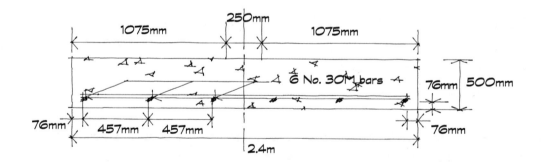

Example M35.2

In a rectangular footing 3.2m x 1.5m the steel required in the short direction is $A_s = 4716mm^2$. Find the distribution of the short bars in the long direction.

A percentage of steel has to be concentrated in a band 1.5m wide under the column.

$B = L/S = (3.2m)/(1.5m) = 2.13$

$p_r = 2/(B + 1) = 2/(2.13 + 1) = 0.64$

Area of steel in band = 0.64(4716mm²) = 3010mm² Use six No. 25M = 3060mm² .

Area of steel at each side of band width = 0.36/2(4716mm²) = 1,698mm².

Use four No. 25M bars = 2,040mm².

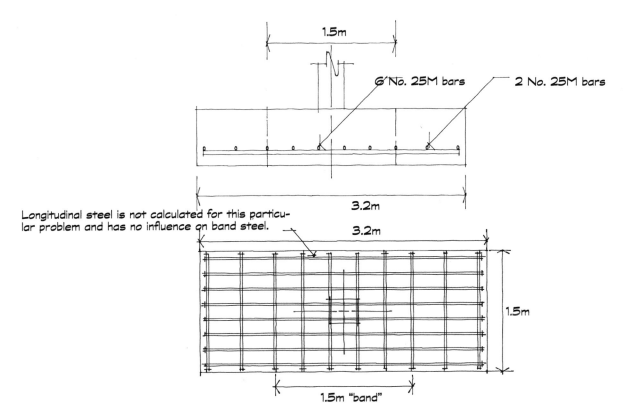

Longitudinal steel is not calculated for this particular problem and has no influence on band steel.

Verify the bar spacing at each side of the band: (850mm)/(457mm o.c.) = 1.86 or 2 spaces. In this case the two spaces are provided by starting at a "bandwidth bar" and providing two additional spaces, which means the two bars we designed meet spacing requirements.

36 COLUMNS

36.1 Construction of R/C columns

Steel reinforcement in concrete columns provides compressive capacity, but its most important role is in controlling the mode of failure. Concrete, when taken beyond yield point, unlike other materials, fails abruptly - even **explosively**; therefore, reinforcement is used even in columns subject only to axial compression to control this potentially dangerous failure mode. The vertical bars work in compression as small steel columns, and even though encased in concrete, their slenderness ratio must be limited by the introduction of horizontal ties.

The most common shapes are rectangular and round, but the **type** of columns is independent from their shape and is determined instead by their reinforcement. The reinforcement can be as follows (Fig. 36.1).

Tied: The vertical bars are tied with looped horizontal ties of smaller diameter, usually No. 3 or 4 bars. The loops are typically rectangular or diamond-shaped.

Spiral: The vertical bars are placed in a circle and are enclosed by a continuous spiral made from steel rod. The circular reinforcement can be placed in a column of virtually any shape. Experience has shown that spiral-reinforced columns are stronger than tied columns with the same concrete and steel area.

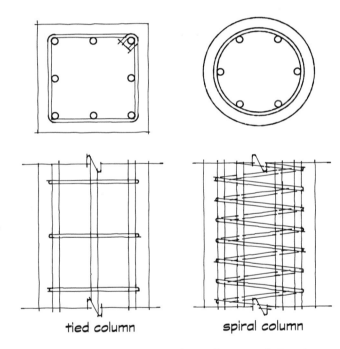

tied column   spiral column

Fig. 36.1 Typical column types (not a function of shape)

The ACI 318-95 Standards require the following column ties:

Longitudinal bar diameter	US tie diameter	Metric tie diameter
No. 10 or smaller [No.32M]	No. 3	No. 10M
Nos. 11, 14, 18 [Nos.35M,45M,57M]	No. 4	No. 13M

Vertical spacing of ties must not exceed:
(a) 16 longitudinal bar diameters, or
(b) 48 tie bar diameters, or
(c) least dimension of the column

Generally, the dimensions or the vertical reinforcing of the column will not be altered for the determination of the ties; consequently, you will determine the most stringent of these two criteria and simply select ties to match that controlling criterion.

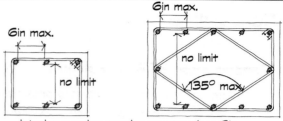

interior spacing may be greater than 6in

Fig. 36.2 Tie criteria for lateral support of bars

Spiral reinforcement must comply with the following requirements:

Minimum diameter of 3/8in [9.5mm]

Clear spacing between spirals 3in maximum,

           1in minimum [76mm max., 25mm min.]

Minimum of six longitudinal bars

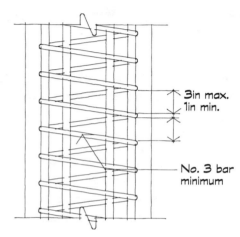

Fig. 36.3 Spiral reinforcement criteria

Example 36.1

A rectangular column 12in x 18in is reinforced with six No. 8 bars; see Fig. 36.4(a). The column is not exposed to the exterior. For No. 8 bars, No. 3 ties are used.

Minimum cover: 1.5 in.

No. 8 diameter: 1.0 in.

Distance of bars from the corner bar:

    Cover: 2(1.5)   = 3.00in

    No. 8 bars: 3(1.0)   = 3.00in

    No. 3 ties: 2(0.375) = 0.75in

                      6.75in

    Spacing = (18in - 6.75in) /2 = 5.6in

See Fig.36.2. Since 6in is the maximum distance from the corner bars, the bar placement is adequate.

Spacing of longitudinal bars:

    $1.5d_b$ or 1.5in,   6in > 1.5in OK!

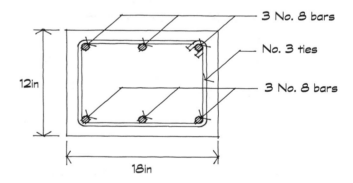

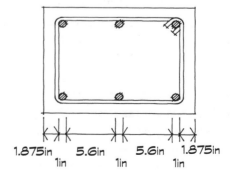

36.2. Design method

Concrete column design is the relatively straightforward addition of the load capacity of the concrete plus the load capacity of the steel. There are, however, complications associated with eccentricity, slenderness, stress distribution, and type of column. These are acknoweldged in the equations as "strength reduction factors" which are listed in Table 36.1

The required strength U of columns (or any concrete member) must not be greater than the design strength $P_{max.}$, or $P_{max.} \geq U$. Remember that the "strength" of a concrete member is based on a load of

$$U = 1.4D + 1.7L$$

where

D = dead load

L = live load

This means that a column designed using this criterion will fail at loads much greater than the actual service live and dead loads. Conceptually, in the strength method, the design strength P_{max} is equal to the combined design strength of the concrete and of the steel, or

$$P_{max.} = P_{concrete} + P_{steel} \quad or \quad P_{max.} = f_c'A_c + F_yA_s$$

where

f_c' = ultimate strength of concrete

A_c = area of concrete

F_y = yield stress of steel

A_s = area of steel reinforcement

Table 36.1 Design strength reduction factors for axially loaded columns

0.85 Stress distribution factor Accounts for the fact that the stress on a cross-section of a member is not uniformly distributed; average stress is taken as a percentage of f_c.

Φ Strength reduction factor Allows for inaccuracies, importance of member, and ductility (e.g., spiral columns are more ductile than tied columns and have a higher Φ).

$\Phi = 0.70$ for tied columns

$\Phi = 0.75$ for spiral columns

k_e Eccentricity factor Allows for accidental eccentricities causing bending stresses.

0.80 for tied columns

0.85 for spiral column

In addition to the above factors, a *slenderness reduction factor* R must be applied to long columns.

Long columns are defined by the slenderness ratio

$$kL/r \geq 22$$

where L is the unbraced length, r is the radius of gyration defined by the formula (see Table 36.2).

$$r = \sqrt{\frac{I}{A}}$$

For columns braced against sidesway (UBC, 1997), K = 1.0. This is a conservative number since columns with fixed ends have a K < 1.0. However, the precise determination of K depends on the effectiveness of the bracing provided at each connection point against sidesway (or its "stiffness").

We are going to assume that there is such a thing as a simple pinned-end concrete column. This is probably not a realistic assumption since the nature of concrete is such that even without providing reinforcement to transfer moment at connections between beams and columns or between slabs and columns, some moment transfer will inevitably occur. The procedure we are outlining will give an accurate answer for "simple pinned" columns (axially loaded) and will give a relatively accurate size and configuration for slightly more complex situations. You should be aware of the built-in error of the assumption and obtain the services of a structural engineer for final calculations of any other than the most elementary columns.

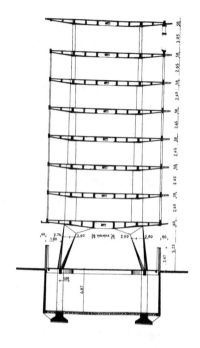

Although the actual r =   holds for simple shapes such as circles and rectangles, this value reduces to the following commonly used approximations.

Fig. 36.4 UNESCO Headquarters section

Table 36.2 Radius of gyrations (r) formulas for common shapes

| Circular column | $r = 0.25D$ | D = diameter |
| Rectangular column | $r = 0.3d$ | d = least dimension |

The **slenderness ratio** of concrete columns is, as usual, KL/r, where K = effective length factor (given in Figure 10.1 for STEEL).

L = actual unbraced length of the column
r = radius of gyration

Columns are defined short when:

$KL/r \leq 22$,   for unbraced columns
$KL/r \leq 34 - 12(M_1/M_2)$ for braced columns

A column is **braced** when it cannot move sideways or buckle when subjected to high loads; this is the typical situation of columns in frames braced by a rigid core or shear walls (Fig. 36.4). M_1 is the smaller of the two moments M_1 and M_2 at the ends due to its action as part of a (rigid) frame.

A case of an unbraced column is a "flagpole" column rigidly connected only at the foundation, but with the top free to rotate and move laterally. Note also in these two examples that the K factor is significantly different (1.0 vs 2.1) (Fig. 36.5).

In a braced column, the end moments M_1 and M_2 must be considered; the ratio M_1/M_2 cannot be taken less than -0.5, even in the case (rare in practice) when the ends can be considered pinned and the moments are zero. If we introduce this minimum value in the formula, braced columns are defined short when

$$KL/r = 34 - 12(-0.5) = 40$$

The slenderness-reduction factor for **unbraced columns** is
for long columns with $KL/r > 22$
$$R = 1.07 - 0.008h/r \leq 1.0$$
for short columns with $KL/r \leq 22$ $R = 1.0$

Fig. 36.5 Johnson Wax offices

where h = length of the column, and r = radius of gyration.

The slenderness-reduction factor for **braced columns** is
for short columns with $KL/r \leq 40$ $R = 1.0$

This assumes NO end moments on the column, which will not generally be the case in a braced situation, but will at least give a reasonable estimate of size for architectural design purposes.

We will adopt a unified formula for short and long unbraced columns where coefficient R is a stress-reduction factor smaller than 1.0 for long columns. This factor takes into account the amplification of the bending moment caused by buckling under axial load (remember Euler's theory?). This procedure will allow us to use basically the same formula for short and long columns. The column formula, giving the ultimate axial compression load, is, therefore:

$$P_{max.} = R\Phi k_e(0.85\, f_c'A_g + F_yA_s)$$

This equation can be written in a different format, more suitable for the use of tabulated values, by introducing the following quantities:

Net area of concrete:
$$A_c = A_g - A_s \text{ (gross area minus area of the steel)}$$

Percentage of steel on the gross area:
$$p_g = A_s/A_g \text{ (area of the steel divided by the gross area)}$$

Substituting these terms into the column equation yields

$$P_{max.} = R\Phi k_e [0.85f_c(A_g - A_s) + F_y A_s]$$

Dividing the column equation by A_g yields

$$P_{max.} = R\Phi k_e A_g [0.85f_c'(1 - A_s/A_g) + F_y A_s/A_g]$$

and since $A_s/A_g = p_g$,

$$P_{max.} = R\Phi k_e A_g [0.85f_c'(1 - p_g) + F_y p_g]$$

This variation in the formula is developed because the code gives some guidance in the use of steel in the form of maximum and minimum percentages. The percentage of steel varies from a minimum of 1% to a maximum of 8% of the gross area, or

$$0.01 \le p_g \le 0.08.$$

In the equation there are three unknowns: R, p_g, and A_g. Generally, an intermediate value of steel is assumed (3% or $p_g = 0.03$) and the equation is solved for A_g. After a preliminary size is determined, the actual value of R can be determined and the percentage of steel (p_g) adjusted to compensate for any loss from R.

Two equations can be written for the two types of column:

Tied columns: $P_{max.} = R(0.70)(0.80)A_g [0.85f_c'(1 - p_g)+ F_y p_g]$

Spiral columns: $P_{max.} = R(0.75)(0.85)A_g [0.85f_c'(1 - p_g)+ F_y p_g]$

These equations are a far cry from the original $P = P_c + P_s$, but if you follow through, you'll see that they consist of that basic idea with a series of applied code restrictions.

Design procedure summary:

1.   Calculate the required strength U.
2.   Find A_g from the column formula with $P_{max.} = U$, assuming R = 1.0 and $p_g = 0.03$. The shape and dimensions of the column consistent with A_g must be determined.
3.   Calculate r, KL/r, and R.
4.   Recalculate p_g from the column equation with the actual R value.
5.   Find $A_s = p_g A_g$; find the required bar sizes corresponding to A_s in the tables.
6.   Determine tie size, spacing, and check the bar spacing.

Example   601

Example 36.2

Design a square tied column 18ft tall with loads LL = 423k and DL = 827k using f'_c = 5,000psi concrete and F_y = 60,000psi steel. We will assume the column is unbraced and has an effective length factor K = 1.0.

1. U = 1.7L + 1.4D = 1.7(423k) + 1.4(827k) = 1876.9k = $P_{max.}$ design strength

2. Tied column formula:

    $P_{max.}$ = R(0.70)(0.80)A_g[0.85(5ksi)(1 - p_g)+ $F_y p_g$]

    R = 1.0

    p_g = 0.03

    1,876.9k = 1.0(0.7)(0.80)A_g[(4.25ksi)(1 - 0.03) + (60ksi)(0.03)]

    A_g = (1,876.9k)/[0.7(0.8)[(4.25ksi)(0.97) + 1.8ksi]] = 566.1 in²
    A_g = 566.1in² = required gross column area
    We have specified a square column; therefore,
    d = $\sqrt{566.1}$ = 23.8in minimum column side.

    We make it 24in; therefore:

    A_g = 24in X 24in = 576in²

3. Radius of gyration:

    r = 0.3(24in) = 7.2in

    Slenderness ratio:

    KL/r = 1.0(18in)(12in/ft)/(7.2in) = 30   The 30 > 22 long column. R must be calculated.

    Slenderness reduction factor:

    R = 1.07 - 0.008(216in)/0.3(24in) = 1.07 - 0.24 = 0.83

4. Since R is not as assumed in step 1, the actual required p_g must be found by solving the column equation.

    1,876.9k = 0.83(0.7)(0.8)(576in²)[4.25ksi(1 - p_g) + (60ksi)p_g]
    1,876.9k = 267.7in²[4.25ksi - 4.25ksi(p_g) + (60ksi)p_g]
    1,876.9k = 1,137.7k + 14,924.27k(p_g)
    p_g = (1,876.9k - 1,137.7k)/(14,924.27k) = 0.049 < 0.08

Steel is between 1% (0.01) and 8% (0.08) of the gross area: OK!

5. $p_g = A_s/A_g$

Required steel area:

$A_s = p_g A_g = 0.049(576\text{in}^2) = 28.2\text{in}^2$

Try with No. 9 bars, A = 1.00in²

Number of bars required = (28.2in²)/(1.00in²/bar) = 28.2 bars

Since this is a large number, we try with No. 11 bars, A = 1.56in²

Number of bars required = (28.2in²)/(1.56in²/bar) = 18 bars

Use 18 No. 11 bars, 9 on each side of the column.

6. No. 11 bar diameter = 1.41in

Minimum spacing = 1.5in, or

$1.5d_b = 1.5(1.41\text{in}) = 2.11\text{in} > 1.5\text{in}$

Ties: with No. 11 bars, the minimum tie diameter d_t is No. 4.

Maximum tie spacing = $16d_b = 16(11/8) = 22\text{in}$ or

$48d_t = 48(4/8) = 24\text{in}$ or

$d_{column} = 24\text{in}$

The least tie spacing value must be used, which is 22in. For lateral bracing, longitudinal bars must comply with the minimum distances from the tie corners as shown in the illustration.

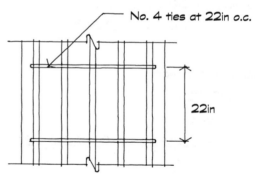

No. 4 ties at 22in o.c.

22in

Placement of the bars:

Tie cover   = 1.5in
Tie diameter = 0.5in
Bar cover   = 2.0in

Check bar spacing, side (a):

Cover: 2(2.0in)	= 4.0in
Bars: 5(1.41in)	= 7.05in
Minimum spacing, 2(3.0)	= 6.0in
Total	= 17.05in

Spacing remaining for the middle bar on both sides = (24in - 17.05in)/2 = 3.47in, which is OK since 2.24in < 3.47in < 6in.

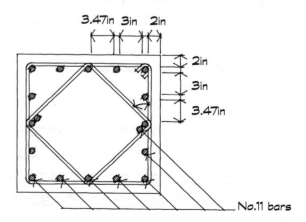

3.47in 3in 2in

2in
3in
3.47in

No.11 bars

On side (b), six bars must be placed. Since each other bar must be tied by a tie corner, the two bars in the middle are bundled and tied together by the 45° tie.

Example 603

A maximum of three bars can be bundled together. The tie angle cannot be smaller than 22.5° (see illustration); in our case, the angle is 45°.

METRIC VERSIONS OF ALL EXAMPLES

Example M36.1

A rectangular column 300mmx460mm is reinforced with six No. 25M bars. The column is not exposed to the exterior. For No. 25M bars, No. 10M ties are used.

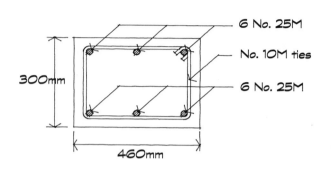

Minimum cover: 38mm

No. 25M diameter: 25.4mm

Distance of bars from corner bar:

Cover: 2(38mm)	= 76mm
No. 25M bars: 3(25.4mm)	= 76mm
No. 10M ties: 2(10mm)	= <u>20mm</u>
Total	= 172mm

Spacing = (460mm - 172mm)/2spaces = 144mm

Since 152mm is the maximum distance from the corner bars, the bar placement is adequate.

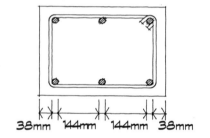

Spacing of longitudinal bars:

$1.5d_b$ or 38mm, 144mm > 38mm OK!

Example M36.2

Design a square tied column 5.5m tall with loads LL = 1881kN and DL = 3678kN using f_c' = 34MPa concrete and F_y = 414MPa steel.

1. $U = 1.7L + 1.4D = 1.7(1,881kN) + 1.4(3,678kN) = 8,347kN = P_{max.}$ design strength

2. Tied column formula:

$P_{max.} = R(0.70)(0.80)A_g[.85f_c'(1 - p_g) + F_y p_g]$

$R = 1.0$

$p_g = 0.03$

$8,347,000N = 1.0(0.7)(.80)A_g[28.9MPa(1 - 0.03) + 414MPa(0.03)]$

Required gross column area:

$A_g = (8,347,000N)/[(0.7(0.8)[(28.9MPa)(0.97) + 12.42MPa]] = 368,500mm^2$

We have specified a square column; therefore,

$d = \sqrt{368,500mr} = 607mm$ minimum column side.

We make it 610mm, therefore: $A_g = 610mm \times 610mm = 372,100mm^2$

3. Radius of gyration:

$$r = 0.3(610mm) = 183mm$$

Slenderness ratio:

$$KL/r = 1.0(5500mm)/(183mm) = 30$$

The 30 > 22 long column. R must be calculated.

Slenderness reduction factor:

$$R = 1.07 - 0.008(5500mm)/(183mm) = 0.83$$

4. Since R is not as assumed in step 1, the actual required p_g must be found by solving the column equation.

$$8,347,000N = 0.83(0.70(0.80)(372,100mm^2)[(28.9MPa)(1 - p_g) + (414MPa)p_g]$$
$$8,347,000N = 172,952mm^2[28.9MPa - (28.9MPa)(p_g) + (414MPa)p_g]$$
$$8,347,000N = 4,998,313N + (66,603815N)p_g$$
$$p_g = (3,348,687N)/(66,603,815N) = 0.05 < 0.08$$

Steel is between 1% (0.01) and 8% (0.08) of the gross area OK!

5. $p_g = A_s/A_g$

Required steel area $A_s = p_g A_g = 0.05(372,100mm^2) = 18,605mm^2$

Try with No. 29M bars, $A = 645mm^2$

Number of bars required = $(18,605mm^2)/(645mm^2/bar) = 28.8$ bars

Since this is a large number, we try with No. 36M bars, $A = 1006mm^2$

Number of bars required = $(18,605mm^2)/(1006mm^2/bar) = 18.5$ bars

Use 18 No. 36M bars, 9 on each side of the column.

6. No. 36M bar diameter = 35.8mm

Minimum spacing = 38mm or

$1.5d_b = 1.5(35.8mm) = 53.7mm > 38mm$

Ties: with No. 36M bars, the minimum tie diameter d_t is No. 13M.

Maximum tie spacing = $16d_b = 16(35.8mm) = 573mm$ or

$48d_t = 48 (12.7mm) = 610mm$ or

$d_{column} = 610mm$

The least tie spacing value must be used, which is 573mm. For lateral bracing, longitudinal bars must comply with minimum distances from tie corners shown in the illustration.

Pacement of the bars:

    Tie cover   = 38mm

    Tie diameter   = <u>13mm</u>

    Bar cover   = 51mm

Check bar spacing, side (a):

Cover: 2(51mm)   = 102mm

Bars: 5(35.8mm)   = 179mm

Minimum spacing: 2(75mm)   = <u>150mm</u>

            Total   = 431mm

Spacing remaining for the middle bar on both sides
= (610mm - 413mm)/2 = 89.5mm, which is OK
since 2 bar diameter < 89.5mm < 152mm.

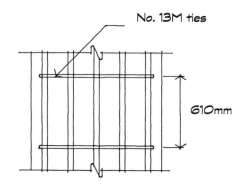

No. 13M ties

610mm

On side (b) six bars must be placed. Since each
other bar must be tied by a tie corner, the
two bars in the middle are bundled and tied together by the 45° tie.

A maximum of three bars can be bundled together. The tie angle cannot be smaller than
22.5° (see illustration); in our case, the angle is
45°.

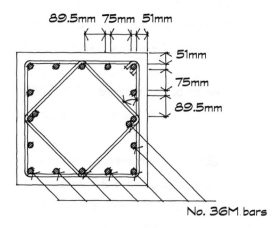

89.5mm 75mm 51mm

51mm

75mm

89.5mm

No. 36M bars

37 WALLS

37.1 Design requirements for vertical loads

You will be happy to know that wall design is a relatively simple variation of column design. In most instances, the capacity of the concrete alone would be sufficient to carry the direct axial load. The reinforcing that will be introduced, will be in the wall primarily to compensate for shrinkage and temperature changes. With this in mind we will look at the unique considerations that apply to walls and the variations in the basic design equation.

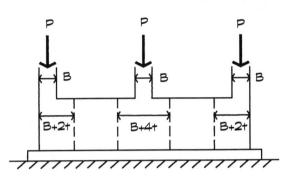

Fig. 37.1 Concentrated load distribution on walls

(a) Loads and stability

Walls can support loads distributed along their length, such as floor structures made of concrete slabs, as well as beams or columns that apply concentrated loads. The distributed loads' compressive stress can be considered applied on a 1ft [1m] long wall section with a horizontal cross-sectional area $A = 12t$ (in²) [mm²], where t is the wall specified thickness. The load must be factored according to the R/C procedure, and the wall section will be designed for U, the required strength, plf (lb/ft of wall length)[N/m].

Concentrated loads could be represented by columns, beams, or joists. Concentrated loads can be considered distributed on length of wall $L = B + 4t$ where B = width of bearing, and t = specified wall thickness, or spacing between loads (Fig. 37.1), whichever is smaller.

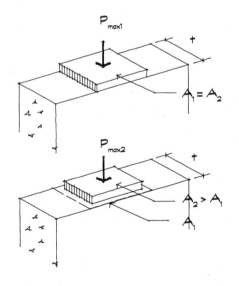

Walls must be anchored to floor structures and vertical bracing structures such as cross walls or moment-resisting frames able to resist overturning under the action of lateral loads.

(b) Bearing strength

Design bearing strength is

$$P_{max.1} = \Phi(0.85f_c)A_1$$

Fig. 37.2 Bearing area

where $\Phi = 0.70$, the strength reduction factor, and A_1 = loaded area as wide as the wall thickness. When the wall is wider than the loaded area,

$$P_{max2} = P_{max1} \sqrt{A_1/A_2}$$

(See Fig. 37.2)

(c) Minimum reinforcement

Minimum reinforcement is given in the UBC as a ratio of the required steel area A_s to gross concrete area A_g, as listed in Table 37.1.

Table 37.1 Minimum reinforcement in concrete walls subject to compressive loads

	A_s/A_G	Reinforcing size/type
1.	**Vertical reinforcement**	
	0.0012	Deformed bars ≤ No. 5 [No.16M], $f_y \geq 60,000$psi [414MPa]
	0.0012	Maximum W31, D31 welded wire fabric
	0.0015	Deformed bars > No. 5 [No.16M]
2.	**Horizontal reinforcement**	
	0.0020	Deformed bars ≤ No. 5 [No.16M], $f_y \geq 60,000$psi [414MPa]
	0.0020	Maximum W31, D31 welded wire fabric
	0.0025	Deformed bars > No. 5 [No.16M] deformed

(d) Placement of bars (Fig. 37.3).

Spacing: Maximum vertical and horizontal spacing = 3t or 18in [457mm] whichever is least, where t is the wall thickness.

Ties: Required as in columns if vertical reinforcing is designed for compression.

$$A_s > 0.01A_g.$$

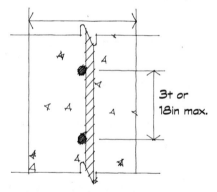

Fig. 37.3 Bar placement

Openings: Minimum two No. 5 [No. 16M] bars around windows, door openings, and extending not less than 24in[610mm] beyond the corners of the openings.

Wall t > 10in [254mm]: Place bars in two layers parallel to the faces of the wall (Fig. 37.4):

Above grade: the interior layer is to have 34% to 50% of the total required vertical and horizontal A_s; the exterior layer is to have 50% to 66% of the total required vertical and horizontal A_s; the interior cover of 3/4in [76mm] minimum, t/3 maximum, (e.g., 4in for a 12in wall) [102mm for a 305mm wall]; and an exterior cover of 2in [51mm] minimum, t/3 maximum (e.g., 4in for a 12in wall) [102mm for a 305mm wall].

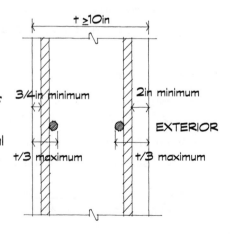

Fig. 37.4 Reinforcement placement equirements for thick walls

37.2. Walls designed as compression members

R/C walls can be designed as columns, whenever the magnitude and type of loads require the use of vertical compressive reinforcing. This could be the case of a concrete shear wall supporting R/C columns (Fig. 37.5). In this case, the procedure illustrated in Chapter 36 should be followed. The UBC, Section 2614(f), allows the use of an empirical procedure for compressive loads that are axial or located within the middle third ("kern" area) of the overall wall thickness. In this procedure, the design requirements listed previously in Sec. 37.1 must be satisfied. The method can be used also for basement and foundation walls.

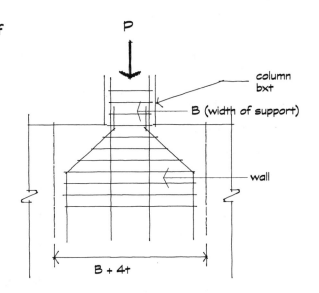

Fig. 37.5 Wall designed as a column

The design axial load strength $P_{max.}$ of a wall is computed by the equation

$$P_{max.} = 0.55\Phi f_c'A_g[1 - (KL_c/32t)^2] \geq U$$

where

Φ = strength-reduction factor for R/C walls = 0.70

kl_c = effective unsupported height of the wall

U = required strength (factored load)

The values of K are the same as those used for any material used for construction. You can refer to Fig. 10.1 in Chapter 10, Steel.

The minimum thicknesses of walls designed with the empirical method are a follows:

Bearing walls: 4in [102mm] or (unsupported length L_c)/25 or (unsupported height h_c)/25

Basement walls and foundations: 7-1/2in [191mm]

Example 37.1

A reinforced-concrete bearing wall supports a roof system consisting of precast single tees spaced 8ft o.c. The stem of each T-section is 8in wide, but the bearing width is taken as 7in to allow for beveled bottom edges. The tees will bear on the full thickness of the wall. The height of the wall is 11ft - 6in and the reaction of each tee due to service loads is 22k DL and 12k LL. Design the wall with f_c' = 4,000psi and F_y = 40,000psi.

On each T-section:

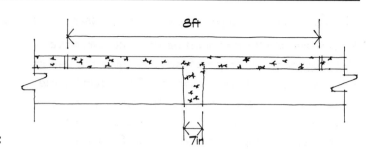

U = 1.4(22k) + 1.7(12k) = 51.2k

Minimum thickness ratio L_c/t = 25

$t = L_c/25 = 11.5ft(12in/ft)/25 = 5.5in$

Make t = 6in

Effective horizontal bearing length of wall:

L = B + 4t = 7in + 4(6in) = 31in

Gross cross-sectional area:

A_g = (6in)(31in) = 186in²

K = 0.8 for a braced cantilever wall

$KL_c/(32t) = 0.8(11.5ft)(12in/ft)/32(6in) = 0.575$

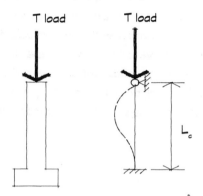

$P_{max.} = 0.55(0.70)(4ksi)(186in^2)[1 - (0.575)^2] =$

$= 191.7k > U = 51.2k \quad OK!$

The capacity of the 6in wall is nearly four times the required strength. Since the thickness cannot be reduced, the only possible increase of efficiency is to use a weaker concrete, or to simplify the connection with the foundation by using a single dowel, making K = 1.0.

37.3 Horizontal forces in basement walls

Soils cause horizontal pressures against basement, foundation, and retaining walls, called **active pressures**, similar to fluids and depending on the nature of the soil. Loose granular soils produce pressures that increase proportionally to the depth, but cohesive soils such as clay do not behave in the same way. If the soil becomes saturated, the water weight will increase the horizontal pressure. It is important to drain retaining walls and foundations in general to prevent the buildup of hydrostatic pressure.

Since the pressures are not equal to the actual weights of the materials, due to the internal friction of the material (the material will stand up due to its own internal resistance or angle of repose), they are expressed in terms of the weight of an equivalent fluid that will produce the given hydrostatic pressure. This is called the fluid weight of the material, and the pressure p_e that it will produce is called the **equivalent fluid pressure**. This can be calculated with the equation.

$$p_e = 0.3gh$$

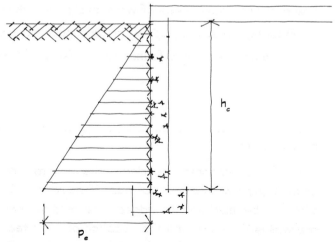

Fig. 37.6 Soil pressure distribution on a basement wall

where

> g = density of the soil (weight of the unit of volume) in lb/ft³ [N/m³]
>
> h = clear height of the wall in ft [m].

The distribution of the pressure is shown in Fig. 37.6; the wall is subject to beam action between the two floor structures. The loads from floors and walls above must be added as well. Therefore, this is potentially a case of combined axial load and bending.

Example 37.2

A basement wall is 12ft high between horizontal supports. No vertical loads are supported on this side. g = 100lb/ft³; f_c' = 3,000psi and F_y = 50,000psi

By using a 1ft-wide strip (as in slab design), the value of the maximum soil pressure is

p = 0.3(100lb/ft³)(12ft) = 360plf on a 1ft-wide vertical strip of wall.

This represents the service live load, and since there are no horizontal dead loads:

Required strength U = 1.7(360plf) = 612plf

Resultant U = (612plf)(12ft)/2 = 3,672lb

Maximum shear = support reaction V = (3,672lb)(4ft)/(12ft) = 1,224lb

Maximum moment M = 0.1283Uh²/2 = 0.1283(612plf)(12ft)²/2 = 5,652ftlb

Required depth for the moment (d_m):

d_m is critical at the face of the support (face of the wall):

$$d_m = \sqrt{\frac{5652ftlb(12in/ft)}{0.9(462.8psi)12in}} = 3.68in$$

Required depth for shear (d_{vb}):

d_{vb} is critical at d inches from the face of the support (face of the wall):

$$f_v' = 0.85 \; 2\sqrt{3000} = 93.1psi$$

If we assume a uniform distribution of force for simplicity, the value of shear V at d inches from the bottom end of the wall is

$$V_{@d} = V_{max.} - U$$

$$\Delta u = \frac{u_{max}(d)}{h} = \frac{612plf(d)}{12ft} = 51.0plf(d) = \frac{51.0plf(d)}{12in/ft} = 4.25pli(d)$$

$$93.1(12)d = 1224 - 4.25d^2$$

$$4.25d^2 + 1,117.2d = 1224$$

$$d^2 + 262.9d = 288$$

$$(d + 131.4)^2 = 288 + 131.4^2 = 17,501.4$$

$$d + 131.4 = 132.3$$

d = 0.90; this is d_{vb}, but we've just used d to simplify the calculations.

Moment controls at d_m = 3.68in

Actual depth:

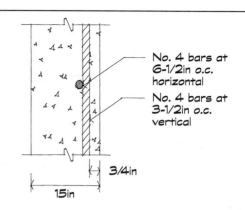

No. 4 bars at 6-1/2in o.c. horizontal

No. 4 bars at 3-1/2in o.c. vertical

3/4in

15in

Controlling d:	= 3.68in
Bars: No. 4	= 0.25in
Cover: interior walls	= 0.75in
Total	= 4.68in

For ease of concrete placement and reinforcement location, use a minimum T of 6in; this will make d = 5in.

Steel area:

$R = (5,652ftlb)(12in/ft)/[0.9(12in)(5in)^2] = 251.2$

From Table 30.4, we can find the value for p_g

$p_g = 0.0053$

$A_s = 0.0053(12in)(5in) = 0.318in^2/ft$

From Table 28.4

No. 4 bars spaced at 7-1/2in o.c. gives $A_s = 0.32in^2/ft$ vertical reinforcing.

Shrinkage steel:

$A_s = 0.002 (12in)(6in) = 0.144in^2/ft$

From Table 28.4

No. 4 bars spaced 17in o.c. gives $A_s = 0.14in^2/ft$ horizontal reinforcing.

37.4 Retaining walls

Retaining walls are cantilever structures supported on a footing and loaded by horizontal soil pressure. They are designed as cantilever slabs, with the additional provisions of minimum horizontal reinforcement for walls as given in Table 37.3.

Table 37.3 Minimum horizontal reinforcement in retaining walls

A_s/A_G	Reinforcing size/type
0.0020	Deformed bars ≤ No. 5 [No.16M], f_y ≥ 60,000psi [414MPa]
0.0020	Maximum W31, D31 welded wire fabric

Retaining walls have to be designed for

1. Factor of safety against sliding of 1.5

2. Factor of safety against overturning of 1.5

3. Maximum soil pressure under the foundation

4. Shear and bending under lateral loads

Short retaining walls up to a height h = 10ft [3m] can be designed as cantilever walls on a footing width b = h/2, with 2/3 of b against the soil. This allows use of the weight of the soil itself against overturning (Fig. 37.7). The lug below the footing, 8 to 12in [203 to 305mm] thick, is used in cohesive soils (clay) to prevent sliding. On soils with a high coefficient of friction (granular material), this is not necessary. The friction force is proportional to the weight; therefore, this force can be increased by making the footing larger so that it is loaded with a larger volume of soil.

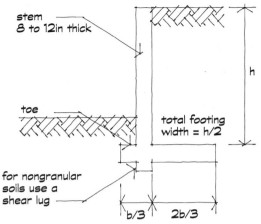

Fig. 37.7 Retaining wall proportions

Tall retaining walls may need buttressing systems or a thicker section at the base to resist overturning moments (Fig. 37.8). Soil loads should be increased by loads of vehicles or materials on the surface behind the wall. These **surcharges** can be considered as additional soil pressures. The added lateral pressures are proportional to the vertical load according to Poisson's ratio (Fig. 37.9).

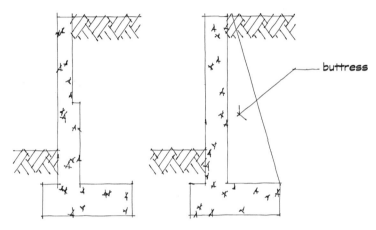

Fig. 37.8 Tall retaining walls use alternative geometries.

For a concentrated load, this pressure is proportional to the distance from the wall and to the height of the wall. Figure 37.9 illustrates the surcharge and its point of application on the wall. For uniform loads, the surcharge is equivalent to an added soil weight. If the earth weight is 100pcf [15,700N/m³] and the load (e.g., parking lot loading) is 300psf [14,362N/m³], the surcharge is equivalent to an extra fill 3ft [1m] deep.

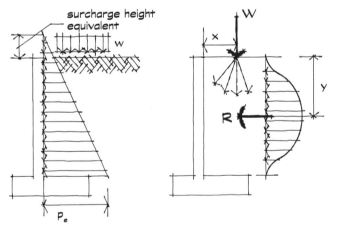

Fig. 37.9 Retaining walls with surcharges

37.4.1 Backfill and drainage

A final concern that is frequently overlooked is how an excavation is backfilled after construction is complete. If material is replaced that was originally removed from the excavation, it is unlikely that it will be adequately compacted or will provide good drainage characteristics. The water behind or against the wall must be able to reach the footing drain for it to

be effective. At the very least, a granular backfill, preferably pea gravel (remember its compaction characteristics) should be used. In some cases, an inverted choke of a layer of large
gravel or stone immediately adjacent to the drain, followed by a layer of smaller stone and
yet another layer of smaller stone, topped of with finish material, is desirable. This sequential
series of stone sizes over the foundation drain will filter out soil particles and keep the drain
from becoming useless. As a last consideration, the finish grade should slope away from the
building at a rate of 1/4-in/ft [20mm/m]. Downspouts and any other source of water should be
directed away from the building to reduce water pressure and infiltration on basement walls.

METRIC VERSIONS OF ALL EXAMPLES

Example M37.1

A reinforced-concrete bearing wall supports a roof system consisting of precast single tees spaced 2.4m o.c. The stem of each T-section is 200mm wide, but the bearing width is taken as 180mm to allow for beveled bottom edges. The tees will bear on the full thickness of the wall. The height of the wall is 3.5m and the reaction of each tee due to service loads is 98kN DL and 53kN LL. Design the wall with f_c' = 28MPa and F_y = 276MPa.

On each T-section U = 1.4(98kN) + 1.7(53kN) = 227.3kN

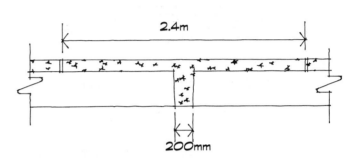

Minimum thickness ratio L_c/t = 25

$t = L_c/25$ = (3500mm)/25 = 140mm

Make t = 150mm

Effective horizontal bearing length of wall:

L = B + 4t = 180mm + 4(150mm)

= 780mm

Gross cross-sectional area:

A_g = (150mm)(780mm) = 117,000mm²

K_e = 0.8 for a braced cantilever wall

$KL_c/32t$ = 0.8(3500mm)/32(150mm) = 0.58

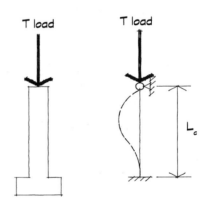

$P_{max.}$ = 0.55(0.70)(28MPa)(117,000mm²)[1 - (0.58)²]

= 836,972N

= 837kN > U = 227.3kN   OK

The capacity of the 150mm wall is nearly four times the required strength. Since the thickness cannot be reduced, the only possible increase of efficiency is to use a weaker concrete, or to simplify the connection with the foundation by using a single dowel, making K_e = 1.0.

Example M37.2

A basement wall is 3.6m high between horizontal supports. No vertical loads are supported on this side. g = 15,708N/m³, f_c' = 21MPa amd F_y = 345MPa

By using a 1m wide strip (as in slab design), the value of the maximum soil pressure is

p = 0.3(15,708N/m³)(3.6m) = 16,967N/m on a 1m wide vertical strip of wall.

This represents the service live load, and there are no horizontal dead loads:

Required strength (distributed load): U = 1.7(16,967N/m) = 28,844N/m

Resultant $U = (28,844 N/m)(3.6m)/2 = 51,919N$

Maximum shear = support reaction $V = (51,919N)(1.2m/3.6m) = 17,306N$

Maximum moment $M = 0.1283Uh^2/2 = 0.1283(28,844N/m)(3.6m)^2/2 = 23,980Nm$

Required depth for moment (d_m):

  d_m is critical at the face of the support (face of the wall):

$$d_m = \sqrt{\frac{23,980Nm(1000mm/m)}{0.9(3.19MPa)(1000mm)}} = 91.4mm$$

Required depth for shear (d_{vb}):

  d_{vb} is critical at d mm from the face of the support (face of the wall):

$$f_v' = 0.85\,(1/6)\sqrt{21MPa} = 0.65MPa$$

If we assume a uniform distribution of force for simplicity, the value of shear V at d inches from the bottom end of the wall is

$$V_{@d} = V_{max} - U$$

$$\Delta u = \frac{u_{max}(d)}{h} = \frac{28.844N/mm(d)}{3600mm} = 0.65MPa(1000mm)d$$

$$17,306N - (0.008N/mm^2)(d^2)$$

$$0.008d^2 + 650d = 17,306$$

$$d^2 + 81,250d = 2,163,250$$

$$(d + 40,625)^2 = 2,163,250 + 40,625^2 = 1,652,553,875mm^2$$

$$d + 40,625 = 40,652$$

d = 27mm; this is d_{vb}, but we've just used d to simplify the calculations.

Moment controls at $d_m = 91.4mm$

Actual depth:

Effective d:	= 90mm
Bars: No. 15M	= 15mm
Cover: interior walls	= <u>19mm</u>
Total	= 124mm

For ease of concrete placement and reinforcement location, use a minimum T of 150mm; this makes d = 150mm - 19mm - 15mm/2 = 108.5mm.

Steel area:

  $R = (23,980Nm)(1000mm/m)/[0.9(1000mm)(108.5mm)^2] = 2.26MPa$

  From Table 30.4 (converting answer from MPa to psi)

$p_g = 0.00705$

$A_s = 0.00705(1000\text{mm})(108.5\text{mm}) = 765\text{mm}^2/\text{m}$

From Table M28.4

No. 15M bars spaced at 230mm o.c. gives $A_s = 768\text{mm}^2$ per meter vertical reinforcing.

Shrinkage steel:

$A_s = 0.002(1000\text{mm})(150\text{mm}) = 300\text{mm}^2/\text{m}$

From Table M28.4

No. 10M bars spaced at 400mm o.c. gives $A_s = 302\text{mm}^2$ per meter horizontal reinforcing.

38 CONNECTIONS

38.1 Connections of footings and vertical structures

Various types of connections are possible depending on the type of superstructure (Fig. 38.1):

(a) Masonry wall on continuous footings:

Use dowels or a key for lateral force.

(b) Concrete or reinforced masonry element:

Use dowels to match vertical bars. This will not only transfer the concentrated loads of the vertical reinforcing, but will facilitate placement of the vertical wall reinforcing during construction.

(c) Anchor bolts for wood plates or steel base plates.

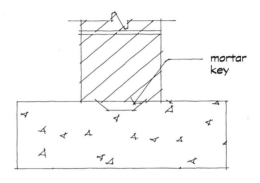

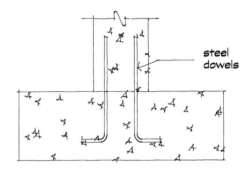

When a reinforced-concrete column rests on a footing, it is usually necessary to develop the vertical compressive stress in the column reinforcing by doweling action into the supporting concrete structure. The dowels help solve the problem of bearing pressures between the column and the footing, which is usually made of a lower-strength concrete (e.g., 5,000 and 3,000psi [33 and 21 MPa], respectively). In this way, the stress at the foot of the column is transferred gradually by the steel into the mass of the footing instead of directly bearing on a small footing area of relatively weak concrete (in much the same way a friction pile transfers its load into the weak subsoil). For reinforcing bars of large size and high yield strength, the distance required for this development length can become considerable and affect the footing design.

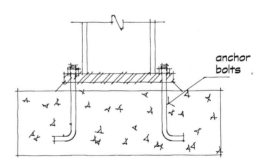

The calculation of the length of dowels is based on the bond strength between concrete and steel. The dowels are typically of the same diameter as the bars of the column or wall resting on the foundation. Multiple smaller dowels can be used in column connections to reduce the required penetration depth. For compression loads, the length of embedment is given by

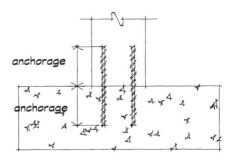

Fig. 38.1 Alternative connection techniques

$$L_d = \frac{0.02F_y d_b}{\sqrt{f'_c}} \geq 0.0003 d_b F_y \qquad \left[L_d = \frac{0.25F_y d_b}{\sqrt{f'_c}} \geq 0.044 F_y d_b\right] \text{ metric version}$$

where

d_b = diameter of bar

F_y = steel yield stress

f'_c = concrete strength

For tension loads (which could be caused by bending), the required length is as follows.

For No. 6 [*No. 19M*] bars or smaller:

$$L_d = \frac{0.04 d_b F_y}{\sqrt{f'_c}} \geq 12\text{in} \qquad \left[L_d = \frac{0.5 F_y d_b}{\sqrt{f'_c}} \geq 305\text{mm}\right] \text{ metric}$$

For No. 7 [*No. 22M*] bars or larger:

$$L_d = \frac{0.05 d_b F_y}{\sqrt{f'_c}} \geq 12\text{in} \qquad \left[L_d = \frac{0.625 F_y d_b}{\sqrt{f'_c}} \geq 305\text{mm}\right] \text{ metric}$$

These are obviously the same criteria for anchorage discussed in Chapter 31, Sec. 31.2.

Precast concrete components require external connecting devices and these usually require additional bolting or welding (Fig. 38.2). Hollowcore precast slab units or precast T's frequently use weld plates that are cast into the unit and provide a connector for attachment to steel beams. If the system is one consisting of both precast beams and precast columns, pockets are precast into the units for connection. If the loads at these points are very high, the pocket may be reinforced by steel plates or brackets. The system in Figure 38.3 is similar to the mortise-and-tenon joint in Fig. 25.1. While technologies have advanced significantly, it is worth looking at solutions from the past for inspiration.

Although the scale is radically different from architectural building systems, classic Lincoln Logs, Tinker Toys, and other children's building kits solve the same problem of connections in a modular way.

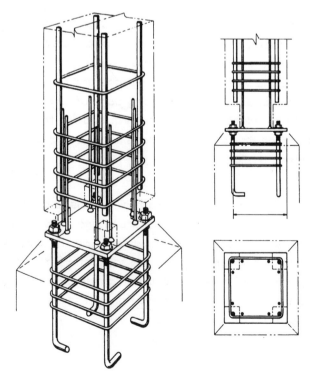

Fig. 38.2 Connection of precast column units

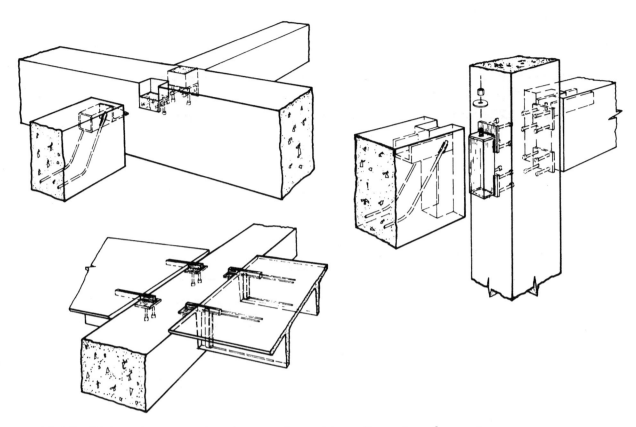

Fig. 38.3 Prefabricated precast units using a mortise-and-tenon-like system of connectors

Prefabricators of contemporary structural systems use the same criteria: ease of fabrication, ease of erection, and repetition, as well as economy of material, to guide their decisions.

Example 38.1

An R/C column, loaded with 210k axial compression, is reinforced with four No. 8 bars. The concrete is 5,000psi for the column and 3,000psi for the footing. The steel is 50,000psi. The footing total depth is 20in. Since the concrete is different in column and footing, two different lengths must be calculated.

In the column:

$$L = \frac{0.02d_bF_y}{\sqrt{f'_c}} = \frac{0.02(1in)(50,000psi)}{\sqrt{5000psi}} = 14.14in$$

$$0.0003\,d_bF_y = 0.0003\,(1in)(50,000psi) = 15in$$

The dowel length in the column must be 15in.

In the footing:

$$L = \frac{0.02d_bF_y}{\sqrt{f'_c}} = \frac{0.02(1in)(50,000psi)}{\sqrt{3000psi}} = 18.25in$$

The dowel length in the footing must be 18.25in.

The required cover is 3in, so the length of the bar in the footing can only be
20in - 3in = 17in. The remaining length 18.25in - 17in = 1.25in has to be provided by bending the
bar for a length = 1.25in. We would probably specify a standard 4in hook.

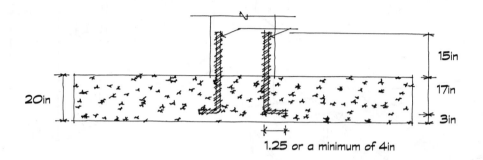

15in

17in

20in

3in

1.25 or a minimum of 4in

METRIC VERSION OF EXAMPLE

Example M38.1

An R/C column, loaded with 35MPa axial compression, is reinforced with four No. 25M bars. The concrete is 35MPa for the column and 21MPa for the footing. The steel is 345MPa. The footing total depth is 400mm. Since the concrete is different in column and footing, two different lengths must be calculated.

In the column:

$$L = \frac{d_b F_y}{4\sqrt{f'_c}} = \frac{25.4mm(345MPa)}{4\sqrt{35MPa}} = 370mm$$

$$0.044\, d_b F_y = 0.044\,(25.4mm)(345MPa) = 386mm$$

The dowel length in the column must be 386mm.

In the footing:

$$L = \frac{d_b F_y}{4\sqrt{f'_c}} = \frac{25.4mm(345MPa)}{4\sqrt{21MPa}} = 478mm$$

The dowel length in the footing must be 478mm.

The required cover is 76mm, so the length of the bar in the footing can only be 400mm - 76mm = 324mm. The remaining length 478mm - 324mm = 154mm has to be provided by bending the bar for a corresponding length.

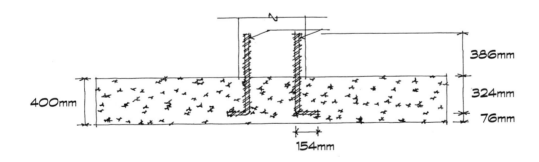

39 PRESTRESSED AND PRECAST CONCRETE

39.1 Prestressed concrete

The majority of concrete structures for multistory buildings are not made entirely of ordinary reinforced concrete but are prestressed. This is particularly advantageous if the design consists of long spans, shallow depths, and light dead loads. Prestressed concrete can save 25% or more concrete and steel and have a depth/span ratio of 1/20 or 1/30 versus 1/12 for conventionally reinforced beams.

Prestressed beams are often produced in a manufacturing plant, and they are best used as simply supported beams for roof or floor structures. Long-span beams develop very large reactions and require carefully designed bearing devices in order to spread the load on the support and to provide the possibility for movement for the larger thermal expansion and shrinkage associated with long spans. Roof or floor slabs and planks, as seen in Chapter 29, are usually produced with straight prestressing reinforcement and used as simply supported members on medium or long spans. Prestressing is also used in frames made of cast-in-place or precast members to create rigid, moment-resisting connections between columns and beams. In these cases, the stressing compression is applied in place and is referred to as posttensioning.

The basic principles of prestressing are shown in Fig. 39.1. In bending, concrete in a normal cast-in-place R/C beam will have hairline cracks in the tension zones. However, in a prestressed section, the normal tensile area of the beam is placed in a state of high compression. Since the reinforcing is placed in the "lower" (tension) portion of the beam, the eccentricity of this force will potentially cause tension in the "upper" (compression) area of the section (lower and upper referring to a simply supported situation). After placement of the beam in its final structural application, loads are introduced that will create the normally anticipated compression and tensile stresses. These stresses are met with "opposite" prestresses, and the net result of the combination of these stresses is such that the entire cross-section of the beam subjected to bending may be in compression. Consequently, the bottom fibers will not crack: no tension, no cracking. If you'll remember our discussion at the end of Chapter 30 explaining why columns are typically a more cost-effective use of concrete than typical cast-in-place beams, it's now easy to see why prestressed or posttensioned sections are an extremely effective use of concrete.

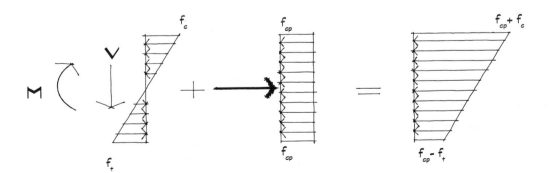

Fig. 39.1 Resultant stresses in a prestressed beam

To illustrate this principle, we can represent a cracked beam with a row of books held up horizontally between two hands (Fig. 39.2). The hands compress the books **at the top**, and the bottom is cracked. If an additional load is applied vertically or if the load is applied too high, the system will explode downward. This beam (a) will offer very little resistance to applied load and may fail under its own weight (similar to a prestressed beam that is lifted improperly or is allowed to lie on its side. Now (b) compress the books **on the bottom** of the beam: Cracks may initially appear at the top, but as we place an additional load over it, (c) the entire beam will be in compression and will carry the load.

This example corresponds to the application of an eccentric compression (Fig. 39.2(b)) producing a combination of direct stress and bending, opposite to the one created by the gravity loads. The resulting stress in working condition is (c) uniform compression. Figure 39.3 illustrates the principle using structural notation.

Compression is applied in the beam with special steel reinforcement bars or cables, as shown in Fig. 39.4. In Fig. 39.4(a), the tension in the steel will cause a uniform compression P and a moment P(e) equal to the maximum bending moment M = $wL^2/8$ due to the uniformly distributed vertical loads. At the ends of the beam, though, the bending moment due to w decreases, and the beam section will not be uniformly compressed. If the steel is placed with a decreasing eccentricity toward the ends (draped), the prestressing moment will also decrease, following the variation of the external bending moment. This variation is parabolic for uniformly distributed loads. In this way, the stress near the ends will be essentially the precompression $f_{ci} = P/A$. The steel acts

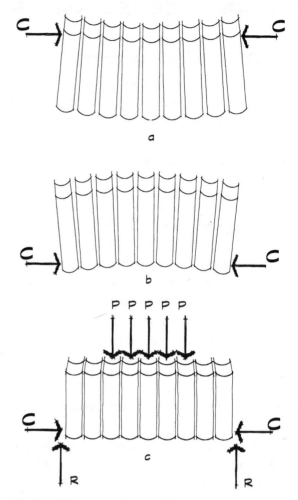

Fig 39.2 Example of load capacity resulting from application of compression

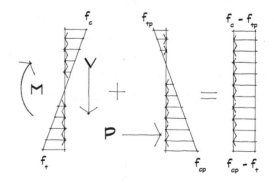

Fig. 39.3 Stresses resulting from eccentric prestressing force

similarly to the cable of a suspension bridge. It is possible to produce the required variable eccentricity by using more than one cable with placements in the shape of increasingly steeper parabolas. In this case, the induced bending stress at any section will be the resultant of all the stress diagrams produced by each cable. With the straight cable system, several beams may be cast simultaneously with bond breaks between them, and the very long cables need be anchored only at the two ends, making it constructionally efficient, but somewhat structurally inefficient. If the parabolic cable system is used, multiple anchor points are necessary to maintain the parabolic configuration; consequently, it may be constructionally difficult, but structurally efficient.

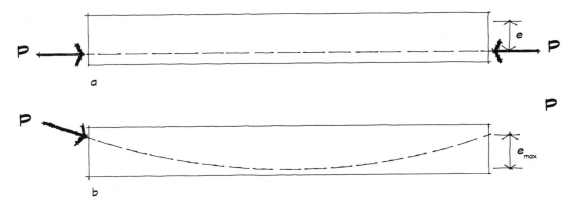

Fig. 39.4 Placement of prestressing reinforcement in simply supported beams

Whereas prestressed sections are efficient uses of both steel and concrete, it should be noted from the "books" example cited earlier that stress reversals within the member are difficult to respond to. A member optimally designed to span between two supports can be quite effectively constructed. If a cantilever is added to this same member (overall a good structural technique), it causes the beam to have tension in the bottom between the supports, and in the top at the support. This will generally necessitate additional nonprestressed steel to be placed in the tensile zone for the cantilever or you would need to prestress both the top and the bottom of the beam, creating a beam with redundant steel in many locations. The construction method of prestressed beams requires a relative simple, straight-line placement of prestressing rods/cables.

39.2 Construction techniques

As we have mentioned, the stressing can be applied in the following two ways.

39.2.1 Prestressed concrete

If the reinforcement is placed under tension **before** the concrete is cast, as in the case of prestressed concrete manufactured in a plant, the term **pretensioned** is used. Pretensioned steel is made of high-yield steel **strands** of 6 to 12 thin wires twisted together. The strands are placed under tension with jacks anchored at the ends of the formwork or on abutments at the end of a casting bed where a series of members can be made with the same strands (Fig. 39.5). The concrete is poured on the steel and steam-cured for 24 hours to reduce the curing time.

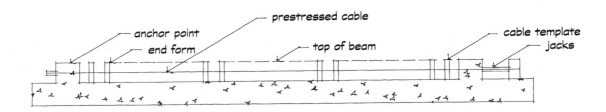

Fig 39.5 Casting bed for prestressed beams allowing multiple beams to be cast simultaneously

The strands are then cut, and as they are released, they compress the concrete. The compression is created as a reaction of the concrete to the tension of the strands along their contact surface; it is essential to ensure a good bond between the two materials, and the twisted surface of the strands helps to ensure this. This pressure in the concrete makes it shorten, which relieves part of the strain and stress of the steel. Therefore, the steel loses some of its initial tension. The uplift of the eccentric compression makes the removal from formwork automatic. A camber is created that must be taken into account in the architectural design.

39.2.2 Posttensioned concrete

When the stressing is applied after the concrete is cast, the process is called posttensioning. The member is made with some ordinary steel reinforcing, and the tensioning steel is greased and wrapped or left loose inside protective ducts that prevent contact with the concrete. The tension is then applied with jacks at the ends of the beam, and the bars or tendons are blocked with steel wedges in the anchoring devices. Grout can be injected in the ducts to protect the tendons against corrosion, or tendons can be left unbonded with a suitable protective coating. When the required compression of the concrete is attained, the steel does not lose tension. This is an obvious advantage of posttensioning. Other advantages are that the efficient draped configuration can be easily used because the draped form is not under tension until after the concrete has cured sufficiently, and in posttensioned systems the dead load of the structure itself can be used to resist the potential tensile uplift forces. These forces can be a problem in prestressed systems during handling. If a member is allowed to change orientation and the dead load no longer resists the tensile forces in the top of the beam, it can simply fold up. Posttensioning systems can use wires, strands, or bars, as listed in Table 39.1.

Table 39.1 Prestressing/posttensioning materials

System	Reinforcement description	Minimum strength	
Wire	0.25in [6mm] diameter steel wire	240,000psi	[1655MPa]
Strand	Seven wires wrapped together	270,000psi	[1890MPa]
Bar	5/8 to 1-3/8in [16 to 35mm] diameter bars	145,000psi	[1015MPa]

The steel used is much stronger than in conventionally reinforced concrete, where bars have strengths of 40,000 to 60,000psi [276 to 414 MPa]. The prestress $f_{ci} = P/A$ has a value of 150 to 250ksi [1050 to 1750 MPa] for slabs and 200 to 500ksi [1380 to 3450 MPa] for beams. In both systems, shrinkage and creep take place, but obviously some of the total shrinkage will take place before compression is applied in posttensioning, and in pretensioning it reduces compression. Higher values of prestressing cause excessive creep.

Creep (plastic flow) of concrete also causes the tension of the reinforcing to decrease; this combines with steel yield or relaxation, anchorage seating loss, elastic shortening of concrete, and friction loss in posttensioning tendons. In the design of prestressed members, these losses of tensions must be taken into account so that the correct initial tension can be specified. Loss of prestress due to creep, shrinkage, and other causes can be as high as 20 to 25%.

The main benefits of prestressing are as follows:

1. Additional capacity of beams for gravity loads, since the bending stress due to the weight of the beam is balanced by the eccentric precompression stress.

2. Increased shear resistance, since the entire section is compressed. This allows us to design prestressed beams with thin webs, thereby reducing the dead load considerably.

3. Elimination of cracks with a consequent increase of durability, particularly in a corrosive atmosphere.

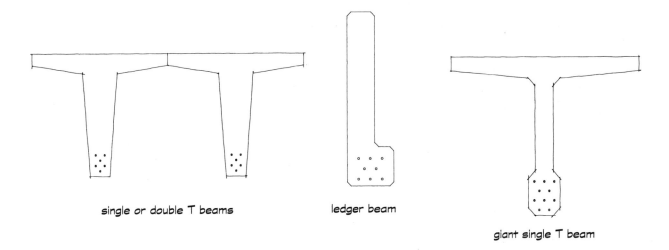

single or double T beams   ledger beam

giant single T beam

Figure 39.6 Typical shapes of prestressed beams

39.3 Precast concrete shapes

The most common precast members are simply supported flexural members such as floor planks, but columns and beams for multistory frames are widely used with ordinary reinforcement or more often with prestressing. It is possible to connect precast members and transfer vertical and horizontal forces and bending moments by joining reinforcing bars or by posttensioning. Steel bars can be joined by welding, using threaded ends or bolted steel plates. The completion of the structure usually involves pouring some concrete on-site to cover the steel connections or to form a concrete topping over prestressed floor systems. If connections are created in "pockets" left in the member that need to be patched after installation, the creation of these pockets can have a significant impact on the appearance of the finished structure.

The camber produced by prestressing is not precise enough to allow for a finished surface to be applied directly over the member, creating the need for either structural or nonstructural topping slabs.

In addition to the service conditions, precast members must be designed for handling, transportation, and erection, often requiring additional reinforcement for these stresses. Storing and lifting of precast members has to be done carefully. In some (most?) cases, the members must be lifted at the same location as their final support to ensure that tensile stresses will not be created in unexpected locations. Threaded inserts or lifting lugs may be cast into the member to ensure attachment at the correct locations. In many cases, the prestress manufacturer will also be the erection contractor to ensure proper handling of the material.

The advantages of prestressed concrete manufactured under plant-controlled conditions are better quality control, reduced shrinkage on-site, faster site work, and the possibility of complex but efficient shapes with potential savings in weight.

The disadvantages of prestressed concrete are shipping or transportation limitations for long or large members, sophisticated erection procedures, and architectural or construction problems created by camber.

In the design of prestressed structures, consistent adjacent bay sizes are important. If, for instance, a 60ft [18m] span is adjacent to spans of 30ft [9m], the maximum camber (and deflection) for the 60ft [18m] span will occur at midspan, while this is the location of "0" camber and ultimately "0" deflection in the adjacent 30ft [9m] span support point. This is an issue in all materials, but the combination of initial camber during construction and deflection after completion can be crucial.

PART FOUR

MASONRY

Otaniemi Technical College, Otaniemi, Finland (photo by Michele Chiuini)

40 MATERIALS AND PROPERTIES

40.1 Masonry units

Masonry is built by bonding masonry units (brick, block, stone, or other) with mortar. The structures primarily addressed by current codes are made of clay or concrete units bedded in mortar. The structural properties of masonry depend on the combination of these two materials.

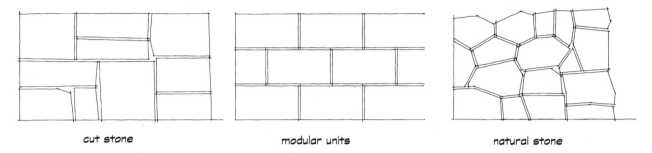

cut stone   modular units   natural stone

Fig. 40.1 Masonry construction can be expressed by a variety of different materials.

Masonry units are classified according to the material. Table 40.1 lists the primary classifications based on structural characteristics.

Table 40.1 Criteria of classification of concrete masonry units

Weight	Normal stone concrete mix, 144pcf [22,620N/m³]
	Medium concrete mix, with some air entraining or lightweight aggregate,
	    110pcf [17,270N/m³]
	Lightweight concrete mix, with air entraining or lightweight aggregate,
	    80pcf [12,560N/m³]

Strength	The compressive strength of the concrete mix used to manufacture the unit Comparable to the strength of normal or precast concrete, 3,000psi [21MPa] or greater
Shape	Solid: cross-sectional area is 75% to 100% solid. Hollow: voids may be more than 25% of the cross-sectional area.
Permeability	Grade N may be exposed to water in below grade or exposed applications. Grade S may be used only when protected from moisture.

Masonry units should be dry. Concrete units in particular can have shrinkage problems if laid when wet, and should be protected from rain during construction.

Calcium silicate bricks are made with a sand and lime (in place of cement) mixture and have a lower initial strength. Their strength will increase over time as the lime continues calcification when exposed to air.

Clay or shale units include bricks and hollow clay blocks (called tiles) for structural use. Some hollow clay blocks have been used for nonbearing partitions and as fillers for concrete slabs. The classification criteria are similar to those used for concrete units, with the obvious exception of shape and sizes. Up to 40% cores are permitted for hollow clay units.

There are different ways of indicating the sizes of units for the purpose of design:
(a) Specified size is the size required in design.
(b) Nominal size is the specified size plus the thickness of the joints.
(c) Actual size is the specified size plus or minus the production tolerance.

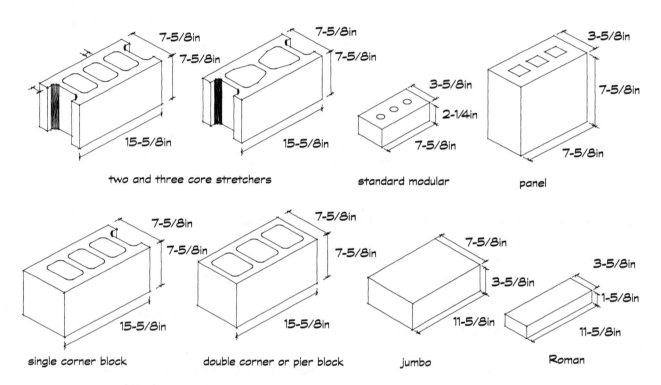

two and three core stretchers   standard modular   panel

single corner block   double corner or pier block   jumbo   Roman

Fig. 40.2 Various common modular masonry units

There are three types of mortar joints (Fig. 40.3):
(a) Bed joint
(b) Head joint
(c) Collar joint

Bed and head joints can have the following thickness: 1/4in (min.), 3/8in, 1/2in, and 5/8in (max.) [6mm, 9mm, 13mm, and 16mm]. It is common to use 3/8in [9 or 10mm] joints for both brickwork and concrete masonry units, unless thicker joints are required to embed reinforcement or anchors.

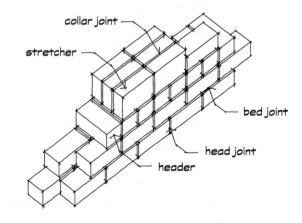

Fig. 40.3 Joint types and unit names as a function of orientation

40.2 Mortar and grout

Mortar is a mixture of cementitious materials, additives (but not in all cases), aggregate, and water.

The mixing of cementitious materials, aggregate, and water is done in volume, in the proportions indicated in Table 40.1. For instance, type M cement-lime mortar is obtained by adding 1 part of Portland cement and 1/4 of lime to 3(1 + 1/4) = 3-3/4 parts of sand.

For proper workability, mortar must have a plastic consistency. Lime retains water and is added to the mixture for good workability, but takes a long time to reach its maximum strength, and its proportion must be limited. Mortar or grout mixed at the job site should be mixed for a period of time not less than 3 minutes and not more than 10 minutes in a mechanical mixer. It should be used within 2-1/2 hours after the initial mixing water has been added, since the cement has started setting by this time. An important issue regarding mortar color is related to when the joint is tooled. If the joint is tooled early, then the mortar color will be light; if tooled late, the same mortar will be dark. When the architectural sample panel is laid, it is important to note the joint tooling time to ensure that the finished wall will match the panel.

Mortar should not be used in freezing conditions. Mortar temperature should be between 40°F and 120°F [*4°C and 49°C*]. When the mean daily air temperature is 40°F to 32°F [*4°C and 0°C*], masonry should be protected from rain or snow for 24 hours with a weather-resistant membrane. Below the freezing point, insulating blankets and supplementary heating (if the mean temperature is below 20°F [*minus 7°C*]) are to be added.

If mortar properties are specified by mortar proportions, then test information is not generally required; however, if the strength properties are specified, then tests will need to be administered to ensure these properties are provided. Water can be added to mortar already mixed (this is referred to as **tempering**) if water is lost because of evaporation. Depending on the mix and on the site conditions, the mortar will set after 1 hour or more. Water should not be added after the mortar has begun to harden, since it will not be part of the chemical reaction and will evaporate, leaving a porous and **friable** (crumbly) mortar.

Masonry walls can be built with two leaves or wythes with the space in between filled with grout. Grout's purpose is to create a solid masonry system or to bond the masonry with the reinforcing bars. **Grout is NOT mortar**; grout is a mixture of cementitious materials: lime, and Portland cement (or a combination of the two) with sand and sufficient water to flow in cavities. Its minimum compressive strength must be 2000psi [*14MPa*].

40.3 Mortar strength

Masonry mortars are available in five designations: M, S, N, O, and K. They range in strength from M as strongest to K as weakest.

Type M mortars are generally used in earthquake zones where high compressive loads may be encountered and high bond strengths are required.

Types S and N are the typical structural mortars, with N being the general-purpose mortar used for above-grade and S for below-grade applications.

Type O can be used for nonbearing partitions, but cannot be subjected to freezing.

Type K mortar is generally used only for tuckpointing (repairing mortar) on older buildings.

You might think you should always specify the strongest mortar and have a strong building. Unfortunately, the stronger mortars are less flexible and are more likely to crack under freezing and thawing cycles. It's a good general rule to use a weaker mortar if it's sufficiently strong and not overspecify strength.

The combination of units and mortar of specified strengths results in different specified compressive strengths; f_m' of masonry is shown in Table 40.2. For instance, 8,000psi [56MPa] brick with type S mortar gives $f_m' = 3,350$psi [23MPa]. These values are not used directly in design, but they are the basis for the calculation of the allowable stresses for compression, tension, shear, and flexure.

Table 40.2 Compressive strength[a] of masonry f_m'

Compressive strength of clay masonry units	Specified compressive strength of masonry, f_m	
	Type M or S mortars	Type N mortar
14,000psi +	5,300 psi	4,400psi
12,000psi	4,700psi	3,800psi
10,000psi	4,000psi	3,300psi
8,000psi	3,350psi	2,700psi
6,000psi	2,700psi	2,200psi
4,000psi	2,000psi	1,600psi
Compressive strength of concrete masonry units		
	Type M or S mortars	Type N mortar
4,800psi +	3,000psi	2,800psi
3,750psi	2,500psi	2,350psi
2,800psi	2,000psi	1,850psi
1,900psi	1,500psi	1,350psi
1,250psi	1,000psi	950psi

[a]Based on the strength of masonry units

Compressive strength of solid clay masonry units is based on gross area. Compressive strength of hollow clay masonry units is based on the minimal net area.

40.4 Construction

Some precautions must be followed for a good execution on-site of masonry structures, in particular for mortar joints.

40.4.1 Joints

The mortar should be sufficiently plastic and units should be placed with sufficient pressure to extrude mortar from the joint and produce a tight joint. The thickness of the joints should be not less than 1/4in [6mm] and not more than 5/8in [16mm]. Head joints must also be filled with mortar, unless the unit has a beveled end (Fig. 40.4). In this case, the unit ends can butt tightly so that the beveled joint can be filled with grout (minimum thickness 5/8in [16mm]).

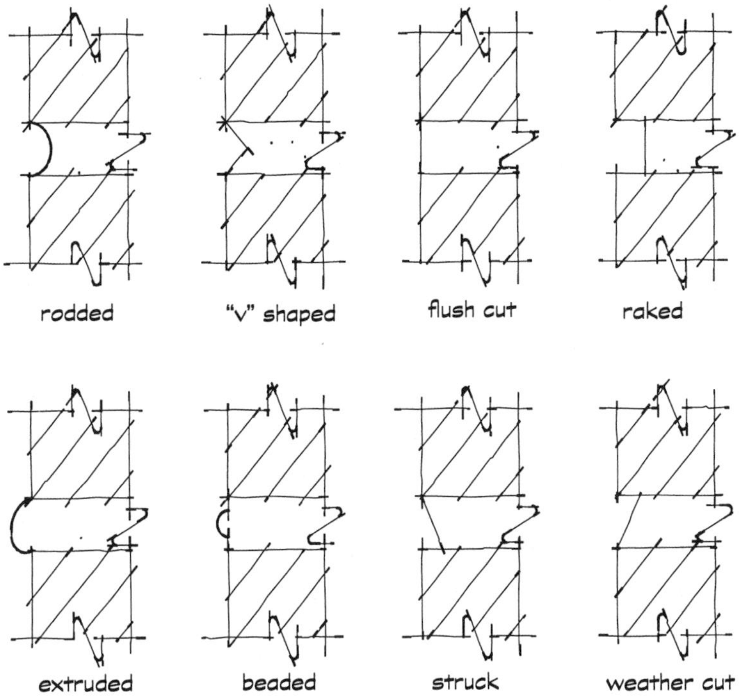

rodded   "v" shaped   flush cut   raked

extruded   beaded   struck   weather cut

Fig. 40.4 Common joint types used in masonry construction

The execution of joints is of paramount importance to achieve the required structural strength. The joint can be:

Troweled: excess mortar struck off with a trowel

Tooled: shaped with a tool

The execution and type of joint affect the durability and long-term strength of the masonry. Troweled joints do not compress the mortar the way tooled joints do; consequently, they tend to be less weather-resistant. The classic **weather** joint, a troweled, outward-sloping joint has the right geometry, but doesn't compact the mortar, which makes it less weather-resistant than a traditional tooled joint.

40.4.2 Grout

In grouted masonry, grout must fill all cavities, and the spaces to be grouted must be cleaned of mortar droppings. Minimum size of spaces or cells is given in Table 40.1. For pours over 5ft [1.5m] and lower if necessary, cleanouts are left at the bottom course 32in [813mm] o.c. or less. The cleanouts are then sealed after inspection and cleaning for grouting. Between grout pours, a horizontal key is formed by stopping the grout at 1-1/2in [38mm] below a mortar joint. **Remember, grout is NOT mortar**; it is basically a high slump (8 to 11in) [203 to 279mm] concrete. As you may remember from our concrete discussion (humor us and say you do), the water-cement ratio governs the strength, so you might want to use a stiffer grout. Unfortunately, the dry block will tend to draw moisture from the grout, so the high moisture content is necessary both for placement in tight places and to ensure proper hydration (curing) of the grout.

Grout should be placed within 1-1/2 hours from initial contact with water, before it starts setting. The supervising architect or engineer can require testing of grout prisms; a minimum of one test prism should be taken for every 5000ft² [465m²] of wall area, in addition to a test prism prior to construction. A test prism is made in a way to simulate the grout in the masonry, so the mold is made up of clay or concrete masonry units of the same type used in the project, and kept moist with damp paper or cloth until the test. The specimen is usually as small as a 2X2in [51X51mm] cross-section and is tested in a laboratory after 48 hours.

40.5 Design requirements

This chapter deals with two structural types: walls and columns.

Walls and columns can be designed as solid walls and columns, either bonded or grouted, and cavity walls, either double wythe with ties or multiwythe with ties.

All these types of construction must provide structural continuity. In solid brick walls, bond courses are provided every sixth course with headers (common bond) or in similar ways. An adequate grout joint can achieve the same result. Metal or plastic ties can also be used, but they do not transfer vertical structural loads. The wythes of cavity walls or grouted walls can be built with different materials; this has to be taken into account in the structural calculations. It is common, for instance, to design a cavity wall with a CMU inner wythe and a brick outer wythe; with the air space in between partially filled with insulation (Fig. 40.8). In this case, however, the outer wythe is usually not load-bearing. In some cases both wythes are bearing or working structurally (in compression, shear, or flexure).

Metal ties, shelf angles, and anchors are used to

(a) bond together two or more masonry wythes
(b) tie a masonry wall to a different structural element (e.g., a brick wall to a steel column)
(c) tie a different structural member to a masonry structure (e.g., a timber plate to a brick wall).

Metal ties and anchors are made of steel with a minimum yield stress of 30,000psi [207MPa]. If not completely embedded in mortar, they must be protected with zinc or other corrosion-resistant coating. The best material is stainless steel or plastic (polyethylene), since other metals will corrode after 10 to 20 [10 to 20] years.

Remember that when anchoring a brick wythe to a concrete masonry or concrete wall the brick expands and the block and concrete shrink, so any anchors must provide for this movement in opposite directions. Figure 40.8 shows the minimum requirements for placement of ties. The minimum tie size is 3/16in [4.8mm]. The requirements for placement of ties in cavity walls is as shown in Table 40.3.

When horizontal prefabricated reinforcement is placed in joints, the following requirements apply:

(a) Maximum vertical spacing = 16in [406mm]
(b) Minimum cover and bar spacing = 5/8in [16mm]
(c) Maximum bar diameter = No. 2 diameter = 0.25in [No. 6M = 6mm], with a joint thickness = 2 × bar diameter.

40.5.1 Solid walls

Solid walls must be bonded by masonry headers extending not less than 3in [76mm] into the backing and constituting 4% of the wall surface of each face. Full-length headers can be spaced at 24in [610mm] maximum, horizontally and vertically. It is common to lay header courses every sixth or seventh course, depending on the size of the units; this corresponds to a vertical spacing of 16in [406mm], which allows modular coordination, for instance, between facing bricks and CMU backing (Fig. 40.5). If hollow units are used, the vertical spacing of header courses on stretcher courses must not exceed 34in [864mm]. In stone masonry, the bond stones must be uniformly distributed and correspond to an area of at least 10% the wall area. Rubble stone 24in [610mm] thick must have bond stones at 3ft [0.9m]o.c; if more than 24in [610mm] thick, one bond stone is re-quired for 6ft² [0.56m²] of wall (Fig. 40.6).

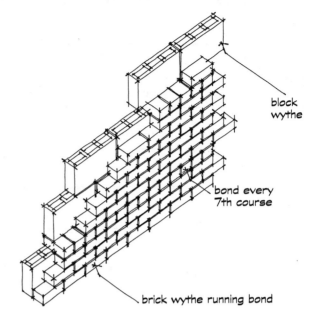

Fig. 40.5 Bonded brick and block wall

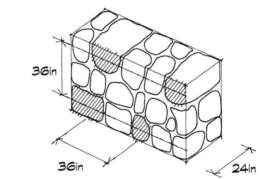

Fig. 40.6 Bonded stone wall

40.5.2 Cavity walls

Requirements for ties:

> Minimum tie diameter of 3/16in [4.8mm]
> One tie for each 4-1/2ft² [0.42m²] of wall
> Maximum spacing of 24in [610mm] horizon-tal, and 36in [914mm] vertical
>
> Tie shape: Z shape, 2in 90° hook in solid units and rectangular shape for hollow units
>
> Three feet [0.9m] o.c. around openings, 12in [254mm] from openings

Requirements for prefabricated reinforce-ment:

> One No. 9 gage cross wire, each 2.67ft² [0.25m²] of wall area
> Maximum vertical spacing of 16in [406mm]

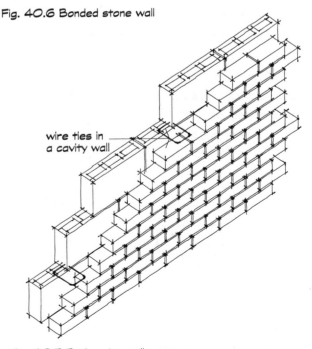

Fig. 40.7 Tied cavity wall

Table 40.3 Tie placement in cavity walls

Number of ties		Cavity size/type	
1 per 4-1/2ft²	[1 per 0.42m²]	< 3in	[<76mm]
1 per 3ft²	[1 per 0.28m²]	3 - 4-1/2in	[76mm to 114mm]
1 per 2ft²	[1 per 0.19m²]	multiwythe	

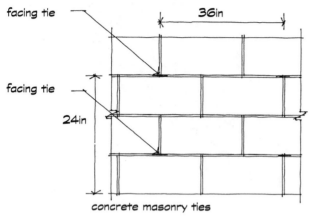
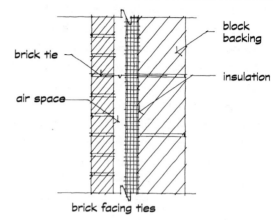

Fig. 40.8 Tie placement in walls

40.6 Expansion/control joints and reinforcing

It is important to note that two very different types of joints are used in brick and concrete masonry construction due to the nature of movement expected. Expansion joints are primarily used in brick construction since the brick will expand or grow about 5/8 inch for every 100 feet of wall (or 0.0005 in/in) [0.52mm/meter], while concrete masonry units will shrink as they age and require cracking "control" joints. The facing and backing of multiple-wythe walls must be bonded together with headers or ties. Any reinforcing should be held back 5/8in [16mm] from the face of the wall to ensure adequate mortar protection.

40.6.1 Anchorage
Intersecting walls can be bonded with different methods.
Alternating 3in [76mm] overlaps.
    Twenty-four-inch [610mm] long steel connectors, and 4ft [1.2m] of vertical spacing
    Thirty-inch [762mm] long No. 9 gage joint reinforcement

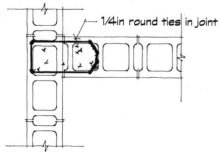

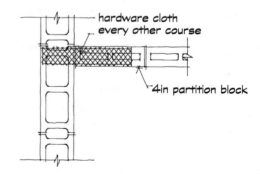

Fig. 40.9 Anchorage of intersecting walls

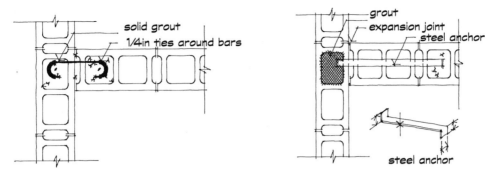

Fig. 40.10 Anchorage of intersecting walls

Other types of anchorage include:

Floor joists of steel or wood: steel anchors 6ft [1.8m] o.c. on all sides

Roof structures: 1/2in [12.7mm] diameter bolts, 6ft [1.8m] o.c., embedded 15in [381mm] into masonry or connected to reinforcement 6in [152mm] below the top of the wall

Structural framing: 1/2in [12.7mm] diameter bolts, 4ft [1.2m] o.c., embedded 4in [102mm] into masonry

Ladder or trussed-wall reinforcing should not be used spanning between concrete masonry and brick wythes. As discussed earlier, these materials move in opposite directions over time; consequently, a "flexible" connection should be made between them.

40.7 Flashing

Moisture penetration into a wall is a major cause of "failures" in masonry walls. The moisture can corrode ties, lintels, steel studs, and even structural steel. As internal moisture evaporates out through the masonry unit, dissolved chemicals are carried to the surface, where the moisture is evaporated, leaving the chemical residue, which causes the wall to bloom. This bloom (efflorescence) is the white substance you see on the surface of many masonry buildings. It can be cleaned, but if moisture continues to penetrate the wall, it will return. Flashing materials should penetrate out slightly beyond the exterior surface of the wall. Architects hate this, and frequently stop flashing before it **surfaces**. Unfortunately, this allows water to move around the end and reenter the wall. The ends of flashing are also major points of water penetration. If the flashing merely stops and doesn't have an **end dam**, some moisture will enter the wall at these points. Unfortunately, this is frequently the bearing point for lintel angles, so the first point of corrosion is at a critical point in the structural system. Remember also that corrosion is an oxidation of the steel and causes expansion of the material. This means the rusting component may also cause additional mortar failure and allow even more moisture to penetrate.

You may be tempted to use aluminum flashing since it is resistant to corrosion and relatively inexpensive, but, unfortunately, it reacts with cement and will quickly deteriorate. OOPS!

Remember: KEEP THE WATER OUT!

41 STRUCTURAL SYSTEMS

41.1 Masonry construction

Modern masonry construction consists of bricks, concrete masonry units, and stones bonded together by mortar. Stone structures, still common in the first part of the twentieth century, are now rare. Unfortunately (nostalgia speaks here), vaulting over large spaces has been replaced by less labor-intensive solutions, although it may still be an appropriate technology for countries that have inexpensive labor forces (Fig. 41.1). Load-bearing masonry is used today mostly for low-rise structures such as housing.

In modern masonry construction, a major evolution took place with the introduction of steel reinforcement. Masonry systems are now classified as either plain or reinforced. Masonry buildings have problems in seismic areas, due to the poor tensile strength of mortar joints. To respond to this weakness, reinforced masonry is used almost exclusively in seismic zones to provide the ductility necessary to resist high, reversing lateral forces. Masonry is currently used primarily in load-bearing walls and columns, although reinforced beams and lintels are occasionally found. These simple vertical systems are used in combination with other materials for horizontal spanning systems. We are not considering a filled lintel block as a "masonry" system since the block acts as formwork for the grouted structural interior.

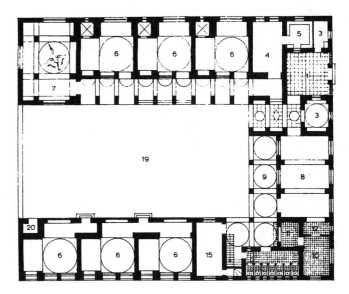

Fig. 41.1 Traditional masonry structures in Egypt

Unreinforced brick structures were used for tall buildings until the beginning of this century, and traditional stone masonry is still employed in the reconstruction of monuments. An understanding of traditional masonry structures is certainly important due to the large number of historic buildings present in all countries, where the architect may be called in to intervene for restoration or alteration work. In California, Frank Lloyd Wright experimented with textile block houses, made of concrete blocks and reinforcing bars in the joints (Fig. 41.2). Today, there are a great many types of block and brick used in low-rise residential and commercial construction. Lightweight concrete blocks with a high thermal insulation value have been in use in Sweden for many years; in the United States, the National Concrete Masonry Association (NCMA) has created a product research-and-development division that has devised adhesive bonds to save labor and create posttensioned walls. Prefabricated masonry panels are now being more commonly used to reduce the amount of on-site construction. These panels are produced by fully automated machines that lay head and bed joints, ensuring both very accurate placement and high strength.

Fig. 41.2 "Textile" block construction

The convenience of this material may also be its weakness; that is, the small modular dimension that makes it versatile and of human scale means that it does not require heavy construction equipment at the construction site but, on the other hand, it is labor-intensive. As with any other structural system, the decision to use masonry will depend on a balance of many factors, ranging from architectural philosophy to functional and cost considerations. Masonry is still one of the best materials to use for the construction of curved forms. Although reinforced concrete does not inherently have a "form," its forms are limited by the materials that are used to create its formwork. Even masonry vaults and arches are typically built on a formwork of wood that is removed after the vault or arch is complete and can be self-supporting. Masonry buildings can be built easily with relatively tight curves in plan since the material is produced in modules of small dimension. Examples of ideas range from the Jefferson wall of the University of Virginia to the houses of Mario Botta (Fig. 41.3).

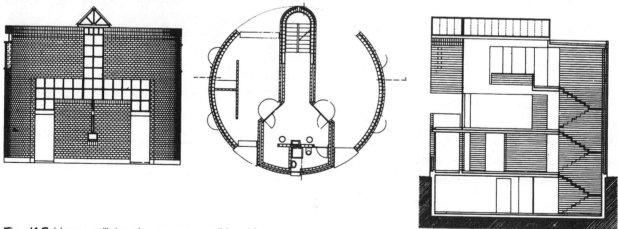

Fig. 41.3 House utilizing the curves possible with masonry

Now the downside: One of the major concerns with masonry, either as a single- or multiple-wythe system, is heat and cold transfer. Insulation within the cores of concrete masonry units is a beginning, but only that. Even with excellent core insulation, the webs of the units allow for direct transfer through the wall. In a multiple-wythe system, a space for the introduction of insulation and for moisture drainage is available, but the outer wythe on an exterior bearing wall becomes largely decorative. To use this wythe as a structural component, load must be applied through a bond beam, bearing plate, or some other technique that will distribute load to both wythes. If you accomplish this, you create a localized point of heat or cold transfer. This is a detailing problem that will eventually be solved, but at the present there is no clear, simple solution. On interior bearing walls, the thermal differential doesn't exist and the problem disappears.

41.2 System selection

We warned you! The good news is that this is the last time you'll see this section, so read it carefully and say a fond farewell.

Selection of a masonry system involves a complex set of interrelated issues, so there is no single answer. Since masonry construction for the most part is a vertical structural system, many of the horizontal issues are not as relevant, but nonetheless should be reviewed. You need to at least consider the following.

Spatial requirements

What are the volumetric requirements of the function that the space will house?

Is there a planning module or a leasing module that should be column-free or could be easily isolated? Parking garages, office structures, schools, hospitals, and retail establishments all have identifiable modules that are more desirable. Unfortunately, these are not absolute, perhaps with the exception of parking facilities, and must be identified during programming and schematic design stages.

Expansion

Is the spatial system likely to require frequent remodeling?

What is the likelihood of expansion of the system?

Will the material be available in the future if needed for expansion or remodeling?

> Even brick produced from the same material by the same plant may have some variation in color from the original. To assure a color match for future changes, the face brick should be purchased at the time of initial construction and stored. This, unfortunately, is an expensive proposal and is unlikely to happen. Creating details that hide or at least deemphasize the potential material difference is the key.

Integration of systems

Will the environmental/lighting systems require frequent change?

Does the spatial system require isolated volumes of space: for temperature, sound, security, utilities management, and so on?

Ease of erection/construction

What is the availability of skilled labor?

Are material or component production facilities easily accessible?

Fire resistance

What are the fire-rating requirements of the building as a whole?

What are the isolation characteristics of zones of the building?

What are the patterns of access required for emergency egress?

Soil conditions

What is the nature of the soil on the site?

> Chemical reactions with masonry products?
>
> Bearing capacities? Masonry structures are relatively heavy structures. This may be a disadvantage for foundation design and an advantage for thermal balancing. Masonry products do not have a good inherent insulation quality, so you will have significant thermal transfer through structural components. A masonry structure will generally weigh more than its equivalent in steel and about the same as its concrete counterpart. This could be a major consideration if the building is supported by weak soils that would require a significant investment in foundations.

41.3 Masonry systems

We will focus on the design of simple masonry walls and columns as the most likely structures that you might be asked to design in ordinary construction. Masonry is used in the United States for residential foundation and basement walls, for load-bearing walls in low-rise construction, and for non-load-bearing walls subjected to lateral loads. These are common in commercial structures that use masonry as infill around or facing on a steel frame.

Although masonry beams are seldom used in construction, walls will bend under lateral load and consequently flexure is sometimes an important consideration in the design of wall systems. Even the simplest building will need to resist lateral loads, and these loads can be a major determinant in the structural configuration. With a wall-bearing system using horizontal diaphragms (floor or roof structures) to transfer horizontal forces, the lateral loads are resisted at the top of the walls. The walls, depending on their height-to-width ratio, will behave like a short or long cantilever (see Fig. 41.4(a) and 41.4(b)), where the controlling factor in the design can be bending or shear as opposed to the expected compression due to gravity loads. These shear walls may be part of the exterior enclosure system, part of the interior spatial definition, or part of the vertical circulation cores in a multistory solution.

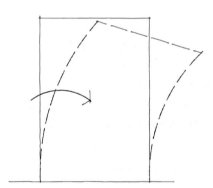

Fig. 41.4a Long cantilever as a bending member

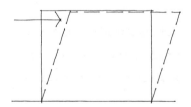

Fig. 41.4b Short cantilever as a shear member

Tall masonry buildings are rare, but they can be economically built even in earthquake areas. Residential buildings and hotels over 10 stories have been built with load-bearing masonry. The advantage in these building types is the use of masonry load-bearing walls for structure, acoustic insulation, and fire resistance. In high-rise masonry buildings, that may be relatively narrow in relation to their height, it is important of create plan organizations which provide sufficient shear wall length parallel to the potential direction of lateral force. Interior shear walls may allow some of the exterior walls to become nonstructural, permitting greater architectural freedom on the facade. Other design considerations that increase effectiveness of shear walls are returns at the ends of walls that act like flanges of beams in bending situations and grouping openings in walls in concentrated areas allowing a greater continuous solid wall length. When a shear wall is interrupted by openings, the openings effectively split the wall into two separate cantilevers. Since these cantilevers are acting as beams, they will be significantly more effective if attached together, similar to the advantage gained by structurally connecting plywood floor sheathing to a wood joist. The two cantilevers can be connected by stiff beams or spandrels to achieve "composite" action. These beams or spandrels act like the nails in a composite wood system to resist the horizontal shear that would be created between multiple, small cantilevered beams (walls).

The need to provide lateral stability in both directions with cross walls or buttresses has to be considered in the architectural organization. A tubelike configuration, similar to that used for tall steel structures, can provide high torsional and bending resistance with the wall only at the perimeter, providing an open plan. A combination of exterior shear walls with simple interior columns has a similar advantage of opening up the interior and creating more flexible architectural space.

42. EMPIRICAL DESIGN OF WALLS

42.1 Limitations of empirical design

The empirical method is a simplified procedure provided by the code, which usually gives a very conservative and sometimes uneconomical result. It was developed by "experience" over the years and used for relatively small-scale masonry bearing walls and partitions. It is useful to illustrate it in order to highlight some of the issues that apply to the design of a masonry structural system. This procedure can be used for buildings meeting the following requirements.

Buildings in seismic zones 0 and 1

With a wind speed less than 80mph [130km/h] or a pressure of 25psf [1190Pa]

A maximum building height of 35ft [10.7m]

This procedure is probably the safest, most conservative approach that an architect can use for preliminary design. Later, we'll compare the two solutions of a single wall system and you'll see the level of conservatism you're accepting.

Fig. 42.1 Contemporary bearing wall construction at Nottingham, England

The empirical design procedure is based on three main considerations:

1. Designs do not exceed the working compressive stress.

2. Lateral stability criteria are met.

3. Unsupported height or length requirements are not exceeded.

42.2 Working stress

Working stress is $f_m = P/A \leq$ allowable f_m

where

P = axial live and dead load

A = gross cross-sectional area of the wall based on specified (not nominal) dimensions

The allowable stresses are given in Table 42.1. No acknowledgment of reinforcement is included in this method.

Table 42.1 Allowable compressive stresses for empirical design of masonry

Compressive strength of unit based on gross area		Allowable compressive stress using			
		Type M or S mortar		Type N Mortar	
Solid masonry of brick (clay or concrete)					
8,000+ psi	[55.1+ MPa]	350psi	2.4MPa	300psi	2.1MPa
4,500psi	[31.0MPa]	225psi	1.6MPa	200psi	1.4MPa
2,500psi	[17.2MPa]	160psi	1.1MPa	140psi	0.97MPa
1,500psi	[10.3MPa]	115psi	0.79MPa	100psi	0.69MPa
Grouted masonry (clay or concrete)					
4,500+ psi	[31.0+MPa]	225psi	1.6MPa	200psi	1.4MPa
2,500psi	[17.2MPa]	160psi	1.1MPa	140psi	0.97MPa
1,500psi	[10.3MPa]	115psi	0.79MPa	100psi	0.69MPa
Solid masonry of solid concrete units					
3,000+ psi	[20.7+MPa]	225psi	1.6MPa	200psi	1.4MPa
2,000psi	[13.8MPa]	160psi	1.1MPa	140psi	0.97MPa
1,200psi	[8.3MPa]	115psi	0.79MPa	100psi	0.69MPa
Hollow laod-bearing units					
2,000+ psi	[13.8+MPa]	140psi	0.97MPa	120psi	0.83MPa
1,500psi	[10.3MPa]	115psi	0.79MPa	100psi	0.69MPa
1,000psi	[6.9MPa]	75psi	0.52MPa	70psi	0.48MPa
700psi	4.8MPa	60psi	0.41MPa	55psi	0.38MPa
Hollow walls (cavity or bonded) solid units[a]					
2,500+ psi	17.2+MPa	160psi	1.1MPa	140psi	0.97MPa
1,500psi	10.3MPa	115psi	0.79MPa	100psi	0.69MPa
Hollow units		75psi	0.52MPa	70psi	0.48MPa
Stone ashlar masonry					
Granite		720psi	5.0MPa	640psi	4.4MPa
Limestone or marble		450psi	3.1MPa	400psi	2.8MPa
Sandstone or cast stone		360psi	2.5MPa	320psi	2.2MPa
Rubble stone masonry					
Course, rough, or random		120psi	0.83MPa	100psi	0.69MPa
Unburned clay masonry		30psi	0.21MPa	-----	

Linear interpolation may be used for units with compressive strengths between those listed in the table.

[a] In hollow walls, the net width of masonry carrying load is used. (Gross thickness minus the cavity, or if bearing only on a single wythe, use that single-wythe thickness.)

42.3 Lateral stability

The lateral loads acting on the wall are transferred to the horizontal supports at the top and bottom at the wall, and the vertical supports parallel to the direction of the lateral load (see Chapter 43, 43.3). These vertical supports are usually other masonry walls that resist lateral loads predominantly in shear, moment, or with a combination of moment and shear, and are referred to as "shear walls." The spacing of shear walls is determined by the L/t ratio according to Table 42.2 and by the shear wall spacing requirements given in Table 42.3.

Table 42.2 Wall lateral support requirements

Type of construction	Minimum L/t or h/t
Bearing walls	
Solid or grouted units	20
All other units	18
Nonbearing walls	
Exterior	18
Interior	36

Table 42.3 Shear wall spacing requirements

Floor or roof construction	Maximum ratio of shear wall spacing to shear wall length
Cast-in-place concrete	5:1
Precast concrete	4:1
Metal deck with concrete fill	3:1
Metal deck with no fill	2:1
Wood diaphragm	2:1

42.4 Unsupported height or length

The wall height or length is limited by the L/t ratios in Table 42.3, where L is the net length between vertical supports (cross walls) or between horizontal supports (floor structures). This is a limitation of the slenderness and of the clear span of the wall, which would potentially bend between supports under lateral loads. Since bending resistance (tensile capacity) of masonry is poor, the empirical method limits the span in relation to the thickness of the wall.

Both the height and length of the wall are referred to as the "length" L in Table 42.3. Note that the code requires the L/t ratio requirement be met by **either** the wall length **or** the wall height, not both.

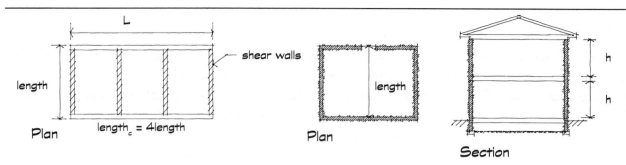

Fig 42.2 Cumulative length of shear walls   Fig. 42.3 Unsupported length of walls

Table 42.4 Maximum unbalanced fill for various wall thicknesses

Foundation wall construction	Nominal thickness	Maximum unbalanced fill
Masonry of ungrouted hollow units	8in 10in 12in	5ft 6ft 7ft
Masonry of solid units	8in 10in 12in	5ft 7ft 7ft
Masonry of hollow units (fully grouted)	8in 10in 12in	7ft 8ft 8ft

Masonry of reinforced hollow units
   No. 4 bars /grout @ 24in o.c.
   Located not less than 4-1/2in
   from the pressure side of wall
	8in	7ft

Table 42.5 Minimum thickness of masonry walls

Structural Type		Thickness	
Load-bearing masonry	Multistory	8in nominal	[203mm]
	One-story, 9ft + 6ft gable	6in nominal	[183mm]
Cantilever wall	Solid masonry	h/t = 6	
	Hollow masonry	h/t = 4	
Parapet		h/t = 3; 8in [203mm] nominal	
Stone (natural or cast)			
load-bearing		16in	[406mm]
Glass block		3in	[76mm]
non-load-bearing		3in	

Example 42.1

A 10ft high wall made of hollow CMU f_m' = 1000psi and S mortar, nominal thickness 8in, and will support a floor load (DL+LL) = 4200lb/ft of length. From the manufacturer's catalog, weight of an 8in wall = 60psf and specified thickness t = 7-5/8in.

Load at the base of the wall P = W + Q   (where Q is the superimposed load from the floor system)

Weight of the wall W = (60psf)(10ft) = 600lb/ft of wall

P = 600lb + 4,200lb = 4,800lb

Working stress f_m = P/A = (4800lb)/(12in)(7.625in) = 52.5psi [*0.36MPa*]

Allowable stress (Table 42.1) = 75psi [*0.52MPa*] OK!

The h/t ratio (Table 42.3) is 18 (bearing wall of hollow units), so the minimum thickness of the 10ft wall is t = h/18 = (10ft)(12in/ft)/18 = 6.67in [*169mm*], which is smaller than the specified thickness, so it is also OK!

42.5 Lateral support and shear walls

The spacing of shear walls is determined by both the L/t ratio and the shear wall spacing requirements in Table 42.5. The spacing acknowledges the idea that the shear walls are supporting the floor or roof structural planes, which behaves like a beam loaded by the lateral loads acting on the walls. To avoid excessive deflection, the plane's depth must be larger for weaker materials. With equal spans, a wood floor must have a depth or width equal to half the span. A more rigid concrete floor system can have a depth of 20% of the span (Fig. 42.4). In each case, this will be the length of the shear wall as well. A sufficiently rigid floor diaphragm will safely support the walls under lateral loads.

The shear walls must meet the following requirements:

Minimum thickness = 8in [*203 mm*] nominal

Cumulative length of shear walls

$L_c \geq 0.4L$

where L is the long side of building perpendicular to shear walls

The maximum spacing of shear walls can be found in Table 42.3.

Example 42.2

A row of three townhouses is designed with 28ft-long party walls spaced 20ft apart. The floor structure consists of engineered wood joists with plywood sheathing at an 8ft clear height. The wall construction is grouted and made of concrete units, 8in nominal thickness. Are the shear walls adequate?

In the direction of loads perpendicular to party walls, the L/d ratio is (28ft)/(20ft) = 1.4:1, assuming the floor of each residential unit acts independently. In the other direction L/d is even smaller. The cumulative length of shear walls is 28ftX4 = 112ft and the total length L = 3 × 20ft = 60ft.

The cumulative length of the walls of 112 ft is > 0.4 × 60ft = 24ft   OK!

The minimum thickness of the walls is determined by h/t = 20, or t = (8ft)(12in/ft)/20 = 4.8in. Since we have an 8in thickness, this is OK as well. In fact, we could use ungrouted construction, with a minimum thickness of t = (8ft)(12in/ft)/18 = 5.33in.

Problem 42.1

A 10ft high brick wall has to be designed around a park for security reasons. Calculate the spacing of the brick piers forming the vertical supports if the brick wall is single wythe construction or two-wythe solid construction with a grouted collar joint. Which solution is more economical in terms of quantity of materials if the piers are 8in × 8in nominal in both cases?

Answer:

Single wythe: piers at 6ft o.c.
Double wythe: piers at 12ft o.c.
The horizontal cross-sectional area of a 12ft long wall is 704in² nominal for the single wythe and 1216in² nominal for the double wythe, consequently, the single wythe wall is more economical.

Problem 42.2

A foundation wall supported on a concrete footing is 5ft high and carries a load of 950plf. If it is built using 8,000psi hollow concrete blocks and N mortar, what thickness should the wall be?

Solution:

(5ft)(12in/ft)/18 = 3.33in; t = (950plf)/(120psi)(12in) = 0.7 in; t = 8in minimum with a 5ft backfill; therefore, use an 8in block.

Problem 42.3

A 12in nominal grouted concrete wall, 20 ft high, supports a precast concrete floor system. At which spacing should the shear walls be placed if the shear walls are 23ft long? What is the load-bearing capacity of the wall if it is built with 3,000psi units and S mortar?

Solution:

Spacing: length = 4:1; therefore, spacing = 4(23ft) = 92ft
P_a = (225psi)(11.625in)(12in) = 31,387plf

METRIC VERSIONS OF ALL EXAMPLES AND PROBLEMS

Example M42.1

A 3m-high wall made of hollow CMU $f_m' = 6.9MPa$ and S mortar, nominal thickness 203mm, will support a floor load (DL+LL) = 61,295N/m of length. From the manufacturer's catalog, weight of an 203mm wall = 293kg/m² = 2874N/m² and specified thickness t = 194mm.

Load at the base of the wall P = W + Q   (where Q is the superimposed load from the floor system)

Weight of the wall W = (2874N/m²)(3m) = 8622N/m of wall

P = 8622N + 61295N = 69917N

Working stress f_m = P/A = (69,917N)/(0.194m²) = 360,397N/m² = 0.36MPa

Allowable stress, (Table M42.1) = 0.52 MPa   OK!

The h/t ratio (Table M42.3) is 18 (bearing wall of hollow units), so the minimum thickness of the 3m wall is t = h/18 = (3,000mm)/18 = 167mm, which is smaller than the specified thickness, so this is also OK!

Example M42.2

A row of three townhouses is designed with 8.5m-long party walls spaced at 6.1m. The floor structure consists of engineered wood joists with plywood sheathing at a 2.4m clear height. The wall construction is grouted and made of reinforced concrete units, with 203mm nominal thickness. Are the shear walls adequate?

In the direction of loads perpendicular to party walls, the L/d ratio is (8.5m)/(6.1m) = 1.4:1, assuming the floor of each residential unit acts independently. In the other direction, L/d is even smaller. The cumulative length of shear walls is 4(8.5m) = 34m, the total length L = 3(6.1m) = 18.3m. The cumulative length of the walls = 34m > 0.4(18.3m) = 7.32m OK!

The minimum thickness of the walls is determined by h/t = 20 or t = (2,400mm)/20 = 120mm. Since we have a 203mm nominal thickness, this is OK as well. In fact, we could use ungrouted construction, with a minimum thickness of t = (2,400mm)/18 = 133mm.

Problem M42.1

A 3m high brick wall has to be designed around a park for security reasons. Calculate the spacing of the brick piers forming the vertical supports if the brick wall is single wythe construction, or two-wythe solid construction with a grouted collar joint. Which solution is more economical in terms of quantity of materials if in both cases the piers are 203mm × 203mm?

Answer:

Single wythe: piers at 1.8m o.c.

Double wythe: piers at 3.6m o.c.

The horizontal cross-sectional area of a 3.6m-long wall is 0.45m² nominal for the single wythe and 0.77m² nominal for the double wythe; consequently, the single wythe wall is more economical.

Problem M42.2

A foundation wall supported on a concrete footing is 1.5m high and carries a load of 13,864N/m. If it is built using 55.1MPa hollow concrete blocks and N mortar, what thickness should the wall be?

Solution:

(1,500mm)/18 = 83mm; t = (13,864N)/(0.83MPa)(1000mm) = 16.7mm; t = 203mm minimum with 1.5m backfill; therefore, use a 203mm nominal block.

Problem M42.3

A 305mm nominal grouted concrete block wall, 6m high, supports a precast concrete floor system. At which spacing should the shear walls be placed if the shear walls are 7m long? What is the load-bearing capacity of the wall if it is built with 31MPa units and S mortar?

Solution:

Spacing : length = 4:1, therefore spacing = 4(7m) = 28m

P_a =F(A) = (1.6MPa)(295mm)(1000mm) = 472kN/m

43 WORKING-STRESS METHOD

43.1 Unreinforced masonry

With the working-stress method, reinforcement can be taken into account in the design for all types of stresses: compression, tension, bending, and shear. There is a fundamental difference in this method between unreinforced and reinforced masonry. In unreinforced masonry, there is no steel reinforcement (did you already guess that?), or if there is, it is not included in the calculations. This makes sense if you think about a common masonry wall that includes some reinforcing to improve the ductility and reduce cracking; however, this steel may be placed in a way that does not contribute to the bending resistance of the wall. In this case, the working-stress method allows us to use the tensile strength of the mortar to resist bending. On the other hand, if a wall is designed as a reinforced bending member, the steel can be included in calculations of tensile bending stresses, and no tensile resistance of the masonry is considered.

The working-stress method takes into account the slenderness of the wall in the stress formula by introducing a factor h/r, where h is the unbraced height and r is the radius of gyration.

There are two allowable axial compression-stress equations depending on the slenderness ratio of the section. You'll note that these are similar in basic form to those used for other axially loaded members.

$$F_a = 0.25f'_m\left(1 - \left(\frac{h}{140r}\right)^2\right) \quad \text{for short columns h/r < 99}$$

$$F_a = 0.25f'_m\left(\frac{70r}{h}\right)^2 \qquad \text{for long columns h/r > 99}$$

where

f'_m = masonry strength   (Table 40.2)
h = effective height

The compressive stress, calculated as F = P/A, must not exceed the allowable stress F_a. The r values can be easily calculated using the approximations from R/C design:

for a rectangular section, r = .3 × minimum dimension
for a circular section, r = 0.25 × diameter

43.2 Concentrated loads on walls

When loads are not uniformly distributed on walls or columns, the first consideration is to ensure that the bearing stress is not excessive. This may occur at locations where beams or columns bear on a masonry wall.

The allowable masonry bearing stresses:
When a member bears on the full area of a masonry element is
$$F_p = 0.26f'_m$$
When a member bears on 1/3 or less of a masonry element is
$$F_p = 0.38f'_m$$

The actual bearing stress may be calculated by using our old friend F = P/A, where A is the direct bearing area (Fig. 43.1).

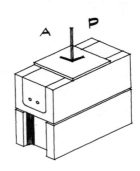

When concentrated loads are spaced on a wall, the effective area of wall supporting each load is $A_e = B + 4t$, where B is the length of the support, and t is the actual thickness of the wall (Fig. 43.2). These look suspiciously like the ones you saw in concrete walls in Chapter 37.

Fig. 43.1 Bearing area

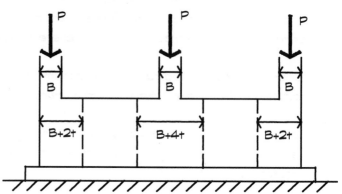

Fig. 43.2 Concentrated load distribution on walls

43.3 Design of shear walls

Shear walls take the lateral loads from the horizontal diaphragms (floors and roofs) and transfer them to the foundations. The shear wall behaves like a cantilever. A short cantilever is generally controlled by shear stresses in the masonry, and this is the common case of a shear wall that is long in relation to the height, as shown in Fig. 43.3.

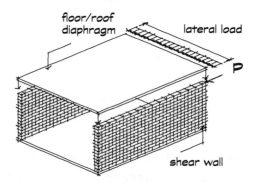

Fig. 43.3 Masonry walls acting in pure shear

The behavior of a shear wall becomes more complex in the case of a tall building or in walls having significant openings (Fig. 43.4a).

The location of shear walls affects the behavior of the building in torsion, which is critical in high seismic zones (Fig. 43.4b).

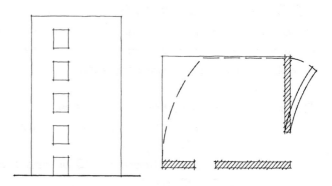

Fig 43.4a Tall shear walls   Fig. 43.4b Plan distorsion

43.4 Reinforced walls

Walls of reinforced masonry are effectively designed as columns in a manner similar to a reinforced concrete wall. To do this, consider a column section 1ft long or a width equal to B = 4t, as in Section 43.3 and use the procedure explained for columns, or compression members in general, for wall design.

Reinforced walls include vertical bars for axial and flexural stress, and horizontal bars for shrinkage. Ties are not required, even if walls are effectively columns when loaded in compression.

43.5 Column and pilaster construction

Columns built with solid or hollow masonry units have to be grouted and reinforced with a minimum of four bars and a minimum steel area A_s of 4% and 0.25% of the net column area:

$$0.0025A_n < A_s < 0.04 A_n$$

The net masonry area A_n is generally equal to the actual column cross-section.
These considerations are reminiscent of the reinforced-concrete column requirements.

The lateral ties must also meet the following requirements:
  Minimum tie diameter 0.25in [6.4mm] or a No. 2 [No. 6M] bar
  Vertical spacing of ties is the same as for reinforced concrete columns.
    Maximum spacing = 16 × bar diameter
          = 48 × tie diameter
          = the minimum column dimension

Columns in general have minimum requirements:
  Minimum dimension = 8in [203mm]
  Maximum L_e/d = 25, where L_e = effective height, and d = least dimension
  Columns should be designed for a load eccentricity equal to 0.1 × the column dimension on both sides. In reinforced-concrete columns, we simply included a numerical factor to compensate for eccentric loading.

Figure 43.5 shows the construction of a column or pilaster. Pilasters are columns connected to walls; their purpose is to carry concentrated loads and to resist lateral loads in bending. In this respect, they are really vertical beams. The walls connected to the pilaster can be considered as acting in composite action with the pilaster in resisting loads, both in compression and bending, if some provisions are satisfied:
  The masonry must be used in running bond.
  The width of the wall on each side of the pilaster that can be considered effective in

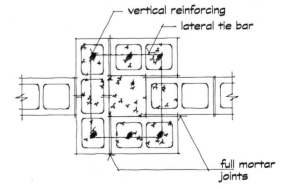

Fig. 43.5 Column/pilaster construction

resisting loads is not greater that six times the thickness of the wall. We must be coming to the end, since this is also a consideration for both wood and concrete for composite action.

43.6 Column design

Walls and columns can be designed with the working-stress method by the following equations giving the allowable stress for axial loads.

Reinforced masonry columns:

$$P_a = (0.25f'_m A_n + 0.65 A_s F_s)\left[1 - \left(\frac{h}{140r}\right)^2\right] \text{ for short columns } h/r < 99$$

$$P_a = (0.25f'_m A_n + 0.65 A_s F_s)\left(\frac{70r}{h}\right)^2 \qquad \text{for long columns } h/r > 99$$

where

f'_m = masonry strength (Table 43.1)
h = effective height
A_n = effective area of masonry
A_s = area of steel
$F_{sc} = 0.4F_y$ allowable compressive stress
for steel; maximum of 24,000psi [165MPa]

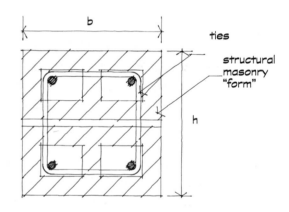

ties

structural masonry "form"

b

h

You should note that this equation is simply the formula for the working-stress method of an unreinforced wall with a component added to account for the inclusion of reinforcing. This is much like the reinforced-concrete column approach, which develops a formula adding the concrete load capacity and the steel load capacity with an adjustment for the slenderness ratio.

The bending stress is calculated considering a moment arm (eccentricity) of 0.1t. Therefore:
$$M = P \times 0.1t$$
and the bending stress is calculated using $f_b = Mc/I$. The combination of bending stress in compression and compression due to axial load must not exceed $F_b = f'_m/3$. For members with $h/r > 99$, it must be $P < 0.25P_e$, where

$$P_e = (p^2 \, EMI/h^2)(1 - 0.577e/r)^3$$
(a type of Euler's stress)

The combined compression stress has to satisfy a condition that we have already seen in other

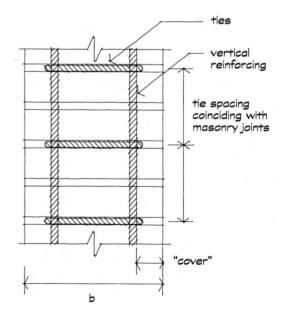

ties

vertical reinforcing

tie spacing coinciding with masonry joints

"cover"

b

combined axially loading and bending members:

$$\frac{f_a}{F_a} + \frac{f_b}{F_b} \leq 1.0$$

where the actual stresses f_a and f_b can be calculated for preliminary purposes, neglecting the contribution of the steel reinforcement. The column cross-section can include the transformed steel area in the section modulus or moment of inertia calculation, although this is also often disregarded.

Example 43.1

With the empirical method illustrated in Chapter 42, an 8in solid wall made of 10,000psi brick in M mortar can have a maximum $h = 20t = 160\text{in} = 13\text{ft-4in}$. No limit therefore applies to length. Allowable load per foot of wall length :

$F_a = 350\text{psi}$ (Table 42.1)
$P_a = F_a A_e = 350\text{psi}(7\text{-}5/8\text{in})(12\text{in}) = 32,025\text{lb}$

Working-stress method for the same wall
Allowable stress formula for $h/r \leq 99$:

$\quad r = 0.3t = 0.3(7.625\text{in}) = 2.29\text{in}$
$\quad h/r = (160\text{in})/(2.29\text{in}) = 69.9 < 99$, so it's a short column.

$$F_a = 0.25 f'_m \left[1 - \left(\frac{h}{140r} \right)^2 \right]$$

$f_m' = 4,000$ psi   (Table No. 42.1)

$$F_a = 0.25(4,000\text{psi}) \left[1 - \left(\frac{160\text{in}}{140(2.29\text{in})} \right)^2 \right] = 751\text{psi}$$

$P_a = (751\text{psi})(12\text{in})(7.625\text{in}) = 68,710\text{lb}$, or over twice that allowed by the empirical method. It's easy to see that the empirical method is very conservative, but don't be tempted to overload the wall to take advantage of that conservative value. Calculating load capacity using working-stress design is a simple task, so even if you initially design the wall using empirical values, it is worth checking in the end using working-stress techniques. Then you'll know!

Example 43.2

A 20ft-high column is built with 3,750psi CMUs in S mortar with minimum reinforcing consisting of four vertical bars, $f_y = 60,000\text{psi}$, placed in grout with the same strength of the CMUs (not in the S mortar, which has a strength of 2,100psi: see Sec. 40.3).

The h/d ratio must be not greater than 25:
$h/d = (20\text{ft})(12\text{in/ft})/(15.625\text{in}) = 15.36$ OK!

Allowable load:
$\quad f_m' = 3,000\text{psi}$ (Table 40.2).
$\quad$ Design the steel reinforcement:
$\quad$ Gross column area:
$$A_e = (15.625\text{in})^2 = 244\text{in}^2$$
$\quad$ Minimum steel reinforcement:

$A_s = 0.0025(244in^2) = 0.6in^2$

$A_{bar} = 0.6in^2/4bars = 0.15in^2 = No. 4$

Use four No. 4:

$A_s = 4(0.20in^2) = 0.80in^2$

Net column area:

$A_n = 244in^2 - 0.80in^2 = 243in^2$ (hardly worth it!)

$F_{sc} = 0.4(60,000psi) = 24,000psi = maximum$

Find the allowable axial load:

$h' = (20ft)(12in/ft) = 240in$

$t = 15.625in$

Radius of gyration $r = 0.3d$ (same as R/C)

$r = 0.3(15.625) = 4.687in$

$h/r = (240in)/(4.687in) = 51.2 < 99$ (short column)

$$P_a = [0.25(20.7MPa)(157,093mm^2) + 0.65(516mm^2)(165.6MPa)]\left[1 - \left(\frac{6000mm}{140(119mm)}\right)^2\right] =$$

$P_a = 755,854N$

Tie spacing $= 16d_b = 16(0.5in) = 8in$ with No. 2 ties.

Example 43.3

The 12in nominal solid brick column, 16ft high, is built with 14,000psi brick, M mortar, and 40ksi steel. The bars are four No. 4 with No. 2 ties at 8in o.c. The column must carry an axial load of 63k. Check if the column design is adequate.

Masonry strength $f_m' = 5,300psi$

Allowable stress in steel $F_{sc} = 0.4(40,000psi) = 16,000psi$

Radius of gyration $r = 0.3 (11.5in) = 3.45in$

$h = (16ft)(12in/ft) = 192in$

$h/r = (192in)/(3.45in) = 55.6 < 99$ (short column)

The allowable axial load is

$$P_a = 0.25(5,300psi)(11.5in \times 11.5in) + 0.65(0.8in^2)(16,000psi) = 183,552lb$$

As we continue to ultimately check combined stresses, remember our discussion of steel; we look at the numbers as we generate them. If at any point the percentage of either component exceeds 1.0, it is foolish to continue. So here we go:

Axial stress:

$$F_a = 0.25(5,300psi)\left(1 - \left(\frac{192in}{140(3.5in)}\right)^2\right) = 1,116psi$$

$$P_a = 1,116psi(11.5in)(11.5in) = 147,591lb > 63,000lb$$

$$f_a = \frac{P}{A} = \frac{63,000 \text{lb}}{(11.5 \text{in} \times 11.5 \text{in})} = 476 \text{psi} \leq 1116 \text{psi} \qquad \text{OK, so far!}$$

Bending with accidental eccentricity:

$$S = \frac{Bt^2}{6} = \frac{(11.5 \text{in})(11.5 \text{in})(11.5)}{6} = 253 \text{in}^3$$

$$f_b = \frac{M}{S} = \frac{0.1(P)}{S} = \frac{0.1(11.5 \text{in})(63,000 \text{lb})}{253 \text{in}^3} = 286 \text{psi}$$

$$F_b = \frac{f'_m}{3} = \frac{5,300 \text{psi}}{3} = 1,767 \text{psi}$$

Checking the interaction equation:

$$\frac{f_a}{F_a} + \frac{f_b}{F_b} = \frac{476 \text{psi}}{1116 \text{pis}} + \frac{286 \text{psi}}{1767 \text{psi}} = 0.59 \leq 1.0 \qquad \text{OK, we're done! finis! hasta la vista! good-bye!}$$

Example 43.4

A 22ft-high load-bearing wall is built with 8in nominal CMSs, 1,000psi concrete, and fully grouted. Design the minimum reinforcement required and calculate the load-bearing capacity of the wall in pounds per foot of length, assuming a perfectly axial load.

Considering a 1ft length of wall:

$A_n = (7.625 \text{in})(12 \text{in}) = 91.5 \text{in}^2$

The minimum reinforcement for columns (or walls in this case) is $0.0025 A_n$.

By using the least expensive 40ksi steel,

$F_{sc} = 0.4(40,000 \text{psi}) = 16,000 \text{psi}$

The bars must be placed consistent with the masonry unit module, which is 8in or a multiple of 8in. From the table of bar areas for concrete, a No. 5 bar at 16in o.c. (one every block) provides

$A_s = 0.23 \text{in}^2$ OK! The steel area per foot of length is, therefore,

$A_s = (0.23 \text{in}^2)(12 \text{in})/(16 \text{in}) = 0.172 \text{in}^2$

Calculate the allowable load:

$h/r = (22 \text{ft})(12 \text{in/ft})/(0.3)(7.625 \text{in}) = (264 \text{in})/(2.29 \text{in}) = 115 > 99$ Long column!

Note that the steel reinforcement contributes to the load-bearing capacity with 1,376lb or less than 2% of the masonry's capacity; however, it is very useful to include it because of the increase in ductility. The slenderness, represented by the h/r ratio, reduces the base capacity by almost twothirds, as the $(h/r)^2$ factor is 0.37.

43.7 Bearing

One of the most common considerations in masonry is bearing. In many cases, steel or wood beams will be resting directly on masonry without bearing plates (this is not a recommended practice). In any case, we can calculate the actual bearing capacity of the masonry surface instead of using the conservative UBC or AISC specified values.

Allowable bearing stress:

(a) Load on the full area of masonry

$$F_{br} = 0.26f_m'$$

(b) Load on one third or less of the masonry area

$$F_{br} = 0.38f_m'$$

Example 43.5

The column of Example 43.3 is loaded with 127.7k from a W8X40 steel girder 8in-wide bearing directly on the masonry. Check if the bearing area is adequate.

Bearing area $A_{br} = 8(15.625) = 125\text{in}^2$

$$A_{br}/A_e = (125\text{in}^2)/244 = 0.51$$

Interpolate between two bearing stress values:

$$F_{br} = 0.49(0.38-0.26)f_m'/0.67 + 0.26f_m' = 0.35f_m'$$
$$F_{br} = 0.35(3,000\text{psi}) = 1,050\text{psi}$$
$$P_a = (1,050\text{psi})(125\text{in}^2) = 131.25k > 127.7k \quad \text{OK!}$$

Problem 43.1

How tall can you build a 20in x 12in brick column according to codes?

Answer:

$h/d = 25$; therefore, $h = 25(12\text{in}) = 300\text{in}$ or 25ft

Problem 43.2

An 8inx16in column is 12ft tall and is built with 8,000psi hollow CMUs fully grouted, S mortar. Calculate the allowable axial load by

(a) the empirical method
(b) working-stress method (unreinforced)
(c) assuming a reinforced column with 4 No. 4 bars, 60ksi steel

Assume the effective length = 12ft.

Answers:

(a) 26,807lb
(b) $h/r = 63 < 99$; $F_a = .25(3,000\text{psi})(0.8) = 5,987\text{psi}$;
    $P_a = (119.14\text{in}^2)(5987\text{psi}) = 71,289\text{lb}$
(c) $(0.65)(0.8\text{in}^2)(24,000\text{psi})(0.8) = 9984\text{lb}$; $P_a = 71,289\text{lb} + 9,984\text{lb} = 81,273\text{lb}$.

METRIC VERSIONS OF ALL EXAMPLES AND PROBLEMS

Example M43.1

With the empirical method illustrated in Chapter 42, a 203mm solid wall made of 70MPa brick in M mortar can have a maximum h = 20t = 20(203mm) = 4060mm. No limit therefore applies to length.

Allowable load per foot of wall length :

F_a = 2.4MPa   (Table No. 42.1)

$P_a = F_a A_e$ = (2.4MPa)(194mm)(1000mm) = 464,820N

Working-stress method for the same wall:

r = 0.3(194mm) = 58.2mm

h/r = (4060mm)/(58.2mm) = 69.8 < 99 so it's a short column.

Allowable stress formula:

$$F_a = 0.25f_m' \left[1 - \left(\frac{h}{140r} \right)^2 \right]$$

f_m' = 27.6MPa   (Table No. 42.1)

$$F_a = 0.25(27.6MPa) \left[1 - \left(\frac{4060mm}{140(58.2mm)} \right)^2 \right] = 5.19MPa$$

P_a = (5.19MPa)(1000mm)(194mm) = 1,006,860N,

or about twice that allowed by the empirical method. It's easy to see that the empirical method is very conservative, but don't be tempted to overload the wall to take advantage of that conservative value. Calculating load capacity using working-stress design is a simple task, so even if you initially design the wall using empirical values, it is worth checking in the end using working-stress techniques. Then you'll know!

Example M43.2

A 6m-high column is built with 25.8MPa CMUs in S mortar with minimum reinforcing consisting of four vertical bars, f_y = 414MPa, placed in grout with the same strength of the CMUs (not in the S mortar, which has a strength of 14.5MPa: see Sec. 40.3).

The h/d ratio must be not greater than 25:
    h/d = (6000mm)/(397mm) = 15.12 OK!

Allowable load:
    f_m' = 20.7MPa (Table 42.1).
Design the steel reinforcement.
    Gross column area:
        A_e = (397mm)^2 = 157,609mm^2
    Minimum steel reinforcement:
        A_s = 0.0025(157,609^2) = 394mm^2
        A_{bar} = (394mm^2)/4 = 98.5mm^2 = No. 13M = 129mm^2
Use four No. 13M: A_s = 4(129mm^2) = 516mm^2
    Net column area:
        A_n = 157,609mm^2 - 516mm^2 = 157,093mm^2 (hardly worth it!)
        F_{sc} = 0.4(414MPa) = 165.6MPa = max.

Find the allowable axial load:
    h' = 6000mm
    t = 397mm
    Radius of gyration r = 0.3t (same as R/C)
    r = 0.3(397mm) = 119mm
    h/r = (6000mm)/(119mm) = 50.4 < 99 (short column)

$$P_a = \left[0.25(20.7\text{MPa})(157,093\text{mm}^2) + 0.65(516\text{mm}^2)(165.6\text{MPa})\right]\left[1 - \left(\frac{6000\text{mm}}{140(119\text{mm})}\right)^2\right] =$$

P_a = 755,854N

Tie spacing = 16d_b = 16(13mm) = 208mm with No. 6M ties.

Example M43.3

A 305mm square, nominal solid brick column, 4.88m high, is built with 96.5MPa brick, M mortar, and 276MPa steel. The bars are four No. 13M with No. 6M ties at 203mm o.c. The column must carry an axial load of 280kN. Check if the column design is adequate.

Masonry strength f_m' = 36.5MPa
Allowable stress in steel F_{sc} = 0.4(276MPa) = 110.4MPa
Radius of gyration r = 0.3 (292mm) = 87.6mm
    h = 4880mm
    h/r = (4880mm)/(87.6mm) = 55.7 < 99 (short column)
    A_n = (292mm)^2 = 85,264mm^2 approximately

The allowable axial load is:

$$P_a = 0.25(36.5 \text{MPa})(292 \text{mm} \times 292 \text{mm}) = 778,034 \text{N}$$

$$P_a = 778,034 \text{N}$$

As we continue to ultimately check combined stresses, remember our discussion of steel; we look at the numbers as we generate them. If at any point the percentage of either component exceeds 1.0, it is foolish to continue. So here we go:

Axial stress:

$$F_a = 0.25(36.5 \text{MPa})\left[1 - \left(\frac{4880 \text{mm}}{140(87.6 \text{mm})}\right)^2\right] = 7.68 \text{MPa}$$

$$P_a = 7.68 \text{MPa}(229 \text{mm})(292 \text{mm}) = 655 \text{kN} \geq 280 \text{kN}$$

Bending with accidental eccentricity:

$$f_a = \frac{P}{A} = \frac{280,000 \text{N}}{(292 \text{mm})(292 \text{mm})} = 3.28 \text{MPa}$$

$$S = \frac{Bt^2}{6} = \frac{24,897,088 \text{mm}^3}{6} = 4,149,515 \text{mm}^3$$

$$f_b = \frac{M}{S} = \frac{0.1(292 \text{mm})(280,000 \text{N})}{4,149,515 \text{mm}^3} = 1.97 \text{MPa}$$

$$F_b = \frac{36.5 \text{MPa}}{3} = 12.18 \text{MPa}$$

Checking the interaction equation:

$$\frac{f_a}{F_a} + \frac{f_b}{F_b} = \frac{3.28 \text{MPa}}{7.68 \text{MPa}} + \frac{1.97 \text{MPa}}{12.18 \text{MPa}} = 0.59 \leq 1.0$$

Example M43.4
A 6.7m-high load-bearing wall is built with 203mm nominal CMSs, 68.9MPa-concrete, and fully grouted. Design the minimum reinforcement required and calculate the load-bearing capacity of the wall in N per meter of length, assuming a perfectly axial load.

Considering a 1ft length of wall:

$$A_n = (193.7 \text{mm})(1000 \text{mm}) = 193,700 \text{mm}^2$$

The minimum reinforcement for columns (or walls in this case) is $0.0025 A_n$.
Using the least expensive 276MPa steel,

$$F_{sc} = 0.4(276 \text{MPa}) = 110.4 \text{MPa}$$

The bars must be placed consistent with the masonry unit module, which is 203mm or a multiple of 203mm. From the table of bar areas for concrete, a No. 16M bar at 406mm o.c. (one every block) provides A_s = 199mm² OK! The steel area per meter of length is therefore, A_s = (199mm²)(1000mm/406mm) = 490mm² per meter of wall.

Calculate the allowable load:

$\quad$ h/r = (6700mm)/(0.3×194mm) = 115 > 99 Long column!

$$P_a = [0.25(27.6\text{MPa})(194\text{mm} \times 1000\text{mm}) + 0.65(490\text{mm}^2)(110.4\text{MPa})]\left(\frac{70(58.2\text{mm})}{6700\text{mm}}\right)^2 =$$

P_a = 508,292N per meter.

Note that the steel reinforcement contributes to the load-bearing capacity with 13kN or less than 2% of the masonry's capacity; however, it is very useful to include it because of the increase in ductility. The slenderness, represented by the h/r ratio, reduces the base capacity by almost two-thirds, as the (h/r)² factor is 0.37.

Example M43.5

The column of Example M43.3 (f_m' = 36.5MPa) is loaded with 568kN from a W200X59 steel girder 205mm wide bearing directly on the masonry. Check if the bearing area is adequate.

Bearing area:

$\quad A_{br}$ = (205mm)(397mm) = 81,359mm²

$\quad F_{br}$ = 0.25f_m' = 0.25(36.5MPa) = 9.1MPa

$\quad P_a$ = 9.1MPa(85,264mm²) = 775,902N = 776kN > 568kN OK.

Problem M43.1

How tall can you build a 508mmx305mm brick column according to codes?

Solution:
h/d = 25; therefore, h = 25(305mm), = 7625mm or 7.6m

Problem M43.2

A 203mmx406mm column is 3.6m tall and is built with 55.1MPa hollow CMUs fully grouted, S mortar. Calculate the allowable axial load by:
(a) the empirical method
(b) the working-stress method (unreinforced)
(c) assuming a reinforced column with four No. 13M bars, 414MPa steel
Assume the effective length = 3.6m

Answers:

(a) 119kN

(b) h/r = 63 < 99; F_a = 0.25(20.7Mpa)(0.8) = 4.14MPa;

$\quad$ P_a = (76,875mm²)(4.14MPa) = 318.3kN

(c) (0.65)(516mm²)(165MPa)(0.8) = 44.3kN;

$\quad$ P_a = 318.3kN + 44.3kN = 362.6kN

CONVERSION FACTORS FROM U.S. customary units to SI metric units

The pound-inch system currently used in the United States, also known as "English" or "U.S. customary," is gradually being replaced in industrial production and in architectural practice by the international system of weights and measures called SI (Systeme International) based on the metric system and commonly referred to as "metric." Since January 1994, all federal construction work in the United States has been designed and built with SI measures, and starting in 2004, the architecture registration examinations will be metric. This is why most measures in the text have the SI equivalent between brackets. To obtain the SI equivalent of U.S. measures, multiply the U.S. measure by the factor in Table A.1.

Example: Convert 1108.ft² into m²

$$(1108.7ft^2)(0.092903) = 103m^2$$

In order to obtain the reverse conversion, multiply the metric measure by the inverse conversion factor; for example, to turn square meters to square feet, multiply the number in m² by 1/ 0.092903.

Example:

$$103m^2(1/0.092903) = 1108.7sqft$$

If a job has to be done in metric, make sure that all factors and coefficients contained in the equations are consistent with the units used. The most important codes (e.g.,the ACI Code Requirements for Reinforced Concrete, for example) are published with the SI versions, but other manuals and codes may not be. So, in some cases, such as wood connections, you have to do all calculations in U.S. customary units, using the NDS tables, and then convert the end result into metric. See the list of references after Table A.1 for additional information.

You should also remenber that in the SI units there is a distinction between mass and force (all bodies have a mass, but their weight is a force acting on buildings, depending on the acceleration of gravity). So, whereas densities of materials are given in grams, weights are expressed in newtons (mass x acceleration of gravity = force).

Table A.1 Conversion factors from U.S. customary units to SI metric units

Length

U.S. unit	multiply by	to obtain	which is
ft	0.305	m	meter
in	25.4	mm	millimeters

Area

U.S. unit	multiply by	to obtain	which is
sqft, ft²	0.0929	m²	square meters
sqin, in²	645.16	mm²	square millimeters

Volume, section properties

U.S. unit	multiply by	to obtain	which is
in³	16,387.1	mm³	cubic millimeters
ft³	0.0283	m³	cubic meters
yd³, cuyd	0.7645	m³	cubic meters
in⁴	416 231.4	mm⁴	

Weight (on Earth), forces on buildings, mass

U.S. unit	multiply by	to obtain	which is
lb (mass)	0.453	kg	kilograms [a]
lb (force)	4.448	N	newton
k (kip or kips)	4.448	kN	kilonewtons
lb/ft, plf	14.594	N/m	newtons/meter
kips/ft, klf	14,594	kN/m	kilonewtons/meter
psf/in	1.885	(N/m²)/mm	
psf, lb/sqft	47.88	N/m² or Pa	newtons/square meter
pcf, lb/cuft (mass)	16	kg/m³	kilograms/cubic meter
pcf, lb/cuft	157.08	N/m³	newtons/cubic meter

[a] To convert kg to N, multiply kg by 9.81

Moments of forces

U.S. unit	multiply by	to obtain	which is
ftlb	1.3566	Nm	newton-meters
inlb	112.979	Nmm	newton-millimeters
ftk	1.3566	kNm	kilonewton-meters

Stress, pressure

U.S. unit	multiply by	to obtain	which is
psi, lb/sqin	0.006894	MPa(N/mm²)	megapascals (newtons/square millimeter)
ksi, kips/sqin	0.006894	Pa(N/m²)	pascals (newtons/square meter)
	6.894	MPa(N/mm²)	megapascals

Temperature

U.S. unit	multiply by	to obtain	which is
°F -32°	5/9	°C	degrees Celsius

Example: 68°F = (68°-32°)5/9 = 20°C; reverse: (20°C)9/5 +32°F = 68°F

Velocity, speed, acceleration

U.S. unit	multiply by	to obtain	which is
mph	1.609	km/hr	kilometers/hour
ft/sec², ft/sec/sec	0.305	m/sec²	meters/square second

Materials properties: see specific sections on steel, wood, concrete, and masonry

Typical code loads (the SI quantities are rounded up for convenience of use)

10psf = 480N/m²
20psf = 960N/m²
30psf = 1440N/m²
40psf = 1920N/m²
50psf = 2400N/m²

REFERENCES

General

International Conference of Building Officials. *1997 Uniform Building Code.* Whittier, CA: ICBO, 1997. (Dual units).

Building Officials and Code Administrators International. *BOCA National Code.* Country Club Hills, IL: BOCA, 1996.

American Society for Testing and Materials. ASTM E621, *Standard Practice for the Use of Metric (SI) Units in Building Design and Construction.* Philadelphia: ASTM, 1991.

National Institute of Building Sciences. *Metric Guide for Federal Construction.* Washington, DC: NIBS,1991.

U.S. General Services Administration. *Metric Design Guide.* Washington, DC: USGSA, 1993.

Steel

American Institute of Steel Construction. *Metric Properties of Structural Shapes with Dimensions According to ASTM A6M.* Chicago: AISC, 1992.

Steel Joist Institute. *Standard Specifications, Load Tables and Weight Tables for Steel Joists amd Joist Girders,* Myrtle Beach, SC: SJI, 1995 (SI tables included).

Wood

American Forest and Paper Association. *Wood Products Metric Planning Package.* Washington, DC: AFPA, 1994.

Canadian Wood Council. *Wood Design Manual.* Ottawa: CWC, 1995.

Concrete

American Concrete Institute. *Metric Building Code Requirements for Reinforced Concrete and Commentary, ACI 318-95.* Detroit: ACI, 1996.

American Society for Testing and Materials. ASTM A615 M-96a and A706 M-96a. *Standard Metric Reinforcing Bar Sizes.* Philadelphia: ASTM, 1996.

Masonry

American Concrete Institute. *Building Code Requirements for Masonry Structures and Commentary.* Detroit: ACI, 1996 (Dual units).

BIBLIOGRAPHY

Publications are listed in alphabetical order by author.

A. Steel structures

American Institute of Steel Construction. *Manual of Steel Construction - Allowable Stress Design*, 9th ed. AISC, Chicago, 1991.

American Institute of Steel Construction. *Steel Design Guide Series*. AISC, Chicago, 1991.

Crawley and Dillon. *Steel Buildings. Analysis and Design*. (4th ed.). John Wiley & Sons, New York, 1993.

Engel, Irving. *Structural Steel in Architecture and Building Technology*. Prentice Hall, 1988.

Hart, Henn, and Sontag. *Multi-story Buildings in Steel*. (2nd ed.). Nichols, New York, 1985.

McCormack, J. C. *Structural Steel Design: ASD Method* (4th ed.). Harper Collins, New York, 1992.

Otto, Frei (ed.). *Tensile Structures - Design, Structure and Calculation*. MIT Press, 1973.

Steel Joist Institute. *Standard Specifications, Load Tables and Weight Tables for Steel Joists and Joist Girders*. SJI, 1996.

Steel Deck Institute. *Design Manual for Composite Decks, Form Decks and Roof Decks*.

B. Wood structures

American Forest & Paper Association. *National Design Specifications for Wood Construction (NDS). 1991 edition*. AFPA, Washington DC, 1993.

American Forest & Paper Association. *Permanent Wood Foundation System. Design fabrication installation manual*. Washington DC, 1987.

American Institute of Timber Construction. *Timber Construction Manual*. (4th ed.). John Wiley & Sons, New York, 1994.

American Institute of Timber Construction. *AITC 104-84: Typical Construction Details*. AITC, Vancouver, WA.

American Institute of Timber Construction. *Laminated Timber Design Guide*. AITC, Vancouver, WA, 1995.

American Institute of Timber Construction. *Glued Laminated Timbers for Residential and Light Commercial Construction*, AITC, 1992.

APA - The Engineered Wood Association (American Plywood Association), Tacoma, WA. *Plywood Design Specifications*. August 1986, and Supplements:
1- *Design and Fabrication of Plywood Curved Panels*, March 1990.
2- *Design and Fabrication of Glued Plywood-lumber Beams*, July 1992.
3- *Design and Fabrication of Plywood Stressed-skin Panels*, August 1990.
4- *Design and Fabrication of Plywood Sandwich Panels*, March 1990.
5- *Design and Fabrication of All-Plywood Beams*, March 1990.

APA - The Engineered Wood Association (American Plywood Association), Tacoma, WA.
APA Design/Construction Guide:
- *Diaphragms, 1995.*
- *Residential and Commercial, 1996.*

APA - The Engineered Wood Association (American Plywood Association), Tacoma, WA.
Product Guide.
- *Oriented strand board, 1996.*
- *Grades and specifications, 1995.*

American Wood Council. *Permanent Wood Foundation. Guide to Design and Construction.* AFPA, Washington DC, 1988.

American Wood Systems. *Product and Application Guide: Glulams.* Tacoma, WA, 1995.

Breyer, D.E. *Design of Wood Structures.* (3rd ed.). McGraw-Hill, New York, 1993.

Canadian Wood Council. *Wood Design Manual.* Canadian Wood Council, Ottawa, 1995.

Canadian Wood Council. *Wood Reference Handbook.* Canadian Wood Council, Ottawa, 1991.

Dietz, Albert. *Dwelling House Construction.* MIT Press, 1991.

Engineered Wood Systems. *Data File - Glued Laminated Beam Design Tables.* Tacoma, WA, 1996.

Faherty, K., and T. Williamson, *Wood Engineering and Construction Handbook,* (2nd ed.). McGraw-Hill, New York, 1995.

Gotz, Hoor, Mohler, and Natterer. *Timber Design and Construction Sourcebook.* McGraw-Hill, New York, 1989.

Halperin, D., and G.T. Bible, *Principles of Timber Design for Architects and Builders.* John Wiley & Sons, New York, 1994.

Stalnaker, J., and E. Harris, *Structural Design in Wood.* Van Nostrand Reinhold, New York, 1989.

Southern Pine Council. *Southern Pine Use Guide.* Southern Forest Products Association, Kenner LA, 1994.

C. Concrete structures

Ambrose, James. *Simplified Design of Concrete Structures* (7th ed.). John Wiley & Sons, New York, 1996.

American Concrete Institute (ACI). *Building Code Requirements for Reinforced Concrete and Commentary, ACI 318-95.* ACI, Detroit, 1995.

Leet, K., and D. Bernal, *Reinforced Concrete Design.* (3rd ed.). McGraw-Hill, New York, 1997.

McCormack, Jack. *Design of Reinforced Concrete* (3rd ed.). Harper Collins, New York, 1993

Nawy, Edward. *Prestressed Concrete - A Fundamental Approach* (2nd ed.). Prentice Hall International, Upper Saddle River, NJ, 1996.

Nervi, Pier Luigi. *The Works of Pier Luigi Nervi.* Frederick A. Praeger, New York, 1957.

Nilson, A., and G. Winter, *Design of Concrete Structures* (11th ed.). McGraw-Hill, New York, 1991.

Precast Concrete Institute. *PCI Design Handbook. Precast and Prestressed Concrete* (4th ed.). PCI, Chicago, 1992.

Shaeffer, R.E. *Reinforced Concrete - Preliminary Design for Architects and Builders.* McGraw-Hill, New York, 1992.

Wang, C., C. and Salmon, *Reinforced Concrete Design* (5th ed.). Harper Collins, New York, 1992.

D. Masonry structures

Ambrose, J. *Simplified Design of Masonry Structures.* John Wiley & Sons, New York, 1991.

National Concrete Masonry Association. *TEK Manual for Concrete Masonry Design and Construction.* NCMA, Herndon VA, 1997.

Panarese, Kosmatka, and Randall. *Concrete Masonry Handbook.* PCI, Skokie, IL., 1991.

Schneider, R.R., W.L. and Dickey, *Reinforced Masonry Design.* Prentice-Hall, 1980.

F. Selected books on structural design and building codes

Ambrose, James. *Building Structures* (2nd ed.) John Wiley & Sons, New York, 1993.

Ambrose, James. *Simplified Design of Building Structures* (3rd ed.). John Wiley & Sons, New York, 1995.

Ambrose, J. *Simplified Design of Building Foundations* (2nd ed.). John Wiley & Sons, New York, 1994.

Ambrose, J., and Vergun. *Simplified Building Design for Wind and Earthquake Forces* (3rd ed.). John Wiley & Sons, New York, 1994.

Ambrose, James. *Design of Building Trusses.* John Wiley & Sons, New York, 1994.

Building Officials and Code Administrators International (BOCA). *National Building Code.* BOCA, Country Club Hills, IL, 1996.

Council of American Building Officials (CABO). *One- and Two-family Dwelling Code.* CABO, Falls Church VA, 1989.

Dowrick, David. *Earthquake Resistant Design for Engineers and Architects* (2nd ed.) John Wiley & Sons, Chichester, 1977.

International Conference of Building Officials. *Uniform Building Code, 1997 ed.* ICBO, Whittier CA, 1997.

Lin, T.Y., and Sidney Stotesbury, *Structural Concepts and Systems for Architects and Engineers* (2nd ed.). Van Nostrand Reinhold, New York, 1988.

Sandaker, B.N., and A.P. Eggen, *The Structural Basis of Architecture.* Whitney, 1992.

Schueller, Wolfgang. *Horizontal-span Building Structures.* John Wiley & Sons, New York, 1983.

Schueller, Wolfgang. *The Vertical Building Structure.* Van Nostrand Reinhold, New York, 1983.

Schueller, Wolfgang. *The Design of Building Structures.* Prentice Hall, Upper Saddle River, NJ, 1996.

Southern Building Code Congress International. *Standard Building Code.* SBCCI, Birmingham, AL, 1988.

G. Selected books on structural systems in architecture

Buchanan, P. *Renzo Piano Building Workshop. Complete Works.* Phaidon Press, London, 1993.

Davies, Colin. *High Tech Architecture.* Rizzoli International, New York, 1988.

Dini, Massimo. *Renzo Piano: Projects and Buildings 1964-1983.* Rizzoli International, New York, 1983.

Drew, Philip. *Frei Otto: Form and Structure.* Westview Press, Boulder, CO, 1976.

Ehrenkrantz, Ezra. *Architectural Systems.* McGraw-Hill, New York, 1989.

Fischer, Robert (ed.). *Engineering for Architecture.* Architectural Record Book / McGraw-Hill, New York, 1980.

Foster, Norman. *Foster Associates - Buildings and Projects.* Watermark, Hong Kong, 1989.

Komendant, August. *18 Years with Architect Louis Kahn.* Aloray, 1975.

Lipman, J. *Frank Lloyd Wright and the Johnson Wax Buildings.* Rizzoli International, New York, 1986.

Mainstone, Rowland. *Developments in Structural Form.* Allen Lane and Penguin Books, London, 1975.

Ramsey/Sleeper. *Architectural Graphic Standards.* (9th ed.). John Wiley & Sons, New York, 1991.

Singer. *A History of Technology.* Oxford at Clarendon Press, 1958.

Nabokov, P., and R. Easton, *Native American Architecture.* Oxford University Press, New York, 1989.

Suzuki, K. *Early Buddhist Architecture in Japan.* Kodansha International, New York, 1980.

Wilkinson. *Supersheds.* Butterworth, 1991.

Yoshida, Tetsuro. *Japanese House and Garden.* Praeger, New York, 1969.

Zannos, Alexander. *Form and Structure in Architecture - The Role of Statical Function.* Van Nostrand Reinhold, New York, 1987.

Illustrations and tables credits

Fig. 42.1   Michele Chiuini

Tables 42.1, 42.2, 42.3, 42.4
                    Reproduced from the 1997 edition of the Uniform Building
                    Code, copyright © 1997, with the permission of the
                    publisher, the International Conference of Building Officials.